Global Emerging Innovation Summit (GEIS-2021)

Edited by

Dharam Buddhi

Rajesh Singh

&

Anita Gehlot
Lovely Professional University
Phagwara
India

Global Emerging Innovation Summit (GEIS-2021)

Editors: Dharam Buddhi, Rajesh Singh and Anita Gehlot

ISBN (Online): 978-1-68108-901-0

ISBN (Print): 978-1-68108-902-7

ISBN (Paperback): 978-1-68108-903-4

©2021, Bentham Books imprint.

Published by Bentham Science Publishers – Sharjah, UAE. All Rights Reserved.

First published in 2021.

need for a court order if at any point you breach any terms of this License Agreement. In no event will any delay or failure by Bentham Science Publishers in enforcing your compliance with this License Agreement constitute a waiver of any of its rights.

3. You acknowledge that you have read this License Agreement, and agree to be bound by its terms and conditions. To the extent that any other terms and conditions presented on any website of Bentham Science Publishers conflict with, or are inconsistent with, the terms and conditions set out in this License Agreement, you acknowledge that the terms and conditions set out in this License Agreement shall prevail.

Bentham Science Publishers Ltd.
Executive Suite Y - 2
PO Box 7917, Saif Zone
Sharjah, U.A.E.
Email: subscriptions@benthamscience.net

BENTHAM SCIENCE

CONTENTS

FOREWORD

Research and Innovation are the prime factors for the creation of knowledge and entrepreneurship. Innovation also ensures the long-term sustainability of institutions. Innovation is a process of conversion from ideation to the creation of a new product or process by applying uniqueness & novelty.

Lovely Professional University organized a Global Emerging Innovation Summit 2021(GEIS-2021) on April 9-10, 2021 in collaboration with The National Agency for New Technologies, Energy and Sustainable Economic Development (ENEA). GEIS-2021 was the most comprehensive conference focused on the various aspects of the Global Emerging Innovations. The Conference provided a platform to academic and industry professionals to discuss recent innovations in the area of Engineering and Sciences. The researchers from academia and industry as well as practitioners share ideas, problems, and solutions relating to the multifaceted aspects of Innovations. It also provided opportunities for the different research area delegates to exchange new ideas to establish research relations which may be able to mobilize future collaborations among academic institutions, industries and entrepreneurs globally.

Authors from academics, government agencies and industries, who contribute 189 papers truly make the conference a unique platform for a productive, interdisciplinary exchange of ideas. Out of 189 papers submitted, 59 manuscripts passed the peer review process and were presented during the conference. The success of this conference cannot be achieved without the involvement of number of individuals. I came to know that a good number of international keynote speakers and participants attended the summit.

I congratulate the GEIS organizers and editors for coming out with the proceedings of the summit, which will be an immense help to innovators.

Giacobbe Braccio
Head of Bioenergy, Biorefinery and Green Chemistry Division
Italian National Agency for New Technologies, Energy and Sustainable Economic
Development
Research Centre of Trisaia
Matera
Italy

PREFACE

I am pleased to present you the Proceeding of the Global Emerging Innovation Summit (GEIS 2021) held on April 9th & 10th 2021. The conference was organized by Lovely Professional University (LPU) and supported by ENEA - the National Agency for New Technologies, Energy and Sustainable Economic Development with Publication partner Bentham Science.

GEIS-2021 is the most comprehensive conference focused on the various aspects of the Global Emerging Innovation Summit-2021 related areas. The Conference provides a platform for academic and industry professionals to discuss recent innovations in the area of engineering and Sciences. The goal of this conference is to bring together the researchers from academia and industry as well as practitioners to share ideas, problems, and solutions relating to the multifaceted aspects of Innovations in Engineering and Sciences. It provides opportunities for the different research area delegates to exchange new ideas to establish research relations to find global partners for future collaboration.

Authors from academics, government agencies and industries, who contribute 189 papers truly make the conference a unique platform for a productive, interdisciplinary exchange of ideas. Out of 189 papers submitted, 59 manuscripts are accepted and presented during the conference.

The success of this conference couldn't be achieved without the involvement of number of individuals. We would therefore, express our sincere gratitude to all of the organizing committees, keynote speakers, reviewers, session chairs, presenters, participants, and all contributors who have put their finest effort into making this conference a big success.

Dharam Buddhi

Rajesh Singh

&

Anita Gehlot
Lovely Professional University
Phagwara
India

List of Contributors

Abhinav Jain	Hamdard Institute of Medical Sciencesand Research Associated HAHC Hospital,, New Delhi, India
Abhinav Srivastava	Lovely Professional University,, Phagwara, India
Abukar Ahmed Muse	Lovely Professional University,, Phagwara, India
Aditee Singh	Lovely Professional University,, Phagwara, India
Aditi Singh	Lovely Professional University,, Phagwara, India
Aeimy Mary Jose	Lovely Professional University,, Phagwara, India
Akshay Mohan	Lovely Professional University,, Phagwara, India
Amit Joshi	Lovely Professional University,, Phagwara, India
Amit Kumar Thakur	Lovely Professional University,, Phagwara, India
Amit Sachdeva	Lovely Professional University,, Phagwara, India
Amita Mane	PCCOER, Pune, India
Amol Kalage	Sinhgad Institute of Technology, Lonavla, India
Amrish Jangra	Gurukul Kangri Vishwavidyalaya, Haridwar, India
Anita Choudhary	Guru Teg Bahadur Institute of Technology, New Delhi, India
Anita Gehlot	Lovely Professional University, Phagwara, India
Anjuvan Singh	Lovely Professional University, Phagwara, India
Ankush Sharma	Guangzhou Rising Dragon Recreation Industrial Co. Ltd., Guangzhou, China
Anshuman Das	Lovely Professional University, Phagwara, India
Anuja Ajit More	Lovely Professional University, Phagwara, India
Apash Roy	Lovely Professional University, Phagwara, India
Atul Malhotra	Lovely Professional University, Phagwara, India
Atul Sharma	Lovely Professional University, Phagwara, India
Atul Upadhyay	Thapar University, Punjab, India
Avnish Tiwari	Lovely Professional University, Phagwara, India
Azher Ashraf Gadoo	Lovely Professional University, Phagwara, India
B. Arun Kumar	Lovely Professional University, Phagwara, India
Balachandra Posa	Lovely Professional University, Phagwara, India
Balpreet Singh	Lovely Professional University, Phagwara, India
Bankuru Gowthami	Lovely Professional University, Phagwara, India
Bharath Chandra	CETPA Infotech Pvt. Ltd., Uttar Pradesh, India
Bhawna Bhawna	Lovely Professional University, Phagwara, India
Bhupender Thakur	Lovely Professional University, Phagwara, India

Chandini Kumari	Lovely Professional University, Phagwara, India
Charanjeet Singh	Lovely Professional University, Phagwara, India
Cherry Bhargava	Lovely Professional University, Phagwara, India
Chirag Chopra	Lovely Professional University, Phagwara, India
Daljeet Singh Dhanjal	Lovely Professional University, Phagwara, India
Deepak Parmar	Lovely Professional University, Phagwara, India
Devesh Kumawat	Lovely Professional University, Phagwara, India
Dibbyo Bhattacharjee	Lovely Professional University, Phagwara, India
Dharam Buddhi	Lovely Professional University, Phagwara, India
Dushyant Kumar Singh	Lovely Professional University, Phagwara, India
E. Pavan Kumar	Lovely Professional University, Phagwara, India
Enjeti Amareswar	Lovely Professional University, Phagwara, India
Farhan Alam	Lovely Professional University, Phagwara, India
Frank Bruno	Lovely Professional University, Phagwara, India
Gade Vivek	Lovely Professional University, Phagwara, India
Gopalchetty Brahma	Lovely Professional University, Phagwara, India
Gourav Singh	Lovely Professional University, Phagwara, India
Gurbakash Phonsa	Lovely Professional University, Phagwara, India
H. Pal Thethi	Lovely Professional University, Phagwara, India
Harjeet Kaur	Lovely Professional University, Phagwara, India
Harjit Singh	Lovely Professional University, Phagwara, India
Harpreet Singh Bedi	Lovely Professional University, Phagwara, India
Himanshu Singh	Lovely Professional University, Phagwara, India
Indresh Kumar Agnihotri	Lovely Professional University, Phagwara, India
Jagannadha Varma Mandhapati	Lovely Professional University, Phagwara, India
Jagjit Kaur	Lovely Professional University, Phagwara, India
Jahangeer Sofi	Lovely Professional University, Phagwara, India
Jasvinder Pal Singh	Lovely Professional University, Phagwara, India
Jyoti Ganai	Jamia Hamdard, New Dehli, India
Kanksha Kaur	Lovely Professional University, Phagwara, India
Kapil Kumar	BIT, Meerut, India
Kesu Manoj Kumar	Lovely Professional University, Phagwara, India
Koushik Barman	Lovely Professional University, Phagwara, India
Krishna Uday K.V.	Lovely Professional University, Phagwara, India
Kunwar Shahbaaz Singh Sahi	Lovely Professional University, Phagwara, India

Amity University Yadav	Uttar Pradesh, India
Liang Bo	Guangzhou Rising Dragon Recreation Industrial Co. Ltd., Guangzhou, China
Lokeshwar Reddy Lingala Bayannagari	Lovely Professional University, Phagwara, India
Mahendra Joshi	Lovely Professional University, Phagwara, India
Mahvish Qaiser	Jamia Hamdard, New Dehli, India
Manali Gupta	Gautam Buddha University, Uttar Pradesh, India
Manish Kumar	Lovely Professional University, Phagwara, India
Manisha Yadav	Lovely Professional University, Phagwara, India
Manjit Kaur	Lovely Professional University, Phagwara, India
Manoj Kumar Jena	Lovely Professional University, Phagwara, India
Manoj Sindhwani	Lovely Professional University, Phagwara, India
Manukonda Manoj Kumar	Lovely Professional University, Phagwara, India
Maries Tahaab	Lovely Professional University, Phagwara, India
Martin Belusko	Mondial Advisory, Malvern, Australia
Max Bhatia	Lovely Professional University, Phagwara, India
Mayank Singh Chouhan	Lovely Professional University, Phagwara, India
Md. Iman Ali	Lovely Professional University, Phagwara, India
Megha Khatri	Lovely Professional University, Phagwara, India
Megha Malhotra	Amity University, Uttar Pradesh, India
Mekapotula Bhuvan Sundhar Reddy	Lovely Professional University, Phagwara, India
Ming Liu	University of South Australia, Mawson Lakes, Australia
Mohit Arora	Lovely Professional University, Phagwara, India
Nahid Khan	Jamia Hamdard, New Dehli, India
Namra Samin	Amity University, Uttar Pradesh, India
Narbada Prasad Gupta	Lovely Professional University, Phagwara, India
Narendra Gali	Lovely Professional University, Phagwara, India
Neelesh Kumar Gupta	AKGEC, Ghaziabad, India
Nelavelli Chandu	Lovely Professional University, Phagwara, India
Nitin Tandon	Lovely Professional University, Phagwara, India
Ovass Shafi Zargar	Lovely Professional University, Phagwara, India
P. Heam Kumar	Lovely Professional University, Phagwara, India
P.B. Karandikar	Army Institute of Technology, Pune, India
Paila Akhil	Lovely Professional University, Phagwara, India
Pallala Akhil Reddy	Lovely Professional University, Phagwara, India

Pankaj Dhaiya	Delhi Technological University, New Delhi, India
Pankaj Jain	Lovely Professional University, Phagwara, India
Parulpreet Singh	Lovely Professional University, Phagwara, India
Pooja Pathak	GLA University, Mathura, India
Pranay Raju Konduru	Lovely Professional University, Phagwara, India
Pratibha Kaushal	Lovely Professional University, Phagwara, India
Pratyaksh Kumar Singh	Lovely Professional University, Phagwara, India
Prem Kapur	Hamdard Institute of Medical Sciences and Research Associated HAHC Hospital, New Delhi, India
Priyanka Shukla	Lovely Professional University, Phagwara, India
Raghav Gupta	Lovely Professional University, Phagwara, India
Rajesh Singh	Lovely Professional University, Phagwara, India
Rajeshwar Singh	Doaba Group of Colleges, Punjab, India
Ram Gopal Nagaboina	Lovely Professional University, Phagwara, India
Ramanpreet Kaur	Guangzhou Rising Dragon Recreation Industrial Co. Ltd., Guangzhou, China
Raminder Kaur	Lovely Professional University, Phagwara, India
Rashi Kaushik	Lovely Professional University, Phagwara, India
Ravi Prakash	Lovely Professional University, Phagwara, India
Reena Singh	Lovely Professional University, Phagwara, India
Rhys Jacob	Mondial Advisory, Malvern, Australia
Rohit Saini	Lovely Professional University, Phagwara, India
Rongsennungsang Jamir	Lovely Professional University, Phagwara, India
Roshi Saxena	Gautam Buddha University, Uttar Pradesh, India
Runjhun Tandon	Lovely Professional University, Phagwara, India
Sahil Uttekar	Army Institute of Technology, Pune, India
Samia Mushtaq	Lovely Professional University, Phagwara, India
Sanjay Kumar Sharma	Gautam Buddha University, Uttar Pradesh, India
Sanjeev Singh	Lovely Professional University, Phagwara, India
Santosh Kumar Nanda	Techversant Infotech Pvt. Ltd, Kerala, India
Sarthak Patra	Amity University, Uttar Pradesh, India
Saurabh Satija	Lovely Professional University, Phagwara, India
Shailza Roy	Amity University, Uttar Pradesh, India
Shalabh Dwivedi	Lovely Professional University, Phagwara, India
Shama Kakkar	Lovely Professional University, Phagwara, India
Shamik Chatterjee	Lovely Professional University, Phagwara, India

Shane Sheoran	University of South Australia, Mawson Lakes, Australia
Shamik Chatterjee	Lovely Professional University, Phagwara, India
Shiva Prasad A.	Lovely Professional University, Phagwara, India
Shubhangi Das	Lovely Professional University, Phagwara, India
Soheila Riahi	University of South Australia, Mawson Lakes, Australia
Sudarshan Sharma	Lovely Professional University, Phagwara, India
Sukhkirandeep Kaur	Lovely Professional University, Phagwara, India
Sumit Bhattacharjee	Lovely Professional University, Phagwara, India
Sunil Krishnan G.	Lovely Professional University, Phagwara, India
Suresh Kumar Sudabattula	Lovely Professional University, Phagwara, India
Tawseef Ahmed Teli	Higher Education Departments, Amar Singh College, J&K
Teetla Anand	Lovely Professional University, Phagwara, India
Tina Bakshi	Lovely Professional University, Phagwara, India
Vankudoth Dinesh	Lovely Professional University, Phagwara, India
Vartika Bhadana	GLA University, Mathura, India
Vikas Kaushik	Lovely Professional University, Phagwara, India
Vikrant Sharma	Lovely Professional University, Phagwara, India
Vinay Thakur	Lovely Professional University, Phagwara, India
Vineeth Changarangath	Lovely Professional University, India
Vishal Agrawal	Lovely Professional University, Phagwara, India
Warsha Jagati	Amity University, Uttar Pradesh, India
Wenhui Xiao	Guangzhou Rising Dragon Recreation Industrial Co. Ltd., Guangzhou, China
Yarra Raviteja	Lovely Professional University, Phagwara, India
Yash Shendokar	Army Institute of Technology, Pune, India
Yedida Venkat Lakshmi	Lovely Professional University, Phagwara, India
Yogesh Kumar	Amity University, Uttar Pradesh, India

CHAPTER 1

Power and Energy Density Analysis of Various Propulsion Systems

Sahil S. Uttekar[1,*], Yash V. Shendokar[1], Aanchal Gupta[1], Poorva Aparaj[1] and **Parashuram B. Karandikar[1]**

[1] Department of Mechanical Engineering, Army Institute of Technology, Pune, India

Abstract: In this paper, the energy and power density of vehicles, energy-storing devices, and motors have been revisited. Hybrid propulsion systems like amphibious vehicles are considered along with normal propulsion modes. Power density reflects the amount of power delivered, and energy density reflects the range of systems. Power and energy densities of various transportation such as Electrically powered vehicles (EVs) Internal Combustion Engine (ICE) vehicles, Hybrid Vehicles (HVs), underwater vehicles, airplanes, ships, and hybrid mode transportation systems, are analyzed. The energy and power density of multi-mode propulsion are also calculated to explore the possibility of converting existing automobiles to multimode transportation systems as a part of retrofitting in the future. Power density matching for energy sources, ICE, motor, and coupling can lead to better operation of the propulsion system. This concept is parallel to the maximum power transfer theorem. Power density matching of all components associated with power flow can lead to better stability, lesser vibrations, and reduced maintenance of the system components. It reduces the need for frequent replacements of power components due to power stress. Energy density matching of all power flow components is equally important. A detailed study by considering various vehicles is presented in this paper.

Keywords: Amphibious Vehicles, Battery, Electric Vehicle, Energy density, Engine, Fuel Cell, Hybrid Vehicles, Motor, Power density, Retrofitting, Seaplanes, Submarines, Super-Capacitor, Ultra-Capacitor.

INTRODUCTION

The electric vehicle was first developed in the 1830s [1]. The principal benefits of these electrically powered vehicles are that these are pollution-free, easy to drive, have more smart features than internal combustion engines, *etc.* The adoption of these vehicles is very less due to the electricity problem. Also, the recharging time of these vehicles is more as compared to Internal Combustion Engine (ICE)

*** Corresponding author Sahil S. Uttekar:** Department of Mechanical Engineering, Army Institute of Technology, Pune, India; E-mail: sahiluttekar4441@gmail.com

Dharam Buddhi, Rajesh Singh and Anita Gehlot (Eds.)

vehicles. The cost of these vehicles and range are not specific. Still, there are many confusions to select an Electric Vehicle's (EVs). Even though E-mobility is a cleaner alternative for the ICE vehicles, they are still not as good as in terms of power density and issues on refueling. The first batteries of lithium-ions were commercially produced by Sony in the early 1990s. These batteries were originally used specifically for small-scale consumer items such as cell phones, wall clocks, remotes, *etc* [2]. and with the new technologies and research. EV industries are taking bigger strides to become a common household commodity in the coming decades. There is a need to relook at the energy and power densities of newer propulsion systems. Gasoline is the abundantly used form of energy, mostly because of its transferability and ease of storage. Whereas the energy extracted from natural sources is not continuous and mainly relies on weather conditions and seasons, As renewable energy sources now become increasingly significant, there has been a growing curiosity among researchers to store this clean energy. Thus, different energy-storing devices are being developed; besides, their performance has also been enhanced extremely.

The Ragone plot named D.V. Ragone is a very famous guideline for energy storing devices [3]. The Ragone plot is commonly used for energy storing device's development for differentiating, blending, and anticipating future energy storing devices. The Ragone plot provides intuitive and systematic information of devices such as fuel cells, capacitors, *etc* [4]. The various EV brand manufacturers are providing battery packs that provide energy density of up to 200W-h/kg and the industry expects increasing electric density over to 500W-h/kg [5], but the power density of batteries still is around 500 W/kg [6]. Whereas the supercapacitors have an energy density of 20 W-h/kg and power density of 10 kW/kg. The ultracapacitors (super-capacitors) can withdraw very high power, but their energy-storing capacity is minimal. The energy density of petrol is to be 100 times greater than that of a battery, but do not reveal any implications that EVs will outweigh conventional vehicles to be able to run the same distance [7].In EV or Hybrid Electric Vehicles (HEVs) energy density and power density need to be investigated from a retrofitting point of view. Vehicle weight reduction also impacts the energy efficiency of a vehicle's configuration. In conventional ICE applications, a one-tenth weight reduction may result from a 6%–8% improvement in fuel economy. Both the conventionally powered vehicles and EVs vehicles are benefitting due to weight-optimization using new materials. Firstly, materials are often want to reduce the load of the vehicle, and secondly, materials can enable higher efficiency engines and power trains [8]. Energy density, power density mismatching in power train components, and energy storage devices need to be investigated. This is a major research gap found by the authors. Such analysis can give direction to society about the reuse of vehicles in different modes. This paper aims at discussing the Energy and Power density of various automobiles, Eelectrical storage devices and Amphibious Vehicles and Seaplanes followed by concluding remarks.

Energy Density and Power Density of Various Automobiles

Heat engines convert thermal energy into a temperature gradient to produce mechanical work on a rotating shaft. Whereas electric motors use electrical energy to produce any mechanical work. The power to weight ratio or the power density in our case is used to measure the actual performance of any vehicle's engine. It is calculated to enable the comparison of the vehicle's performance with each other. If a system or a vehicle is said to have a high specific power, it means that it can withdraw a large amount of energy based on its volume or its weight. Since the energy is released in a small-time, a high-power density system can also recharge rapidly. The energy density can be calculated by using the enthalpy of the combustion of fuel. Table **1** contains the enthalpy of combustion of different fuels. This energy is calculated by multiply with the fuel capacity of a vehicle and thus taking its ratio with the weight of a vehicle. The higher the specific energy of a system, the greater will be the amount of energy stored in its body, and hence, the greater will be the range of the vehicle.

Table 1. Energy density of various fuels [9 - 11].

Fuel type	The energy density of fuel W-h/L)	The energy density of fuel (W-h/Kg)
Gasoline	9,500	12,888
CNG	2,500	14,888
Jet fuel	9722.22	11944.44
Diesel	10,700	12,666
Hydrogen Gas,1atm, 250C	2.8	33,313

Power density gives power swept per volume or brake horsepower per cm^3 (Fig. **1**). It is based on the internal energy capacity of the engine. Power to fuel capacity ratio is used to compare the efficiency of the vehicles based on their maximum distance covered in one full consumption cycle of the full fuel tank (Fig. **2**). This is the basic assumption for the comparison of various vehicles. For calculating parameters of various power densities of different kinds of two-wheelers, four-wheelers, heavy-duty motor cargos, and locomotive engines, aircraft, and watercraft, their engine displacements, maximum power, fuel capacity, and gross weight were collected [12].

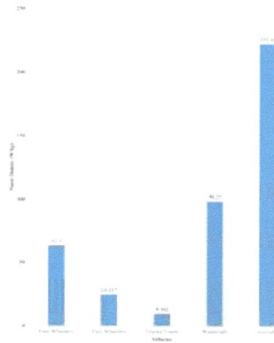

Fig. (1). Power to Gross Weight Ratios (Power density) of various vehicles [13 - 15].

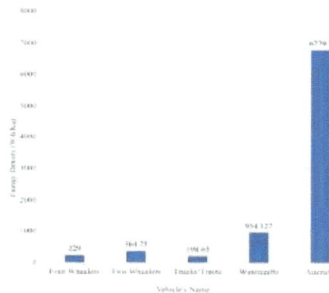

Fig. (2). Power to fuel tank capacity ratios (Energy Density) of various vehicles [13 - 15].

E.D. and P.D. of Electrical Storage Devices

The storing of energy and power density in electrical devices is based on some factors like cycle efficiency, life cycles, charge/discharge characteristics, *etc*. Also, there are some constraints related to space and the existing allowable weight. In the ICE the energy storage is at the flywheel, secondary batteries or taken from the fuel cells. In electrical devices, battery, capacitor, fuel cells, supercapacitor, or ultra-capacitor are keywords for storing energy. In the market, there are various types of batteries like lead-acid, lithium-titanate, lithium-ion batteries, *etc*. The usage of lithium-ion batteries is versatile, and these can be found in a laptop, electric vehicles, *etc*. Whereas, the lead-acid battery is also used in electric vehicles and emergency lighting. Due to some factors, the storing of energy and power density in batteries differ. The main parameter in the battery is the life cycle; if it increases, it increases the power and energy density. The below chart represents different batteries comparison of energy and power density. The Power density is denoted by the orange bar and energy density by the blue bar, as shown in Fig. (**3**).

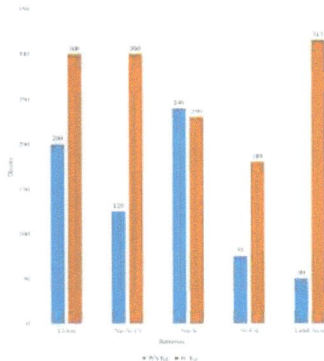

Fig. (3). Comparison of various batteries in terms of energy and power density [16].

In this transportation world, fuel cells have also contributed by producing zero emissions. Recently, Toyota announced "MIRAI" as their first hydrogen fuel cell vehicle which will be launched in 2020 [17]. The use of fuel cells in the automotive field is in the beginning stage though it has some advantages and disadvantages. Much research had contributed to the interest in the storage of hydrogen fuel cells [18]. Many types of fuel cells are used as per their application. Some of those fuel cells are presented in below Table **2** with their respective power densities.

Table 2. Energy and Power densities of various fuel cells [19].

Types of Fuel cells	Energy Density (W-h/kg)	Power Density (W/kg)
Proton exchange membrane fuel cells (PEMFC)	54	18
Direct Methanol fuel cells (DMFC)	290	15
Alkaline fuel cells (AFC)	160	11
Phosphorous Acid fuel cells (PADC)	95	32

When it comes to electric energy storage, supercapacitors or ultra-capacitor are also used in between capacitors and batteries. The biggest advantage of supercapacitors is that their life cycle and power density are high as compared to others. Due to this advantage, the storage of energy is sufficient. It is stated that the power density and energy density of fabricated supercapacitor is 496 W/kg and 206 W-h/kg at 0.25 A/g current density and at 50 A/g current density, the P.D is 32 kW/kg and E.D is 9.58 W-h/kg [20].

Amphibious Vehicles and Seaplanes

Amphibious Vehicles are flat-bottomed motor vehicles that can operate on both lands and in water with an opposed pair of flotation side hulls and a rear hull. They can be further classified into two kinds: that travel on air cushions (hovercraft or Air Cushion Vehicle (ACVs)), and those that are not designed to broaden an off-road capability to all-terrain capabilities. One simple way to convert land vehicles into amphibious vehicles is to install them with a waterproof or hydrophobic hull and a propeller, which is feasible only because of its displacement. Amphibious vehicles include the categories of ATVs, cars, combat vehicles, boats, *etc.* are presented in Table **3** and seaplanes in Table **4**. The increasing utilization of these vehicles by military forces to secure the sea borders with the increasing water sports activities and water transportation are accelerating the amphibious vehicles market growth.

Table 3. Amphibious Vehicles with their respective ratios [21 - 23].

Amphibious vehicles	Power to Gross Weight Ratio (W/kg)	Power to Fuel Capacity Ratio (W-h/kg)
AAVP7A1 Assault Amphibious Vehicle	11	460
Panther Car-boat	170	61500
Kailai	12	1040

Table 4. Seaplane Models with their respective ratios [25 - 27].

Seaplanes models	Power to Gross Weight Ratio (W/kg)	Power to Fuel Capacity Ratio (W-h/kg)
Cessna 180	135	697
1985 CESSNA 152	107	833
Cessna 207	134	908

Seaplanes are powered flying machines that are designed in a way such that they take off and land on the water bodies. They come in the subclass of the amphibious vehicles and use fluid dynamical lift to glide off the water body while moving above a surface at a pace. The fuselage of a floatplane is not intended to touch the water. A flying boat is a fixed-winged seaplane that can land on the water body surface with the help of a hull, which generally does not have any type of landing gears that is commonly used operation on the plane's landing. These machines are stable on a water surface due to their wing-shaped projection known as sponsons from the main body. These can generally handle rougher rougher water and are more stable than floatplanes on the water surface, as shown in Fig. (4).

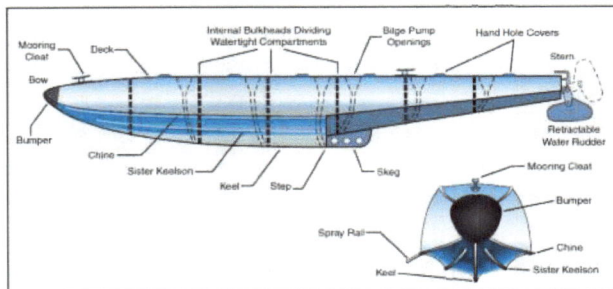

Fig. (4). Float components of a seaplane [24].

Some vehicles are used underwater. One of the vehicles is an Autonomous underwater vehicle (AUV). It is mostly used by the marine industry and by the military. These vehicles are controlled by the computer [28]. As per the researcher, power density is 11 W/kg, and Energy density is 110 W-h/kg [29]. Another underwater vehicle is the submarine which is widely used to cover aspects like research and development in the marine industry. Military activities are also carried out by submarines [30]. The submarine's key features are GPS and radar navigation which are widely used. The Power density of a typhoon class submarine is 9.47 W/kg, and Energy density is 28,0512 W-h/kg [31].

Table **5** gives the approximate data on the comparison of various systems in power trains with continuous power density. This power density should not be misunderstood as peak power density, about five times the continuous power density. It is found that motors' power density is high as compared to the battery. By increasing the power density of the battery, we can match its power density with motors. Similarly, matching the power density of motor and IC engines is very important in hybrid electric vehicles as well as hybrid mode-operated vehicles. Thus, all components of the power train should have continuous power density in the same range to balance power stresses.

Table 5. Comparison of Power Densities of various major components of the power train [32 - 34].

Sources	Types	Power density W/kg(approx.)
Motor	Permanent magnet synchronous motor	920
	BLDC Motor	590
	Single AC permanent magnet motor	2300

(Table 5) cont.....

Sources	Types	Power density W/kg(approx.)
IC Engine	4 Stroke (Diesel)	75
	4 Stroke (petrol)	170
Battery	Lithium-ion	315
	Lead-acid	300
	Ni-Cd	300

CONCLUSION

We are approaching a world in which the conventional internal combustion engine may become partially obsolete by replacing it with EVs. This research analyses different ICE-powered vehicles, also various non-traditional vehicles in terms of power and energy density in their respective engines. This study also gives insights into the different available battery technologies in terms of energy and power density. The terms energy density and power density provide an important platform to compare these two different power-producing devices *i.e.* Batteries, fuel cells, and supercapacitors. Motor and IC engine's energy density and power density should match with the source from which energy is obtained for propulsion. Despite the efforts in research and advancement of new technology, there is a significant gap between the energy density and power density produced by the batteries and engine. Therefore, this study provides a starting point for the upcoming researchers to aim to achieve these values to design batteries for vehicles with the variation of the application of transportation in air, water, and land. Lightweight two-wheelers and boats can be redesigned to make affordable multi-mode propulsion systems. Energy density matching of sources and consumer components will reduce thermal stresses on system components.

CONSENT FOR PUBLICATION

Not applicable.

CONFLICT OF INTEREST

The authors declare no conflict of interest, financial or otherwise.

ACKNOWLEDGEMENTS

The authors would like to thank the management of the Army Institute of Technology, Pune, for encouraging this kind of innovative student.

REFERENCES

[1]   https://www.iedconline.org/clientuploads/Downloads/edrp/IEDC_Electric_Vehicle_Industry.pdf

[2]   https://www.sony.net/sonyinfo/corporateinfo/History/sonyhistory/2-13.html

[3]   D. Ragone, "Review of Battery Systems for Electrically Powered Vehicles", *SAE Technical Paper,* p. 680453, 1968.
        [http://dx.doi.org/10.4271/680453]

[4]   S. Lee, and W. Jung, "Analogical understanding of the Ragone plot and a new categorization of energy devices", *Energy Procedia,* vol. 88, pp. 526-530, 2016.
        [http://dx.doi.org/10.1016/j.egypro.2016.06.073]

[5]   https://www.iea.org/reports/technology-roadmap-electric-and-plug-in-hybrid-electric-vehicles

[6]   W. Zuo, R. Li, C. Zhou, Y. Li, J. Xia, and J. Liu, "Battery-Supercapacitor Hybrid Devices: Recent Progress and Future Prospects", *Adv. Sci. (Weinh.),* vol. 4, no. 7, p. 1600539, 2017.
        [http://dx.doi.org/10.1002/advs.201600539] [PMID: 28725528]

[7]   R. Vijayagopal, K. Gallagher, D. Lee, and A. Rousseau, *Comparing the Powertrain Energy Densities of electrical and Gasoline Vehicles.*, 2016, pp. 533-540.

[8]   https://www.drivealuminum.org/research-resources/impact-of-vehicle-weight-reduction--n-fuel-economy-for-various-vehicle-architectures/

[9]   IOR Energy, *"List of common conversion factors (Engineering conversion factors)",* 2008.

[10]   "College of the Desert", *Module 1, Hydrogen Properties,* 2001.

[11]   "Characteristics of Petroleum Products Stored and Dispensed", *Petroleum Products Division.* 122-135.

[12]   N.G. Zacharof, and G. Fontaras, *Review of in use factors affecting the fuel consumption and CO_2 emissions of passenger cars* vol. 1. , 2016, pp. 7-170.

[13]   http://www.motorbikesindia.com/types-of-two-wheelers/

[14]   https://www.marineinsight.com/guidelines/a-guide-to-types-of-ships/

[15]   https://www.caranddriver.com/shopping-advice/g26100588/car-types/

[16]   L. Jorissen, and H. Frey, *Energy Storage,* vol. 1, pp. 215-231, 2009.

[17]   https://www.toyota.com/mirai/fuel.html

[18]   H.T. Hwang, and A. Varma, "Hydrogen storage for fuel cell vehicles", In: *Current Opinion in Chemical Engineering,* vol. 5. , 2014, pp. 42-48.

[19]   http://earthsci.org/mineral/energy/fuelcell/fuelcell.html#AFC

[20]   J.R Rani, R Thangavel, Y.S Lee 2, and J.-H. Jang, "An Ultra-High-Energy Density Supercapacitor; Fabrication Based on Thiol-functionalized Graphene Oxide Scroll", *Nanomaterials (Basel).,,* vol. 9, no. 24, pp. 1-12, 2019.

[21]   https://fas.org/man/dod-101/sys/land/aavp7a1.html

[22]   https://www.autoblog.com/2013/06/25/watercar-panther-looks-to-wrangle-the-road-and-open-water/

[23]   https://cncltruck.en.alibaba.com/product/62584339650-807395128/China_supply_Amphibious_rescue_engineering_vehicle_for_reasonable_price.html

[24]   *Principles of seaplanes.*, 2004, pp. 2-4.

[25]   https://www.seaplanepilotsassociation.org/resources/the-largest-seaplane-resou-ce-collection/seaplanes-specs/cessna-180/

[26]   https://www.manualslib.com/manual/1471611/Cessna-152.html

[27]   www.flugzeuginfo.net/acdata_php/acdata_cessna_207_en.php

[28]   J.G. Bellingham, "Platforms", *Autonoums Underwater Vehicles,* vol. 1, pp. 473-484, 2009.

[29]   https://www.imas.utas.edu.au/antarctic-gateway-partnership/autonomous-underwater-vehicle-auv-specifications

[30]   C. Gong, X. Zhu, F. Wang, P. Jiang, and W. Yu, *Research on Submarine Straight-Line Track Control Underwater Based on Nonlinear Proportional Development,* vol. 1, pp. 1-7, 2016.

[31]   http://www.military-today.com/navy/typhoon_class.htm

[32]   https://www.electricmotorsport.com/me1616-brushless-65hp-liquid-cooled-ipm-motor-24-120v.html

[33]   https://www.servovision.com/hub%20motor/Motor%2010%20KW/index.html

[34]   https://www.evwest.com/catalog/product_info.php?cPath=8&products_id=297&osCsid=8vth5vkrdvpl m6k44hscisgd30

CHAPTER 2

Classification of P2P-VoIP (Video) Traffic Using Heuristic-Based and Statistical-Based Technique

Max Bhatia[1,*] and **Vikrant Sharma[1]**

[1] Department of Computer Science Engineering, Lovely Professional University, Punjab, India

Abstract: In recent years, VoIP technology has become very popular among internet users as it allows the users to transfer their voice & video over an IP network. Its advantage is that it is cost-effective, works over traditional telephone networks, and is also compatible with the public switched telephone network. It is based on P2P architecture and allows users to communicate with each other in real-time through audio or video conferencing over the network. It consumes a lot of network bandwidth. That is why, the classification of VoIP traffic is important from the viewpoint of monitoring, prioritizing, or blocking such traffic by the ISPs or network administrators. Traditional classification techniques such as port-based and payload-based are ineffective since modern VoIP applications make use of random port numbers, encryption, or proprietary protocol to obfuscate their traffic. In this paper, we focus on classifying VoIP (video) traffic by utilizing a 2-step classification process. The 1st step uses a packet-level process, where P2P-port based technique is utilized to classify VoIP traffic. The traffic which remains un-classified as VoIP is then fed to the 2nd step classification process (*i.e.* flow-level process), which combines proposed heuristic rules (with a unique packet size distribution of VoIP traffic) and statistical-based technique to classify VoIP traffic. The experiments have been conducted on real traffic traces using offline datasets and results show that the proposed technique not only achieves high classification accuracy of over 98.6% but also works with both TCP & UDP protocols and is not affected even if traffic is encrypted.

Keywords: Heuristic-Based Classification, Statistical-Based Classification, Traffic Classification, P2P Port-Based Classification, P2P-VoIP Traffic.

1. INTRODUCTION

Voice over Internet Protocol (VoIP) is an internet technology that provides the ability to transfer voice and media sessions over the IP networks. In recent years, Peer-to-Peer (P2P) Voice over Internet Protocol (VoIP) applications have become

* **Corresponding author Max Bhatia:** Department of Computer Science Engineering, Lovely Professional University, Punjab, India; E-mail: max.16870@lpu.co.in

Dharam Buddhi, Rajesh Singh and Anita Gehlot (Eds.)

very popular among individuals and enterprises due to high bandwidth connectivity and low cost in comparison to traditional Public Switched Telephone networks (PSTN) [1]. With the evolution of such applications, it provides better quality of voice & video, free communication between users, and can circumvent the restrictive network environments such as NAT (Network Address Translation) and firewalls.

A VoIP infrastructure typically consists of VoIP clients and signaling servers for call establishment, authentication, and associated services. In addition to that, it may also include additional servers for accelerating media transport, achieving traversal of the media path, and interfacing with PSTN and mobile networks. Recently, there is tremendous growth in VoIP traffic as it is becoming a major communication service for individuals and enterprises [2]. Classifying VoIP traffic can help ISPs and enterprises to prioritize such type of traffic in the network and enforce policies for network monitoring, load balancing, flow control, managing network bandwidth, providing quality of service, enforcing intrusion detection & prevention services, and auditing.

There are several challenges in classifying VoIP traffic accurately since many applications such as Skype, Google-meet, *etc.*, obfuscate/hide their traffic by making use of random port numbers, encryption, or proprietary protocol for communication. Conventional techniques include port-based and payload-based techniques for classifying network traffic. The port-based technique is the oldest & simplest technique to classify the traffic by using well-known port numbers ranging between 0-1023, which are assigned by IANA [3] to various protocols such as FTP, DNS, HTTP, *etc*. But, this technique is ineffective in classifying the traffic, which uses random or dynamic port numbers for communication. The payload-based technique (also known as Deep Packet Inspection) relies on packet payload and is the most accurate technique in classifying the traffic. It examines the packet payload to search for application-specific signatures and maps it with the database containing the signatures of previously stored application protocols. However, this technique also suffers from various limitations such as a) unable to deal with encrypted traffic, b) involves a lot of processing load and complexity, c) infeasible in high-speed networks, d) need to find application signatures every time as new application protocol emerges, e) leads to breach of some organization privacy policies by direct inspection of the packet payload, *etc.* [4, 6]. Therefore, traditional classification techniques *i.e.*, port-based and payload-based techniques, are ineffective in classifying VoIP traffic, and since they are conventional techniques, so its related work is referred to in [7]. Currently, modern techniques (known as Classification in the Dark) are being employed to classify traffic which makes use of statistical/heuristic-based techniques. Statistical-based techniques classify traffic using statistical features calculated from the traffic, such as packet length, flow duration, the number of packets sent, number of packets received, inter-arrival time of packets, *etc* [4]. On the other hand, the heuristic-based technique classifies traffic using a pre-defined set of rules by observing the behavioral patterns of the traffic; such as the number of out-going connections of a host, host acting as both client & server, number of ports used by a host, *etc*.

The main objective of this research work is to classify P2P-VoIP (video) traffic. Many modern VoIP applications have the functionality to make voice calls, video calls, file transfer, and chat, but we specifically focus on classifying video traffic which is generally used for video conferencing or conducting online meetings. For this purpose, we propose a 2-step hybrid classification approach which is categorized into packet-level and flow-level classification processes. Packet-level classification process uses of P2P-port based classification technique whereas flow-level classification process uses a combination of heuristic-based and statistical-based techniques for classifying VoIP (video) traffic. The experiments have been conducted on 3 popular VoIP applications, namely Skype, Zoom & Google-meet and the results show that the proposed technique not only achieves high classification accuracy (*i.e.* 98.6%) but also works with both TCP & UDP protocols and is not affected even if traffic is encrypted.

The remaining paper is organized as follows. Section 2 discusses the related work. Section 3 discusses the proposed methodology to classify VoIP (video) traffic. Section 4 discusses evaluation criteria and experimental results. Finally, Section 5 concludes the research work.

2. RELATED WORK

Jiang *et al.* [8] analyzed the network structure of Skype and conducted experiments to observe its new communication pattern after its acquisition by Microsoft. They designed a methodology to detect Skype users in real-time by analyzing the log-in & log-out phases of Skype in each traffic flow. The authors evaluated their technique using 11 hosts (where 8 hosts were running Skype) in an actual network environment and claimed that Skype users could be accurately & quickly identified using this approach. Yuan *et al.* [9] used an automated packet-sequence signature construction system to construct packet sequence signatures from the application payloads and discovered the sequence of signatures generated by Skype UDP flows. Their technique utilized the combination of login signal detection (to search for '0x02' string present in packet payloads during Skype login session) and destination IP-address lookup (which is one of the destination IP-address used for authentication purposes from the list of IP-addresses used by Skype servers) to identify Skype traffic. The experimental results achieved 98.93% precision and 99.54% recall, but their approach relied on a payload-based technique which has various limitations. Lee *et al.* [10] classified

Skype traffic by combining pattern-based and signature-based techniques. It consists of 3 modules which were applied to the traffic in the following sequence: a) login detection (which used pattern-based detection method), b) list-based detection (which used list of IP-port information of Skype detected clients fetched during login detection process) and c) signature-based detection (which used IP correlation and IP-based detection). The experimental results achieved a detection rate of 95%. Saqib *et al.* [11] proposed a hybrid technique based on behavioral and statistical analysis to detect and classify VoIP voice packets over IP networks. The 1st step uses behavioral analysis to separate voice and non-voice packets. The 2nd step employed a proposed voice detection algorithm that uses statistical traffic features to confirm further and classify VoIP traffic. But, the proposed technique focused on classifying VoIP voice traffic, and experimental results achieved true positive rates of 93.6% & 95% for offline & online traffic traces, respectively. Munir *et al.* [12] performed an analysis on VoIP and non-VoIP traffic and proposed a statistical-based technique that can classify unencrypted, encrypted, and tunneled VoIP-voice traffic. For this purpose, they formulated rules based on threshold values of 9 statistical parameters (*i.e.* packet rate, mean packet size, standard dev*ia*tion of time difference, *etc.*) of traffic flows and achieved a detection rate of 97.165%. Datta *et al.* [13] proposed a technique to classify Google Hangouts traffic by observing the application behavior. They extracted a set of 7 statistical features by analyzing the connection behavior of Google Hangouts and used them in 3 different ML algorithms (*i.e.* Naïve Base, AdaBoost & J48) to assess the classification performance. The authors performed dataset collection and evaluation in a controlled network environment. The experimental results found that J48 performs comparatively better in classification and achieved recall rates ranging between 99.99% - 100%.

The technique proposed in this paper focuses on classifying P2P-VoIP (video) traffic specifically since various government organizations and enterprises are currently employing video conferencing to run their businesses [2]; hence such traffic is contributing a lot to the overall P2P traffic on the internet. For this purpose, a combination of heuristic-based and statistical-based techniques have been utilized to classify VoIP traffic.

3. P2P-VOIP TRAFFIC CLASSIFICATION METHODOLOGY

A 2-step classification process (*i.e.*, packet-level & flow-level) has been employed to classify VoIP traffic, as shown in Fig. (**1**). The packet-level process utilizes the P2P-port- based technique to classify VoIP traffic flows, whereas   the   flow-level

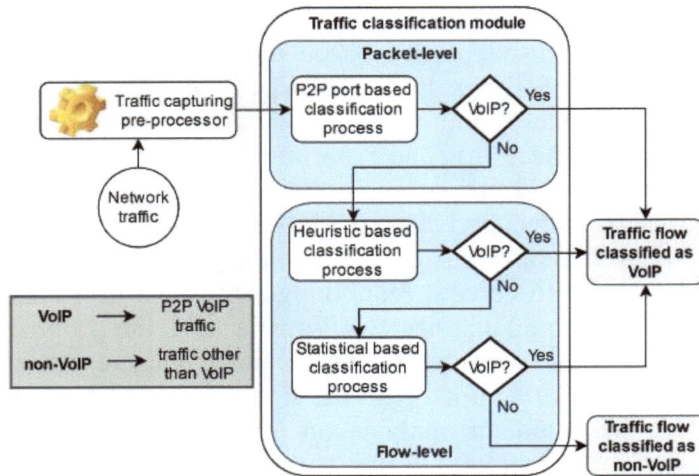

Fig. (1). P2P-VoIP traffic classification technique.

process utilizes the combination of heuristic-based & statistical-based techniques to classify the remaining traffic flows, which could not be classified as VoIP in the 1st step.

A traffic flow is generally defined as the combination of 5-tuples (*i.e.* source-IP, destination-IP, source-port, destination-port, protocol). When 2 hosts are involved in VoIP communication, packets travel in both directions (*i.e.* from source to destination and vice-versa). Therefore, during the classification process, the hash-key of packets is calculated by concatenating 5-tuple flow information (as shown in Fig. **2**), so that packets associated with the same traffic flow (traveling in either direction) can be identified; since they

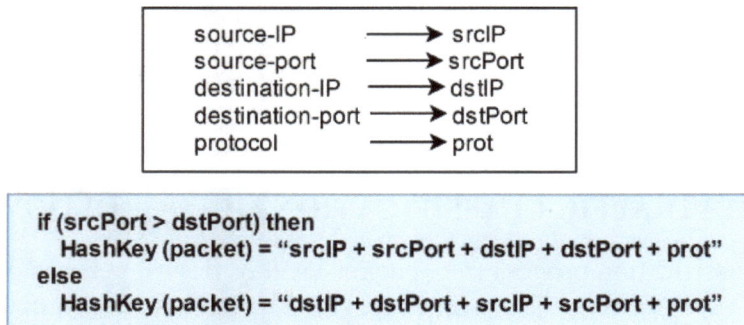

```
if (srcPort > dstPort) then
    HashKey (packet) = "srcIP + srcPort + dstIP + dstPort + prot"
else
    HashKey (packet) = "dstIP + dstPort + srcIP + srcPort + prot"
```

Fig. (2). Calculation of hash-key of the packet.

will have the same hash-key. This hash-key is mainly used to find whether the flow of a packet is already classified as VoIP or not. Here, a P2P flow table is used to store both the flow details & packet-hash information of those flows which have already been classified as VoIP.

In the packet-level classification process, P2P-port based technique is employed to check if a traffic flow is using VoIP default-port numbers for communication [14 - 16] as shown in Table **1** . For this purpose, the transport-layer port number is extracted from the packet header and mapped with the default port numbers used by various VoIP applications. If a match is found, then the corresponding flow (with which the packet is associated) is classified as VoIP, and flow details are added in the P2P flow table. However, these applications don't need to use default port numbers for communication (*e.g.* Skype may use other UDP ports in the range 50000 – 60000 also). Various VoIP applications mostly use random port numbers (or masquerade well-known port numbers such as 80, 443, *etc.*) for communication, thus making port-based technique ineffective for classification; however, this technique is still employed here so that if any VoIP application uses default port number for communication, then its flow can be classified at an early stage. In the packet-level classification process, only default UDP port numbers (used by various VoIP applications) have been considered for classification purposes to avoid false-positive cases.

Table 1. Default port numbers are used by various VoIP applications.

VoIP application	Default ports used	
	TCP	**UDP**
Google-meet	443	19302 - 19309
Zoom	443, 80, 8801, 8802	3478, 3479, 8801, 8802
Skype	443	3478 - 3481

The traffic that remains un-classified as VoIP in the 1st step undergoes further analysis in the flow-level classification process;, a heuristic-based technique is employed for classification. By analyzing the behavior of various VoIP applications, a set of heuristic rules have been proposed for classifying VoIP (video) traffic, which is described below:

a) Usage of TCP & UDP (heuristic_1): It has been observed that some P2P-VoIP applications (such as Skype) utilize both TCP & UDP protocols simultaneously for communication; where UDP could be used for transferring the data between the peers and TCP could be used for establishing/maintaining the connection between them [17, 18]. Therefore, if source-IP simultaneously uses both TCP & UDP to communicate with the destination-IP, then such traffic flow can be considered VoIP.

b) UDP datagram-length ratio (heuristic_2): It has been observed that VoIP applications generally make use of UDP protocol for video communication amongst the peers and the packet size (*i.e.* UDP datagram-length) distribution of the majority of the packets lie in between the range: 23-289 bytes & 1037-1222 bytes. Hence, we define the packet-size-distribution ratio (psd_ratio) of two communicating peers as mentioned in Equation 1.

$$psd\_ratio = \frac{\text{total\_UDP\_datagram\_length}}{\text{total\_number\_of\_packets}} \tag{1}$$

Here, total_UDP_datagram_length refers to the total number of packets whose datagram-length lies in the range: 23-289 bytes & 1037-1222 bytes. Hence, if psd_ratio of a UDP flow is found to be greater than threshold-ratio, then the flow is classified as VoIP (video) traffic. For experimental purposes, the threshold ratio considered here is 0.75.

c) Usage of ephemeral port numbers (heuristic_3): It has been observed that various P2P-VoIP applications utilize ephemeral port numbers (*i.e.* above 1023) for communication. Hence, in a traffic flow, if both source-IP & destination-IP use ephemeral port numbers for communication, then it can be considered as VoIP.

d) Data transfer between peers (heuristic_4): It has been observed that during VoIP (video) communication between two peers, both source & destination peers transfer a large number of bytes to each other. This is because both source & destination peers simultaneously upload/download the video data to/from each other. Therefore, in a traffic flow, if it is found that several bytes (*i.e.* data) transferred in both directions is greater than the threshold value, then such traffic flow can be considered as VoIP. For experimental purposes, the threshold value considered here is 10MB. Such kind of traffic differs from FTP traffic in a way that VoIP traffic transfers data in both directions simultaneously, whereas FTP traffic transfers data in a single direction only.

It is to be noted that heuristic_1, heuristic_3 & heuristic_4 (discussed above) alone are not sufficient to verify whether a traffic flow belongs to VoIP or not since similar behavior can be seen in P2P file-sharing applications (such as BitTorrent) as well. In this research work, heuristic-based classification employs the proposed heuristics rules in the Algorithm-1 (as shown in Table **2**) to classify the traffic flow either as VoIP or non-VoIP. Here, we analyse that if both source & destination IPs of a flow use TCP & UDP simultaneously for communication (*i.e.* heuristic_1 == true) and their psd_ratio >= 0.75 (*i.e.* heuristic_2 == true) then the flow is classified as VoIP. Otherwise, we analyse that if both source &

destination IPs of a flow either use ephemeral port numbers (*i.e.* heuristic_3 == true) or data transfer between them is greater than threshold-value (*i.e.* heuristic_4 == true) and their psd_ratio >= 0.75 (*i.e.* heuristic_2 == true), then the flow is classified as VoIP.

Table 2. Algorithm for performing Heuristic-based traffic classification.

Legends:	Algorithm 1: Heuristic-Based Classification

src → source peer
dst → destination peer
psd_ratio → ratio of all packets with a *datagram-lengt--range to the total number of packets of a flow (*datagram-length-range = 23-289 bytes & 1037-1222 bytes)
data(src_to_dst) → data transferred from source to destination
data(dst_to_src) → data transferred from destination to source
data(threshold) → 10MB
	Begin
	while(trafficFlows_not_finished)
	{
	flow = fetch_next_trafficFlow()
	if(flow.heuristic_1() == true)
	{
	if(flow.heuristic_2() == true)
	{ //flow classified as VoIP
	write: flow → P2P_flowTable
	}
	}
	else if((flow.heuristic_3() == true) || (flow.heuristic_4() == true))

heuristic_1 → both src & dst uses TCP & UDP simultaneously
heuristic_2 → psd_ratio >= 0.75
heuristic_3 → both src & dst use ephemeral ports
heuristic_4 → data(src_to_dst) & data(dst_to_src) > data(threshold)
	{
	if(flow.heuristic_2() == true)
	{ //flow classified as VoIP
	write: flow → P2P_flowTable
	}
	}
	}
	}
	End

The traffic flow, which remains un-classified as VoIP undergoes further analysis and is fed to the statistical-based classification process, where the ML algorithm (C4.5 decision tree) is applied to the statistical properties of the traffic flow. Therefore, if any VoIP traffic flow goes undetected in the previous processes, then it gets classified at this stage.

4. DATASETS, VALIDATION, AND EXPERIMENTAL RESULTS

The proposed technique classifies the traffic flow either as P2P-VoIP or non-VoIP. The technique is implemented in java using jNetPcap library [19] and weka [20] to validate its classification performance. The metrics used for measuring the classification performance are accuracy, false positive (FP), and false-negative (FN) rates. The experiments have been conducted using offline traffic traces, where two individual datasets consisting of P2P and non-P2P traffic flows have been employed, as shown in Table **3** . Here, P2P traffic flows consist of both VoIP and file-sharing traffic.

Table 3. The number of flows in Dataset-1 and Dataset-2.

Dataset	P2P flows	non-P2P flows	Total
Dataset-1	7599	71399	78998
Dataset-2	20821	10967	31788

Dataset-1 (D-1) is a publicly available dataset that belongs to the University of Brescia [21, 22]. It consists of both P2P & non-P2P traffic, where P2P traffic traces consist of file-sharing applications (eDonkey, BitTorrent, *etc.*) and a VoIP application (Skype). The Dataset-2 (D-2) consists of real-traffic traces which are captured in the campus area network (using Wireshark [23]), which consists of a mixture of P2P (including VoIP) and non-P2P traffic. The data capturing was accomplished in a controlled environment where various popular VoIP applications named Google-meet, Skype & Zoom were executed on individual systems for analyzing their pattern of communication, and all other applications were stopped from being executed during this period. Therefore, it was well known in advance regarding the flows which were associated with VoIP traffic and hence were labeled accordingly for ground-truth verification. In addition to that, another system was also made to generate non-VoIP traffic consisting of P2P file-sharing applications (*e.g.* BitTorrent) & non-P2P applications (*e.g.* HTTP, HTTPS, DNS, *etc.*). Hence, overall Dataset-2 consists of both P2P (VoIP & non-VoIP) and non-P2P applications.

During the experiment, a detailed analysis has been conducted on Dataset-2, which shows that packet-level classification process can achieve the classification accuracy of 56.47% only (as shown in Fig. (**3a**)) and has FP & FN rates of 1.23% & 42.30%, respectively (as shown in Fig. (**3b**)). This is because only some of the traffic flows used default VoIP port numbers (shown in Table **1** in the previous section) during connection establishment; thereafter, random port numbers were used for communication. It is to be noted that if this initial communication is not captured (or does not contain default VoIP port numbers), then the performance of the packet-level classification process will become very poor. Here, the main purpose of the packet-level classification process is to classify VoIP (video) traffic at an early stage (if some traffic flows are found to be using default VoIP ports) which will reduce the amount of traffic that needs to be analyzed at the flow-level classification process.

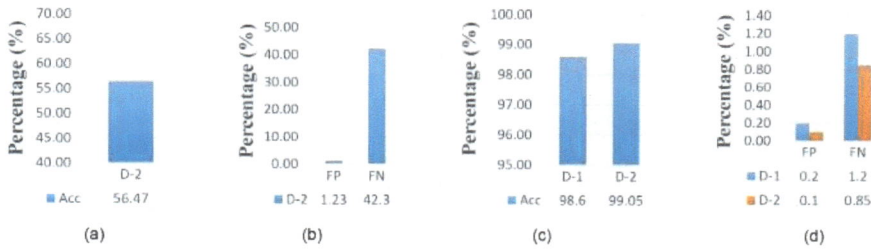

Fig. (3). (a) Accuracy of packet-level classification process, (b) FP & FN rates of the packet-level classification process, (c) Overall classification accuracy of the proposed technique, (d) Overall FP & FN rates of the proposed technique.

The combination of packet-level and flow-level processes achieved overall classification accuracy ranging between 98.6% - 99.05% (as shown in Fig. (3) (c)). In addition to that, it has FP & FN rates ranging between 0.1 – 0.2% & 0.85 – 1.2%, respectively (as shown in Fig. (3d)). It can be seen that heuristic rules used in the proposed technique work equally with both TCP & UDP traffic and are not affected by encrypted traffic. In addition to that, the proposed technique has low overhead (since it does not depend upon the DPI technique, which is computationally expensive), and works with encrypted traffic as well.

CONCLUSION

P2P-VoIP applications have become very popular in recent years and are currently being used extensively by various companies, enterprises, and government organizations globally to run their businesses. VoIP traffic which involves video communication, consumes a lot of network bandwidth, and hence it needs to be classified so that it can be managed/prioritized by the ISPs and network administrators. In this paper, a 2-step hybrid approach is employed, which combines heuristic-based & statistical-based techniques to classify VoIP (video) traffic. The experimental results show that the proposed technique achieves high classification accuracy of over 98.6%. In addition to that, it classifies both TCP & UDP traffic flows and is not affected even if traffic is encrypted. However, there are certain limitations of the proposed technique: a) It may produce some false positives (during VoIP-port based classification process) if network traffic includes malicious applications using the default port numbers which could be used by VoIP applications for communication also and b) It does not perform fine-grained classification to classify VoIP traffic into specific applications. Therefore, our future work will focus on classifying various other VoIP applications such as MS-Teams, WhatsApp, *etc*., and enhance the technique to perform VoIP-voice & fine-grained classification.

CONSENT FOR PUBLICATION

Not applicable.

CONFLICT OF INTEREST

The authors declare no conflict of interest, financial or otherwise.

ACKNOWLEDGEMENT

Declared none.

REFERENCES

[1] "Global Internet Phenomena," Sandvine,", *[Online]*, 2019.https://www.sandvine.com/phenomena

[2] https://www.businesswire.com/news/home/20200416005739/en/Impact-of-COVID-19-on-the-Video-Conferencing-Market-2020---ResearchAndMarkets.com

[3] "Service Name and Transport Protocol Port Number Registry", *[Online]*.https://www.iana.org/assignments/ service-names-port-numbers/service-names-port-numbers.xhtml

[4] J.V. Gomes, P.R. Inacio, M. Pereira, M.M. Freire, and P.P. Monteiro, "Detection and classification of peer-to-peer traffic: A survey", *ACM Comput. Surv.,* vol. 45, no. 3, p. 30, 2013. [CSUR]. [http://dx.doi.org/10.1145/2480741.2480747]

[5] S-M. Liu, and Z-X. Sun, "Active learning for P2P traffic identification", *Peer-to-Peer Netw. Appl.,* vol. 8, no. 5, pp. 733-740, 2015. [http://dx.doi.org/10.1007/s12083-014-0281-3]

[6] J.V. Gomes, P.R. Inacio, M.M. Freire, M. Pereira, and P.P. Monteiro, "Analysis of peer-to-peer traffic using a behavioural method based on entropy", *Computing and Communications Conference,* 2008 [http://dx.doi.org/10.1109/PCCC.2008.4745138]

[7] M. Bhatia, and M.K. Rai, "Identifying P2P traffic: A survey", *Peer-to-Peer Netw. Appl.,* vol. 10, no. 5, pp. 1182-1203, 2017. [http://dx.doi.org/10.1007/s12083-016-0471-2]

[8] Q. Jiang, H. Hu, and G. Hu, "Real-Time Identification of Users under the New Structure of Skype", *in 2016 IEEE International Conference on Sensing, Communication and Networking (SECON Workshops),* 2016. [http://dx.doi.org/10.1109/SECONW.2016.7746809]

[9] Z. Yuan, C. Du, X. Chen, D. Wang, and Y. Xue, "Skytracer: Towards fine-grained identification for skype traffic *via* sequence signatures", *International Conference on Computing, Networking and Communications (ICNC),* 2014. [http://dx.doi.org/10.1109/ICCNC.2014.6785294]

[10] S-H. Lee, Y-H. Goo, J-T. Park, S-H. Ji, and M-S. Kim, "Sky-Scope: Skype application traffic identification system", *in 2017 19th Asia-Pacific Network Operations and Management Symposium (APNOMS),* 2017. [http://dx.doi.org/10.1109/APNOMS.2017.8094123]

[11] N.A. Saqib, "Y. SHAKEEL, M. A. Khan, H. MEHMOOD and M. Zia, "An effective empirical approach to VoIP traffic classification", *Turk. J. Electr. Eng. Comput. Sci.,* vol. 25, no. 2, pp. 888-900, 2017. [http://dx.doi.org/10.3906/elk-1501-126]

[12] S. Munir, N. Majeed, S. Babu, I. Bari, J. Harry, and Z.A. Masood, "A joint port and statistical analysis

based technique to detect encrypted VoIP traffic", *Int. J. Comput. Sci. Inf. Secur.,* vol. 14, no. 2, p. 117, 2016.

[13]   J. Datta, N. Kataria, and N. Hubballi, "Network traffic classification in encrypted environment: a case study of google hangout", *Twenty First National Conference on Communications (NCC),* 2015. [http://dx.doi.org/10.1109/NCC.2015.7084879]

[14]   "Prepare your network for Meet video calls", *[Online],* 2020.https:// support.google.com/a/answer/1279090?hl=en

[15]   "Network Firewall Settings for Meeting Connector", *[Online],* 2020.https://support.zoom.us/hc/en-us/articles/202342006-Network-Firewall-Settings-for-Meeting-Connector. [Accessed September 2020].

[16]   "Ports need to be open to use Skype", *[Online].*https://support.skype.com/en/faq/FA148/which-port--need-to-be-open-to-use-skype-on-desktop

[17]   J.M. Reddy, and C. Hota, "Heuristic-based Real-Time P2P Traffic Identification", *2015 International Conference on Emerging Information Technology and Engineering Solutions,* 2015. [http://dx.doi.org/10.1109/EITES.2015.16]

[18]   P. Velan, M. Cermak, P. Celeda, and M. Drasar, "A survey of methods for encrypted traffic classification and analysis", *Int. J. Netw. Manage.,* vol. 25, no. 5, pp. 355-374, 2015. [http://dx.doi.org/10.1002/nem.1901]

[19]   "jNetPcap", *[Online].*.https://sourceforge.net/projects/jnetpcap/

[20]   https://www.cs.waikato.ac.nz/ml/weka

[21]   F. Gringoli, L. Salgarelli, M. Dusi, and N. Cascarano, "F. Risso and others, "Gt: picking up the truth from the ground for internet traffic", *Comput. Commun. Rev.,* vol. 39, no. 5, pp. 12-18, 2009. [http://dx.doi.org/10.1145/1629607.1629610]

[22]   M. Dusi, F. Gringoli, and L. Salgarelli, "Quantifying the accuracy of the ground truth associated with Internet traffic traces", *Comput. Netw.,* vol. 55, no. 5, pp. 1158-1167, 2011. [http://dx.doi.org/10.1016/j.comnet.2010.11.006]

[23]   "Wireshark", *[Online].*https://www.wireshark.org. [Accessed 2020].

CHAPTER 3

Comparative Study of FxLMS, Flann, PSO, and GA for Active Noise Controller

Balpreet Singh[1], H. Pal Thethi[1] and **Santosh Kumar Nanda[2,*]**

[1] *Department of ECE, SEEE, Lovely Professional University, Punjab, India*

[2] *Techversant infotech Pvt Ltd,Trivandrum, Kerala, India*

Abstract: This paper presents a comparison between the different adaptive algorithms for designing a cost-effective Anti-Noise Control (ANC) system. Typically, the ANC systems use linear FIR-adaptive filter Filtered- Least Mean Square (FxLMS) configuration. FxLMS is straightforward in hardware implementation but, its efficiency substantially degrades in case time-varying and non-linear acoustic environment and the probability that it will converge to local minima. Meanwhile, the implementation of Trigonometric Functional-Linear Adaptive Neural Network (FLANN) enhances ANC performance for non-linear noise signals. However, at the same time, it increases the complexity of hardware implementation and is unable to solve the problem of local minima convergence. Whereas evolutionary algorithms -Genetic Algorithm (GA) and metaheuristic algorithm Particle Swarm Optimization (PSO) increase the robustness and stability in non-linear, and the time-varying acoustic environment with absolute zero probability to converge to local minima. This paper briefly discusses on implementation of FxLMS, FLANN, PSO, and GA-based ANC systems. Further, simulation compares Mean Square Error, BER, and PSNR to provide computational efficiency of these algorithms.

Keywords: Active Noise Control, Adaptive algorithm, Filtered-Least Square Method (FxLMS), Functional Link Adaptive Neural Network (FLANN), Genetic Algorithm (GA), MSE, Particle Swarm Optimization (PSO), PSNR, SNR.

INTRODUCTION

The enigma of eliminating undesired sound, commonly referred to as noise, generated by automobiles, airplanes, industries, giant constructions machinery, and other sources, is not only a prevalent research problem but also causing hearing impairment, stress, anxiety, and mental illness. Moreover, disturbance due to noise in the background damping focus and resulted in inefficiency in productivity. To diminish noise efficiently, the two predominant techniques are;

---

* **Corresponding author Santosh Kumar Nanda:** Techversant infotech Pvt Ltd,Trivandrum, Kerala, India; E-mail: santoshnanda@live.in

Dharam Buddhi, Rajesh Singh and Anita Gehlot (Eds.)

Passive Filters and Active Filters. Passive filters utilize the conventional method to use physical materials, which act as silencers, enclosers, and barriers to absorb and stop the undesired sound from reaching the eardrums. However, designing passive filters for a wide range of frequencies is expensive and ineffective to reduce low frequencies roughly below 500Hz. Consequently, active filter or Active Noise Controller (ANC) is cost-effective as well as efficient for low frequencies. It generates anti-noise sound of the identical amplitude and frequency but of an inverse phase of noise in an acoustic medium which results in destructive interference with noise and eliminating noise. Further, the active noise control system is a self-driven adaptive system that can adapt according to the continuously changing noise source and the acoustic environment on its own. The ANC system uses Adaptive Feedforward and Adaptive Feedback system configuration. The feedforward ANC [1] system comprises of reference microphone, anti-noise control system (usually adaptive filter), speaker, and error microphone. In the feedforward system, a reference microphone feeds the incoming noise to the controller to produce the desired anti-noise through the loudspeaker. The controller adapts to time-varying and non-stationary noise signals and acoustic mediums and drives the speaker to minimize acoustic pressure at the error microphone. Whereas, Feedback ANC configuration is built of an error microphone and adaptive controller [1, 2] The error microphone provides an error signal to a controller for updating its output and turns the speaker to generate anti-noise sound in an acoustic environment for diminishing noise [3].

This paper briefly discusses Active Noise Control algorithms in subsequent sections. Section II discusses details of Fx-LMS algorithms, FLANN algorithm., Particle Swarm Optimization (PSO), and Genetic Algorithm (GA) based adaptive algorithms in detail. Lastly, Section III presents the MATLAB Simulation results, comparison, and conclusion.

Section II: Algorithms

Filtered -Least Mean Square Algorithm (FxLMS)

In standard LMS, when electrical delay (time controller takes to compute anti-noise) is more significant than acoustic delay (time noise signal takes to reach speaker from reference microphone), it leads to non-casualty of system. Hence, performance sustainably degrades in diminishing broad-band noise; it can only reduce narrow-band/periodic noise [4]. The controller followed by the secondary path transfer function for compensating acoustic superposition of primary noise and anti-noise in the secondary path is a potential solution for the non-casualty problem [5]. However, this results in the non-alignment of the error signal with the reference signal in time cause instability [6, 7]. Morgan *et al.* [8] proposed to use an inverse filter in series with a secondary-path transfer function or use an inverse filter with the reference signal for updating the weight of the LMS adaptive algorithm, which is also known as the Filtered-X-LMS (FxLMS) algorithm. Fig. (**1**) presents the block diagram of FxLMS.

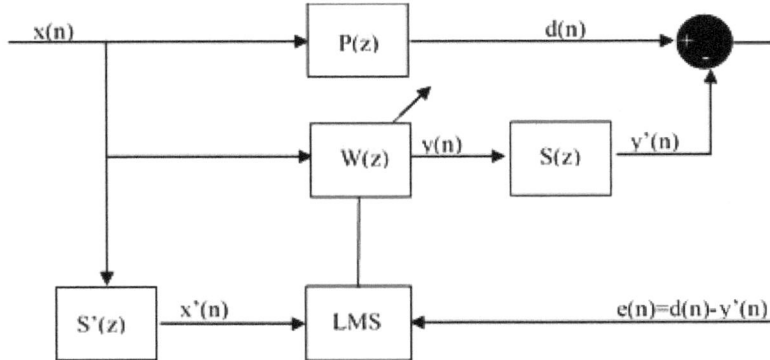

Fig. (1). Block diagram for FxLMS.

whereas, x(n)= Reference Signal; p(n)=impulse-response of primary path; s(n)=impulse-response of secondary path; w(n)=weight of the filter of length L $[w_0(1),w_1(2),…………w_{L-1}(n)]$ and $w^T(n)$ is transpose of w(n);

FxLMS algorithms first estimate the secondary path transfer function S'(z), which leads to divergence of the algorithm due to estimation error. Additionally, the secondary path is usually time-varying in nature, and gradient-based decent method of weight update led to convergence at local minima [8].

Functional Link Artificial Neural Network (FLANN)

The LMS and FxLMS have shown exceptional performance in the case of linear noise, but in non-linearity, both algorithms lose efficiency [9,10]. The non-recursive, and shift-invariant filters such as Volterra-filtered-x LMS (VFxLMS) [11], neural network-based controllers [12] and functional link artificial neural networks (FLANN) based filtered-s LMS (FsLMS) [13] are potential solution for non-linear input noise signal. The FLANN uses a trigonometric function to calculate the output. The expression is as follows:

$$y(n) = \sum_{i=0}^{Na-1} a_i x(n-i) + \sum_{j=0}^{Nb-1} a_j y(n-j) + \sum_{j=0}^{Ne-1} e_j \cos[\pi y(n-j)] + \sum_{j=0}^{Ne-1} g_j \sin[\pi y(n-j)] \qquad (2)$$

Particle Swarm Optimization

Particle Swarm Optimization is a meta-heuristic evolutionary algorithm and is most preferred for an optimization problem. Kennedy *et al.* 1 [13] first proposed PSO based on the collective behavior of birds flock flying together in search of food. During their search, they continuously modify their path and velocity to find an optimum place. PSO applied to all problems of optimization, Krusienski *et al.* [14] has optimized mean square error for adaptive filters. Xia *et al.* [15] developed PSO-based Adaptive noise canceler, and Modares *et al.* developed a novel technique to adapt the weight of multilayer neural network for nonlinear ANC algorithm. Rout *et al.* used a bank of PSO adaptive filter to collectively minimize the mean square error and also proposed a PSO algorithm to identify the primary path and secondary path efficiently for ANC [16]. The foremost importance in PSO is to identify the cost function that is needed to be minimized. In the case of the ANC system, cost function in mean square error.

Genetic Algorithm

Chang *et al.* [17] proposed the Genetic Algorithm based ANC system, which utilizes GA instead of FxLMS algorithm. The proposed algorithm is with GA is an inherent solution for nonlinear noises and the problem of local minima as compared to FxLMS. A genetic algorithm is an evolutionary algorithm, which mimics natural selection to optimize the solution. It involves evaluation of cost function, *i.e.* in ANC mean square error, reproduction of new solutions, crossover with potential solutions, mutation to generate a solution, and replacement [18]. Adaptive Genetic Algorithm has implemented without estimation of secondary path hence improved stability. Moreover, due to property global searching algorithm of GA prevented the local minima problem.

Section III: Simulation Result and Conclusion

This section presents and discuss the MATLAB R2018a simulation result of FxLMS, FLANN, PSO and GA algorithm, for this purpose we have consider primary path P(z)=[0.01z-1 +0.25 z-2 +0.5 z-3 +1 z-4 +0.5 z-5 +0.25 z-6 +0.01z-7]; and secondary path as S(z)=0.25*P(z) [1]. Uniform random white noise with zero mean with T=1000 sample is taken as a reference signal.

FxLMS

FxLMS algorithm starts with the estimation of the impulse response of the secondary transfer function. The result is evaluated for different filter lengths of L =8,10,12, and 16 and mu=0.1. The error of estimation is shown in Fig. (**2**) Error signal for secondary system estimation with time and semilogy plot of error with time

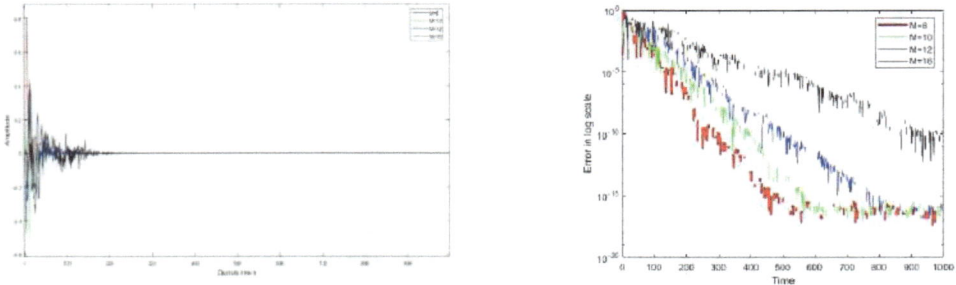

Fig. (2). Error amplitude with iterations and Semilog plot of error with iteration.

It has been observed that as filter length increases, the time required to estimate the impulse response of secondary path also increases; for instance, M=16 required more than 1000 samples to identify transfer function efficiently.

Table 1. Comparison of Fx-LMS with different Filter lengths.

Filter Length	N=1000			N=2000		
	Mean Square Error(MSE)	SNR dB	PSNR dB	MSE	SNR dB	PSNR dB
M=8	-10.6148	10.481	12.39	-13.516	14.8058	13.624
M=10	-10.3621	10.391	11.792	-13.07	15.680	13.606
M=12	-10.7624	10.536	11.691	-13.429	15.3100	13.287
M=16	-8.3594	10.266	11.869	11.844	14.9788	12.874

As per Table **1** in FxLMS the error is significantly dependent upon the length of the filter when N=1000, but for larger samples, the SNR and PSNR do not change significantly.

FLANN

Figs. (**3** and **4**) presents the result of the FLANN algorithm. Error amplitude in Fig. (**4**) with FLNN filter of order P=2 and harmonics as 15 with a step size of 0.01. Whereas, Fig. (**3**) and Table **2**, show the MSE for N=10,15,20 number of harmonics, and it has been observed that increasing harmonics degrades the efficiency of FLANN.

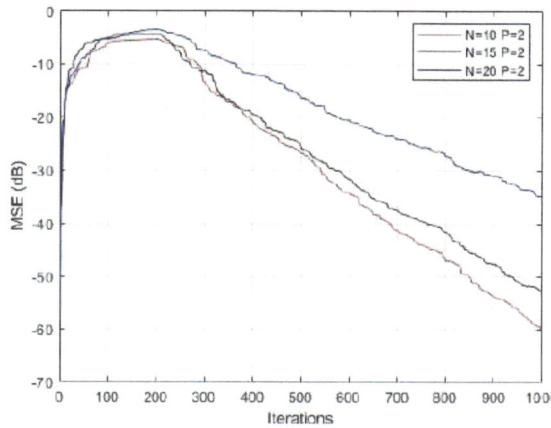

Fig. (3). MSE for different filter lengths.

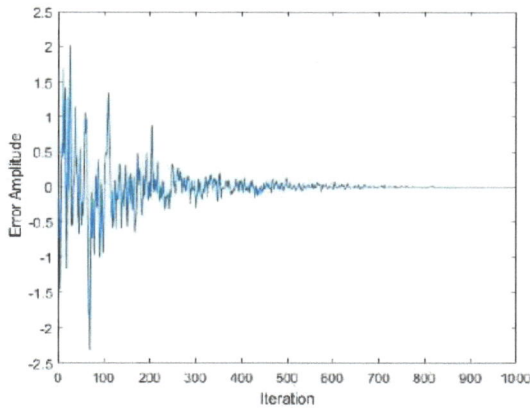

Fig. (4). Error Amplitude for P=2, N=15.

Table 2. Comparison of FLANN on filter length.

Configuration	PSNR in dB	SNR in dB	MSE in dB
P=2, M=10	12.2747	14,49	-87.300
P=2 M=15	11.1322	13.29	-52.014
P=2 M=20	9.9758	11.83	-52.5966

Particle Swarm Optimization

PSO is an evolutionary algorithm performed with 200 populations for 1000 iterations. The learning factors are taken as rand and damping coefficient as 0.99.

The PSO is significantly reducing error and MSE. PSNR is 8.2683 dB, SNR is 8.4120 dB and MSE= -145.6149 dB. Fig. (**5**) represent the semilog plot Mean Square Error with iteration and is observed that error is reduced significantly, whereas Fig. (**6**) present the mean square amplitude concerning iteration and it has been observed that it approaches nearly zero after 100th iterations.

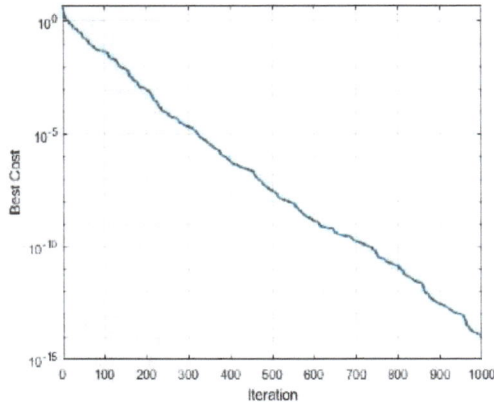

Fig. (5). Semilog plot of Mean square error with iteration.

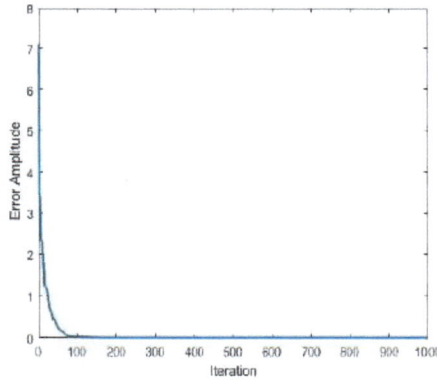

Fig. (6). Error Amplitude plot with iteration.

Genetic Algorithm

A genetic algorithm-based ANC system has been implemented with a Population size of 50 for 1000 iterations. Fig. (**7**) shows the simulation results. MSE is -26. 8074 dB, PSNR is 8.7628 dB and SNR is 8.8894 dB for a population of 50, and the Number of populations 100 MSE has significantly improved to -34.8131 dB

PSNR to 10.2872 dB and 10.4269 dB. Fig. **(7)** represent the semilog plot Mean Square Error with iteration and is observed that error is reduced significantly, whereas Fig. **(8)** present the mean square amplitude concerning iteration and it has been observed that it approaches nearly zero after 200th iterations.

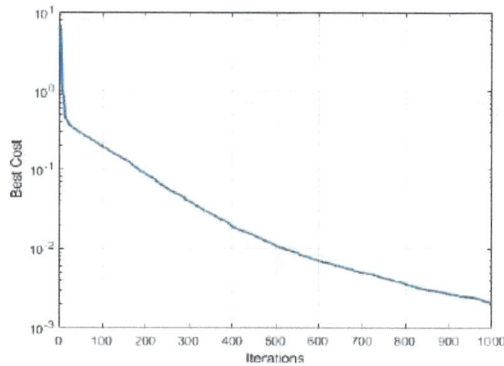

Fig. (7). Semilog plot of Mean square error with iteration.

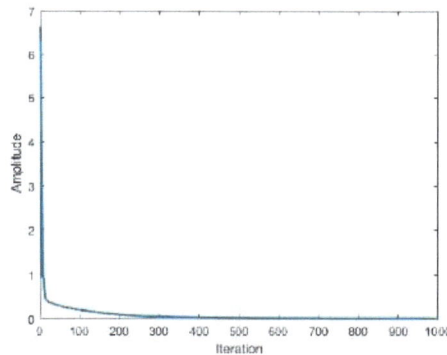

Fig. (8). Error Amplitude plot with iteration.

Comparison of FXLMS, FLANN, PSO, and GA

Table **3** presents the comparison of FxLMS, FLANN, PSO, and GA in terms of MSE, SNR, and PSNR and Fig. **(9)** shows the MSE of a different algorithm for nonlinear noise. Notably, the performance of PSO is sustainably effective as compared to its counterparts. As per Fig. **(9)** GA converges fast but the efficiency of convergence is better in PSO.

Table 3. Comparison of FxLMS, FLANN, PSO, and GA.

Filters	MSE in dB	SNR in dB	PSNR in dB
FxLMS (M=16, N=1000)	-8.3594	10.2668	11.8696
FLANN(P=2, N=15)	-52.014	13.29	11.1322
PSO(P=50, N=1000)	-145.6149	8.4120	8.2683
GA(P=50, N=1000)	-26. 8074	8.8894	8.7628
GA(P=100, N=1000)	-34.8131	10.4269	10.2872

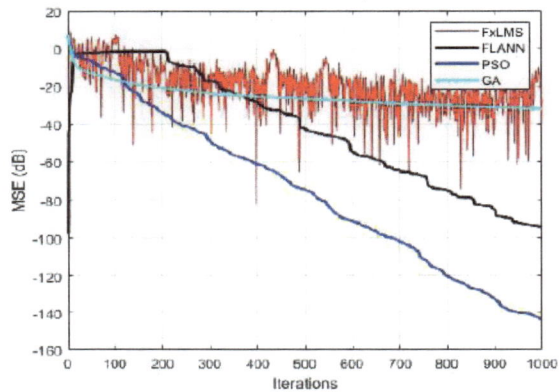

Fig. (9). MSE in(dB) of FxLMS, FLANN, PSO and GA.

CONCLUSION

This paper discusses and compares the simulation result of four algorithms for designing an Active Noise Cancellation system. Where, FxLMS is simple and easy to implement, but the performance substantially degrades in the case of non-linearity in noise as well as has the problem of local minima. The FLANN has shown better performance than Fx-LMS, but its complexity increases with an increase in the sinusoidal component. As compared to FLANN, and Fx-LMS evolutionary algorithms are showing efficient results in non-linear and non-uniform random noise. But comparatively, PSO has emerged as the most efficient algorithm for the implementation of ANC. It not only converges to a minimum faster than GA but also required less time complexity.

CONSENT FOR PUBLICATION

Not Applicable.

CONFLICT OF INTEREST

The authors declare no conflict of interest, financial or otherwise.

ACKNOWLEDGEMENT

Declared none.

REFERENCES

[1]   S.M. Kuo, and D.R. Morgan, *Active Noise Control Systems—Algorithms and DSP Implementations.* Wiley: New York, 1996.

[2]   J.G Proakis, and D.G Manolakis, "Digital Signal Processing", *Pearson Publication, NJ,* 2007.

[3]   C. H. Hansen, "Understanding Active Noise Cancellation", *Taylor & Francis group,* 2001.

[4]   B.B. Widrow, and M.E. Hoff, *Adaptive Switching Crcuit.* IRE WESCON Convention Record, 1960, pp. 96-104.
      [http://dx.doi.org/10.21236/AD0241531]

[5]   S.D. Snyder, *Active Noise Control Primer.* Modern Acoustics and Signal Processing, 2000.
      [http://dx.doi.org/10.1007/978-1-4419-8560-6]

[6]   C.Y.C.K.K. Shyu, "Active noise cancellation with a fuzzy adaptive filtered-X algorithm", *Institute of Electical Engineering and Circit System,* vol. 150, no. 5, pp. 416-422, 2003.

[7]   D. Zhou, and V. DeBrunner, "Efficient Adaptive Nonlinear Filters for Nonlinear Active Noise Control", *IEEE Trans. Circuits Syst. I Regul. Pap.,* vol. 54, no. 3, pp. 669-681, 2007.
      [http://dx.doi.org/10.1109/TCSI.2006.887636]

[8]   D.R. Morgan, "A hierarchy of performance analysis techniques for adaptive active control of sound and vibration", *J. Acoust. Soc. Am.,* vol. 89, no. 5, pp. 2362-2369, 1991.
      [http://dx.doi.org/10.1121/1.400925]

[9]   G.L. Sicuranza, and A. Carini, "A Generalized FLANN Filter for Nonlinear Active Noise Control", *IEEE Trans. Audio Speech Lang. Process.,* vol. 19, no. 8, pp. 2412-2417, 2011.
      [http://dx.doi.org/10.1109/TASL.2011.2136336]

[10] V. Patel, V. Gandhi, and S. Heda, "Design of Adaptive Exponential Functional Link Network-Based Nonlinear Filters", *IEEE Trans. Circuits Syst. I Regul. Pap.,* vol. 63, no. 9, pp. 1434-1442, 2016.
      [http://dx.doi.org/10.1109/TCSI.2016.2572091]

[11] D.J. Krusienski, and W.K. Jenkins, "A particle swarm optimization - Least mean squares algorithm for adaptive filtering", *Conf. Rec. Asilomar Conf. Signals Syst. Comput.,* vol. 1, no. 1, pp. 241-245, 2004.
      [http://dx.doi.org/10.1109/ACSSC.2004.1399128]

[12] L. Xia, "G. Hui and. L. Jinfeng, "Adaptive noise canceller based on PSO algorithm", *2008 IEEE International Conference on Automation and Logistics,* 2008.

[13] H. Modares, A. Ahmadyfard, and M. Hadadzarif, "A PSO approach for non-linear active noise cancellation, in 6th WSEAS International Conference on Simulation, Modelling and Optimization, Lisbon, Portugal", September 22-24, 2006.

[14] N.K. Rout, D.P. Das, and G. Panda, "Particle Swarm Optimization Based Active Noise Control Algorithm Without Secondary Path Identification", *IEEE Trans. Instrum. Meas.,* vol. 61, no. 2, pp. 554-563, 2012.
      [http://dx.doi.org/10.1109/TIM.2011.2169180]

[15] D.E. Goldberg, *Genetic Algorithms in Search, Optimization and Machine Learning.* Addison-Wesley Longman Publishing Co.: USA, 1989.

[16]   Z. Michalewicz, *Genetic Algorithms + Data Structures = Evolution Programs.* Springer-Verlag Berlin Heidelberg, 1996.
[http://dx.doi.org/10.1007/978-3-662-03315-9]

[17]   M. Mitchell, *An Introduction to Genetic Algorithms.* MIT Press: Cambridge, MA, 1998.
[http://dx.doi.org/10.7551/mitpress/3927.001.0001]

[18]   F.M. Janeiro, and P.M. Ramos, "Impedance Measurements Using Genetic Algorithms and Multiharmonic Signals", *IEEE Trans. Instrum. Meas.,* vol. 58, no. 2, pp. 383-388, 2009.
[http://dx.doi.org/10.1109/TIM.2008.2005077]

CHAPTER 4

Application of Computational Methods for Identification of Drugs against *Tropheryma Whipplei*

Amit Joshi[1] and **Vikas Kaushik**[2,*]

[1] *Department of Biochemistry, Lovely Professional University, Punjab, India*

[2] *Department of Bioinformatics, Lovely Professional University, Punjab, India*

Abstract: *Tropheryma whipplei* causes severe malady termed as Whipple's disease, a multisystemic lethal problem and we still require modified best regimens. To treat it successfully, 3 medications were distinguished in this investigation by using in-silico methods. 2-amino7fluoro5oxo5Hchromeno[2,3b]pyridine3carboxamide(2APC), Nicotinamide mononucleotide (NMN), and Riboflavin Monophosphate (RFMP) were seen as putative medications. 2APC and NMN restrain DNA Ligase catalytic activity for *Tropheryma whipplei* and compelling in impeding genomic copying and repairing mechanisms, RFMP shows the inhibitory impact on Chorismate synthase that drives hindrance in metabolic biosynthesis of amino acids. Our investigation used modern advanced in-silico assemblies. BLAST, CDART, CD-HIT were utilized to choose target catalytic biomolecules of a bacterium. Phyre2, dependent on HMM calculation, was applied to discover the best auxiliary models of chosen biocatalysts. AutoDock-Vina assembly was utilized for molecular docking and scoring restricting energies of these medications with catalytic proteins of the bacterium. 2 APC and NMN hindering DNA Ligase show - 8.3 and - 8.2 kcal/mol individually while RFMP represses Chorismate synthase - 7.3 kcal/mol binding energy. Sub-atomic re-enactment or simulative mechanistic analysis gives further approval to concluding 2APC as impeccable inhibitory medication having remedial activity against *T. whipplei*. This escalated and novel examination is simple, quick, and valuable in anticipating drugs by incorporating computational insights in medicinal sciences.

Keywords: Biomolecules, Computational approach, Drug Discovery, Molecular dynamics Simulation, *Tropheryma whipplei*.

---

[*] **Corresponding author Vikas Kaushik:** Department of Bioinformatics, Lovely Professional University, Punjab, India; E-mail: vikas31bt@gmail.com

Dharam Buddhi, Rajesh Singh and Anita Gehlot (Eds.)

INTRODUCTION

Whipple's malady is an uncommon multisystem disorder affecting the gastro-intestine and central nervous systems. *Tropheryma whipplei*, Gram **+ve** actinobacterium about which little is known. The total DNA arrangement of its genome consists of 925 938, not complete genes set with an absence for significant biosynthesis-pathways and a diminished limit concerning vitality in metabolic mechanisms [1]. T. whipplei normally contaminates sewage laborers in Caucasian populations and among little children under 7 years in poor unclean surroundings [2]. Still, no data about the transmission of this microscopic organism distinguishing that this bacterium is associated with Homo sapiens and transferred from other animals. T. whipplei make duplicates inside macrophages, living in mucous and monocytes of blood vasculature in patients. Like Mycobacterium tuberculosis, it gradually develops under culture. Currently, hydroxychloroquine, doxycycline, and trimethoprim/sulfamethoxazole drugs, and also an injection of ceftriaxone given to patients with *T. whipplei* but the problem with this medication approach was very long and time taking up to 2 to 3 years with lifetime follow up [3]. Modern computational approaches can be useful in predicting drugs for the rare bacterium that causes severe health effects like *T. whipplei* [4]. In this study, we used genomic and proteomic analysis along with molecular docking and simulation analysis to find out possible drugs. Evaluation of pharmacokinetic characteristics in computer-aided drug designing is integral for hit-to-lead improvements. Exceptional unpredictability of the present research design for medication search, scientists strongly sought molecular docking and simulation mechanistic models to characterize patterns in ADMET information to develop practical insights [5]. This investigation deployed ADMET analysis along with 2D and 3D interaction of drugs to biocatalysts and it was found to be very successful in drug predictions.

MATERIAL AND METHODS

Enzyme Selection Bias

Proteomic sequences of Tropheryma whipplei were retrieved from NCBI-Genbank for two significant compounds to be specific, DNA Ligase (AAO44511) and Chorismate synthase (WP_011096348), and these enzymatic edifices are engaged with DNA replication, biosynthesis of amino acids individually (Table 1).The choice of these fundamental proteins depends on DEG (database of basic qualities) server investigation [6].

Table1. Selected proteins information for *Tropheryma whipplei*, that are found to be drug targets by screening NCBI- Genbank database and identification by KAAS (KEGGAutomatic annotation server) for pathway analysis of crucial genes.

NCBI-Genbank Accession no.	Identified Protein	KEGG Orthology Number	Functionality
AAO44511	DNA Ligase	K01972	Replicative and reparative Mechanisms of DeoxyriboNucleicAcids
WP_011096348	Chorismate synthase	K01736	Peptide and amino acid biosynthetic mechanisms

Protein Drug Analysis

After this CD-HIT server [7] was utilized to distinguish paralogs, it depends on a fast heuristic examination approach and accommodating in deciding likeness investigation between peptide stretches. Basic local search alignment was utilized for deciding homology [8] between considered proteins of Tropheryma whipplei and proteomic spaces of Homo sapiens. This gives more approval to the determination from escalated proteomic sets of the bacterium. To examine pharmacogenetics or medication capacity of considered proteins, a drug bank web-server (http://www.drugbank.ca/) was applied. To recognize the space homogeneity of protein groupings, 2 WebServers conserved-domain-architectu-e-retrieval-tool (CDART) & Pfam was utilized. KEGG automatic annotation server (KAAS) assisted in distinguish metabolic pathways for selected biocatalysts and here *T. whipplei*Twist and *T. whipplei*TW08/27 strains were browsed from the NCBIgenbank organisms list at the time of the investigation.

Structural Analysis: Docking & Simulation

The selected proteins, after intensive investigation and KEGG annotation, was exposed to homology displaying using Phyre2, it is a hidden Markov model-based server for structural predictions of catalytic enzymes. A quick overview of medications acting on chosen proteins and their 3D structure was acquired by utilizing Pubchem web server and RCSB- PDB databank. Sub-atomic docking was led using Autodock-vina assembly to examine the interaction energies of ligand-protein docked structures. SwissADME tool was deployed to examine biochemical properties like pharmo-kinetics, drug-likeness, and inhibitory action on cytochrome P450 isoforms, structural properties, bioavailability, and synthetic accessibility. Molecular simulation studies were performed for 40ns by using GROMACS ver.2019 simulation suite.

RESULT AND DISCUSSION

The selected proteins listed in Table **1** indicate KEGG annotation and NCBI-Genbank accession no. along with their known functionality (based on DEG and KEGG server) in *Tropheryma whipplei* twist strain. Phyre2 itself is an HMM algorithm-based server deployed to determine the protein structure of DNA Ligase and Chorismate synthase enzymes. DNA ligase is a crucial enzymatic assembly that is used by bacterium for repair and copying DNA sequences and inhibited by 2 amino7fluoro5oxo5Hchromeno [2,3b]pyridine3carboxamide (2APC) and Nicotinamide mononucleotide (NMN). While ChorismateSynthase performs dephosphorylation of 5O(1carboxyvinyl)3phosphoshikimate to chorismate. This enzyme does not exist in *Homo sapiens*. Chorismate is a precursor for aromatic-ring containing amino acids. This enzyme interacts with riboflavin monophosphate (as per the Drug-Bank database). Riboflavin monophosphate (RFMP) is a potent oxidizing agent which has been used in the food processing industry as a colorant. In earlier studies, Riboflavin has been seen to effectively regulate *Staphylococcus aureus* colonization when used in conjunction with antibiotics. Similar studies deployed to eradicate *T. whipplei* infection and promote its novel treatment strategy. Pubchem database was deployed to retrieve the structure of drugs interacting with selected proteins. Pubchem CID and the name of drugs were mentioned in Table **3** with Swiss-ADME characteristics. In Fig. (**1**) structure of drugs is represented and for better visualization of pharmacophore analysis, pymol software is used. The structure of enzymatic complexes was retrieved from phyre2 server, based on a detailed homology report of modeled structure of DNA ligase and Chorismate synthase enzymatic assembly of *Tropheryma whipplei*. The best model result was retrieved to obtain the PDB file of their structure. Out of 120 best structures, one was finalized for DNA ligase, while out of 99 models, one was finalized for Chorismate synthase. Docking studies reveal binding energies for docked complexes and perfect binding energies for all docked complexes represented in Table **2**. The perfectly docked complex of inhibitory drugs and enzymatic assembly was represented in Fig. **2(A,B and C)**. These results satisfy the perfect interaction of complexes suggests that 2amino7fluoro5oxo5Hchromeno [2,3 b]pyridine3carboxamide as well as Nicotinamide mononucleotide interacts with DNA Ligase while Riboflavin monophosphate can interact with Chorismate synthase. And these selected chemicals can be used as putative drug candidates. Drug 2D interaction pattern with enzymes based on ligPlot v2.2 software represented in Fig. **2(P,Q and T)**. Mostly all the selected drugs not only show better binding scores but 3D and 2D interaction pattern reveals hydrogen bonds interaction with considered enzymes in their binding pocket. In Table **3** drug physicochemical parameters were represented based on the SwissADME server (www.swissadme.ch). Lipinski rule was also considered during drug analysis and found to show zero violations for 2APC. This indicates that all drugs have good inhibitory properties against selected enzymes. All of the drugs don't show blood-brain barrier permeability also any inhibition for CYP3A4 inhibition; these results suggest effective drug clearance in the body after effective action. 2APC, NMN and RFMP show logKp(cm/s) values -6.97, -8.26 and 10.9 respectively. TPSA values for 2APC, NMN, and RFMP were found 112.21, 165.61, and 217.9, respectively. 2APC also shows high GI absorption. MD simulation for 40 nanoseconds (ns) was conducted by deploying GROMACS software ver.2019. Constant pressure and temperature (NPT) ensembles were used to set the equilibration measures. At a normal temperature of 300 K and a pressure level of 1.013 bar, MD simulations were run. The MD trajectories were calculated using the perfectly docked-complexes to determine the root mean square de*via*tion (RMSD) and root mean square fluctuation (RMSF) for a timescale of 40-ns. It was found the best-docked complexes were exhibiting perfect characteristics based on RMSD and RMSF plots (Fig. **3**). The simulation study revealed that drug-enzyme complexes did not face any alterations in their binding patterns. RMSD values for docked complexes were found in the range of 1- 3 Angstrom, while RMSF values for docked complexes 1 – 4 Angstrom. These values indicate perfect interaction between considered drugs and enzymes of *T.whipplei.*

Table 2. AutoDock-vina docked results: Binding energies of best-docked complexes.

Best Docked Model	Binding-Energy (Kcal/mol)
DNA Ligase and 2APC	-8.3
DNA Ligase and NMN	-8.2
Chorismate Synthase and RFMP	-7.3

Table 3. -Drug characteristics analysed by Pubchem database and SwissADME.

Molecule	Formula	MW	Lipinski #violations
2-amino-7-fluoro-5- oxo-5H-chromeno[2,3- b]pyridine-3-carboxamide (CID- 10038928)	C13H8FN3O3	273.22	0
Nicotinamide mononucleotide (CID- 14180)	C27H29NO9	511.52	1
Riboflavin monophosphate (CID- 643976)	C17H21N4O9P	456.34	2

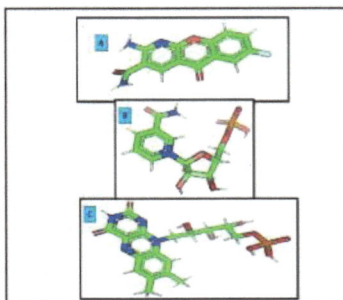

Fig. (1). Chemical structure of drugs obtained from Pubchem database and analyzed in Pymol **A.** 2amino7fluoro5oxo5Hchromeno[2,3-b]pyridine3carboxamide   (CID-10038928),   **B.**   Nicotinamide mononucleotide (CID- 14180), **C.** Riboflavin monophosphate (CID- 643976).

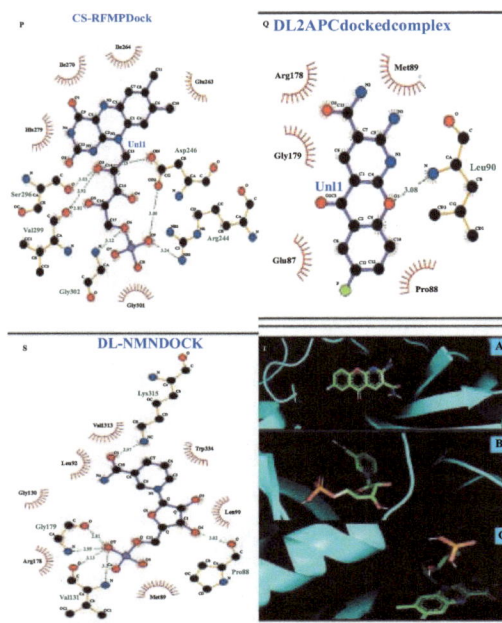

Fig. (2). Molecular interactions between drugs and receptor: **P)** Chorismate synthase interaction with RFMP: Val299, Gly 302, Ser 296, Asp 246, Arg 244 interacts with RFMP drug *via* hydrogen bond (2.00 to 3.50 Å). **Q)** DNA Ligase interaction with 2APC drug: Leu at 90 position interacts with 2APC drug *via* hydrogen bond of strength 3.08Å. **S)** DNA Ligase interaction with NMN drug: Lys at 315 position interacts with NMN drug *via* hydrogen bond of strength 2.97Å, also Gly 179 & Val 131 show hydrogen bonding with the NMN drug. **T)** AutoDock vina docking results of drugs interacting with proteins- **A.** DNA Ligase with 2amino7fluoro5oxo5H chromeno [2,3b]pyridine3carboxamide **B.** DNA Ligase with Nicotinamide mononucleotide **C.** Chorismate synthase with Riboflavin monophosphate.

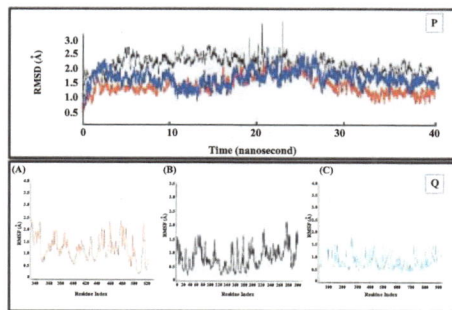

Fig. (3). Molecular simulation analysis of drugs complexed with enzymes of *T. whipplei*: **P)** RMSD plot: Black color-(RFMP-Chorismate synthase complex), Red Color- (2APC-DNA Ligase complex), Blue color-(NMN-DNA Ligase complex); **Q)** RMSF plots: **A.** RFMP interaction with Chorismate synthase, **B.** 2APC interacting DNA Ligase **C.**NMN drug interacting DNA Ligase.

CONCLUSION

This in-silico investigation discovers that 2-APC, NMN, and RFMP as possible medications to treat Whipple's disease and can be used for animal testing and clinical trials. All the pharmaco-kinetics depict that these medications would perfectly interact with bacterial biocatalysts to hamper their activity. This approach was found to be rapid for predicting drug candidates and even effective against harmful organisms like *Tropheryma whipplei* having a reduced genome.

CONSENT FOR PUBLICATION

Not Applicable.

CONFLICT OF INTEREST

The author declares no conflict of interest, financial or otherwise.

ACKNOWLEDGEMENTS

All the authors A.J. and V.K. conducted this research study and also manuscript preparation under the Department of bioinformatics, LPU, Punjab India.

REFERENCES

[1]   S.D. Bentley, M. Maiwald, L.D. Murphy, M.J. Pallen, C.A. Yeats, L.G. Dover, H.T. Norbertczak, G.S. Besra, M.A. Quail, D.E. Harris, A. von Herbay, A. Goble, S. Rutter, R. Squares, S. Squares, B.G. Barrell, J. Parkhill, and D.A. Relman, "Sequencing and analysis of the genome of the Whipple's disease bacterium Tropheryma whipplei", *Lancet,* vol. 361, no. 9358, pp. 637-644, 2003. [http://dx.doi.org/10.1016/S0140-6736(03)12597-4] [PMID: 12606174]

[2]   F. Fenollar, M. Célard, J-C. Lagier, H. Lepidi, P-E. Fournier, and D. Raoult, "Tropheryma whipplei endocarditis", *Emerg. Infect. Dis.,* vol. 19, no. 11, pp. 1721-1730, 2013. [http://dx.doi.org/10.3201/eid1911.121356] [PMID: 24207100]

[3]   A. Joshi, and V. Kaushik, "In-Silico Proteomic Exploratory Quest: Crafting T-Cell Epitope Vaccine

Against Whipple's Disease", *Int. J. Pept. Res. Ther.,* vol. 27, no. 1, pp. 1-11, 2020.
[http://dx.doi.org/10.1007/s10989-020-10077-9] [PMID: 32427224]

[4]   N. Palanisamy, "Identification of putative drug targets and annotation of unknown proteins in Tropheryma whipplei", *Comput. Biol. Chem.,* vol. 76, pp. 130-138, 2018.
[http://dx.doi.org/10.1016/j.compbiolchem.2018.05.024] [PMID: 30005292]

[5]   L.L.G. Ferreira, and A.D. Andricopulo, "ADMET modeling approaches in drug discovery", *Drug Discov. Today,* vol. 24, no. 5, pp. 1157-1165, 2019.
[http://dx.doi.org/10.1016/j.drudis.2019.03.015] [PMID: 30890362]

[6]   R. Zhang, H.Y. Ou, and C.T. Zhang, "DEG: a database of essential genes", *Nucleic Acids Res.,* vol. 32, no. Database issue, pp. D271-D272, 2004.
[http://dx.doi.org/10.1093/nar/gkh024] [PMID: 14681410]

[7]   Y. Huang, B. Niu, Y. Gao, L. Fu, and W. Li, "CD-HIT Suite: a web server for clustering and comparing biological sequences", *Bioinformatics,* vol. 26, no. 5, pp. 680-682, 2010.
[http://dx.doi.org/10.1093/bioinformatics/btq003] [PMID: 20053844]

[8]   M. Johnson, I. Zaretskaya, Y. Raytselis, Y. Merezhuk, S. McGinnis, and T.L. Madden, ""NCBI BLAST: a better web interface," Nucleic Acids Research, vol. 36, no", *Web Server,* no. May, pp. W5-W9, 2008.

CHAPTER 5

Optimizing the Power Flow in Interconnected Systems Using Hybrid Flower Pollination Algorithm

Megha Khatri[1,*], Pankaj Dahiya[2], Amrish [3] and Anita Choudhary[4]

[1] School of Electronics and Electrical Engineering, Lovely Professional University, Jalandhar, Punjab, India

[2] Department of Electronics and Communication Engineering, Delhi Technological University, New Delhi, India

[3] Department of Electronics and Communication Engineering, Gurukul Kangri Vishwavidyalaya, Haridwar, India

[4] GuruTegBahudar Institute of Engineering and Technology, New Delhi, India

Abstract: To standardize the active power flow in the interlinked power systems, the controller must be crafted in such a way that it would improve the system's stability. With this objective, the paper is dedicated to the hybrid flower pollination algorithm: a metaheuristic optimization algorithm (hFPA) applicable to optimize the parameters of the PI and PID structure-controller incorporated in the two-area power systems. The projected algorithm is compared with the articles published where the PI and PID controller structure parameters upgraded with hFPA are compared with the enhanced strength Pareto differential evolution and grey wolf optimization algorithms to demonstrate its robustness for a vast span of system parameters and varying load situations. The superiority of the technique has been presented in respect of peak undershoot, settling time, peak overshoot, tie-line power, frequency divergence, *etc.*

Keywords: Automatic generation control , Hybrid flower pollination algorithm , PI controller , PID controller .

INTRODUCTION

Modernization in interconnected power systems takes the responsibility of providing uninterrupted power, which is economical, efficient, sustainable, and safe. The power system is subjected to short time and long-time disturbances such as imbalance, reduction in generated power concerning the consumer demand, internal and external losses, *etc*. In addition, the power exchange in the interconn-

---
*** Corresponding author Megha Khatri:** School of Electronics and Electrical Engineering, Lovely Professional University, Jalandhar, Punjab, India; E-mail: meghakhatri7@gmail.com

Dharam Buddhi, Rajesh Singh and Anita Gehlot (Eds.)

eted systems may result in frequency deviations, due to trouble inflow of active power, which needs to be addressed meticulously. The frequency variations in power generation and load side affect the overall power flow, stability, and performance of the system. However, the interconnected power systems have their separate controllers, which are synchronized to manage the frequency variations and load demand. While these conventional controllers with fixed gains or parameters are unable to adjust with the real-time situations, which may lead to system failure.

Therefore, to ensure continuous circulation of quality power, the system frequency must be controlled to a nominal operating range during random load perturbations. The interconnected power system with various uncertainties inactive power delivery, load variations, system modeling errors is studied by researchers. The control strategies that combine understanding, expertise and methodologies from several sources are presented in the literature [1]. However to cope up with the system nonlinearities and inconsistent behavior the net changes are required in the controller parameters according to the variations in consumer power demand.

In literature, several controller structures have been discussed for the interdependent power units in the power system [2]. Recently, the biologically inspired control optimization algorithms such as genetic algorithm [3] chaotic quantum genetic algorithm [4], artificial immune system algorithm [5], plant growth simulation algorithm [6], particle swarm algorithm [7], artificial bee colony algorithm [8], ant colony algorithm [9], grey wolf optimization (GWO) algorithm [10], Jaya algorithm [11], *etc.*, have been developed for designing different controllers incorporated to several single and multi-area power systems.

In this article, an effort has been made to demonstrate the advantages of applying hybrid flower pollination algorithm (hFPA)for PI and PID controller structures over Enhanced strength Pareto differential evolution (ESPDE) [10] and GWO [10], in maintaining the system frequency under scheduled range.

An optimization problem is formulated using two thermal power systems and hFPA is commissioned to showcase the efficacy of the proffer technique with the considered controller structures for power flow control and system stability. The algorithm is providing superior damping characteristics when the system is put through the different variations discussed in this work. The proposed parameter control mechanism, can monitor the parameter deviations, take decisions and take action to minimize the variations, hold the system frequency within specified values and streamline the power distribution. The work is divided as; Section 2 illustrates the mathematical representation of interconnected thermal generation system with controllers; Section 3 represents the proffer hFPA for controller parameter optimization and Section 4 explains the simulations and the results obtained followed by the conclusions, respectively.

POWER SYSTEM MODELLING

A dynamic model [10] of an interconnected system under investigation is shown in Fig. (1). In the figure, R_i (p.u. Hz) is the governor speed regulation variables; B_i is the frequency bias constants; ΔP_{Vi} are deviations in governor valve positions; ΔP_{Ti} are the variations in turbine power out-turn; K_{PSi} are the gains of the power system; T_{Gi} are the governor time constants; ΔF_i are frequency shifts; T_{Ti} is the turbine time constants; T_{PSi} are the system time constants; The nominal parameters taken for the simulations can be seen in [10].

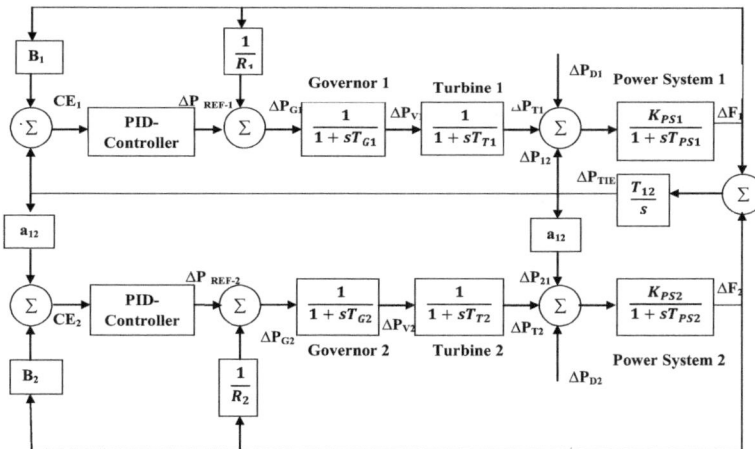

Fig. (1). Transfer function representation of 2-interdependent thermal systems.

An objective function is necessary to establish based on the required constraint, specifications and has to be selected for the optimization of the controller specifications. The function used in this work is integral time absolute error (ITAE) and can be described as:

$$ITAE = \int_0^{t_s} t(\sum_{i=1}^{2}|\Delta F_i| + |\Delta P_{TIE}|)dt \qquad (1)$$

PROPOSED ALGORITHM

Flower pollination algorithm (FPA) and its evolved forms are applicable in almost every field of engineering such as chemical, civil, electronics, and communication, computer science, power system, *etc.* developed in [12] and extended for multi-objective optimization problems [13]. Generally, plants are flowering using their species for breeding *via* a pollination process that implies a transfer of pollen between flowers. Biotic pollination is observed as global pollination because the pollinators can travel long distances and show Lévy flight behavior and mathematically formulated using Lévy distribution to get the random solutions. On the contrary, abiotic is viewed as local pollination, moreover mathematically formulated for effective convergence of the solutions [14]. While following Lévyflight behavior, for global pollination the flower pollination algorithm is generalized as

$$X_i(t+1) = \gamma \left(G_* - X_i^t(t) \right) L(\rho) + X_i(t) \tag{2}$$

As the closer flowers are presumably pollinate due to local pollination while the distant flowers have fair chances of pollinated globally. In this work, a standard FPA is combined with the differential evolution to improve the searching accuracy. Here, a modified hybrid algorithm using differential evolution techniques along with an optimization flower pollination algorithm is applied for dynamic and customizable weights to improve the system performance. It also eliminates the use of switch probability, which often causes disturbances and further enhances the stability of the presented interconnected system. The weights W_1 and W_2 are directing the algorithm for fast response and effective convergence to the global solution is defined as:

$$W_1 = W_{max} - t \frac{W_{max} - W_{min}}{T} \tag{3}$$

$$W_2^i = \frac{\left| \min\left(fit_i^t, mean\left(fit^t \right) \right) \right|}{\left| \max\left(fit_i^t, mean\left(fit^t \right) \right) \right|} \tag{4}$$

where W_{max} and W_{min} are weights limits decided by the end-user, T = iterations and t is current generation. The weight W_1 is computed every time and W_2 is calculated adaptively based on the likely solution. Now the modified flower pollination generalized equation become:

$$X_i(t+1) = W_1 \gamma L(\rho)(G_* - X_i^t(t) + W_2^i f\left(X_j^t(t) - X_k^t(t) \right) + X_i(t) \tag{5}$$

From the above equation, it is apparent that W_1 is utilized for controlling Lévy flight steps and W_2 is used for rapid and intelligent convergence.

SIMULATIONS AND RESULTS

In this section, simulations of interconnected thermal systems incorporated with controllers like PI and PID are carried out. In the simulations step load disturbance of 10% p.u. is taken into account for one source. For the tuning of PI and PID, gains hFPA is employed with 100 iterations.

Case 1

The power system fitted with PI controller is simulated with the gains adjustment through hFPA, EPSDE [10], and GWO [10]. The variations in controller gains, performance indices *i.e.* PU, PO, and ST are presented in Table **1**. It has been observed that the variations in controller parameters when the proffer algorithm is applied are lesser than with the other algorithms. The objective function is minimized *i.e.* ITAE value is 0.8092 to achieve convergence of system frequencies to steady-state faster than any other technique *i.e.* ΔF_1 = 4.8485 and ΔF_2= 6.9979 respectively along with the tie-line power stability *i.e.* ΔP_{tie} = 5.8150. In Fig. (**2**), the relative performance of the dynamic responses with the selected 10% step disturbance applied on Area-1 is shown. The results acquired from the profferhFPA are better in terms of ST and ITAE.

Table 1. Comparative Examination of 2-area system fitted with PI controller.

Parameter	Optimization Techniques			Parameter	Optimization Techniques		
	hFPA	*EPSDE*	*GWO*		*hFPA*	*EPSDE*	*GWO*
K_{P1}	-0.1658	0.0145	0.0630	**ITAE**	0.8092	1.3529	1.221
K_{I1}	0.6618	0.8502	0.5565	**PO(ΔF_1)**	0.0342	0.0628	0.0184
K_{P2}	0.8025	0.0478	0.0730	**PO(ΔF_2)**	0.0254	0.0579	0.0189
K_{I2}	-0.4284	0.0334	0.0199	**PO(ΔP_{tie})**	0.0020	0.0092	0.0019
PU(ΔF_1)	-0.2289	-0.2134	-0.2130	**STΔF_1)**	4.8485	8.9299	7.8914
PU(ΔF_2)	-0.1811	-0.1570	-0.1595	**ST(ΔF_2)**	6.9979	9.8026	8.7536
PU(ΔP_{tie})	-0.0612	-0.0569	-0.0572	**ST(ΔP_{tie})**	5.8150	9.3920	8.3841

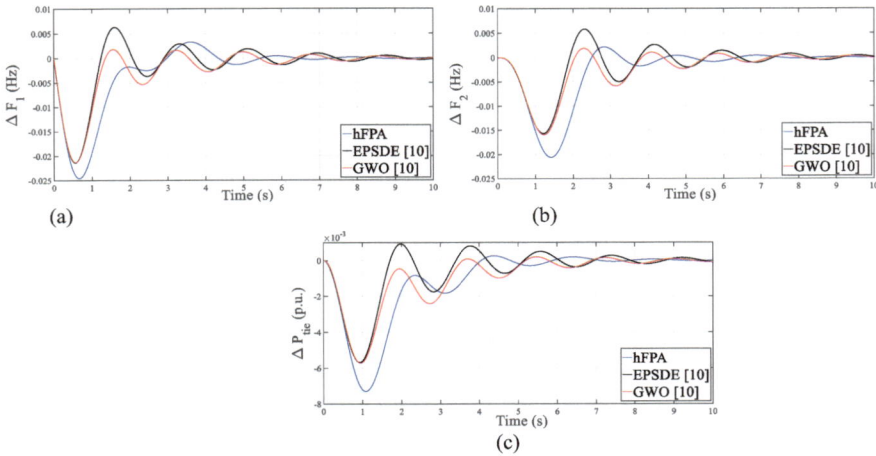

Fig. (2). Frequency shifts in Area-1(a), Area-2 (b) and tie-line power (c) incorporated with PI type controllers.

Case 2

In this case, the two area system is incorporated with the PID controllers, and the controller gains obtained using hFPA along with the various performance indices are presented in Table **2**, whereas the supremacy of hFPA tuned controller over EPSDE and GWO algorithm applied to the PID control structure has been depicted in terms of minimization of objective function *i.e.* ITAE which also represented the reduction in deviations. The responses with 10% step load perturbation are shown in Fig. (**3**). The ST for frequency divergence of systems 1 and 2 and tie-line power is found to be as $\Delta F_1 = 4.8485$, $\Delta F_2 = 6.9979$, and $\Delta P_{tie} = 5.8150$, respectively. The results show that the hFPA tuned controller is again a better performer with performance indices *i.e* ITAE, ST, PU, and PO as depicted from Table **2**.

Table 2. Comparative Examination of 2-area system fitted with PID controller.

Parameter	Optimization Techniques			Parameter	Optimization Techniques		
	hFPA	*EPSDE*	*GWO*		*hFPA*	*EPSDE*	*GWO*
K_{P1}	1.2472	0.8599	1.0569	PU (ΔP_{tie})	-0.0201	-0.0236	-0.0215
K_{I1}	2.4591	1.7733	1.9107	ITAE	0.1024	0.1494	0.1339
K_{D1}	0.4714	0.3883	0.4221	PO (ΔF_1)	0.0073	0.0028	0.0020
K_{P2}	1.2472	1.0411	1.7486	PO (ΔF_2)	$8.3699 * 10^{-5}$	$2.8146 * 10^{-4}$	$9.3176 * 10^{-5}$
K_{I2}	2.4591	0.1650	0.0400	PO (ΔP_{tie})	$1.5104 * 10^{-4}$	$4.5515 * 10^{-5}$	$2.1873 * 10^{-5}$

(Table 2) cont.....

K_{D2}	0.4714	1.0110	1.1988	ST (ΔF_1)	2.6486	2.8741	1.0589
PU (ΔF_1)	-0.1058	-0.1179	-0.1123	ST (ΔF_2)	2.5749	2.9713	3.1751
PU (ΔF_2)	-0.0592	-0.0612	-0.0560				

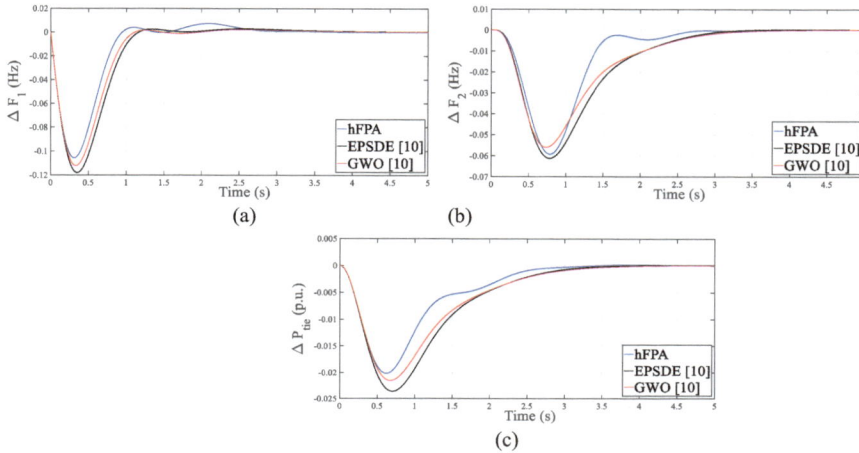

Fig. (3). Frequency shifts in Area-1(a), Area-2(b) and tie line power (c) incorporated PID type controllers.

CONCLUSIONS

An interconnected non-reheat thermal power system with various controller structures optimized with proposed hybrid flower pollination algorithm (hFPA) for adjusting the power flow and stabilizing the operation of the complete system. The proffer algorithm applies to the considered controller's structure for their gains optimization and minimizing the deviations. The PI and PID type controller structure optimized gains are compared with EPSDE and GWO tuned controllers. The robustness of the presented algorithm is judged on the PU, PO, ST, and ITAE performance indices with varying load conditions. The results obtained with the proffer algorithm achieved a finer dynamic response compared to other algorithms.

CONSENT FOR PUBLICATION

Not applicable.

CONFLICTS OF INTEREST

The authors declare no conflict of interest, financial or otherwise.

ACKNOWLEDGEMENT

Declared none.

REFERENCES

[1]   S.K. Pandey, S.R. Mohanty, and N. Kishor, "A survey on load–frequency control for conventional and distribution generation power systems", *Renew. Sustain. Energy Rev.,* vol. 25, pp. 318-334, 2013. [http://dx.doi.org/10.1016/j.rser.2013.04.029]

[2]   H.H. Alhelou, M-E. Hamedani-Golshan, R. Zamani, E. Heydarian-Forushani, and P. Siano, "Challenges and opportunities of load frequency control in conventional, modern and future smart power systems: A comprehensive review", *Energies,* vol. 11, no. 10, p. 2497, 2018. [http://dx.doi.org/10.3390/en11102497]

[3]   H. Golpira, and H. Bevrani, "Application of GA optimization for automatic generation control design in an interconnected power system", *Energy Convers. Manage.,* vol. 52, no. 5, pp. 2247-2255, 2011. [http://dx.doi.org/10.1016/j.enconman.2011.01.010]

[4]   G-C. Liao, "Solve environmental economic dispatch of Smart MicroGrid containing distributed generation system–Using chaotic quantum genetic algorithm", *Int. J. Electr. Power Energy Syst.,* vol. 43, no. 1, pp. 779-787, 2012. [http://dx.doi.org/10.1016/j.ijepes.2012.06.040]

[5]   R. Syahputra, and I. Soesanti, "Power System Stabilizer Model Using Artificial Immune System for Power System Controlling", *Int. J. Appl. Eng. Res.,* vol. 11, no. 18, pp. 9269-9278, 2016.

[6]   C. Wang, and H.Z. Cheng, "Optimization of network configuration in large distribution systems using plant growth simulation algorithm", *IEEE Trans. Power Syst.,* vol. 23, no. 1, pp. 119-126, 2008. [http://dx.doi.org/10.1109/TPWRS.2007.913293]

[7]   Y. Del Valle, G.K. Venayagamoorthy, S. Mohagheghi, and R.G. Harley, "PSO: basic concepts, variants and app. in power systems", *IEEE Trans. Evol. Comput.,* vol. 12, no. 2, pp. 171-195, 2008. [http://dx.doi.org/10.1109/TEVC.2007.896686]

[8]   A.A. El-Fergany, and A.Y. Abdelaziz, "Artificial bee colony algorithm to allocate fixed and switched static shunt capacitors in radial distribution networks", *Electr. Power Compon. Syst.,* vol. 42, no. 5, pp. 427-438, 2014. [http://dx.doi.org/10.1080/15325008.2013.856965]

[9]   C.F. Chang, "Reconfiguration and capacitor placement for loss reduction of distribution systems by ant colony search algorithm", *IEEE Trans. Power Syst.,* vol. 23, no. 4, pp. 1747-1755, 2008. [http://dx.doi.org/10.1109/TPWRS.2008.2002169]

[10]   D. Guha, P.K. Roy, and S. Banerjee, "Load frequency control of interconnected power system using grey wolf optimization", *Swarm Evol. Comput.,* vol. 27, pp. 97-115, 2016. [http://dx.doi.org/10.1016/j.swevo.2015.10.004]

[11]   S.P. Singh, T. Prakash, V.P. Singh, and M.G. Babu, "Analytic hierarchy process based automatic generation control of multi-area interconnected power system using Jaya algorithm", *Eng. Appl. Artif. Intell.,* vol. 60, pp. 35-44, 2017. [http://dx.doi.org/10.1016/j.engappai.2017.01.008]

[12]   D. Chakraborty, S. Saha, and O. Dutta, "DE-FPA: a hybrid differential evolution-flower pollination algorithm for function minimization", In: *in 2014 international conference on high performance computing and applications (ICHPCA),* 2014, pp. 1-6. [http://dx.doi.org/10.1109/ICHPCA.2014.7045350]

[13]   N. Jayakumar, S. Subramanian, S. Ganesan, and E.B. Elanchezhian, "Grey wolf optimization for combined heat and power dispatch with cogeneration systems", *Int. J. Electr. Power Energy Syst.,* vol. 74, pp. 252-264, 2016.

[http://dx.doi.org/10.1016/j.ijepes.2015.07.031]

[14]   N.C. Patel, M.K. Debnath, B.K. Sahu, and P. Das, "2DOF-PID controller-based load frequency control of linear/nonlinear unified power system", *In International Conference on Intelligent Computing and Applications,* 2019, pp. 227-236.
[http://dx.doi.org/10.1007/978-981-13-2182-5_23]

CHAPTER 6

A Comparative Study on Various Data Mining Techniques for Early Prediction of Diabetes Mellitus

Ovass Shafi[1,*], S. Jahangeer Sidiq[1], Tawseef Ahmed Teli[2] and Majid Zaman[3]

[1] School of Computer Applications, Lovely Professional University, Punjab, India

[2] Department of Computer Applications, Amar Singh College, J and K, India

[3] Directorate of IT and SS, University of Kashmir, Srinagar, India

Abstract: Diabetes mellitus is a deadly disease that affects people all over the globe. An early prediction of diabetes is very beneficial as it can be controlled before the onset of the disease. Various data mining classification techniques have proven fruitful in the early detection and prediction of multiple diseases like heart attack, depression, kidney-related diseases, and many more. This paper discusses and compares various data mining techniques for the prediction of Diabetes Mellitus. Also, three widely used data mining techniques *via* Artificial Neural Networks (ANN), K-nearest neighbor (K-NN), and Support Vector Machine (SVM) have been implemented in Matlab and the results are compared based on accuracy, recall, true negative rate, and precision.

KEYWORDS: ANN, Classification, Data Mining, Detection, Diabetes, KNN, Prediction, SVM.

1. INTRODUCTION

Among various types of Diabetes, diabetes mellitus is the most common type that affects humans around the globe. The main reason for the disease is a change in the lifestyle of the common man. It is a kind of metabolic disorder with hyperglycemia which means a high concentration of glucose found in the blood. The reason can be a defect in the insulin secretion or improper action of insulin in the body or both in some cases. Due to diabetes mellitus, the metabolism of the body is affected which causes a high level of sugar in the blood. Chronic hyperglycemia is dangerous as it may lead to the dysfunction of various organs and ultimately in organ failure. Neuropathy, Nephropathy, Retinopathy, Cardiac

---

* **Corresponding author Ovass Shafi:** School of Computer Applications, Lovely Professional University, Punjab, India; E-mail: owaisfour03@gmail.com

Dharam Buddhi, Rajesh Singh and Anita Gehlot (Eds.)

Disorder, and Blood vessels compression are some of the main illnesses caused by this disease.

Long-lasting diabetes has been classified broadly into two major types: A) Type I diabetes B) Type II diabetes. The first one is the pathogenesis of TYPE I and is caused when the pancreas secretes damaged β-cells. The damaged β-cells prevent the pancreas to reduce glucose levels in the blood. The second type of diabetes mellitus is known as TYPE 2 or non-insulin-dependent diabetes mellitus that is caused due to resistance of insulin and deficiency in insulin secretion. This is one of the serious diseases that may result in death if not treated properly. To save the lives of people, the diagnosis of disease before its onset is the need of the hour. The use of DM (Data Mining) techniques in the early prediction and diagnosis of various deadly diseases like heart attack, breast cancer, Alzheimer's disease, *etc.* has proven useful for saving the lives of several people. These DM techniques may also be used to improve the prediction probability of diabetes mellitus so that it can be treated on time.

The medical dataset taken for the study of diabetes mellitus contains several attributes and the study of all the attributes is time consuming, inefficient, and cumbersome process. Therefore for the efficient, accurate, and correct prediction of diabetes mellitus, only significant features are selected that includes: (NTP) Number of Times Pregnant, (BDI) Body Mass Index, (2HSI) 2-Hr Serum Insulin, (TSFT) Triceps Skin Fold Thickness, Age and (DPF) Diabetic Pedigree Function and Class (Yes or No).

2. LITERATURE REVIEW

The researchers [1] develop five prediction models with 9 input and one output variable. They provide a comparative performance analysis of multiple classification methods for the prediction of diabetes mellitus by using vector measurements. The authors in [2] studied the data of 768 women and selected nine attributes for their study. They propose a model for the detection of diabetes mellitus using data mining classification algorithms using Clementine 12 software. The authors in [3] make use of invasive and non-invasive features to identify diabetes patients with high risk. They use the J48 decision tree with WEKA version 3.9.2 along with linear regression. Researchers in [4] studied various data mining algorithms for predicting diabetes mellitus using the dataset obtained from UCI Machine on WEKA software. The authors found decision tree records the highest prediction accuracy. The authors in [5] proposed a MADG algorithm using various mining techniques for the prediction of diabetes mellitus with three parameters viz Oral Glucose Tolerance Test, Body Mass Index, and Diastolic Blood Pressure for rule extraction. Researchers in [6] worked on a study to check if sex, age, socio-economic status, parental education level, DKA at early diagnosis of type 1 diabetes mellitus, infection, psychological problems, poor metabolic control can be used for the prediction of DKA in type 1 diabetes mellitus patients. The authors in [7] form multiple data groups of various individuals and tested each group with K nearest algorithm, SVM, RF, NB, and ANN models and use negative prediction, sensitivity, specificity, and precision for assessing model performance. Authors in [8] studied the harmonization of biomarkers like immunoassays for the prediction of diabetes. Researchers in [9] studied various blood count parameters between healthy pregnant women and pregnant women suffering from Gestational Diabetes Mellitus to find the relation between the various blood count parameters and Gestational Diabetes Mellitus. The researchers studied [10] the role of interleukin-6 for early diagnosis of Gestational Diabetes Mellitus and found that assessment of serum interleukin 6 levels can be studied and used for early prediction of diabetes mellitus. Attempts to make an early prediction of diabetes more accurate and a novel model using D-M technique for the prediction of Type 2 diabetes was proposed by [11]. The researchers in [12] developed a technique for the diagnosis of gestational diabetes mellitus during the time of early pregnancy using the artificial neural network and decision tree. The metabonomics technique to see the changes of metabolites and metabolic trails pre and post-diabetes for the efficient and accurate treatment of diabetes was studied by [13]. The researchers studied [14] the possibility of various data mining classification techniques to classify persons suffering from diabetes based on various non-invasive and early collected clinical features. Authors in [15] studied 942446 U.S. persons without any sign of diabetes,>-3 RPG in baseline year and >-1 primary care visit/ year for the consecutive 5 years. The researchers in [16] studied the probability of heart attacks in patients with type 2 diabetes mellitus. The authors in [17] proposed a rule extraction algorithm enhanced STAD model in order to make accurate prediction of diabetes by developing accurate, interpretable and concise classification rules and then apply the rules on dataset for early diagnosis of diabetes mellitus. Researchers in [18] gave the comparison of various data mining classification algorithms based on experimental study along with the merits and demerits of each algorithm. Authors in [19] presented a review and explains the various challenges for diagnosis of gestational diabetes mellitus and identifies the various biomarkers that may help in the identification of gestational diabetes mellitus. Researchers in [20] studied five machine learning techniques for prediction of diabetes mellitus patients [21]. A study estimated diabetes using major characteristics and also gives the categorization of the relation between contradictory characteristics.

3. CONVENTIONAL MACHINE LEARNING TECHNIQUES

As diabetes is a kind of long-lasting disease, a huge volume of data is generated during the treatment of patients suffering from diabetes like cardiovascular disease [22] and the early prediction could prove lifesaving. Various ML (machine learning) methods can be used for the diagnosis and prediction of diabetes mellitus; however, these techniques can either be supervised or unsupervised forms of learning.

A. Supervised Learning

In supervised learning methods [23 - 25], the researchers set a goal function that defines the model and then fine-tune arguments of the classification technique by the application of a group of recognized sample categories. The set of known output variable values is used to compose training data. Supervised learning has two types of learning tasks, classification, and regression. Some of the commonly used supervised learning techniques include DT (Decision Trees), ANN, and SVM.

1) Decision Tree (DT): A decision tree is a flow chart type of classification algorithm that starts with the root node and then contains intermediary nodes and the leaf nodes. The labels are allotted to leaf nodes. Different test conditions are assigned at the root node and internal nodes. The selection of the root node depends on the process of information gain. The advantage of the Decision Tree is a precise and understandable decision-making process that provides a strong method for the classification and prediction of diabetes.

2) Support Vector Machine (SVM): SVM transforms data from input space to multidimensional feature space and then searches for a separating hyper-plane using a kernel function. Various kernel functions that are commonly used in SVM are linear polynomial, PUK, and radial basis functions.

3) Artificial Neural Network (ANN): ANN is one of the most powerful tools used in the analysis [26, 27] of complex clinical datasets. ANNs use known data in the training process and then identify the complex relationship between input and output variables, followed by the prediction of output values from a given set of input values. The ANN is suitable for incomplete, complex, and non-linear datasets.

4) K-nearest neighbor (KNN): The KNN algorithm was developed by Thomas Cover for regression and classification. It is a non-parametric mechanism that comprises K-closest training examples for input in the feature space and the output is dependent upon whether it is used for classification or Regression. This technique produces class membership as output which is an object classified on plurality vote. The object that is the most common among the neighbors is allotted to the class. If k=1, the object is simply allotted to the class of that single nearest neighbor. The k-NN regression produces the property value of the object as output that is the mean of the values of k nearest neighbors.

B. Unsupervised Learning

Enormous data is produced by the medical industry, but some data doesn't possess corresponding labels. As a result of which the use of supervised learning techniques is not suitable. In such cases, unsupervised learning can be useful to discover the relation between hidden structures of unlabeled data. Some of the commonly used unsupervised learning methods include clustering techniques and association rule techniques

1) Clustering Techniques: The clustering techniques can be used for searching for useful patterns in unorganized datasets.

2) Association Rule Learning: The general form of an association rule is {X2, X2,....., Xn}$\rightarrow$Y. An association rule is one where the two entities support degree and trust degree reach a threshold value which is already specified. Commonly used association rule discovery algorithm includes Apriori.

C. Deep Learning

The traditional machine learning algorithms have limitations in handling raw natural data. To deal with such type of data the researchers use the notion of representation learning in which there is a set of methods that are used by a machine and provided with input in the form of raw data and it learns the representations required for the classification. One of the widely used representations learning methods is Deep-Learning that has many layers of representation, obtained by a combination of simple nonlinear modules that helps to transform a simple level of representation into a higher, more abstract representation. The datasets generated by the medical industry are multidimensional, heterogeneous, and sparse. Therefore the application of Deep Learning methods for the classification of such datasets is more suitable and has applications in the prediction of diabetes.

1) Deep Belief Network: DBN is another classification mechanism that specifically consists of Restricted Boltzmann Machines with hidden layers interconnected with a visible layer of the next RBM. Deep Belief Network has also produced promising results in the classification of diseases and can be advantageous in the detection of diabetes.

2) Deep Recurrent Neural Network D-RNN: Another kind of unsupervised deep learning algorithm is D-RNN in which the connections of units generate a directed cycle and can work on temporal data, time-series data, or sequential data.

3) Deep Convolutional Neural Network D-CNN: D-CNN is another kind of deep learning method that gives better results for the study of medical datasets. It is an interleaved collection of feed-forward layers applying filters that are followed by the process of reduction and then the rectification of pooling layers.

4. EXPERIMENT AND RESULTS

The three most widely used data mining techniques *via* ANN, K-NN Algorithm, and SVM were implemented using Matlab. The dataset used for this experiment was taken from the National Institute of Diabetes and Digestive and Kidney Diseases popularly known as the PIMA dataset. Table **1** shows the Attributes Chosen for Study.

Table 1. Attributes Chosen for Study.

Sr. No.	Attribute Name	Sr. No.	Attribute Name
1.	Diastolic Blood Pressure	6	Triceps Skin Fold Thickness
2.	Plasma Glucose Concentration	7	Age
3.	Number of Times Pregnant	8	Diabetic Pedigree Function
4.	Body Mass Index	9	Class (Yes or No)
5.	2-Hr Serum Insulin	-	-

1) Artificial Neural Network: From the dataset consisting of 768 records, a total of 550 samples were taken randomly out of which 384 records were used as training data, 83 records were used for validation, and 83 for testing. A total of 10 layers were used.

The confusion matrix of the ANN Algorithm is given in Table **2**.

Table 2. Confusion Matrix Of ANN.

	Training Confusion Matrix				Validation Confusion Matrix		
0	224 58.3%	56 14.6%	80.0% 20.0%	**0**	54 65.1%	14 16.9%	79.4% 20.6%
1	31 8.1%	73 19.0%	70.2% 29.8%	**1**	2 2.4%	13 15.7%	86.7% 13.3%
2	87.8% 12.2%	56.6% 43.4%	77.3% 22.7%	**2**	96.4% 3.6%	48.1% 51.9%	80.7% 19.3%
	0	**1**	**2**		**0**	**1**	**2**
	Target Class				**Target Class**		
	Test Confusion Matrix				**All Confusion Matrix**		
0	42 50.6%	19 22.9%	68.9% 31.1%	**0**	320 58.2%	89 16.2%	78.2% 21.8%
1	5 6.0%	17 20.5%	77.3% 22.7%	**1**	38 6.9%	103 18.7%	73% 27.0%
2	89.4% 10.6%	47.2% 52.8%	71% 28.9%	**2**	89.4% 10.6%	53.6% 46.4%	76.9% 23.1%
	0	**1**	**2**		**0**	**1**	**2**

From the Confusion Matrix of the ANN algorithm, it is clear that the algorithm gives an accuracy rate of 76.9%.

2) K-Nearest Neighbour: The KNN was run using all the 768 records of the dataset out of which 278 records were used for testing data and 550 records for training data.

The Confusion Matrix of KNN is given in Table **3**.

Table 3. Confusion Matrix Of KNN (training data, testing data and total).

123	13	46	36
90.4%	9.6%	43.9%	56.1%
72.8%	73.5%	27.2%	26.5%

Predicted Class

TABLE 4. Results Obtained from KNN.

Measure	KNN
Recall	72.781065
True Negative Rate	73.469388
Precision	90.441176
Accuracy	72.935780

When the dataset was used with the K-NN algorithm it gives an accuracy of 72.9% which is slightly less than the accuracy rate of the ANN Algorithm (Table **4**).

3) Support Vector Machine: The SVM was again run using all the 768 records of the dataset out of which 278 records were used for testing data and 550 records for training data.

The Confusion Matrix of SVM is given in Table **5**.

TABLE 5. Confusion Matrix of SVM (Training data, testing data and total).

123	13	37	45
90.4%	9.6%	54.9%	45.1%
76.9%	77.6%	23.1%	22.4%

Predicted Class

TABLE 6. Results Obtained from SVM.

Measure	SVM
Recall	76.875000
True Negative Rate	77.586207
Precision	90.441176
Accuracy	77.064220

The Results obtained from SVM give an accuracy value of 77% which is better than KNN and ANN (Table **6**).

CONCLUSION

Various data mining techniques have been sued by researchers to assist the early detection of Diabetes. In this paper, many such techniques were discussed. Also, this research work carried out the experimentation of three popular data mining techniques used for Diabetes prediction namely ANN, K-NN, and SVM using Matlab and compared the results. The performance of these methods was compared based on these metrics; Accuracy, Recall, True Negative Rate, and Precision. The results showed that SVM gives better accuracy results followed by ANN and KNN.

CONSENT FOR PUBLICATION

Not Applicable.

CONFLICT OF INTEREST

The author declares no conflict of interest, financial or otherwise.

ACKNOWLEDGEMENT

Declared none.

REFERENCES

[1]   A.S Sunge, "Comparison Data Mining Techniques To Prediction Diabetes Mellitus", *Journal of Sustainable Engineering: Proceedings Series,* vol. 1, no. 2, pp. 225-230, 2019.
      [http://dx.doi.org/10.35793/joseps.v1i2.31]

[2]   S.S. Mirzajani, and S. Salimi, *Prediction and Diagnosis of Diabetes by Using Data Mining Techniques,* vol. 6, no. 1, pp. 3-7, 2018.
      [http://dx.doi.org/10.15171/ajmb.2018.02]

[3]   M.N. Sohail, *A hybrid Forecast Cost Benefit Classification of diabetes mellitus prevalence based on epidemiological study on Real-life patient 's data*, 2019.
      [http://dx.doi.org/10.1038/s41598-019-46631-9]

[4]   B. O. Oguntunde, S. A. Arekete, and M. Odim, "Assessment of Selected Data Mining Classification Algorithms for Analysis and Prediction of Certain Diseases University of Ibadan Journal of Science and Logics in ICT Research Assessment of Selected Data Mining Classification Algorithms for Analysis and Prediction of Certain Diseases", *March,* 2020.

[5]   G. Amalarethinam, N. A. Vignesh, and A. V. Arts, "A New Monotony Advanced Decision Tree Using Graft Algorithm to Predict the A New Monotony Advanced Decision Tree Using Graft Algorithm to Predict the Diagnosis of Diabetes Mellitus", vol. 118, no. 6, 2018.

[6]   S. Patria, L Listianingrum, and T. Wibowo, "Predictive factors of ketoacidosis in type 1 diabetes mellitus", *PI [Internet],* vol. 59, no. 4, pp. 169-4, 2019.
      [http://dx.doi.org/10.14238/pi59.4.2019.169-74]

[7]   G. Cardozo, S. Cossul, J. Luis, and B. Marques, "Prediction of Glycated Haemoglobin Based on Routine Blood Count Tests to Support the Diagnosis of Diabetes Mellitus Method", pp. 1-27.

[8]   S. Hörber, P. Achenbach, E. Schleicher, and A. Peter, *Harmonization of immunoassays for biomarkers*

in diabetes mellitus, 2020.
[http://dx.doi.org/10.1016/j.biotechadv.2019.02.015]

[9]   K. Dogan, R.K. Calıskan, D.E. Iliman, I. Karaca, and L. Yasar, *Mpv for Predicting Gestational Diabetes Mellitus.*, 2019, pp. 1-4.
[http://dx.doi.org/10.31487/j.JDMC.2019.01.04]

[10]   A. Amirian, M. B. Mahani, and F. Abdi, "Role of interleukin-6 (IL-6) in predicting gestational diabetes mellitus", *Obstet. Gynecol. Sci.,* vol. 63, no. 4, pp. 407-416, 2020.
[http://dx.doi.org/10.5468/ogs.20020]

[11]   H. Wu, S. Yang, Z. Huang, J. He, and X. Wang, "Informatics in Medicine Unlocked Type 2 diabetes mellitus prediction model based on data mining", In: *Informatics Med. Unlocked,* vol. 10. , 2018, pp. 100-107.
[http://dx.doi.org/10.1016/j.imu.2017.12.006]

[12]   S. Rouhani, *Data Mining Approach for the Early Risk Assessment of Gestational Diabetes Mellitus,* April, 2018.
[http://dx.doi.org/10.4018/IJKDB.2018010101]

[13]   Q. Ma, Y. Li, M. Wang, Z. Tang, T. Wang, C. Liu, C. Wang, and B. Zhao, "Progress in Metabonomics of Type 2 Diabetes Mellitus", *Molecules,* vol. 23, no. 1834, 2018 .
[http://dx.doi.org/10.3390/molecules23071834]

[14]   D. Pei, Y. Gong, H. Kang, C. Zhang, and Q. Guo, *Accurate and rapid screening model for potential diabetes mellitus,* vol. 3, pp. 1-8, 2019.

[15]   M.K. Rhee, Y.-L. Rhee, M.K. HO, S. Raghavan, J.L. Vass, K. Cho, D. Gagnon, DL.R. Staimez, C.N. Ford, P.W.F. Wilson, and L.s. Phillips, "Random plasma glucose predicts the diagnosis of diabetes", *PLoS ONE,* vol. 14, no. 7, p. e0219964, 2019.
[http://dx.doi.org/https://doi.org/10.1371/journal.pone.0219964]

[16]   B.A. Williams, and S.S. Shetty, "A risk prediction model for heart failure hospitalization in type 2 diabetes mellitus", In: *Clinical Cardiology*, 2019, pp. 1-9.
[http://dx.doi.org/10.1002/clc.23298]

[17]   N.A. Vignesh, "Available Online at", www.ijarcs.info*International Journal of Advanced research in computer sciencea new stad model to predict the diabetes mellitus,* 2020.

[18]   R. Patil, and S. Tamane, *A Comparative Analysis on the Evaluation of Classification Algorithms in the Prediction of Diabetes,* vol. 8, no. 5, pp. 3966-3975, 2018.
[http://dx.doi.org/10.11591/ijece.v8i5.pp3966-3975]

[19]   N. Rodrigo, and S. J. Glastras, *The Emerging Role of Biomarkers in the Diagnosis of Gestational Diabetes Mellitus,* 2018.
[http://dx.doi.org/10.3390/jcm7060120]

[20]   V. Rawat, "A Classification System for Diabetic Patients with Machine Learning Techniques", 2020.
[http://dx.doi.org/10.33889/IJMEMS.2019.4.3-057]

[21]   P. Tiwari, and V. Singh, *J. Phys. Conf. Ser.,* vol. 1714, 2021.012013
[http://dx.doi.org/10.1088/1742-6596/1714/1/012013]

[22]   T. A. Teli, "Fuzzy Logic And Medicine With Focus On Cardiovascular Disease Diagnosis", *5th IEEE International Conference on Computing for Sustainable Global Development,* 2018.

[23]   M. Ashraf, M. Zaman, and M.A. Butt, "An Empirical Comparison of Supervised Classifiers for Diabetic Diagnosis Available Online at", www.ijarcs.info*An Empirical Comparison of Supervised Classifiers for Diabetic Diagnosis,,* 2017.

[24]   M. Ashraf, M. Zaman, and M.A. Butt, "Knowledge Discovery in Academia : A Survey on Related Literature Available Online at ", www.ijarcs.info*Knowledge Discovery in Academia : A Survey on Related Literature,,* 2017.

[25]   M. Ashraf, M. Zaman, and M.A. Butt, "An empirical comparison of supervised classifiers for diabetic diagnosis", *Int J Adv Res Comput Sci,* vol. 8, no. 1, pp. 311-315, 2017.

[26]   K. Neha, and S.J. Sidiq, "Analysis of Student Academic Performance through", *Expert Syst.,* vol. 02, no. 09, pp. 48-54, 2020.

[27]   M. Ashraf, M. Zaman, and M.A. Butt, "A Binarization Approach for Predicting Box-Office Success", *Solid State Technol.,* vol. 63, no. 5, pp. 8652-8660, 2020.

CHAPTER 7

Systematic Review of BYOD Cyber Forensic Ecosystem

Md. Iman Ali[1,*] and **Sukhkirandeep Kaur**[2]

[1] Department of Computer Application, Lovely Professional University, Phagwara, Punjab, India

[2] School of Computer Science and Engineering, Lovely Professional University, Punjab, India

Abstract: Since the inception of Bring Your Own device (BYOD) a series of the study conducted in various segments of BYOD. The exponential growth of the adoption of BYOD technology increased the demand for research. As a result, academic researchers created a good amount of abstracts with new techniques and methods. This study article offers a systematic study of existing techniques and development on the BYOD cybersecurity ecosystem. The primary goal of this systematic review is to identify the existing research specifically on BYOD Cyber forensic ecosystem and grouping the techniques developed in various areas and summarized the findings. Post analyzing 8519 articles this study identifies the potential 18 research which contributes to enhancing the cybersecurity forensic ecosystem. Limitations of existing research are also identified which organizations need to mitigate to build a cyber secured Forensic BYOD environment. Finally concluded that BYOD Cyber secured ecosystem is indeed needed for the organization for enabling BYOD services so that due to the impact of cyberattacks in the BYOD environment business ecosystem does not get fragmented.

Keywords: BYOD, Cyberthreat, Cyber Crime, Cyber Security, Digital Forensic, Mobile Security.

1. INTRODUCTION

Periodic incremental research on Bring Your Device (BYOD) in various areas is conducted continuously. As BYOD becomes a fundamental need for the organization to enable an extended work environment. This is also witnessed that providing BYOD services to the employee increased productivity, business agility while employee satisfaction was a key requirement. While providing BYOD services to the employee for business growth and employee personal activities but this service becomes one of the major business disruption and survival reasons

---

* **Corresponding author Md. Iman Ali:** Department of Computer Application, Lovely Professional University, Phagwara, Punjab, India; E-mail: mdiman@rediffmail.com

Dharam Buddhi, Rajesh Singh and Anita Gehlot (Eds.)

due to the excessive cyber threat landscape [1]. Various secured authentication mechanisms [2] of providing BYOD access implemented and even also malicious traffic detection mechanism are framed [3] but limitations on authentication technique and threat detection model is always an important aspect.

Previous researchers have done lots of development in the area of security threats and systematic review was also performed. An advanced level of in-depth systematic review results are very unique and findings are very directive of existing research. As per this study where more than 100 indicative papers were grouped and analyzed and results are like security attack 32%, where research related to malware was 29%, data leakage focus was 14%, stolen device focused research was 12% and DDOS related research was 5% and unauthorized access and bandwidth related focused research was only 1% [4]. This advanced-level study indicates the demand for more abstract in providing a cyber secured BYOD environment.

The key goal of this review is to identify the available research technique in BYOD malicious traffic detection and protection mechanism to reduce cyber threats. Post identification of existing research, categorization, and finding out the gap and limitation in the existing research are key focused areas. Summarization of the existing studies and finding the development so far in the BYOD malicious traffic detection model and identifying the area of the development to reduce the cyber threat are key goals of this research.

To complete the systematic review rest of the papers are organized as follows

Section 2 described the Review Techniques and methods, Section 3 Review the Planning and conducting of the review, section 4 described the Result and Analysis, and finally, Section 5 include the Conclusion.

2. REVIEW TECHNIQUE

The outcome of the existing study in the BYOD threat detection model was the main target phenomena of this systematic review. Finding out the result, the Conclusion of the existing phenomena in the BYOD traffic threat detection model needs a robust exploration of the existing research in this area. Defining the process of how to conclude this was a tedious task, To finalize the process of Systematic review, key rationale mechanism, and process used during this research. A standard practice of Systematic Literature Review (SLR) was followed during this study. For structuring the process of this study standard defined strategies followed as per the study "Procedures for Performing Systematic Reviews" [5]. SLR protocols are defined as per the guideline of "A

Guide to Conducting a Systematic Literature Review of Information Systems Research" [6]. All the procedures are followed for the systematic review for Planning of the systematic assessment, performing the review, and reporting process. Major components of the systematic review which contain 6 subsections are covered. Finding the source of data, exploration strategy, review selection, review inclusion criteria and, review exclusion principles which are the key SLR questions.

SLR procedure's major 3 phases are considered as per the best practice during this review as mentioned in below Fig. (**1**).

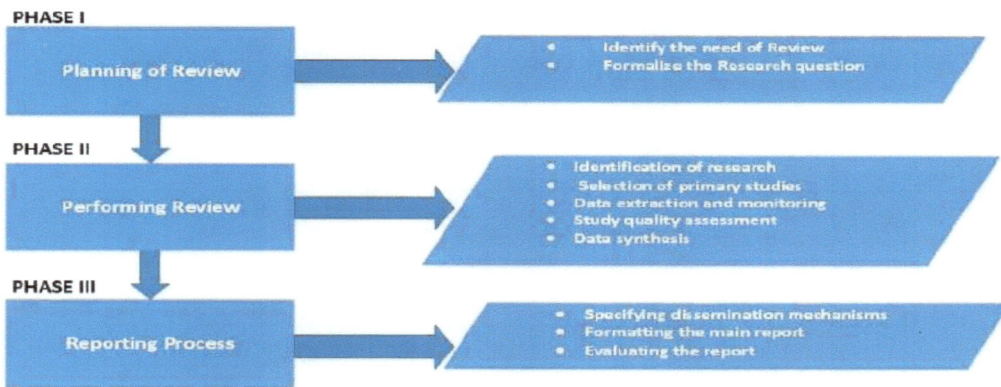

Fig. (1). Procedures followed for Performing Systematic Reviews are mentioned.

As shown in Fig. (**1**), this review was completed in 3 main phases. In phase 1, the Need for the review was identified then the research questions were formalized.

In phase 2 during conducting the research main goal was the identification of the research then selection of the primary studies followed by Data extraction and study quality assessment and at last data combination was done.

Finally, in phase 3 which is the Reporting process specification of distribution mechanism, the formatting of primary report and evaluation was done.

3. REVIEW PLANNING AND CONDUCTING LITERATURE REVIEW

Identify the Need for Research:

As already an advanced level in-depth analysis done from quality high impact factor journal that only 1% focus was towards unauthorized access [4], basis this, further incremental study conducted with the only key focused to cyber threat modeling due to unauthorized access due to certificate-based authentication backdoor.

The analysis was done based on 8519 papers from different journal databases.

3.1. Research Questions

Finalizing the SLR question concisely to get the specific result is a critical task. To derive the contextual result after granular level analysis SLR questions were designed as per the formalized standard criteria by Petticrew and Roberts [7] and PICO [8] paradigm. Table **1** shows the scope and criteria of the Research Questions formalized.

Table 1. Illustrations of the detailed scope and criteria of the research question are formalized.

Criteria	Scope
Population	Finding out the total number of research development so far in the academic and industrial area from 2009 till 2019 in the Cybersecurity area of BYOD related to authentication backdoor or authentication issues in certificate-based authentication model
Intervention	Analysis and finding out the existing, assessment, models, techniques identified by researchers
Comparison	For the determination of techniques used to meet the objective and solve the problem
Outcome	Finding out the limitation in existing research methods, technique models

Based on the Criteria and scope research questions formalized as below

Research Question 1:

What is the total number of research articles published in the area of BYOD authentication-related open loophole/backdoor issues from 2009 till 2019?

Research Question 2:

Techniques and methods used so far to mitigate the need for BYOD cyber forensic ecosystem?

Research Question 3:

What is BYOD key area addressed so Far?

Research Question 4:

What are the limitations of existing research-developed techniques/methods?

3.2. Data Source

Finding the source of data was an important task. For a detailed study, previously published papers were hunted from ten different online databases. These databases are IEEE Explorer, Taylor & Francis, SpringerLink, ScienceDirect, Ethos (UK Thesis), Australian Digital Thesis (ADT), Emerald, ProQuest (USA Thesis), and Scopus.

3.3. Search Strategy

As a known fact that the development of research conducted so far will be available and understandable by searching the data based on Title, abstract, and metadata. So the search string was prepared and executed in the digital library based on this parameter.

Search strings were prepared on BYOD technology in the different areas related to the key objective of the research. 26 different Preliminary search strings are constructed to apply in the different databases as mentioned in Table **2**.

Table 2. Search string on BYOD research based on journal-title.

BYOD Malicious traffic detection	BYOD Suspicious traffic	BYOD Authentication loophole
BYOD Certificate authentication threat	BYOD Certificate-based authentication	BYOD Certificate threat
BYOD unauthorized users	BYOD unauthorized traffic	BYOD authentication
BYOD authentication mechanism	BYOD authentication Threat	BYOD Threat
BYOD Threat detection	BYOD attack	BYOD Attack detection
BYOD cyber forensic technique	BYOD forensic technique	BYOD cybersecurity Method
BYOD cyber Security Mechanism	BYOD Cyber Security threat detection	BYOD Cyber Security
BYOD Cyber forensic	BYOD Cyber Threat	BYOD security
BYOD Forensic	BYOD traffic forensic	

Post this, strings are also constructed with Boolean "OR" and "AND" and different keyword variants, so that all related research can be captured for study.

3.4. Study Selection

During the study, selection process ranking was done of the papers from different sources highest precedence to lowest precedence as journal papers are considered as highest-ranking followed by conference papers. Then proceeding further to the selection process technical reports, thesis and at last books and magazine articles were considered.

3.4.1. Inclusion and Exclusion Criteria

During the inclusion and exclusion criteria, multiple parameters are considered. The first criteria was all articles written in the English language were considered for the study. Secondly, papers published between 2009 till 2019 are included for review. In the 3rd criteria, only articles related to BYOD technology-related study are considered. Any specific articles not related to BYOD and not meeting the objective of this study are excluded.

4. RESULT AND ANALYSIS

Research Question 1:

What is the total number of research articles published in the area of BYOD authentication related to open loophole/backdoor issues from 2009 till 2019?

To derive the conclusion of the first research question, different databases are filtered out to extract the articles. During the initial search result from a different database, a total of 8519 papers were filtered out as mentioned in below Table **3**.

Table 3. The number of research articles found in the initial search result.

Library Search string	Emerald	Science Direct	Scopus	Australian Digital Thesis (ADT)	Springer Link	ProQuest (USA Theses)	IEEE Explorer	Ethos (UK Thesis)	Taylor and Francis
BYOD Malicious traffic detection	3	37	47	2	41	40	1	0	9
BYOD Suspicious traffic	2	12	6	0	18	18	1	0	2

(Table 3) cont.....

Library Search string	Emerald	Science Direct	Scopus	Australian Digital Thesis (ADT)	Springer Link	ProQuest (USA Theses)	IEEE Explorer	Ethos (UK Thesis)	Taylor and Francis
BYOD Authentication loophole	0	5	4	0	4	8	0	0	4
BYOD Certificate authentication threat	3	27	14	9	24	38	0	0	6
BYOD Certificate based authentication	4	30	158	21	40	48	1	0	7
BYOD Certificate threat	4	43	10	9	35	49	0	0	15
BYOD unauthorised users	18	46	16	3	88	135	4	0	23
BYOD unauthorised traffic	4	17	6	3	35	43	0	0	8
BYOD authentication	17	117	4	25	133	163	28	0	21
BYOD authentication mechanism	11	50	29	21	90	95	5	0	14
BYOD authentication Threat	14	86	5	9	79	111	7	0	14
BYOD Threat	41	192	1	7	195	334	40	1	55
BYOD Threat detection	14	79	42	7	88	117	7	0	28
BYOD attack	23	174	3	6	179	230	31	2	33
BYOD Attack detection	11	86	43	5	88	103	7	0	22
BYOD cyber forensic technique	2	26	41	0	15	22	0	0	3
BYOD forensic technique	3	35	29	1	33	36	1	0	6

Library Search string	Emerald	Science Direct	Scopus	Australian Digital Thesis (ADT)	Springer Link	ProQuest (USA Theses)	IEEE Explorer	Ethos (UK Thesis)	Taylor and Francis
BYOD cyber security Method	18	80	69	7	80	133	3	0	18
BYOD cyber Security Mechanism	14	53	67	7	53	99	4	0	15
BYOD Cyber Security threat detection	9	57	79	5	42	76	0	0	13
BYOD Cyber Security	20	135	47	9	122	193	21	1	23
BYOD Cyber forensic	2	28	30	0	19	31	1	0	4
BYOD Cyber Threat	18	119	13	7	89	159	7	0	17
BYOD security	75	268	44	27	369	586	141	5	102
BYOD Forensic	5	44	18	1	42	52	3	0	7
BYOD traffic forensic	1	13	24	0	22	20	0	0	5
SUM=>	336	1859	849	191	2023	2939	313	9	474

Out of these 8519 papers again filtered out more specific papers based on tiles abstract and outcome. More specific papers are filtered based on inclusion and exclusion criteria. For evidence, more specific papers are considered for further analysis which is very relevant to the hunting criteria of the topic. Again, excluded those papers where no novel approach was found related to the objective. however, the excluded paper might be more relevant for other specific studies. Finally, the potential 18 papers were selected for in-depth analysis as shown in Table **4**.

Research Question 2:

Techniques and methods used so far to mitigate the need for a BYOD cyber forensic ecosystem?

Various research conducted so far and the findings are also unique. The outcome of this research is documented.

Table 4. Analysis of finding from existing research.

S. No.	Paper Title	Findings/Problems Addressed
N1	Bring your device in organizations: Extending the reversed IT adoption logic to security paradoxes for CEOs and end users [9]	This research focuses on BYOD adoption and security risk with BYOD and reverses IT adoption logic. Conceptually this paper has studied the security risk of BYOD. Data security concerns is been addressed in this study. The adoption of a tool for BYOD security is been addressed.
N2	A Generic Digital Forensic Readiness Model for BYOD using Honeypot Technology	This study is conducted on Digital Forensic readiness using honeypot, a deception technology. This study analyzed and shown an approach on how to conduct digital forensic analysis after an incident and detection mechanism. A security model in DFR has been proposed in this study. This study has concluded a DFI (Digital Forensic Investigation model using honeypot in BYOD deployment. DFR model has been proposed.
N3.	Improving Forensic Triage Efficiency through Cyber Threat Intelligence	A study was conducted on the extended approach of DFR (Digital Forensic readiness) in BYOD infrastructure to matured the DRF and reduce time and cost in the investigation after an incident. In this study accuracy of the DFI is conducted, to minimize the cost and time for investigation. The maturity level is increased by collecting patterns of malicious activities from different Threat Intelligent platform
N4.	BYOD Authentication Process (BAP) Using Blockchain Technology [10]	This paper studied a model of BYOD authentication technique using Blockchain, which is considered as a secured model. The paper concluded with BYOD blockchain technique authentication with a self-service portal. This covered the data leakage threat due to unauthorized access. The use of blockchain in the authentication process helps to reduce data leakage and secured the core network from attack.
N5	A Novel Approach on Mobile User Authentication for the Internet of Things [11].	The user authentication validation technique studied, the Two-factor authentication method is the area of the paper. The dual-factor authentication method has been analyzed for secure communication. Using the Scyther Tool, the computerized dual-factor authentication mechanism is tested as an automatic verification tool with a secured approach.
N6	Enhanced Bring your Device (BYOD) Environment Security based on Blockchain Technology [12]	The blockchain model of BYOD authentication is the area of research. Record keeping model of authentication is the approach. The multi-factor blockchain secured model is studied. A secured model of blockchain cryptographic authentication model is suggested.
N7	A Study on Improvement of Blockchain Application to Overcome Vulnerability of IoT [13] Multiplatform Security	Blockchain an improved model using spectrum chain is studied. A radix of blockchain core algorithm is proposed, to reduce the security challenges which is weakest in the IoT environment. PKI environment is used.

(Table 4) cont.....

S. No.	Paper Title	Findings/Problems Addressed
N8	Certificate-based hybrid authentication for Bring Your Device (BYOD) in Wi-Fi enabled Environment [14]	Certificated-based authentication is proposed in the hybrid model. During authentication, a potential attack can strike and can create a cyber-attack. This paper studied the authentication model using a certificate. Certificate-based authentication method proposed in general. Attribute-based certificate signing is proposed.
N9	Bring Your Disclosure: Analysing BYOD Threats to Corporate Information [15].	In this paper, the Stride-based BYOD threat model is proposed, analyzed threat interaction in BYOD. BYOD internal and external threat interaction to the corporate network is analyzed so that security and forensic threats in BYOD can be understood.
N10	Security risk analysis of bringing your device (BYOD) system in a manufacturing company at Tangerang [16]	This paper studied the cybersecurity aspect of the BYOD environment to improve security systems with a risk analysis of internal data. This paper concluded with a framework of BYOD with ISO27002:2013. This paper conducted a study on the BYOD security framework on cybersecurity activities to protect the data with Identity, protect, detect, respond, and recover.
N11	Diverging deep learning cognitive computing techniques into cyber forensics [17]	The Paper is based on Deep learning using a machine learning technique for forensic investigation after the crime. Deep learning technique, cognitive computing technology for the forensic investigation. A deep learning Cyber forensic framework has been proposed.
N12	An Enterprise Security Architecture for Accessing SaaS Cloud Services with BYOD [18]	This paper's concept is the security concern on the cloud with BYOD infrastructure. An enterprise security architecture while accessing SaaS through BYOD. This paper studied the "Sharewood Applied Business Security Architecture". This paper has concluded the information security architecture of BYOD while accessing SaaS applications on the cloud.
N13	Implement Network Security Control Solutions in BYOD Environment [19]	This paper studied the BYOD users' authentication process. Monitoring and performance parameter of BYOD is focused. In this paper, the authentication process is well framed with the 802.1x model. In this study 2, VLAN has been created to segregate the traffic. In this study storage capacity was also highlighted for logging.
N14	Byod Implementation in University: Balancing Accessibility And Security	The paper focused more on the approach of BYOD security guidelines. Study on university campus environment BYOD internet access. This study proposed a clear solution to be in place for BYOD infrastructure. This study needs extended research on BYOD secured infrastructure to build with proper security policies.
N15	Cybersecurity Threats Analysis for Airports [20]	This study is related to Cybersecurity on BYOD in Airport security systems. New-gen E-enabled airport Cyber Security studied. Potential Cyberattack in Airport and aircraft system has been studied. Security approach is required to incorporate in BYOD model. Airport critical infrastructure cyberattack risk is highlighted.

(Table 4) cont.....

S. No.	Paper Title	Findings/Problems Addressed
N16	SmartPhone Triggered Security Challenges-Issues, Case Studies and Prevention [21]	Mobile security challenges have been studied. Preventive measurement to improve mobile security is explored. Mobile-related crimes are explored. The cybersecurity risk is addressed with a set of policies that need to implement.
N17	Future Challenges for Smart Cities: Cyber-Security and Digital Forensics [22]	The smart city components are part of IoT. All devices, humans are connected. A network of the networks where all components are part of the network. Digital fraud and Cybercrime happened. Forensic analysis and finding out the root cause analysis is the important view that has been highlighted. Intelligent artificial is also one of the major points which is been pointed here in this study. Identification of the security threats. This study especially gives the idea of data security on the cloud. Stored data in the cloud does not have enough control, and location of the data. Since cloud data is not an enterprise control data center, so data accessed by the unauthorized entity is a risk. Data integrity is an important parameter for forensic analysis post-incident
N18	The Future of BYOD in organizations and higher Institution of Learning [23]	This paper study the use of BYOD for learning purpose, benefit, and flexibility. This paper demonstrated how BYOD is helping the higher education system. This helps in the learning process as explored in this study.

Research Question 3:

What is BYOD key area addressed so Far?

In this phase, all relevant articles are evaluated in detail and documented in major focus areas. Different research addressed the different areas of the BYOD ecosystem. Various techniques and methods are proposed to mitigate the identified gap. Below Table **5** shows the key area of BYOD major gap mitigation using different techniques.

Table 5. Various research outcomes from potential research.

Paper ID	Security Risk	Data Security	Deception Technology	Detection Mechanism	Protection Mechanism	Forensic Model	Data Leakage Framework	Secure Authentication	Threat Modeling
N1	yes	yes	No	NO	No		No	No	No
N2	No	No	Yes	Yes	No	Yes	No	No	No
N3	No	No	Yes	No	No	Yes	No	No	No
N4	No	No	No	No	No	No	Yes	Yes	No
N5	Yes	No	No	No	No	No	No	Yes	No
N6	Yes	No	No	No	No	No	No	Yes	No
N7	Yes	No	No	No	No	No	No	Yes	No

(Table 5) cont.....

Paper ID	Security Risk	Data Security	Deception Technology	Detection Mechanism	Protection Mechanism	Forensic Model	Data Leakage Framework	Secure Authentication	Threat Modeling
N8	Yes	No	No	No	No	No	No	Yes	No
N9	Yes	No	No	No	No	No	No	No	Yes
N10	Yes	No	No	No	No	No	Yes	No	No
N11	No	No	No	No	No	Yes	No	No	No
N12	Yes	No	No	No	No	No	No	No	No
N13	Yes	No	No	No	No	No	No	Yes	No
N14	Yes	No	No	No	No	No	No	No	No
N15	Yes	No	No	No	No	No	No	No	No
N16	Yes	No	No	NO	No	No	No	No	No
N17	No	Yes	No	No	No	Yes	No	No	No
N18	Yes	No	No	NO	No	No	No	No	No

Research Question 4:

Limitation of these research findings concerning defining objective?

Limitation in these studies concerning the targeted Research question is identified as mentioned in Table **6**.

Table 6. Limitations of the study and gap analysis.

Article ID	Author	Gap Analysis
N2	A Generic Digital Forensic Readiness Model for BYOD using Honeypot Technology [24]	Approached is towards deception technology which is a reactive mechanism. The advanced level protection mechanism is the next phase of the study which is pending for the next level of study.
N3.	Improving Forensic Triage Efficiency through Cyber Threat Intelligence	This study can be extended to detection and protection mechanisms as well. Threat prevention can be the next approach before the incident in endpoint, network, and core infrastructure. Digital forensic investigation and automatic prevention mechanisms can be considered. For further study in BYOD infrastructure which is pending
N4	BYOD Authentication Process (BAP) Using Blockchain Technology [10]	Authentication mechanism can be further studied with certificate-based authentication to automate BAP while keeping the same security level to protect the BYOD infrastructure
N6	Enhanced Bring your Device (BYOD) Environment Security based on Blockchain Technology [12]	Instead of user interaction more, this can be reduced with certificate-based cryptographic authentication using the key, and authentication parameters

(Table 6) cont.....

Article ID	Author	Gap Analysis
N15	Cybersecurity Threats Analysis for Airports [20]	The extended study required in threat hunting modeling in BYOD to build cyber secured BYOD ecosystem which is pending to explore
N8	Certificate-based hybrid authentication for Bring Your Device (BYOD) in Wi-Fi enabled Environment [14]	The certificate-based advance level authentication mechanism proposed is 3 tier capita model, but again next level BYOD authentication loophole mechanism is missed which arises post-authentication expiry date and if a valid certificate exists with the user.

CONCLUSION

Starting from 2009 when Intel adopted BYOD technology [23] incremental progression was observed on various tools and techniques. Various academic researchers and product tools and technology are developed by OEM to enable the services. Research in this area is observed in the key area as data security, cybersecurity framework, forensic ecosystem modeling. However, most of the researchers connect and created the abstract in terms of theoretical and practical approaches of enabling the services with creating the framework, identifying the major risk as well. Most of the research was conducted in a fragmented way concerning data security, secured authentication mechanism, deception technology to detect attacks, forensic investigation. To enable cyber secured citizenship, there is a need of doing a more advanced level study that covers the complete 360 ecosystems including security policies, BYOD technology, secured BYOD onboarding mechanism, malicious activity detection mechanism, protection protocol, and finally cyber forensic ecosystem. Finally to mitigate all these gaps future research is an indication that provides a secured authentication mechanism, a unique detection process that also included a protection process, and a cyber forensic ecosystem to protect the organization from an advanced threat. So that due business ecosystem does not fragment due to enabling BYOD.

CONSENT FOR PUBLICATION

Not applicable

CONFLICT OF INTEREST

The authors declare no conflict of interest, financial or otherwise.

ACKNOWLEDGEMENTS

Heartiest Acknowledged to Ayush Upadhyay, for the support, especially in formatting the paper, grammar check of the article.

DECLARATION

The research, hypothesis, assessments, and analysis articulated in this paper are those of the authors alone and not the organization with whom the authors are associated.

REFERENCES

[1]   M.I. Ali, "Security Challenges and Cyber Forensic Ecosystem in IOT Driven BYOD Environment", *IEEE Access,* vol. 8, pp. 1-1, 2020.
[http://dx.doi.org/10.1109/ACCESS.2020.3024784]

[2]   "BYOD Secured Solution Framework", *Int. J. Eng. Adv. Technol.,* vol. 8, no. 6, pp. 1602-1606, 2019.
[http://dx.doi.org/10.35940/ijeat.F8202.088619]

[3]   I. Ali, "BYOD CYBER FORENSIC ECO-SYSTEM",

[4]   F. Jamal, "Mohd. Taufik, A. A. Abdullah, and Z. Mohd. Hanapi, "A Systematic Review Of Bring Your Own Device (BYOD) Authentication Technique", *J. Phys. Conf. Ser.,* vol. 1529, p. 042071, 2020.
[http://dx.doi.org/10.1088/1742-6596/1529/4/042071]

[5]   B. Kitchenham, "Procedures for Performing Systematic Reviews", *Keele University Technical Report TR/SE-0401,* p. 33, 2004.

[6]   C. Okoli, and K. Schabram, "A Guide to Conducting a Systematic Literature Review of Information Systems Research", *SSRN,* 2010.
[http://dx.doi.org/10.2139/ssrn.1954824]

[7]   M. Petticrew, and H. Roberts, "Systematic Reviews in the Social Sciences: A Practical Guide, Wiley",
[http://dx.doi.org/10.1002/9780470754887]

[8]   X. Huang, and M.L.S.J. Lin, "Evaluation of PICO as a Knowledge Representation for Clinical Questions", *AMIA Annu Symp Proc.,* vol. 2006, pp. 359-363, 2006.

[9]   P. Baillette, Y. Barlette, and A. Leclercq-Vandelannoitte, "Bring your own device in organizations: Extending the reversed IT adoption logic to security paradoxes for CEOs and end users", *Int. J. Inf. Manage.,* vol. 43, pp. 76-84, 2018.
[http://dx.doi.org/10.1016/j.ijinfomgt.2018.07.007]

[10]   F. Jamal, M.T. Abdullah, A. Abdullah, and Z.M. Hanapi, "BYOD Authentication Process (BAP) Using Blockchain Technology", *Control Syst. (Tonbridge),* vol. 10, no. 11, p. 8, 2018.

[11]   "International Journal of Scientific Research in Computer Science, Engineering and Information Technology", p. 7, 2018.

[12]   F. Jamal, M. T. Abdullah, A. Abdullah, and Z. M. Hanapi, "Enhanced Bring your Own Device (BYOD) Environment Security based on Blockchain Technology", *International Journal of Engineering,* vol. 7, no. 4, 2018.

[13]   S-K. Kim, U-M. Kim, and J-H. Huh, "A Study on Improvement of Blockchain Application to Overcome Vulnerability of IoT Multiplatform Security", *Energies,* vol. 12, no. 3, p. 402, 2019.
[http://dx.doi.org/10.3390/en12030402]

[14]   U. Raj, "Certificate based hybrid authentication for Bring Your Own Device (BYOD) in Wi-Fi enabled", *Environment,* vol. 13, no. 12, p. 7, 2015.

[15]   D. A. Flores, F. Qazi, and A. Jhumka, *Bring Your Own Disclosure: Analysing BYOD Threats to Corporate Information,* pp. 1008-15, 2016.
[http://dx.doi.org/10.1109/TrustCom.2016.0169]

[16]   A. Retnowardhani, R.H. Diputra, and Y.S. Triana, "Security risk analysis of bring your own device (BYOD) system in manufacturing company at Tangerang", *TELKOMNIKA,* vol. 17, no. 2, p. 753, 2019. [Telecommunication Computing Electronics and Control].
[http://dx.doi.org/10.12928/telkomnika.v17i2.10165]

[17]   N.M. Karie, V.R. Kebande, and H.S. Venter, "Diverging deep learning cognitive computing techniques into cyber forensics", *Forensic Science International: Synergy,* vol. 1, pp. 61-67, 2019.
[http://dx.doi.org/10.1016/j.fsisyn.2019.03.006] [PMID: 32411955]

[18]   V. Samaras, S. Daskapan, R. Ahmad, and S.K. Ray, "An enterprise security architecture for accessing SaaS cloud services with BYOD", *2014 Australasian Telecommunication Networks and Applications Conference (ATNAC),Southbank, Australia,* 2014 pp. 129-134
[http://dx.doi.org/10.1109/ATNAC.2014.7020886]

[19]   K. AlHarthy, and W. Shawkat, "Implement network security control solutions in BYOD environment", *in 2013 IEEE International Conference on Control System, Computing and Engineering, Penang, Malaysia,* 2013 pp. 7-11 Penang, Malaysia
[http://dx.doi.org/10.1109/ICCSCE.2013.6719923]

[20]   G. Suciu, A. Scheianu, I. Petre, L. Chiva, and C.S. Bosoc, *"Cybersecurity Threats Analysis for Airports,"* in New Knowledge in Information Systems and Technologies., Á. Rocha, H. Adeli, L.P. Reis, S. Costanzo, Eds., vol. Vol. 931. Springer International Publishing: Cham, 2019, pp. 252-262.

[21]   S.R. Srivastava, S. Dube, G. Shrivastaya, and K. Sharma, *"Smartphone Triggered Security Challenges - Issues, Case Studies and Prevention,"* in Cyber Security in Parallel and Distributed Computing., D. Le, R. Kumar, B.K. Mishra, M. Khari, J.M. Chatterjee, Eds., John Wiley & Sons, Inc.: Hoboken, NJ, USA, 2019, pp. 187-206.

[22]   Z.A. Baig, "Future challenges for smart cities: Cyber-security and digital forensics", *Digit. Invest.,* vol. 22, pp. 3-13, 2017.
[http://dx.doi.org/10.1016/j.diin.2017.06.015]

[23]   "O. U. Franklin, M. Ismail Z., and FTMS College, "THE FUTURE OF BYOD IN ORGANIZATIONS AND HIGHER INSTITUTION OF LEARNING", *International Journal of Information Systems and Engineering,* vol. 3, no. 1, pp. 110-128, 2015.
[http://dx.doi.org/10.24924/ijise/2015.11/v3.iss1/110.128]

[24]   V. R. Kebande, N. M. Karie, and H. S. Venter, "A generic Digital Forensic Readiness model for BYOD using honeypot technology", In: *IST-Africa Week Conference*South Africa, 2016, pp. 1-12.
[http://dx.doi.org/10.1109/ISTAFRICA.2016.7530590]

CHAPTER 8

Dengue Viral Protein Interaction Study Derived Immune Epitope for *In-Silico* Vaccine Design

Sunil Krishnan G., Amit Joshi[1] and **Vikas Kaushik[1]**

[1] *Lovely Professional University, Phagwara, India*

Abstract: Dengue viral illness is communicated to humans through the bite of female Aedes aegypti mosquitoes. This disease may become lethal in numerous patients. The availability of an efficacious vaccine makes alarm for public healthcare. The dengue virus multiplied inside the host framework involved many host-viral protein interactions. This immunoinformatics study was designed for the prediction of immune epitopes for T cell-mediated immunity. The epitopes are anticipated from the most connecting viral protein with humans. We utilized a couple of epitope mapping tools for the determination of immunodominant epitope for vaccine design. The physical interaction between epitope ligand and receptor MHC class I alleles was analyzed in the molecular docking study. This study was concluded that two epitopes ('SRAIWYMWL' and 'FLEFEALGF') are suitable for the designing of an efficacious multi-epitope vaccine. The clinical validation is considered necessary for the final confirmation of vaccine potency.

Keywords: Conservancy, Dengue, Epitope, Host, Immunogenicity, Immunoinformatics, Protein-interaction, Vaccine, Virus.

INTRODUCTION

The female Aedes aegypti mosquitoes are the significant offender for the dengue viral illness. Asymptomatic, mild, or fatal dengue disorders are ordinarily announced in dengue patients [1]. All significant four serotypes of this infection show 65% closeness in their genome. The structural and non-structural proteins help infection endurance and increase [2]. This lethal infection has been causing social and financial weight in numerous (> 129) endemic nations [3]. The unavailability of an efficacious vaccine makes the circumstance more frenzy [4], [5]. Conventional vaccines may take long R&D processing time, safety concerns, and are also unaffordable for economically weaker nations. Most countries

 Corresponding author Sunil Krishnan G.: Lovely Professional University, Phagwara, India; E-mail: sunil.41800559@lpu.in

Dharam Buddhi, Rajesh Singh and Anita Gehlot (Eds.)

investigate an option for alternative, affordable, efficacious, and safe vaccines [6]. Reverse vaccinology procedures may help for this excursion to design a peptide-based vaccine [7]. Highly interacting viral protein with human protein distinguished through human viral protein interaction network database. One viral serotype was chosen for protein sequence retrieval. From protein sequence database retrieved sequence and anticipated a couple of T cell epitopes. These epitopes were further analyzed for conservancy among the available four viral serotypes protein sequence. Immunogenic, low antigenic, and non-toxic epitopes chose for molecular docking prediction with respective MHC I- HLA allele. The perfectly docked epitopes are predicted as potent vaccine candidates.

METHODOLOGY

A. Dengue Virus-Host Protein Interaction Data Retrieval and Sequence Mapping

The protein-protein interactions (PPI) network study was performed by utilizing the DenvInt database. This tool was assisted with recognizing PPI between the dengue infection and its human host [8]. This data set investigation assisted with foreseeing the dengue protein which was indicating the greatest number of interactions with human protein and was chosen for additional examinations.

B. Strain identification and Non-structural Protein sequences retrieval

Dengue virus strain and protein sequences dataset was retrieved from the ViPR http://www.viprbrc.org/. The database is a useful resource for strain and sequence analysis [9]. The retrieved protein sequence was the input for epitope forecast.

Computation of MHC Class I Binding Epitopes

The protein sequence of Non-structural protein 5 was used for the MHC I presenting CD8 T cell epitopes. An integrated MHC class I binding epitopes established by the latest version of NetCTL artificial neural network. This tool was accurately evaluated binding affinity and proteasomal cleavage [10].

Epitope Analysis

We used the IEDB web tool for conservancy analysis of linear T cell epitopes. This web tool helps to calculate the epitope linear conservancy [11]. The conserved epitope peptide sequence was tested for immunogenicity score [12]. VaxiJen web tool version 2.0 predicted the antigenicity of epitopes. VaxiJen 2.0 server predicts antigenicity 70% to 89% accuracy [13]. ToxinPred is a quantitative matrix peptide toxicity predictor [14]. This tool was predicted nontoxic epitopes from the antigenic epitope sequence dataset.

Prediction of Epitope 3D Structures and Molecular Docking Analysis

PEP-FOLD3 is a Hidden Markov Model-based *de novo* peptide structure forecast instrument that has been accounted for the capacity to discover 3D designs of 5-50 amino acids lengthen peptides [15]. The three-dimensional structure of the distinguished epitope anticipated the PEP-FOLD server from the amino acid sequences. The MHC-HLA allele three-dimensional structure was obtained from protein data bank (PDB) (https://www.rcsb.org). PatchDock webserver was utilized for molecular docking study of MHC- HLA receptor with an epitope peptide ligand. This tool was worked based on Shape complementarity principles [16]. Both the proteins were docked at 1.5 root-mean-square deviations (RMSD) for accurate output.

RESULTS AND DISCUSSIONS

Dengue Virus and Human Protein Interaction Network Study

The DenvInt data made known that NS5 dengue protein interacts with 169 human proteins with the highest interaction than other dengue viral proteins. This protein was selected for further analysis. Many flavivirus PPI and proteome studies were revealed that NS5 protein suppresses the stimulation of interferon [17].

Strain and Protein Sequence Data Retrieval

The DENV serotype type 1 strain particulars and Gen Bank accession number (AF311956) were retrieved from the ViPR database. The complete genome sequence with a length of 10735 base pairs was obtained. This strain was used for NS5 protein sequence retrieval. The functional and structural features of immune epitope locations in the NS5 protein sequence were also identified. The functional and structural variations in the NS1 protein sequence of the influenza virus are identified [18].

MHC I Presenting CD8 T Cell CTL Epitopes

The NetCTL 1.2 web tool was recognized MHC I presenting CD8 T cell CTL epitopes (n=179). These epitopes were identified from DENV -NS5 protein sequence. The MHC I -HLA supertypes, A3 (n= 52), A2 (n=26), A24 (n=54), B7 (22) and B44 (25) at threshold value 0.75 as prediction parameter. These epitope peptides were 9-mer long (9 amino acids). This webserver was utilized to identify five 2019-nCoV CTL T Cell epitopes for vaccine design [7].

Conservancy, Immunogenicity Antigenicity, and Toxicity Prediction

Immunogenic epitope peptides were an asset for T cell epitope-based vaccine designing and a proficient tool for T cell epitope mapping [8]. The conservancy analysis was predicted conserved epitopes (n=16) were 100 percent conserved in all four DENV serotype's NS5 protein sequences. Conserved peptides checked for immunogenicity and found SRAIWYMWL has the highest immunogenicity score (0.34006) and FLEFEALGF has the second highest (0.29507). These peptide's immunogenicity & antigenicity score and toxicity status are detailed in Table **1** .

Molecular Docking Analysis of Epitope Peptide Ligand and MHC Receptor

The binding affinity analysis of the immunodominant epitope peptides with the MHC 1 HLA- A allele was assessed utilizing the Patch dock molecular docking tool. MHC I HLA-A*2402 allele receptor was selected as the receptor and the identified epitope peptides ('FLEFEALGF' and 'SRAIWYMWL') as the ligand. Most suitable MHC 1 -epitope peptide complex was decided based on the shape complementarity analysis result. The docking analysis output result contains, number of the solution, geometric shape complementarity score (see reference 1 for details). The docked complex of HLA-A*2402 MHC I allele (PDB ID: 2BCK) - epitope peptide 'FLEFEALGF' had 7444 and the 'SRAIWYMWL' epitope peptide - HLA-A*2402 allele complex "7936" geometric shape complementarity score. The visualization of 3D structures of predicted epitopes ligand, MHC receptor, and MHC receptor-epitope ligand docked complexes of predicted epitopes in Fig. **(1)** .

Epitopes & HLA A * 24:02 Receptor and docked complex

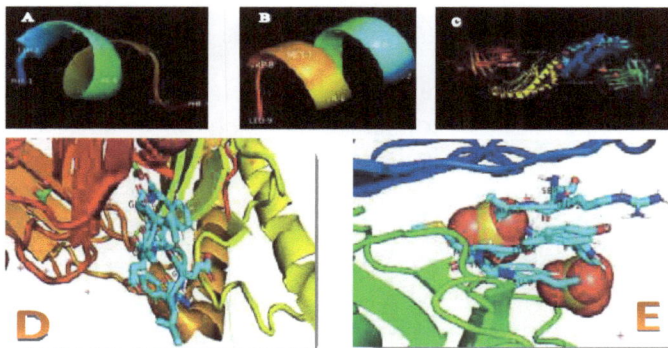

Fig. (1). PyMOL visualization of 3D structures of predicted epitope ligand, MHC receptor, docked complexes of predicted epitopes ligand and MHC receptor: Three-dimensional structures of (A) 'FLEFEALGF' CD8 T Cell CTL epitope peptide (B), SRAIWYMWL' CD8 T Cell CTL epitope peptide (C) HLA A 24:02 with chains, (D) docked complex of MHC I HLA-A*2402 with FLEFEALGF CD8 T Cell CTL epitope (E) Docked complex of MHC I HLA-A*2402 with SRAIWYMWL CD8 T Cell CTL epitope.

Table 1. Epitope analysis details.

S.No.	Epitope Amino Acid Sequence	Number of Amino Acids in Peptide	Conservancy(strain) Percentage	Immunogenicity Score	Antigenicity Score	Toxicity Status
1	AMTDTTPFG	9	100.00% (4/4)	0.1683	0.6509	Non-Toxic
2	PFGQQRVFK	9	100.00% (4/4)	-0.0664	-0.4815	Non-Toxic
3	GQQRVFKEK	9	100.00% (4/4)	0.04138	-1.2200	Non-Toxic
4	CVYNMMGKR	9	100.00% (4/4)	-0.44141	-0.0325	Non-Toxic
5	YNMMGKREK	9	100.00% (4/4)	-0.30152	1.0248	Non-Toxic
6	NMMGKREKK	9	100.00% (4/4)	-0.22568	1.4621	Non-Toxic
7	MMGKREKKL	9	100.00% (4/4)	-0.36935	1.8840	Non-Toxic
8	SRAIWYMWL	9	100.00% (4/4)	0.34006	0.3258	Non-Toxic
9	WYMWLGARF	9	100.00% (4/4)	0.25025	0.7264	Non-Toxic
10	YMWLGARFL	9	100.00% (4/4)	0.24265	0.6068	Non-Toxic
11	WLGARFLEF	9	100.00% (4/4)	0.26011	1.0434	Non-Toxic
12	FLEFEALGF	9	100.00% (4/4)	0.29507	1.7726	Non-Toxic
13	AISGDDCVV	9	100.00% (4/4)	0.0015	-0.4297	Non-Toxic
14	ISGDDCVVK	9	100.00% (4/4)	0.06313	-0.9457	Non-Toxic
15	LMYFHRRDL	9	100.00% (4/4)	0.25346	1.4737	Non-Toxic
16	YFHRRDLRL	9	100.00% (4/4)	0.15474	1.6075	Non-Toxic

CONCLUDING REMARKS

NetCTL server utilized for the forecast of 179 T cell epitopes from DENV-non structural protein 5. The IEDB conservancy web tool anticipated sixteen conserved epitopes in the NS5 protein sequence of all four serotypes with 100% conservancy. Two immunogenic CD8 T Cell CTL epitope with MHC I HLA allele 'Patchdock' docking evaluation uncovered the potential binding affinity of immunodominant epitope peptide with MHC I HLA alleles. The docking results were satisfactory and the 'SRAIWYMWL' 'FLEFEALGF' was selected for the designing of the immunodominant multi-epitope dengue vaccine and the clinical validation studies recommended for further analysis.

CONSENT FOR PUBLICATION

Not applicable.

CONFLICT OF INTEREST

The authors declare no conflict of interest, financial or otherwise.

ACKNOWLEDGEMENTS

Authors SKG, AJ, and, VK are thankful to, Bioinformatics and Biotechnology Department, Lovely professional university, Phagwara, India for providing a well-organized computational and bioinformatics environment for the execution of this research project.

REFERENCES

[1]   K.A. Lintang, E. Ferry, M. Makhfudli, E.H. Eka Misbahatul Mar'ah Has, and E.A. Gading, "Social support and its correlation with '3M plus' behavior in the prevention of dengue hemorrhagic fever", *Indian J. Public Health Res. Dev.,* vol. 10, no. 8, pp. 2681-2685, 2019.
[http://dx.doi.org/10.5958/0976-5506.2019.02274.5]

[2]   M.G. Guzman, S.B. Halstead, H. Artsob, P. Buchy, J. Farrar, D.J. Gubler, E. Hunsperger, A. Kroeger, H.S. Margolis, E. Martínez, M.B. Nathan, J.L. Pelegrino, C. Simmons, S. Yoksan, and R.W. Peeling, "Dengue: a continuing global threat", *Nat. Rev. Microbiol.,* vol. 8, no. 12, suppl. Suppl., pp. S7-S16, 2010.
[http://dx.doi.org/10.1038/nrmicro2460] [PMID: 21079655]

[3]   J.P. Messina, O.J. Brady, T.W. Scott, C. Zou, D.M. Pigott, K.A. Duda, S. Bhatt, L. Katzelnick, R.E. Howes, K.E. Battle, C.P. Simmons, and S.I. Hay, "Global spread of dengue virus types: mapping the 70 year history", *Trends Microbiol.,* vol. 22, no. 3, pp. 138-146, 2014.
[http://dx.doi.org/10.1016/j.tim.2013.12.011] [PMID: 24468533]

[4]   S-F. Wang, W.H. Wang, K. Chang, Y.H. Chen, S.P. Tseng, C.H. Yen, D.C. Wu, and Y.M. Chen, "Severe Dengue Fever Outbreak in Taiwan", *Am. J. Trop. Med. Hyg.,* vol. 94, no. 1, pp. 193-197, 2016.
[http://dx.doi.org/10.4269/ajtmh.15-0422] [PMID: 26572871]

[5]   J. Flipse, and J.M. Smit, "The Complexity of a Dengue Vaccine: A Review of the Human Antibody Response", *PLoS Negl. Trop. Dis.,* vol. 9, no. 6, 2015.e0003749
[http://dx.doi.org/10.1371/journal.pntd.0003749] [PMID: 26065421]

[6]   "In-Silico Proteomic Exploratory Quest: Crafting T-Cell Epitope Vaccine Against Whipple's Disease", *Int. J. Pept. Res. Ther.,* vol. 27, no. 1, pp. 169-179, 2020.

[7]   V. Kaushik, "in silico Identification of Epitope-Based Peptide Vaccine for Nipah Virus", *Int. J. Pept. Res. Ther.,* vol. 26, no. 2, pp. 1147-1153, 2019.
[http://dx.doi.org/10.1007/s10989-019-09917-0]

[8]   L. Dey, and A. Mukhopadhyay, "DenvInt: A database of protein-protein interactions between dengue virus and its hosts", *PLoS Negl. Trop. Dis.,* vol. 11, no. 10, 2017.e0005879
[http://dx.doi.org/10.1371/journal.pntd.0005879] [PMID: 29049286]

[9]   B.E. Pickett, D.S. Greer, Y. Zhang, L. Stewart, L. Zhou, G. Sun, Z. Gu, S. Kumar, S. Zaremba, C.N. Larsen, W. Jen, E.B. Klem, and R.H. Scheuermann, "Virus pathogen database and analysis resource (ViPR): a comprehensive bioinformatics database and analysis resource for the coronavirus research community", *Viruses,* vol. 4, no. 11, pp. 3209-3226, 2012.
[http://dx.doi.org/10.3390/v4113209] [PMID: 23202522]

[10]   M.V. Larsen, C. Lundegaard, K. Lamberth, S. Buus, O. Lund, and M. Nielsen, "Large-scale validation of methods for cytotoxic T-lymphocyte epitope prediction", *BMC Bioinformatics,* vol. 8, no. 1, p. 424, 2007.

[http://dx.doi.org/10.1186/1471-2105-8-424] [PMID: 17973982]

[11] H-H. Bui, J. Sidney, W. Li, N. Fusseder, and A. Sette, "Development of an epitope conservancy analysis tool to facilitate the design of epitope-based diagnostics and vaccines", *BMC Bioinformatics,* vol. 8, no. 1, p. 361, 2007.
[http://dx.doi.org/10.1186/1471-2105-8-361] [PMID: 17897458]

[12] S. Krishnan G, A. Joshi, N. Akhtar, and V. Kaushik, "Immunoinformatics designed T cell multi epitope dengue peptide vaccine derived from non structural proteome", *Microb. Pathog.,* vol. 150, 2021.104728
[http://dx.doi.org/10.1016/j.micpath.2020.104728] [PMID: 33400987]

[13] I.A. Doytchinova, and D.R. Flower, "VaxiJen: a server for prediction of protective antigens, tumour antigens and subunit vaccines", *BMC Bioinformatics,* vol. 8, no. 1, p. 4, 2007.
[http://dx.doi.org/10.1186/1471-2105-8-4] [PMID: 17207271]

[14] S. Gupta, P. Kapoor, K. Chaudhary, A. Gautam, R. Kumar, and G.P.S. Raghava, "in silico approach for predicting toxicity of peptides and proteins", *PLoS One,* vol. 8, no. 9, 2013.e73957
[http://dx.doi.org/10.1371/journal.pone.0073957] [PMID: 24058508]

[15] P. Thévenet, Y. Shen, J. Maupetit, F. Guyon, P. Derreumaux, and P. Tufféry, "PEP-FOLD: an updated de novo structure prediction server for both linear and disulfide bonded cyclic peptides", *Nucleic Acids Res.,* vol. 40, no. Web Server issue, 2012.W288-93
[http://dx.doi.org/10.1093/nar/gks419] [PMID: 22581768]

[16] D. Duhovny, R. Nussinov, and H.J. Wolfson, *Efficient Unbound Docking of Rigid Molecules.* Algorithms in Bioinformatics, 2002, pp. 185-200.

[17] S. Rothenburg, and G. Brennan, "Species-Specific Host-Virus Interactions: Implications for Viral Host Range and Virulence", *Trends Microbiol.,* vol. 28, no. 1, pp. 46-56, 2020.
[http://dx.doi.org/10.1016/j.tim.2019.08.007] [PMID: 31597598]

[18] J.M. Noronha, M. Liu, R.B. Squires, B.E. Pickett, B.G. Hale, G.M. Air, S.E. Galloway, T. Takimoto, M. Schmolke, V. Hunt, E. Klem, A. García-Sastre, M. McGee, and R.H. Scheuermann, "Influenza virus sequence feature variant type analysis: evidence of a role for NS1 in influenza virus host range restriction", *J. Virol.,* vol. 86, no. 10, pp. 5857-5866, 2012.
[http://dx.doi.org/10.1128/JVI.06901-11] [PMID: 22398283]

CHAPTER 9

Significance of Dark Energy and Dark Matter in the Transformation of Cosmological Periods, Focusing on the Evolution of the Universe

Gopalchetty Brahma[1] and **Amit Kumar Thakur[1,*]**

[1] Department of Aerospace Engineering, Lovely Professional University, Phagwara, Punjab-144411, India

Abstract: In this paper, we review some detailed facts and information about dark energy & dark matter and try to examine their existence and nature. We also seek to extend their influence to understand various transformations of cosmological periods that have taken place, the evidence of which can be found in the chronology of the universe, along with the significance of dark energy and dark matter. We study the present-case scenario regarding the evolution of the universe concerning the Friedmann Equations, derived from Einstein's field equations, along with some important cosmological parameters. This analysis also underlines an effort to obtain a clear and crisp picture towards which the universe is heading with time. We also review the various cosmological eras found in the chronology of the universe, which was dominated by radiation, matter, and dark energy respectively, and derive a conclusion of what the universe might have to attain during its evolutionary course.

Keywords: Cosmic Microwave Background Radiation, Dark Energy, Dark Matter, De-Sitter Universe, Einstein's Cosmological Constant, FRW Metric, Λ-CDM Model.

INTRODUCTION

The universe and its law of cause and effect never fail to surprise us. But the most intriguing fact is what inspires us to unveil a bit more, of its greatest secrets. Recent years of our study about the universe have shown striking, yet significant evidence of the detection of Einstein's so-called cosmological constant (Λ), which was considered as his "Biggest Blunder". The observational study of type IA supernovae, presented possible evidence regarding the pace of the universe, which is speeding up. The discovery of the acceleration of the universe (which is

*** Corresponding author Amit Kumar Thakur:** Department of Aerospace Engineering, Lovely Professional University, Phagwara, Punjab-144411, India; E-mail: amit.25010@lpu.co.in

Dharam Buddhi, Rajesh Singh and Anita Gehlot (Eds.)

believed to be following Λ-term), is a huge milestone in our understanding of modern cosmology.

According to (Turner 2002), the mass/energy density of the visible constituents of the universe in terms of percentile is [1]:

Bright Stars: 0.5%

Baryons (total): 4% ± 1%

Neutrinos: at least 0.1% and possibly as large as 5%

From the given data, it can be inferred that, Ωstars=0.005 and Ωbaryons=0.04. Although the value of Ωbaryons is much higher in comparison to the value of Ωstars, the total density of the universe comes close to 1. (Varun Sahni 2004) raised a question about what else can be out there in the universe, contributing to the total density as follows [2]:

$$\Omega_{total} - \Omega_{baryons} = ?$$ **(1)**

Therefore, it was quite clear about an unknown essential component(s) present in the universe, apart from ordinary matter and energy. Until recent years, scientists and astronomers thought that the entire universe was comprised entirely of "baryonic matter" (matter comprising mainly of protons and neutrons). But the results of observational data portrayed that there is something else in the universe also, which is not visible, contributing to the total density, as in the fore mentioned question, and a reason for the non-linear rate of expansion of cosmos.

Cosmic acceleration, and its deep-rooted origin, remains a big mystery. Speaking by the general relativity, if the universe would have been made only of ordinary matter and radiation, then the gravitational force would have slowed down the rate of expansion, that expansion which was set by a powerful explosion, known as the "Big Bang". But, if we observe the chronology of the universe, after the advent of cosmic inflation, the universe still expands at an accelerating rate. Hence, the hypothetical terms "Dark Energy" and "Dark Matter" came into the picture, which was associated with Einstein's cosmological constant, explaining the unrealistic expansion and evolution of the universe.

DARK ENERGY AND DARK MATTER

Dark Energy can be defined as a hypothetical and unknown form of energy, proposed to explain the accelerating rate of expansion of the universe, at the largest scale. It is thought to affect quite opposite to gravity, and similar to repulsion. But, if that's the only reason sufficient, what could have been the reason for the formation of the large-scale structures found in the cosmos? Gravity, being the weakest fundamental force of the universe, couldn't have alone played the role. Therefore, a hypothetical form of matter, known as "Dark Matter" was introduced. Dark Matter could be defined as a hypothetical form of matter, which is believed to interact with visible matter *via* gravitational force. It is referred to be dark because there is no evidence of interaction with electromagnetic radiation, which makes it almost undetectable with current scientific instruments. Hence, the origin of dark forces from the present epoch of the Big Bang was known.

Assuming the standard cosmological model to be correct, and according to observational data procured by (Bolotin *et al.* 2014), the relative densities of main constituents making up the universe, such as dark energy, dark matter, baryonic matter, and radiation are as follows [3]:

$\Omega_{de} = 73\% \pm 4\%$ (dark energy)

$\Omega_{dm} = 23\% \pm 4\%$ (dark matter)

$\Omega_{b} = 4\% \pm 0.4\%$ (baryonic matter)

$\Omega_{r} = 5 \times 10^{-5}$ (radiation)

In the currently accepted model in modern cosmology as discussed by (Peebles and Ratra 2003), the universe is spatially flat, since it's known that [4]:

$$\Omega_{M0} + \Omega_{R0} + \Omega_{\Lambda0} + \Omega_{K0} = 1 \tag{2}$$

where, (Ω_{M0}) is the mean mass density of main baryons and non-baryonic dark matter, (Ω_{R0}) is the measure of present mass in 3-K cosmic microwave background radiation accompanying low-mass neutrinos, $(\Omega_{\Lambda0})$ is the measure of equivalent dark energy, and (Ω_{K0}) is the measure of the curvature of space.

Therefore, we can say that the fate and face of the universe are tied to the nature of dark energy and dark matter and their respective roles. They must have had a significant role in shaping the universe in certain dominations, leading the universe to its current state.

COSMOLOGICAL MODEL AND THE COSMOLOGICAL PARAMETERS

Cosmological Model

There have been numerous views and discussions about cosmological models, which can explain the role of dark energy and dark matter, following the current state of the universe, and also agrees with the cosmological tests and observational evidence.

Albert Einstein, in 1915, proposed his ground-breaking theory named "The General Theory of Relativity", which comprised of field equations, describing gravitation as a result of curved spacetime, being curved by mass and energy, paved the way to a new understanding of gravity. Accordingly, massive bodies distort space-time by making a dent in the 4-D spacetime, and this distorted space-time, in turn, affects the bodies residing nearby.

According to (Walters 2016), Einstein's governing field equation of general relativity can be given as [5]:

$$R_{\mu\nu} - \frac{1}{2} g_{\mu\nu} R = -\kappa T_{\mu\nu} \qquad (3)$$

For a long time, the value of the cosmological constant was believed to be "zero". Following the recent observations, there was a need for a positive value of (Λ), to explain the accelerating rate of expansion of the universe, and also to counteract the gravitational attraction. Therefore, (Rugh and Zinkernagel 2000) modified Einstein's field equation as [6]:

$$R_{\mu\nu} - \frac{1}{2} g_{\mu\nu} R - \Lambda g_{\mu\nu} = \frac{8\pi G}{c^4} T_{\mu\nu} \qquad (4)$$

where, ($R_{\mu\nu}$) as well as (R) refers to the Ricci tensor and scalar curvature respectively, ($g_{\mu\nu}$) refers to the metric tensor, ($T_{\mu\nu}$) is the energy-momentum tensor, (G) and (c) being constants, namely the gravitational constant, and speed of light respectively.

The Λ-CDM Model is the widely accepted, current standard model of cosmology. This model is a huge milestone in modern physics since it is based upon two theoretical models, governing everything in the universe. The first model, being the Standard Model of Particle Physics (SMPP), which describes the physics of very small scales, and the General Theory of Relativity, which describes the physics of large scales. The Λ-CDM Model provides a good account of various properties like the existence of cosmic microwave background radiation (CMBR), large-scale structure formation, and the accelerating expansion of the universe. Also, it's extendable by the addition of cosmic inflation, quintessence, *etc.*

According to the Λ-CDM Model, the universe comprises of three important constituents: the cosmological constant (Λ), associated with dark energy, cold dark matter, which is a slowly propagating dark matter in comparison to the speed of light, mainly responsible for the creation of numerous small-scale structures in the early universe, and ordinary matter which is mostly baryonic matter. The Λ-CDM Model could be further extended by considering cosmic inflation, quintessence (a scalar field form of dark energy), and other elements.

Considering the universe to be spatially homogeneous and isotropic, Friedmann equations were derived from Einstein's field equations by the substitution of the metric. The solution of Einstein's field equations, the FRW metric as given by (Kao and Pen 1991) can be read as [7]:

$$ds^2 \equiv g_{\mu\nu}dx^\mu dx^\nu = -dt^2 + a^2(t)\left[\frac{dr^2}{1-kr^2} + r^2 d\Omega\right] \tag{5}$$

where, $d\Omega = d\theta^2 + \sin^2\theta d\chi^2$ (solid angle) and the values of k=0,-1,(+1) which stands for flat, open, or closed universe respectively, which represents the spatial curvature.

The simplified form of the metric equation, from the FRW metric, as given in the research work carried out by (Zakharov *et al.* 2009) can be given as [8]:

$$ds^2 = c^2 dt^2 - a^2(t)ds_3^2 \tag{6}$$

where (a) is the cosmic scale factor and (ds_3^2) represents a 3-dimensional space metric. These equations, along with the Λ-CDM Model can provide a good account of the transformation of the universe to the current state.

Talking in terms of expansion and scale factor (a), we have from a pair of Friedmann Equations as discussed by (Kalita 2014) as follows [9]:

$$\frac{\ddot{a}}{a} = -\frac{4\pi G}{3}(\rho + 3P/c^2) + \frac{\Lambda c^2}{3} \tag{7}$$

where (ρc^2) is the energy density of matter and $(\ddot{a})$ is the second derivative of scale factor (a), concerning time.

Following the recent observations, the expansion rate of the universe is certainly accelerating, which means $(\ddot{a})$ is positive, implying the first derivative of the scale factor, *i.e.*, $(\dot{a})$ is increasing concerning time. This analogy provides us the mathematical proof of cosmic acceleration.

From the Friedmann Equations, the above analogy was expressed mathematically by (Sapkota and Adhikari 2017) as [10]:

$$\left(\frac{\dot{a}}{a}\right)^2 - \frac{8\pi G}{3}\rho - \frac{\Lambda c^2}{3} = -\frac{kc^2}{a^2} \tag{8}$$

where (k) is the spatial curvature of the universe. Previously, the term containing "cosmological constant" in the LHS was absent in the equation. According to the fore mentioned analogy, the RHS term must be positive and greater than zero. But, according to the observations of CMB, the universe seems to be spatially flat, *i.e.* (k=0). In this case, the RHS term becomes zero. To keep the RHS term positive, another term containing the "cosmological constant", which is associated with dark energy, is added to the LHS.

Cosmological Parameters

There are some important cosmological parameters mentioned in the previous equations and models which are crucial to derive conclusions about the evolution of the universe. They are as follows:

Density Parameter (Ω)

Density Parameter (Ω) is an important parameter, dependent on the constituents of the universe. It is defined as the ratio of average matter density present in the universe to the critical value of that density. The theory of cosmic inflation suggests that the total density of the universe be close to the critical one $(\Omega_{total} \cong 1)$, where $\Omega_i = \rho_i / \rho crit$, which is supported by the available data on the cosmic microwave background radiation (CMB) (Ellis 2003) [11].

Scale Factor (a)

Scale factor (a) is a very important dimensionless key parameter of Friedmann Equations, which is very helpful to obtain an insight into the accelerating expansion of the universe and is also present in various equations. As per the definition, scale factor (a) can be defined as the ratio of the distance (d) between any two objects at a time (t) to the distance (d) between the objects at a reference time (t_0) as follows:

$$a(t) = \frac{d(t)}{d(t_0)}$$

(9)

Therefore, we can say that scale factor (a) is a time-dependent parameter, and its evolution with time is discussed later in the paper.

Hubble's Constant (H_0)

The Hubble's Constant is an important parameter in Hubble's Law. Edwin Hubble, in 1929 gave this law, which states that a galaxy's velocity is proportional to its distance. This

directly tells us that the far the galaxy is from the point of observation, the faster the galaxy is moving away. Mathematically, this can be expressed (Bahcall 2015) as [12]:

$$v = H_0 \, x \, d$$

(10)

Here, (v) denotes the velocity of the galaxy, and (d) denotes the distance between the galaxy and the point of observation. This principle laid the foundations for the cosmological redshift as well as the expansion of the universe.

Deceleration Parameter (q)

Along with scale factor (a), there is another dimensionless measure of cosmic acceleration, known as the deceleration parameter (q). As per the definition, it can be given in relation to scale factor as stated by (Mamon and Das 2017) [13]:

$$q = -\frac{a\ddot{a}}{\dot{a}^2}$$

(11)

where (a) is the scale factor. Since the deceleration parameter also depends on the density parameter as well as matter-energy composition; it can be expressed (Turner 2000) as follows [14]:

$$q_0 \equiv \frac{(\ddot{R}/R)_0}{H_0^2} = \frac{1}{2}\Omega_0 + \frac{3}{2}\sum_i \Omega_i w_i$$

(12)

Here, the subscript (0) denotes respective values at the present epoch, subscript (i) denotes respective components, like matter, energy, *etc*, (H_0) denotes Hubble's constant, and (w) denotes the equation of state. Since it's known that for an accelerating expansion of the universe (a''>0), the value of the deceleration parameter must be negative (q<0), which can be understood from the fore mentioned equation.

Equation of State (w)

The equation of state (w) is a dimensionless parameter, which is defined as the ratio of pressure to the energy density. In case of constant (wi), it can be defined (Frieman *et al.* 2008) as [15]:

$$w_i \equiv \frac{p_i}{\rho_i} = constant, \quad \rho_i \propto (1+z)^{3(1+w_i)}$$

(13)

where (p_i) is the pressure and (ρ_i) is the energy density of the respective component.

Parameter Values of Λ-CDM Model

The values of some important cosmological parameters of the Λ-CDM Model, which were derived from the cosmic microwave background (CMB) fluctuations in temperature are tabulated in Table **1** (Wang *et al.* 2016) [16]:

Table 1. Cosmological Parameters of the Λ-CDM Model.

H0/kms-1Mpc-1	67.74±0.46
$\Omega b,0h2$	0.022 30±0.000 14
$\Omega c,0h2$	0.1188±0.0010
$\Omega \Lambda,0$	0.6911±0.0062
$\Omega K,0$	0.0008-0.0039+0.0040
wd	-1.019-0.080+0.075

SIGNIFICANCE OF DARK ENERGY AND DARK MATTER IN COSMOLOGICAL PERIODS

Cosmological Periods

The true nature of dark energy and dark matter is still a mystery. Observations have implied that there is a constant cosmic war between the two. Although they may seem related in many different ways, their effects are quite different. In brief, dark energy possesses anti-gravitational properties, by the virtue of which it repels matter and favors expansion. On the contrary, dark matter possesses properties similar to gravitation, by the virtue of which it attracts matter. Also, while the effect of dark energy is observed at the largest cosmic scale, dark matter exerts its influence on the celestial bodies residing nearby as well as the universe at large. We can distinguish the various cosmological eras in the chronology of the universe by paying attention to the dominant constituent affecting the dynamics of the universe in certain periods of the evolution of the universe.

Radiation / Energy Dominant Early Universe

Firstly, after the advent of cosmic inflation, we had a universe dominated by energy and radiation. From the study of cosmic evolution, it can be understood that the influence of dark energy and dark matter could be almost negligible, since photons (light particles) and cosmic energy in the form of rays were unaffected by any perturbations and neutrinos, being an example of WIMPS (Weakly Interactive Massive Particles), remain interactive and undisturbed. The dark energy might have had a lesser effect on radiation, which was dominating. Therefore, the tremendous amount of energy released from the Big Bang dominated this period in the form of radiation, whose traces can also be found today in the form of cosmic microwave background radiation (CMBR), like the fingerprints of cosmological creation which provides us with an insight of an early universe.

The expansion rate of a radiation-dominated early universe can be given (Hooper *et al.* 2019) as follows [17]:

$$H^2 \equiv \left(\frac{\dot{a}}{a}\right)^2 = \frac{8\pi G \rho_R}{3} = \frac{1}{4t^2} \qquad (14)$$

where, ρ_R represents the energy density of radiation, and H is the Hubble parameter.

Matter Dominant Medieval Universe

A very significant change took place after the radiation dominant period, where the presence and energy density of matter surpassed both radiation and vacuum energy. This can be accounted for the drop of radiation and conversion of energy into matter, forming elementary particles, although the universe remained optically thick to radiation. This era was quite closely related to the standard cosmological model (Λ-CDM Model) in the beginning. The conditions required for the formation of matter and basic elements such as hydrogen and helium were quite dominant and also favored the formation of matter over anti-matter. Along with gravity, cold dark matter might have had a significant contribution to the formation of various small-scale structures and first-generation stars, while cosmological constant (dark energy) led to a decelerating expansion of space.

Dark-Energy Dominant Modern Universe

This period can probably be considered the final phase of the universe to date. This period has been in tune with the observational shreds of evidence and theories since recent years. Due to the expansion of the universe and hence, due to an increase in its volume, radiation and matter densities drop to low concentrations. Hence, the dynamics of the universe become dependent on the accelerating expansion, caused by dark energy. Also, since scale factor (a) is a time-dependent parameter, it could be useful in determining the expansion of the universe with time in the fore mentioned cosmological periods.

The general equation of scale factor (a) as calculated by (Arun *et al.* 2017) for a dark energy dominant universe can be given as [18]:

$$a(t) = \left[a_0^{3(1+W)/2} + \frac{3}{2}(1+W)\left(\frac{8\pi G \rho_0}{3}\right)^{1/2} t \right]^{2/[3(1+W)]}$$

(15)

Dark Energy and Dark Matter Significance

The various values of cosmological parameters, such as the equation of state, scale factor, and the equivalent densities for different substances in the universe is provided in Table **2** (Tkachev 2018) [19].

Table 2. Various values of cosmological parameters for different substances.

Substance	Equation of state	$\rho(a)$	$a(t)$
Radiation	w=1/3	$\rho = a-4\rho 0$	a=t/t01/2
Matter	w=0	$\rho = a-3\rho 0$	a=t/t02/3
Vacuum	w=-1	$\rho =$const	a=expH0t

where the vacuum energy represents dark energy and (H_0) is the Hubble's constant.

Also, by solving Friedmann Equations with conditions of the respective cosmological periods, we get the dependence of scale factor (a) over time (Cline 2018) for the periods as follows [20]:

$$a \sim \begin{cases} t^{1/2}, radiation \\ t^{2/3}, matter \\ exp\left(t\sqrt{\rho_\Lambda/3}/m_p\right), vaccum\ energy \end{cases} \tag{16}$$

As the value of scale factor (a) keeps increasing at an exponential rate along with time, so as the universe's expansion over time. We can very well observe that the value of (a) kept increasing throughout the evolutionary course of the universe to date. This analogy directs us towards the de-Sitter universe, a universe of dark energy domination.

The density parameter (Ω) is also helpful for the determination of the spatial curvature of the universe. This point towards one among the three possible geometries as discussed by (Copeland *et al.* 2006) of space-time depending upon the values of (Ω) as follows [21]:

- $\Omega > 1\ or\ \rho > \rho_c \rightarrow K = +1$ (Closed or Spherical Universe - Positive Curvature)
- $\Omega = 1\ or\ \rho = \rho_c \rightarrow K = 0$ (Flat Universe - Zero Curvature)
- $\Omega < 1\ or\ \rho < \rho_c \rightarrow K = -1$ (Open or Hyperbolic Universe - Negative Curvature)

where the density parameter is given as $\Omega(t) = \rho(t)/\rho_c(t)$ and the critical density is given as $\rho_c = 3H^2(t)/8\pi G$.

The observational evidence has shown that, within the limits of experimental error, the universe seems to be spatially flat or has a spatially flat geometry $(\Omega \cong 1)$. It can be accounted for by the cosmic inflation of the early universe. But the exact direction towards which the universe is heading remains a mystery. But, from the latest observations and findings, we can certainly say that the expansion of the universe is accelerating. Therefore, it is most likely to evolve close to a de-Sitter universe.

de-Sitter Universe

The de-Sitter universe can certainly be accounted for the evolution of the present-day universe. According to some scientists, we have already entered into a de-Sitter universe. The de-Sitter universe models the universe to be spatially flat, neglecting ordinary matter to a maximum extent. In this case, the dynamics of the universe will be entirely dominated by dark energy, like the inflation field in the early universe. This model wasn't considered appropriate for the observable universe until the models of inflation and dark energy were developed. Quintessence is regarded as a scalar field, which was introduced as an explanation, for the current accelerating rate of expansion of the universe. The evolution of scale factor and expansion rate for a de-Sitter universe is similar to the evolution of scale factor under the vacuum energy as $[a(t) = e^{H_0 t}]$ Where H_0 is the Hubble's constant.

CONCLUSION AND FUTURE SCOPE

From this paper, we can talk about the existence and effect of dark energy and dark matter. We have reviewed the widely accepted Λ-CDM Model, along with various important governing equations, such as Einstein's Field Equations and Friedmann Equations. We also reviewed the nature of dark energy and dark matter and how could they have affected the transformations of various cosmological periods and eras along with their associated parameters, which drove the universe to the present state, along with its spatial curvature. But the true and exact nature of dark energy and dark matter, and how exactly the universe is going to evolve, remains a mystery. We are certainly missing onto some more theories and observations, serving as the missing pieces of the ultimate puzzle which we seek to complete.

With the advancement in astrophysical studies and the latest technologies, there is a lot more to be observed, analyzed, and learned in our never-ending universe. According to multiverse theory, there can be possibilities of many alternative universes with different timelines, different laws of physics, never-ending possibilities, and new concepts. Therefore, this is still a puzzle unsolved by our dedicated scientists of the world, far from the ultimate understanding of what the universe has to offer. There is always a possibility that we may come across various other cosmological phenomena, which will further enhance our understanding of the universe.

CONSENT FOR PUBLICATION

Not applicable.

CONFLICT OF INTEREST

The authors declare no conflict of interest, financial or otherwise.

ACKNOWLEDGEMENTS

We acknowledge Lovely Professional University, Phagwara, Punjab, India.

REFERENCES

[1]   M.S Turner, "Dark Matter and Dark Energy: The Critical Questions", *arXiv Prepr , .* arXiv: astro-ph/0207297

[2]   V. Sahni, *5 Dark Matter and Dark Energy.* Springer: Berlin, Heidelberg, 2004, pp. 141-179.

[3]   Y.L. Bolotin, A. Kostenko, O.A. Lemets, and D.A. Yerokhin, "Cosmological evolution with interaction between dark energy and dark matter", *Int. J. Mod. Phys. D,* vol. 24, no. 3, 2015.
[http://dx.doi.org/10.1142/S0218271815300074]

[4]   P.J.E. Peebles, and B. Ratra, "The cosmological constant and dark energy", *Rev. Mod. Phys.,* vol. 75, p. 559, 2003.
[http://dx.doi.org/10.1103/RevModPhys.75.559]

[5]   A. Einstein, D. Physik, H. Minkowski, and H Weyl, ""How Einstein Got His Field Equations," arXiv", https://arxiv.org/abs/1608.05752

[6]   S. E. Rugh, and H. Zinkernagel, "The quantum vacuum and the cosmological constant problem,", *Stud. Hist. Philos. Sci. Part B - Stud. Hist. Philos. Mod. Phys,* vol. 33, no. 4, pp. 663-705, 2002.
[http://dx.doi.org/10.1016/S1355-2198(02)00033-3]

[7]   W.F. Kao, and U.L. Pen, "Generalized Friedmann-Robertson-Walker metric and redundancy in the generalized Einstein equations", *Phys. Rev. D Part. Fields,* vol. 44, no. 12, pp. 3974-3977, 1991.
[http://dx.doi.org/10.1103/PhysRevD.44.3974] [PMID: 10013875]

[8]   A.F. Zakharov, S. Capozziello, F. De Paolis, G. Ingrosso, and A.A. Nucita, "The role of dark matter and dark energy in cosmological models: Theoretical overview", *Space Sci. Rev.,* vol. 148, no. 99, pp. 301-313, 2009.
[http://dx.doi.org/10.1007/s11214-009-9500-2]

[9]   R. Kalita, "The Nature of Dark Energy and Dark Matter", *Int. J. Astron,* vol. 3, no. 1, pp. 8-21, 2014.
[http://dx.doi.org/10.5923/j.astronomy.20140301.02]

[10]   N. Sapkota, and B. Adhikari, *Dark Matter and Dark Energy: Mysteries of the Universe | IJRRAS.*.http://ijrras.com/dark-matter-and-dark-energy-mysteries-of-the-universe/ (accessed Apr. 20, 2021).

[11]   J. Ellis, *Dark matter and dark energy: Summary and future directions,* 2003.
[http://dx.doi.org/10.1098/rsta.2003.1297]

[12]   N.A. Bahcall, "Hubble's Law and the expanding universe", *Proc. Natl. Acad. Sci. USA,* vol. 112, no. 11, pp. 3173-3175, 2015.
[http://dx.doi.org/10.1073/pnas.1424299112] [PMID: 25784761]

[13]   A. Al Mamon, and S. Das, "A parametric reconstruction of the deceleration parameter", *Eur. Phys. J. C,* 2017.
[http://dx.doi.org/10.1140/epjc/s10052-017-5066-4]

[14]   M.S. Turner, "Dark Matter and Dark Energy in the Universe", *Phys. Scr. T,* 2000.
[http://dx.doi.org/10.1238/Physica.Topical.085a00210]

[15]   J.A. Frieman, M.S. Turner, and D. Huterer, "Dark energy and the accelerating universe", *Annu. Rev. Astron. Astrophys.,* 2008.
[http://dx.doi.org/10.1146/annurev.astro.46.060407.145243]

[16]   B. Wang, E. Abdalla, F. Atrio-Barandela, and D. Pavón, "Dark matter and dark energy interactions: theoretical challenges, cosmological implications and observational signatures", *Rep. Prog. Phys.,* vol. 79, no. 9, p. 096901, 2016.
[http://dx.doi.org/10.1088/0034-4885/79/9/096901] [PMID: 27517328]

[17]   D. Hooper, G. Krnjaic, and S.D. McDermott, "Dark radiation and superheavy dark matter from black hole domination", *J. High Energy Phys.,* 2019.
[http://dx.doi.org/10.1007/JHEP08(2019)001]

[18]   K. Arun, S.B. Gudennavar, and C. Sivaram, *Dark matter, dark energy, and alternate models: A review.* Adv. Sp. Res, 2017.
[http://dx.doi.org/10.1016/j.asr.2017.03.043]

[19]   I. Tkachev, *Cosmology and dark matter,* vol. 5, 2017.

[http://dx.doi.org/10.23730/CYRSP-2017-005.259]

[20]   J. M. Cline, *Tasi lectures on early universe cosmology: Inflation, baryogenesis and dark matter,* 2018. [http://dx.doi.org/10.22323/1.333.0001]

[21]   E.J. Copeland, M. Sami, and S. Tsujikawa, "Dynamics of dark energy", *Int. J. Mod. Phys. D,* 2006. [http://dx.doi.org/10.1142/S021827180600942X]

CHAPTER 10

Magnetic Field in the Solar System – a Brief Review

Rashi Kaushik[1,*] and **Amit Kumar Thakur**[1]

[1] *Department of Aerospace Engineering, Lovely Professional University, Phagwara, Punjab-144411, India*

Abstract: The following review is concerned with the working and development of the magnetic field around the solar system. It combines the various theoretical facts that have been extracted to date emphasizing how the magnetic field is present in the solar system and how it plays a crucial role in protecting it as well as other constituents of the solar system comprising planets. Since the magnetic field is all-pervasive throughout the universe, it can be considered to be a crucial element for the development of any planet or cluster or other celestial and intergalactic entities as well. Through this theoretical review, we will be able to discern the role of the magnetic field in a solar body, whereas through observational data we will be able to understand the practical orientation of the respective magnetic field in different planets throughout the solar system.

Keywords: Dynamo Effect, Heat Flux, Magnetic Field, Magnetosphere, Plasma, Solar Wind.

INTRODUCTION

There have been a lot of speculations regarding the origin, working, and importance of the magnetic field. The magnetic field is not limited to a particular domain. The following review is concerned with the magnetic field of the solar system and its respective planetary bodies. Most of the bodies in and exceeding the solar system possess a magnetic field that acts as a protective shield around their respective atmospheres. Sometimes heavenly bodies like galaxies act as giant magnets. In general, the magnetic fields in-universe are weak. However, in some compact objects, these might be assumed to be of extraordinarily large values. These fields are produced by the gravitational disintegration of massive magnetized bodies. According to [1] it is generally presumed that the magnetic fields in the astronomical structure having different sizes, from the stars $R \sim 10^{11}$

---

* **Corresponding author Amit Kumar Thakur:** Department of Aerospace Engineering, Lovely Professional University, Phagwara, Punjab-144411, India; E-mail: amitthakur3177@gmail.com

centimeters to the galactic clusters R~10^{24} centimeters are generated by the amplification of the preceding weaker magnetic field seed *via* different dynamo. The magnetic fields are often measured with μ, Gauss strength which is abbre*via*ted as G or Gs is the CGS unit for measuring magnetic flux density or magnetic induction or magnetic field [2]. Revealed that the evolution of the magnetic- field is governed by the magnetic-induction equation which is derived by combining both the Maxwell equation while neglecting the current due to displacement as well as a simple form of the ohms law.

The planetary magnetic fields are produced by the interaction between the convection of the molten core and the planet's rotation. Not every planet has a measurable magnetic field. Mercury has a weak magnetic field because of its slow rotation. Venus also doesn't have a measurable field as there is very little convection in its molten core. Earth has a moderately strong field of magnetism which protects it from solar winds as well as facilitates navigation. Mars lost its magnetic field because of the solidification of its interior. Jupiter has an enormous magnetic field for which its core is responsible. Saturn's magnetic field is slowed down as it moves through the ring of particles that litter the orbit of Enceladus. The field of magnetism on Uranus varies from place to place, in the southern hemisphere, its field is only 1/3rd strong as that of earth's while Neptune has a turbulent magnetic field being 27 times powerful to that of earth's and resides at an angle on the planet, varying chaotically as it comes in contact with the solar winds. Comparing with planets, the galaxies have a weak magnetic field. The magnetic field of our Milky Way galaxy is 100 times weaker as compared to that of earth. The fields of magnetism are an important factor in the interstellar medium of barred, irregular, spiral, and other dwarf galaxies as well as in the formation of stars by enabling the removal of angular momentum.

MAGNETISM IN THE SOLAR SYSTEM

Around the Sun

It has been more than 10 decades since magnetic fields have been first measured around the sun. The sun has a quite large and complex field of magnetism [3]. Found that the structural dynamics of the Sun are mainly dominated by the intense field of magnetism which is created by the turbulent plasma present in the convective zone (solar dynamo). The sun's magnetic field is estimated to be twice as strong as that of Earth. The magnetic domain of the Sun, in reality, expands far into space, even beyond Pluto (the farthest planet), and this extension of the Sun's field of magnetism is known as Inter-planetary Magnetic Field or IMF. It is often stated that the sun is a magnetic star itself. It can be understood as a thermo-nuclear furnace burning at 15 million degrees with an excessively heated core that churns the electrically conductive plasma in the exterior third of the sun just like a heated stove that churns the boiling water. The visible disk of the Sun called photosphere contains dark earth-sized sunspots which are the provinces of intensely magnetic domains [4]. explained through this research that numerous missions have been executed for understanding the ejections of coronal mass, their intensity, nature, and extent of the magnetic fields of solar origin taking into concern the 11 years solar activity cycle.

Exploring Mercury

There have been various speculations on the genesis of the magnetic fields in Mercury. The intrinsic field of magnetism in Mercury might be a result of a dynamo effect in a liquefied core, despite the obtuse rotation rate of the planet. The core of mercury is partially in the liquid phase. According to earlier observations, Mercury seems to have an intrinsic dipolar-magnetic domain having a moment of 5×10^{12} Tesla cubic meter. It was conferred that the observations received through a space probe studying magnetosphere-type region around mercury might have been an outcome of reciprocity with the winds of solar origin [5]. Explained the concept of radius and the core-density while there have been variations due to consideration of present conditions and availability of data. Since, Mercury is the only planet to have a notable field of magnetism (revealed by Mariner 10 in 1974-1975) other than Earth, having a field 1% to that of Earth's. It has been a tough task to model the presently observed magnetic field of Mercury in terms of crustal magnetization, while the theoretical studies confer that if the dynamo effect works similarly on Mercury as it does on Earth, then the field of magnetism around it should be ~30% of its terrestrial equivalence.

Exploring Venus

Venus is a rare exception among planets. It cannot be wrong to state that it is a world that doesn't have a magnetic field at all. This exquisite exception in planetary science forces us to speculate that how a planet with no magnetic field can is sustaining itself through the universe. The solar wind may sometimes interact with the atmosphere of the planet without any resistance as there is no magnetic field. However, through recent observations, it has been found that Venus has an induced magnetic domain that protects it, partially though. According to [6] the highest heat flux that can be derived from the core without thermal convection and is given by

$$R_c = k\,\alpha\,a\,t/c_p \tag{1}$$

Where α and k are the thermal-conductivity and thermal-expansivity, a is the gravitational acceleration, t is the core-temperature, and c_p is the specific heat capacity. Thermal convection will halt if the heat that is being derived from the core is less than the R_c.

Exploring the Lively Planet: Earth

It has been over 3.5 billion years since the evolution of Earth's magnetosphere or geomagnetic field, but still, it remains a mystery to many. The earth's field of magnetism is not just a mere domain but is a shield that protects our planet from solar winds [7]. The solar wind does not directly affect Earth as it gets deflected by our magnetic field which is like a tear droplet-shaped cavity in the solar-wind motion. This cavity known as the magnetosphere carries highly diluted plasma whose activity is governed by Earth's magnetic field as well as its reciprocity with the solar winds. The major part of Earth's magnetic field is suspected to have originated deep inside our very planet inside the outer fluidic core. It originates from a self-sustaining dynamo process involving turbulence in the flow of molten iron [8]. Explained the evolution of the magnetic field by the magnetic induction equation which is written as:

$$\frac{\partial M}{\partial T} = \eta \nabla^2 M + \nabla \times (v \times M), \tag{2}$$

Where v is the fluidic velocity; M symbolizes the magnetic-field and $\eta = 1/\sigma\mu$ is the magnetic- diffusivity, which is inversely proportional to the product of electrical-conductivity σ &permeability μ. Also, the term $\partial M/\partial T$ symbolizes the time-derivative of the field; while ∇^2 is the Laplacian operator.

Observing Mars

Similar to Venus, planet Mars also lacks a congenital magnetic field. The solar wind directly impacts the Martian atmosphere leading towards the formation of a magnetosphere. Martian magnetosphere is less extensive and complex in comparison to that of Earth. Since the magnetic field lines from Sun cannot enter through the electrically conductive objects; they drape themselves around the planet generating a magnetosphere, irrespective of the absence of a magnetic field. Despite the reality that Mars has lost its internal dynamo which was capable of generating magnetic fields, similar to that of Earth's, there is evidence that it once had a dynamo. Since it is believed that excessive heating can lead to loss of magnetism, there might be a possibility that Venus lacked magnetic fields due to the stated reason, but the case of Mars is quite odd, as it is colder and has various similarities with Earth [9]. discovered numerous pieces of evidence pointing towards the traces of the presence of magnetic field and Earth-like conditions on

Mars, but still, it remains a mystery.

Around Jupiter

The magnetic field of Jupiter is believed to be about 20,000 times powerful than Earth's. It has a humungous, composite, and fervent field of magnetism, which is thought to have been originated from the electric currents in a swiftly spinning interior consisting of metallic hydrogen. The Jovian field of magnetism is toroidal in shape consisting of colossal versions of Earth's Van Allen belt, which traps high-energy charged atoms. Due to the forces associated with rapid rotations and complex magnetic fields, the belts get flattened and transform into plasma sheets. The plasma torus depicted in the above picture is associated with the orbit of the satellite Io, it is affected by the 9 hours rotation period of the planet along with the satellites Europa, Amalthea, and Ganymede which orbit the region. The magnetosphere of Jupiter is the region dominated by the enormous magnetic field, which interacts with the charged fragments of solar wind all the way long from the Sun as far as 7 million km, forming a bow shock.

Exploring Saturn

Very little was known about Saturn's magnetosphere and its evolution until Cassini's observations completely transformed our understanding of Saturn's magnetosphere. It has been observed that forces prevailing deep inside the planet Saturn create a gigantic bubble consisting of magnetic energy, around the planet, exerting a strong influence over the surrounding space environment. The magnetic field in Saturn is generated in the form of material cycles deep within the fluidic interior of the planet. Whenever, there is an interaction between the solar-charged particles and Saturn's magnetic domain, a gigantic bubble of Saturn's magnetosphere forms. While Sun's magnetic forces dominate the external environment, inside Saturn's magnetosphere its magnetic forces reign. In a similar manner to Earth, the Satian magnetic field also consists of north& south poles, just like a bar magnet. The magnetic field rotates along with the planet. Saturn has its field is perfectly parallel with the planet's rotational axis. Saturn's magnetic field is much weaker in comparison to that of Jupiter's and so are its radio signals that are not powerful enough to be able of getting traced from Earth [10].

Exploring Uranus

The cold planet Uranus comes with a strange magnetic field as its poles are not at all close to its geographic poles. The moons along with their rings are deeply embedded in the planet's magnetosphere, thus due to the huge dipolar tilt, they might have an extreme and diurnally fluctuating influence as told by [11]. The main magnetic field of the planet is tilted by an angle of 58.6° away from its spin axis. The magnetic field of the planet varies from place to place. It is a planet having multipolar magnetic fields. Just like other planetary giants, Uranus generates its magnetic fields with the help of convective currents in an electrically conductive interior. The magnetic center is dislocated from its center by 31%of its radius. This dislodgement is mainly along the rotational axis towards the northern pole. The reason behind the complexity of Uranus's magnetic field is not only its tilt but its relatively massive small-scale elements. Similar to other planets having magnetic fields, Uranus also has a field of magnetism that resists the solar forces consisting of solar wind and charged particles.

Exploring Neptune

Neptune is the farthest planet in our solar system having vivid bluish clouds and a badly behaved field of magnetism along with fierce wind storms and like other planets, it also possesses an intrinsic magnetic field which was first discovered in 1989 by Voyager 2 spacecraft. Just like the magnetic field of Earth. The planet's magnetic field is non-dipolar as rightly stated by [12] in their research survey. The magnetic field of Neptune is tilted from its rotational axis by 47° respectively. Also, the fields are not well centered in the interiors. The planet's field has an offset of 55 percent of its radius and is centered in a portion of the interior that is closer to the cloud tops than to be to the planetary center. The magnetic domain of Neptune is somewhat apple-shaped which the stem end and opposite end placed in the direction of magnetic poles. Most of the trapped particles get absorbed by the moon s and ring's material emptying the magnetosphere from a large fraction of the charged content. Neptune's field is believed to be 27 times stronger than Earth's field of magnetism.

CONCLUSIONS

Through this compact review, we get a basic understanding of how the magnetic field is generated and how it works in different terrenes of our solar system as well as how it acts as a protective shield against erratic behavior of solar forces and their interaction with the respective magnetospheres of the different planets. However, through the different theoretical facts, we understand that the core reason responsible for the initiation of the magnetic field is the dynamo effect which is not only responsible for the spinning motion of a planet but also the initiation of the magnetic field. For finding the reason for universal magnetism the dynamo theory can be seen as the basic step for unleashing the unanswered mysteries of magnetic-field inception in the universe. The magnetic field is a valuable asset in determining the fate of a planet as it can be seen from the above planetary analysis that some of our planets have almost lost their magnetic fields while others still have extremely strong fields. Hence with the help of exquisite models proposed by numerous assiduous scientists and observations made by various spacecraft, it has been possible to reach this far in the understanding of planetary sciences, but there is still a lot of universe left to be explored and more questions to be answered in the line of how magnetic field keeps the universe from falling apart as well as its relation with the universal entropy.

CONSENT FOR PUBLICATION

Not applicable.

CONFLICT OF INTEREST

The authors declare no conflict of interest, financial or otherwise.

ACKNOWLEDGEMENTS

We acknowledge Lovely Professional University, Phagwara, Punjab, India.

REFERENCES

[1]   A. R. Zhitnitsky, "Cosmological magnetic field and dark energy as two sides of the same coin", *Phys. Rev. D,*, vol. 99, no. 10, p. 103518, 2019.
[http://dx.doi.org/10.1103/PhysRevD.99.103518]

[2]   K Subramanian, "Magnetizing the universe", *arXiv,* 2008.
[http://dx.doi.org/10.22323/1.052.0071]

[3]   P. Charbonneau, "Solar dynamo theory", *Annu. Rev. Astron. Astrophys.,* vol. 52, pp. 251-290, 2014.
[http://dx.doi.org/10.1146/annurev-astro-081913-040012]

[4]   M.L. Demidov, "The magnetic Sun from different views: A comparison of the mean and background magnetic field observations made in different observatories and in different spectral lines", *J. Astrophys. Astron.,* vol. 21, no. 3-4, pp. 209-212, 2000.
[http://dx.doi.org/10.1007/BF02702393]

[5]   R.W. Siegfried II, and S.C. Solomon, "Mercury: Internal structure and thermal evolution", *Icarus,* vol. 23, no. 2, pp. 192-205, 1974.
[http://dx.doi.org/10.1016/0019-1035(74)90005-0]

[6]   F. Nimmo, "Why does Venus lack a magnetic field", *Geology,* vol. 30, no. 11, pp. 987-990, 2002.
[http://dx.doi.org/10.1130/0091-7613(2002)030<0987:WDVLAM>2.0.CO;2]

[7]   S. Carolan, A.A. Vidotto, C. Loesch, and P. Coogan, "The evolution of Earth's magnetosphere during the solar main Sequence", *Mon. Not. R. Astron. Soc.,* vol. 489, no. 4, pp. 5784-5801, 2019.
[http://dx.doi.org/10.1093/mnras/stz2422]

[8]   K. Subramanian, "Magnetic fields in the Universe", *Sci. Cult.,* vol. 84, no. 11–12, 2018.

[9]   J.E.P. Connerney, J. Espley, P. Lawton, S. Murphy, J. Odom, R. Oliversen, and D. Sheppard, "The MAVEN magnetic field investigation", *Space Sci. Rev.,* vol. 195, no. 1-4, pp. 257-291, 2015.
[http://dx.doi.org/10.1007/s11214-015-0169-4]

[10]   U.R. Christensen, and J. Wicht, "Models of magnetic field generation in partly stable planetary cores: Applications to Mercury and Saturn", *Icarus,* vol. 196, no. 1, pp. 16-34, 2008.
[http://dx.doi.org/10.1016/j.icarus.2008.02.013]

[11]   N.F. Ness, M.H. Acuña, K.W. Behannon, L.F. Burlaga, J.E. Connerney, R.P. Lepping, and F.M.

Neubauer, "Magnetic fields at uranus", *Science,* vol. 233, no. 4759, pp. 85-89, 1986.
[http://dx.doi.org/10.1126/science.233.4759.85] [PMID: 17812894]

[12]   R. Redmer, T.R. Mattsson, N. Nettelmann, and M. French, "The phase diagram of water and the magnetic fields of Uranus and Neptune", *Icarus,* vol. 211, no. 1, pp. 798-803, 2011.
[http://dx.doi.org/10.1016/j.icarus.2010.08.008]

___

CHAPTER 11

Comparative Study on Identification and Classification of Plant Diseases with Deep Learning Techniques

Aditi Singh[1,*] and **Harjeet Kaur**[1]

[1] Lovely Professional University, Jalandhar, Punjab, India

Abstract: Proper development and growth of crops had always been a major concern and challenge in Agriculture. Proper crop development assures good quality of crops and also bumper harvest. Humans may not always identity all plant diseases accurately at all stages having an automated system for crop disease identification and detection can be a great help for a tiller. This thought inspired me to perform the proposed research work. VGG-16 based learning model achieved an accuracy of 98.74%, ResNet-50 based transfer achieved an accuracy of 98.84%, and ResNet-50 v2 based transfer learning model achieved an accuracy of 98.21%.

Keywords: CNN Achitecture, ResNet, ResNet50-v2, Transfer Learning, VGG.

1. INTRODUCTION

Development from an infant to a fully developed crop is a tedious process that requires close monitoring. Framers depend completely on the crop yield for their living. The plant needs a lot of care in its development process. One of such risks involved includes disease. There are various types of diseases of which a plant can get affected. Crop failure due to disease can have a grave impact not only on farmer's economic status but can also impact the national economy. Automated early detection and identification of plant disease can be a great help to farmers. It has a benefit over the traditional method of making an expert guess based on what farmers perceive from their eyes. Early detection of disease is important to provide scientific manuring.

With the amelioration of artificial intelligence, techniques of plant disease detection with help of leaf images are an emerging research area. The deep

___

* **Corresponding author Aditi Singh:** Lovely Professional University, Jalandhar, Punjab, India; E-mail: aditisingh4614@gmail.com

Dharam Buddhi, Rajesh Singh and Anita Gehlot (Eds.)

learning technique has been proved to be a powerful tool to perform such image classification techniques. In this aspect, deep learning techniques play an indispensable role in form of Convolutional Neural Network techniques. Convolutional neural network feature extraction plays a vital role in model performance.

CNN-based models are broadly classified into two categories: 1) Customized models: made by the developer according to the problem statement. 2) Standard models: These are pre-developed models whose efficiency had already been proved.

In the real world, when plant images are captured in farms the image contains green background, along with the infected leaf. A series of image processing steps need to be performed like image pre-processing, image segmentation, feature extraction. Based on the extracted features models are trained. And disease mapping is done for classification purposes. CNN provides a pre-built architecture for the majority of phase, whereas if one is using a machine learning algorithm need to develop all these phases.

The proposed research focuses on the following key points:

• An experiment is carried on multiple plant datasets to validate the result and avoid overfitting and underfitting conditions.
• Transfer learning techniques helped to achieve a higher result accuracy.

2. RELATED WORK

Konstantinos (2018) [1] presented a convolution neural network model to identify various plant diseases. Images of various plant leaves which include both the infected and healthy leaves images are used to train the model. In this proposed research work, an accuracy of 97% is achieved. The CNN architectures used in the proposed framework include AlexNet, AlexNetOWTBn, GoogLeNet.

Golhani *et al*. (2018) made a detailed review of various deep learning algorithms along with their advantages and disadvantages also their optimization techniques. A comparison has also been made for these techniques about the related work [2].

Sardogan *et al*. in (2018) [3] proposed a model that is a combination of convolution neural network and Linear vector quantization algorithm. The dataset used included 500 images of tomato leaves.

Wallelign *et al*. in 2018 [4] presented a model that was implemented using LeNet one of the efficient and standard CNN architectures for the classification of soybean leaves. The dataset includes 12,763 images. This proposed model gained an accuracy of 99.32%.

Zahid Ullah *et al.*, (2020) [5], in this paper author has explained a detailed methodology for image enhancement. There are various scenarios where some amount of useful information of the image is damaged, distorted, or lost. So, at first, the author has advised filtering the image. A median filter is passed through the image to remove some amount of noise. A good contrast of the image is required for good analysis. Good contrast makes all image object visible. In this paper, the author has suggested a very effective methodology by limiting contrast in histogram equitization. This method is better than the histogram equalization method because it removes the over-amplification issue. And then it is good to do the wavelet transform process. 2D discrete wavelet transform is an efficient tool to do image enhancement.

Pantazi *et al.*, (2019) [6] provides a detailed study of each step of the proposed algorithm. Grabcut algorithm for image segmentation. LBP algorithm for thresholding. One Class Classification method and its hybridization with the SVM algorithm for conflict resolution. Sankaran *et al.* (2010) [7] this paper shows one of the early works in plant disease detection technology. This paper does a comparative analysis of disease prediction in plants with the help of molecular techniques, imaging and spectroscopic techniques, and also volatile organic compound profile study.

Fuentes *et al.* in 2017 [8] provided a framework that needed stage-wise implementation. First meta architectures: Faster RCNN, R-FCN, and SSD are combined to form a single meta-architecture. Then standard CNN architectures like VGG-16, ResNet-50 are used to extract features and train the model.Sharma *et al.*, 2019 [9] in this paper the author has addressed a major problem about the success of the model in the real world. Most of the algorithms proposed work efficiently on training data but fail to achieve the desired output in a real-world scenario. The author made a change in training a convolutional neural network. Instead of training the convolutional neural network with a full image the author trained the model with the help of a segmented image and got a very efficient output. Arivazhagan *et al.*, 2018 [10] proposed a framework to identify the disease with leaf images of mango plat with an accuracy of 96.97%. Oppenheim *et al.*, (2017) [11] proposed a framework using CNN for potato plants from a dataset of size 2465. Tammina (2019) [12], the author has proposed a framework for the use and implementation of the transfer learning model and how to extract necessary data from pre-trained models. Khirade, *et al.*, (2015) [13] proposed a methodology that uses image processing as the initial step for the identification of infected parts followed by image classification.

3. METHODOLOGY

This section is divided into two parts: 1) Dataset description 2) Methodologies for proposed research work. The second part further includes a complete description of VGG-16, ResNet-50, and ResNet-50 v2 architectures used in research.

3.1. Dataset Description

The dataset used for the proposed framework is an open-source dataset taken from Kaggle. The dataset is named the "plant village" dataset. This dataset contains 20,600 images of three crops: potato, pepper, and tomato leaves. These images are divided into 15 categories including both infected and healthy leaves images.

3.2. Methodologies

This research focuses on transfer learning techniques. The existing pre-trained models need to be utilized to carry out the research work. The three pre-trained models used are: ResNet-50, VGG-16, ResNet-50 v1. The next section contains a detailed discussion of these models.

3.2.1. Transfer Learning

Transfer learning is one of the emerging and efficient approaches in deep learning. It can be used to solve various computer vision problems. There are a few things needed to be considered while selecting a model in the transfer learning approach:

- Carefully study the answers to the model knowledge, which is transferred from pre-trained models.
- To identify the correct stage when the knowledge of the pre-trained model needs to be passed.
- How knowledge can be passed on between objective jobs.

3.2.2. VGG-16

VGG-16 is a pre-trained convolutional neural network model having a depth of 16 layers. It alone consists of 138 million parameters which will be covering up a good amount of storage. VGG-16 was a kind of extension or up-gradation to standard pre-existing models AlexNet and LeNet-5. This extension in model performance was by the increasing depth which could be achieved by adding the convolution layers. In the model. VGG-16 had 13 convolution layers, 5 max-pool, 1 flatten, and 3 fully connected layers. After the development of this model, researchers believed that by increasing the depth of a model we can improve the model's performance and developed VGG-19 which consisted of 16 convolution layers. But this model failed to give a considerable amount of efficiency and performance. Also, the increased parameters utilize more amount of space. Fig. (**1**) shows the VGG-16 basic architecture.

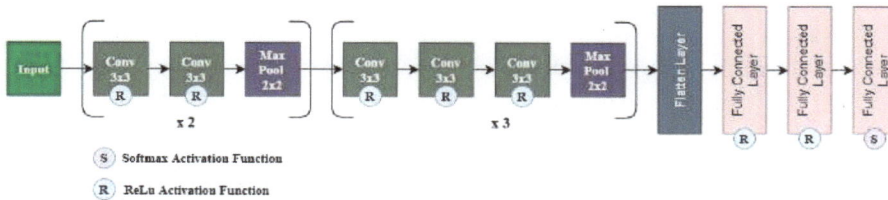

Fig. (1). VGG-16 basic architecture [12].

3.2.3. ResNet-50

This ResNet-50 model is based on the concept of batch normalization. This helps these models to have a stabilized learning process in the training phases of the model. This concept has also made the model time efficient by reducing the number of epochs, that is model is trained with sufficient accuracy with a small number of epochs. The ResNet-50 architecture incorporates 49 convolution layers and 1 fully connected layer and also with 26 million parameters. Given below is the ResNet-50 architecture diagram.

3.2.4. ResNet-20 v2

It alone ResNet-50 v2 and ResNet-50 are almost identical in their functions. The only difference is in the way batch normalization and activation functions are applied. ResNet-50 v2 batch normalization is applied before the weight layer hence, called the pre-activation model. Whereas in ResNet-50 it is applied after the weight layer and hence, called the post-activation model. Fig. (**2**) explains the architectural difference between these two models.

3.2.5. Optimization Algorithm

The main issue while proposing a model in deep learning is to avoid bias and variance problems. Below mentioned are the steps to remove biases of the model:

- Pre-trained deep neural networks are used in the proposed framework.
- Three different models are trained to get the best results.
- Adam optimization algorithm is used.

Below mentioned are the steps are taken to remove high variance from the proposed model:

- Large data set.
- Data regularization using data augmentation.

3.2.6. The proposed research framework can be summarized in the diagram given below

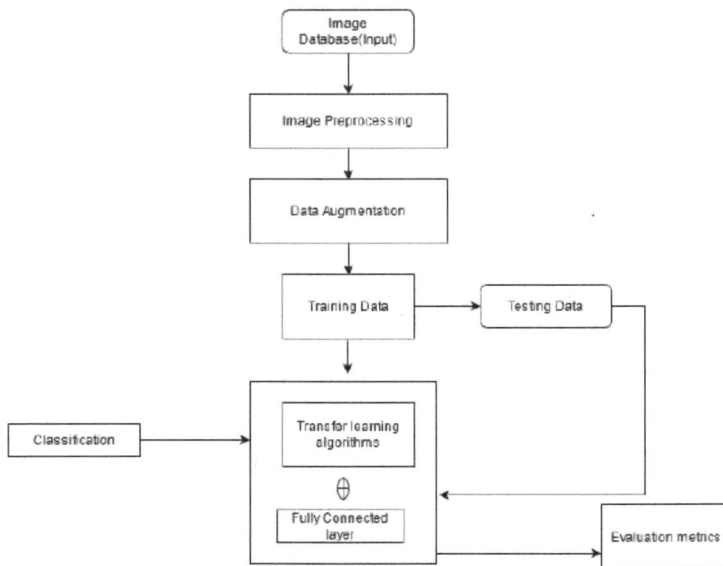

Fig. (2). Flow chart for proposed research methodology [13].

4. RESULT AND DISCUSSION

Table 1 represents the outcome of the proposed framework:

Table 1. Outcome of the proposed framework:

Model	Accuracy (%)	Precision (%)	Recall (%)	F1-score (%)
VGG-16	97.74	97.75	97.74	97.76
ResNet-50	97.84	97.85	97.84	97.65
ResNet-50 v2	97.87	97.65	97.21	97.75

CONCLUSION

Transfer learning is one of the most efficient approaches to solve computer vision problems. It is promising both in terms of efficiency and development effort. This was efficiently demonstrated with the help of the proposed framework. The proposed research framework with the help of VGG-16, ResNet-50, and ResNet-50 v2 models efficiently performed multi-class classification and gave very satisfactory results and accuracy. It could efficiently identify infected and healthy leaves of potato, tomato, and pepper leaves.

CONSENT FOR PUBLICATION

Not applicable.

CONFLICT OF INTEREST

The authors declare no conflict of interest, financial or otherwise.

ACNOWLEDGEMENT

Declared none.

REFERENCES

[1]   K.P. Ferentinos, "Deep learning models for plant disease detection and diagnosis", *Comput. Electron. Agric.,* vol. 145, no. January, pp. 311-318, 2018.
[http://dx.doi.org/10.1016/j.compag.2018.01.009]

[2]   K. Golhani, S.K. Balasundram, G. Vadamalai, and B. Pradhan, "A review of neural networks in plant disease detection using hyperspectral data", *Inf. Process. Agric.,* vol. 5, no. 3, pp. 354-371, 2018.
[http://dx.doi.org/10.1016/j.inpa.2018.05.002]

[3]   M. Sardogan, A. Tuncer, and Y. Ozen, "Plant Leaf Disease Detection and Classification Based on CNN with LVQ Algorithm", *UBMK 2018 - 3rd Int. Conf. Comput. Sci. Eng,* pp. 382-385, 2018.
[http://dx.doi.org/10.1109/UBMK.2018.8566635]

[4]   S. Wallelign, M. Polceanu, and C. Buche, "Soybean plant disease identification using convolutional neural network,", *Proc. 31st Int. Florida Artif. Intell. Res. Soc. Conf. FLAIRS,* 2018, pp. 146-151.

[5]   Z. Ullah, M.U. Farooq, S.H. Lee, and D. An, "A hybrid image enhancement based brain MRI images classification technique", *Med. Hypotheses,* vol. 143, no. April, p. 109922, 2020.
[http://dx.doi.org/10.1016/j.mehy.2020.109922] [PMID: 32682214]

[6]   X.E. Pantazi, D. Moshou, and A.A. Tamouridou, ">Automated leaf disease detection in different crop species through image features analysis and One Class Classifiers", *Comput. Electron.Agric* , vol. 156, pp. 96-104, 2019.
[http://dx.doi.org/10.1016/j.compag.2018.11.005]

[7]   S. Sankaran, A. Mishra, R. Ehsani, and C. Davis, "A review of advanced techniques for detecting plant diseases", *Comput. Electron. Agric.*, vol. 72, no. 1, pp. 1-13, 2010.
[http://dx.doi.org/10.1016/j.compag.2010.02.007]

[8]   A. Fuentes, S. Yoon, S.C. Kim, and D.S. Park, "A robust deep-learning-based detector for real-time tomato plant diseases and pests recognition", *Sensors (Basel)*, vol. 17, no. 9, p. E2022, 2017.
[http://dx.doi.org/10.3390/s17092022] [PMID: 28869539]

[9]   P. Sharma, Y. Paul, S. Berwal, and W. Ghai, "Performance analysis of deep learning CNN models for disease detection in plants using image segmentation", In: *," Inf. Process. Agric*, 2019, pp. 1-9.
[http://dx.doi.org/10.1016/j.inpa.2019.11.001]

[10]   S. Arivazhagan, and S.V. Ligi, "Mango Leaf Diseases Identification Using Convolutional Neural Network", *Int. J. Pure Appl. Math.*, vol. 120, no. 6, pp. 11067-11079, 2018.

[11]   D. Oppenheim, and G. Shani, "Potato Disease Classification Using Convolution Neural Networks", *Adv. Anim. Biosci.*, vol. 8, no. 2, pp. 244-249, 2017.
[http://dx.doi.org/10.1017/S2040470017001376]

[12]   S. Tammina, "Transfer learning using VGG-16 with Deep Convolutional Neural Network for Classifying Images", *Int. J. Sci. Res. Publ.*, vol. 9, no. 10, p. 9420, 2019.
[http://dx.doi.org/10.29322/IJSRP.9.10.2019.p9420]

[13]   S.D. Khirade, and A.B. Patil, *Plant Disease Detection Using Image Processing.*, 2015, pp. 1-4.
[http://dx.doi.org/10.1109/ICCUBEA.2015.153]

CHAPTER 12

Analysis of Heat Sink Consideration Based on their Size

Ramanpreet Kaur[1,*], Wenhui Xiao[1], Ankush Sharma[1] and Liang Bo[1]

[1] *Guangzhou Rising Dragon Recreation Industrial Co. Ltd., Guangzhou, China*

Abstract: Light-emitting diodes have become very popular in general lighting due to their various advantages over other luminaires. Power consumption and LED size also play an important part in their selection over CFL's. LEDs are compact and brighter and can be modelled according to the requirement. Due to the compact size of high-power LEDs, the temperature or heat produced by the LEDs will be more, so some efficient cooling methods must be employed. This paper compares heat dissipation through heat sinks of different sizes. This paper claims that the temperature of high-power LEDs cannot be lowered after reaching a certain point, as a result, an increase in the size of the heat sink at that particular level is useless.

Keywords: Fin Thickness, Heat Dissipation, High-Power LED (Light Emitting Diode), Heat Sink, Thermal Behavior.

INTRODUCTION

The ordinary LEDs consume almost about 0.05W of power and the operating current is equal to 20mA. Whereas high-power LEDs can be of 1W, 2W or maybe up to 100W, and operating current is several hundred milliamperes. The main disadvantage concerned with the LEDs is low light lumen that only 10~20% power transforms into the light which gives rise to the expansion of high-power LEDs. A high-power LED requires that with the stable input voltage, the constant current and the operating temperature below 60°C, can the long-life expectancy of the LED lighting system be achieved. Using a high-power LED, the temperature can be increased to more than 60°C without any cooling system. If the temperature keeps on increasing LED can be burnt.

As we know heat sinks are mechanical devices generally used to dissipate the heat of electronic devices. In an electronic device, the heat increases due to the rise of

*Corresponding author **Ramanpreet Kaur**: Guangzhou Rising Dragon Recreation Industrial Co. Ltd., Guangzhou, China; E-mail: ramansweet5@gmail.com

Dharam Buddhi, Rajesh Singh and Anita Gehlot (Eds.)

temperature of electronic components with an applied voltage which leads to a rise in device temperature. Consequently, high power devices are vulnerable to high temperatures. The heat sink attached to the device helps to spread the heat away from the device to the environment which eventually reduces the device temperature [1, 2].

The heat sinks available in the market are of different shapes and sizes [3]. The construction of a heat sink depends upon the various parameters like a component to be cooled and the size of the device. There are two kinds of heat sinks called active heat sinks and passive heat sinks.

A. An active heat sink uses an electronic device's power supply to connect a fan to dissipate the heat to the environment from the heat sink to make it cooler. Active heat sinks are always employed with passive heat sinks [4].
B. Whereas, passive heat sinks have no mechanical parts hence, making them easy to install within electronics [5]. The generally used material for heat sink design is aluminium because it is economical compared to copper. But the thermal conductivity of copper is more than that of aluminium. Passive heat sinks can spread heat fast through its fins design which allows air to circulate, hence making it cool down faster [6]. Thus, the fins help to spread heat away from the heat sink without employing any active cooling device.

This paper compares the heat dissipated by passive heat sinks. This experiment comprised of two heat sinks of different sizes with no moving parts and the third heat sink with fins is used here without any fan. The first heat sink volume (L x W x H) is of size 5cm x 5cm x 5cm, the second fixed heat sink volume (L x W x H) of 15cm x 15cm x 15cm is used and the third heat sink with fins is of size (L x W) 7.8cm x 7cm with fin height of 4.2cm and fin thickness of 0.19cm is employed. Experiments have been performed over high-power LEDs and with two different wattages. The first LED chosen was around 5 watt and the other one is around 10 watts. The heat dissipation behaviour of all three heat sinks is analysed.

There are several parameters like thermal resistance, the component temperature that is cooled, mass, reliability, application area, size, acquisition and process cost, designated cooling process, and electronic performance that need to be taken into account while choosing a cooling method for high power LEDs [7].

EXPERIMENTAL SET-UP

Here, initially for the first-round measurements were performed for three demonstrations as shown in Fig. (**1**) power LED of 5watts is mounted on the heat sink of size 5cm x 5cm x 5cm which is used to manage the heat dissipated by the high-power LED. 2) the same high-power LED of 5watts is mounted on the other heat sink of different sizes 15cm x 15cm x 15cm. 3) the same high-power LED of

5watts is mounted on the other heat sink with fins of size (L x W) 7.8cm x 7cm with fin height of 4.2cm and fin thickness of 0.19cm.

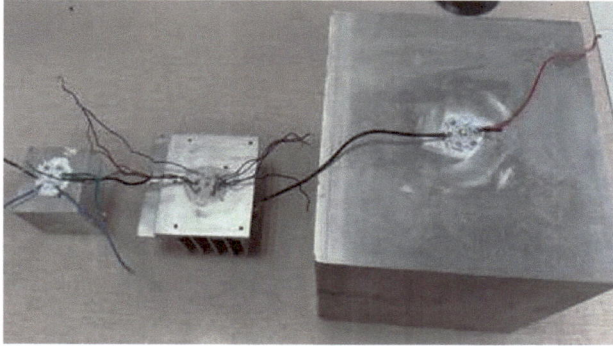

Fig. (1). Different sized heat sinks used in the experiment.

In the second round, measurements were performed for the next three demonstrations: 1) power LED of 10watts is mounted on the heat sink of size 5cm x 5cm x 5cm as shown in Fig. **(2)** . which is used to manage the heat dissipated by the high-power LED. 2) the same high-power LED of 10watts is mounted on the other heat sink of different sizes 15cm x 15cm x 15cm as shown in Fig. **(3)** the same high-power LED of 10watts is mounted on the other heat sink with fins of size (L x W) 7.8cm x 7cm with fin height of 4.2cm and fin thickness of 0.19cm as shown in Fig. **(4)** . The diameter of the LED chip is around 20mm.

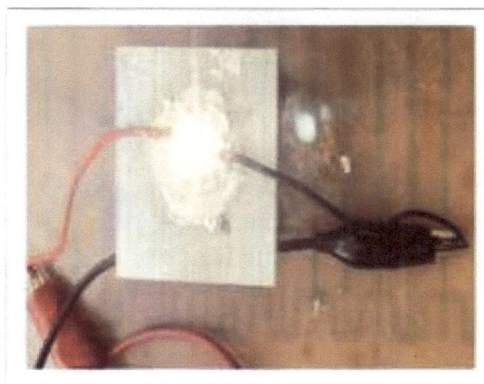

Fig. (2). Heat sink of size 5cm x 5cm.

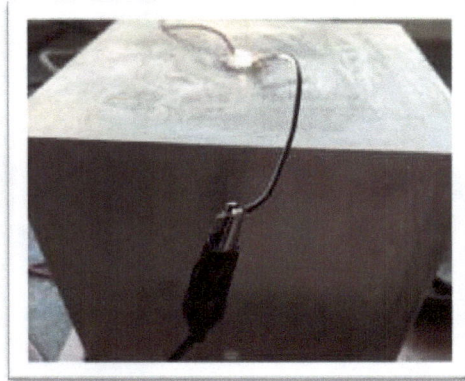

Fig. (3). Heat sink of size 15cm x 15cm.

Fig. (4). Fin Heat sink of size 7.8cm x 7cm.

A thermal conducting paste is used to join the high-power LED with the heat sink. Thermal paste (also known as a thermal compound, thermal grease, thermal interface material (TIM), thermal gel, heat paste, heat sink compound, heat sink paste or CPU grease) is said to be a thermally conductive compound but also an electrically insulating. The thermal paste is used as an interface material between heat sinks and high-power LEDs to reduce the thermal resistance. The main purpose of using the thermal paste is to remove air gaps or spaces from the interface region to increase heat transfer and dissipation. It helps to improve the cooling performance of the heat sink. Thermal paste is an example of thermal interface material.

Fig. (**5**). shows the schematic of a heat sink and the main components can be observed as: the LED and heat sink. A high-power LED that needs to be cooled is connected with the heat sink through thermal interface material. The heat sink attached to the device helps to spread the heat away from the device to the environment which eventually reduces the device temperature.

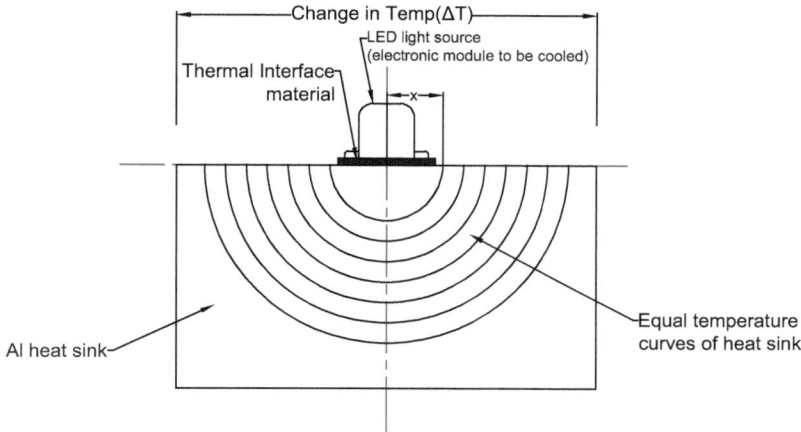

Fig. (5). Schematic of Heat sink.

The temperature curves show the heat distribution along the area of the aluminium heat sink. ΔT is the change in temperature observed over the area from x to x+Δx. Whereas, Δx is the change in distance/change in length of the heat sink.

Thermal Calculations for Power LED

It has been experimentally observed that the change in temperature of the heat sink is directly proportional to the power dissipation across the length of a heat sink, but it is inversely proportional to the surface area or heat transfer area.

To get ΔT (change in Temp./ temperature difference between a heat sink and ambient temperature) from x to x+Δx;

$$\Delta T = \frac{W.dx}{C.A} \tag{1}$$

where w = power dissipation,

c = Heat Conductivity constant = 210 &

$A = 4\pi x^2$, surface area of sphere, x = length in meters

Whereas, dx = change in length of the heat sink through which heat will dissipate)

The constant proportionality c is the *thermal conductivity* of the material. Thermal conductivity c is the measure of a material's ability to conduct heat. Thermal conductivity is defined as the rate of power dissipation through a unit length of heat sink per unit area temperature difference.

We are calculating for semi-circle, so;

$$\Delta T = \frac{W.dx}{210 \ x \ 2\pi x^2}$$

$$= \frac{W}{420\pi} \cdot \frac{dx}{x^2}$$

(2)

By integrating eq. (2);

$$\Delta T = \frac{W}{420\pi} \int_{X0}^{Xmax} \frac{dx}{x^2}$$

(3)

(where, ΔT = change in temperature, in degree celsius)

Case 1: Let's suppose; $W = 100$ watts; $X0 = 10^{-2}$ meters; $Xmax = 0.1$ meters;

$$= \frac{W}{420 \ x \ 3.14} \left[\frac{1}{x}\right]\Big|_{X_0}^{X_{max}}$$

(4)

$$= \frac{W}{420 \ x \ 3.14} \cdot \left(\frac{1}{10-2} - \frac{1}{10-1}\right)$$

$$= \frac{100}{420 \ x \ 3.14} \cdot 90 = 6.824$$

(5)

Case 2: Let's suppose; $W = 100$ watts; $X0 = 10^{-2}$ meters; $Xmax = 0.5$ meters

$$= \frac{W}{420 \text{ x } 3.14} \left[\frac{1}{x}\right]\Big|_{X_0}^{X_{max}}$$

$$= \frac{W}{420 \text{ x } 3.14} \cdot \left(\frac{1}{10-2} - \frac{1}{0.5}\right) \tag{6}$$

$$= \frac{100}{420 \text{ x } 3.14} \cdot 98 = 7.43$$

If W=100 watts, X0 = 10^{-2} (1mm); Xmax = 100meters; then

$$= \frac{W}{420 \text{ x } 3.14} \left[\frac{1}{x}\right]\Big|_{X_0}^{X_{max}}$$

$$= \frac{100}{420 \text{ x } 3.14} \cdot \left(\frac{1}{10-2} - \frac{1}{100}\right) \tag{7}$$

$$= \frac{100}{420 \text{ x } 3.14} \cdot 99.99$$

$$= 7.581$$

If W = 100, ΔT = 7.581

& if W = 200, ΔT = 15.61

From the equations stated above, it can be observed that the change in temperature doesn't change much with the increase in heat sink size. With the increase in heat sink size from 1cm to 100meters from eq. (5) & (7), the temperature change that occurs will be only 0.757°C. This temperature change is not a big change when size is increasing 10,000 folds, thus heat sink of bigger size cannot dissipate much heat as expected.

MEASUREMENTS AND RESULTS

In the first demonstration, LED operates at I_{LED} = 1.6A (optimal current value) with a voltage equal to 3.1V for white light. All measurements were performed with a thermal imaging camera as shown in Fig. (6) and to verify the results multiplex temperature tester was connected directly to the LED base. The ambient temperature recorded at the time the experiment was conducted was 26°C, the experiment then was performed for 9 hours. Thereafter, recorded the temperature value for all three heat sinks *i.e.*, small, large and fin heat sink as well.

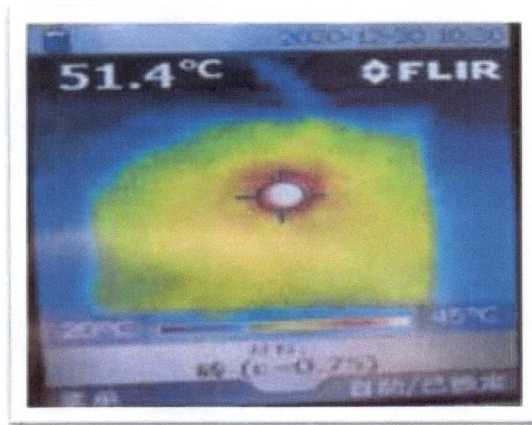

Fig. (6). Thermal imaging camera used for temp. measurement; showing a maximum temperature for a small heat sink with a 10watt power LED.

Temperature curves in Fig. **(7)** . shows the measurement results obtained for 5W LED used with the small heat sink of size 5cm x 5cm x 5cm, the large heat sink of size 15cm x 15cm x 15cm and heat sink with fins of size (L x W) 7.8cm x 7cm with fin height of 4.2cm and fin thickness of 0.19cm. The curve yellow(T1) shows the temperature rise for 15cm x 15cm heat sink whereas the orange curve(T2) represents the temperature rise for 5cm x 5cm heat sink and the curve green(T3) shows the temperature rise for a fin heat sink.

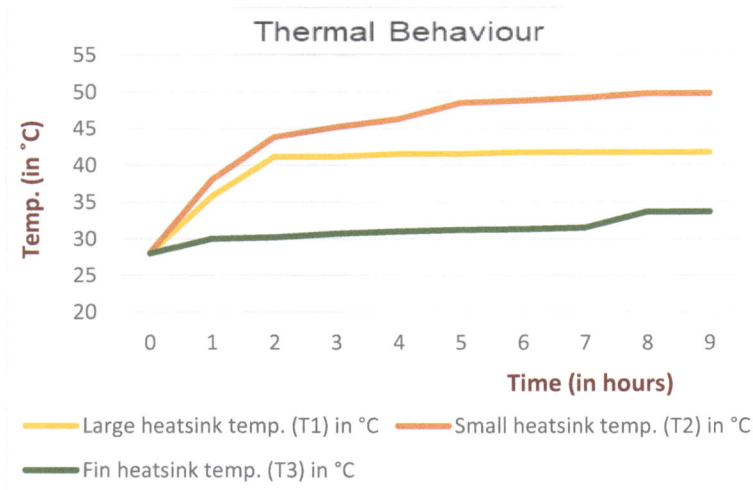

Fig. (7). Measurement results for 5W LED.

This experiment shows that initially the temperature is rising rapidly for the small heat sink but little lesser for the large heat sink and much lesser for the fin heat sink with the same LED power. The capacity to cool down the same power LED is a little higher around 7°C for the large heat sink and around 17°C for fin heat as compared to the small heat sink. After four hours maximum temperature for small heat sink stabilizes at 48.5°C, large heat sink at 41.5°C and fin heat sink at 31.2°C.

In the second demonstration, LED is made to operate at $I_{LED} = 3A$ (optimal current value) with a voltage equal to 3.2V for white light. Thus, it generates power of around 10watts which is the high power for this LED to withstand.

Temperature curves in Fig. (**8**) . shows the measurement results obtained for 10W LED used with the small heat sink of size 5cm x 5cm, the large heat sink of size 15cm x 15cm and heat sink with fins of size (L x W) 7.8cm x 7cm with fin height of 4.2cm and fin thickness of 0.19cm similar to the first demonstration. The curve yellow(T1) shows the temperature rise for 15cm x 15cm heat sink whereas the orange curve(T2) represents the temperature rise for 5cm x 5cm heat sink and the curve green(T3) shows the temperature rise for a fin heat sink.

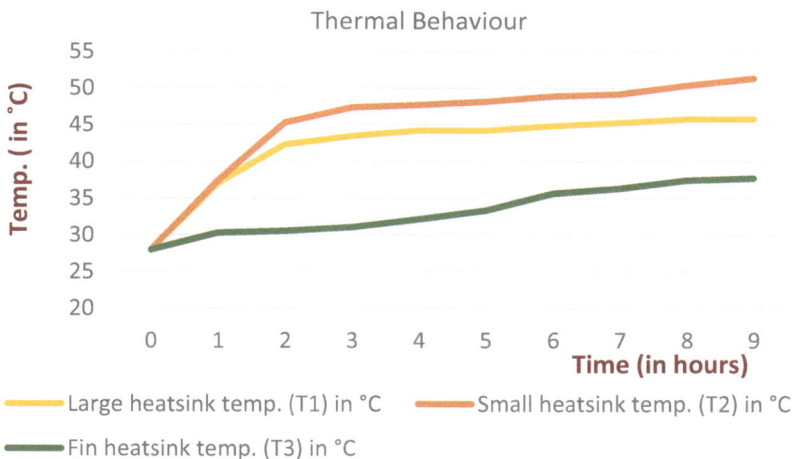

Fig. (8). Measurement results for 10W LED.

Here, in the second experiment, it shows that again temperature rises sharply for the small heat sink but little lesser for the large heat sink and a little rise for the fin heat sink with the same 10watt LED power. The capacity to cool down the same power LED is a little higher around 6°C for the large heat sink and around 14°C for fin heat. After four hours, the maximum temperature for the small heat sink stabilizes at 48.2°C, the large heat sink at 44.2°C and the fin heat sink at 33.4°C.

RESULTS AND DISCUSSION

If we increase the power, temperature increases. But if we increase the size of the heat sink; it actually, triples the existing heat sink size then, temperature decreases. But, this decrease in temperature is small as compared to the increase in heat sink size. In the experiment, we can notice the temperature of a small heat sink is 49.8°C when we increase the size of the heat sink to three folds then the temperature comes down to 41.7°C which means we need to increase the size to thrice to get rid of 7°C rise in temperature. However, increasing the size is not a solution. The size of the heat sink is again associated with its weight. The small heat sink is of size 5cm x 5cm x 5cm square cube of weight around 700grams whereas the large heat sink is of size 15cm x 15cm x 15cm square cube weighs around 10kgs which is huge and not feasible to use with light products. Hence, we need to think about some more practical solutions to get rid of large size and overweighing heat sinks to avoid heat dissipation problems. Therefore, in this experiment, we used the fin heat sink which shows better results.

CONCLUSION

The thermal behaviour of a device plays an important role in the design of an electronic device. This is because the high-power electronic circuits generate a high amount of heat thus, require some cooling devices. Two types of heat sinks are available in the market one is active and another is passive. Active devices are better than passive devices but are very expensive and incorporate complex arrangements. Due to this, active devices are not generally used. The passive heat sinks are normally less expensive, take less instalment time, and provides higher reliability.

In this experiment, it is found that the size of the heat sink does not contribute much towards solving the problem of heat dissipation of high power LEDs. However, this problem can be resolved by changing the design of the heat sink *i.e.*, by using a fin heat sink design rather than a heat sink cube. It has been observed that the temperature of the fin heat sink is 16.1°C and 13.6 °C lower than the small heat sink for 5watts and 10watts LED respectively, whereas approximately 8°C lower than large heat sink for both 5watts and 10watts LED.

CONSENT FOR PUBLICATION

Not applicable.

CONFLICT OF INTEREST

The authors declare no conflict of interest, financial or otherwise.

ACKNOWLEDGEMENT

Declared none.

REFERENCES

[1]   N.B. Drăghici, P. Svasta, and N.D. Codreanu, "Comparison of synthetic air jet vs. passive heat sinks for cooling COB LED modules,", *2017 40th International Spring Seminar on Electronics Technology (ISSE), Sofia,* 2017 pp. 1-4.
        [http://dx.doi.org/10.1109/ISSE.2017.8000897]

[2]   N. Bădălan, and P. Svasta, "Fan *vs.*passive heat sink with heat pipe in cooling of high-power LED", *2017 IEEE 23rd International Symposium for Design and Technology in Electronic Packaging (SIITME), Constanta ,* 2017, pp. 296-299.
        [http://dx.doi.org/10.1109/SIITME.2017.8259911]

[3]   H. Hsu, S. Wu, J. Li, J. Su, J. Huang, and S. Fu, "Thermal design for high power arrayed LED heat-dissipating system", *Assembly and Circuits Technology Conference (IMPACT),* 2013, pp. 222-225 Taipei
        [http://dx.doi.org/10.1109/IMPACT.2013.6706685]

[4]   M. Saini, and R.L. Webb, "Heat rejection limits of air-cooled plane fin heat sinks for computer cooling," ITherm 2002", *Eighth Intersociety Conference on Thermal and Thermomechanical Phenomena in Electronic Systems (Cat. No. 02CH37258), San Diego, CA, USA,* 2002, pp. 1-8.
        [http://dx.doi.org/10.1109/ITHERM.2002.1012431]

[5]   R. Singh, A. Jalilvand, K. Goto, K. Mashiko, Y. Saito, and M. Mochizuki, "Direct impingement cooling of LED by Piezo fan", *2014 International Conference on Electronics Packaging (ICEP),* 2014, pp. 1-5 Toyama.
        [http://dx.doi.org/10.1109/ICEP.2014.6826650]

[6]   Amer Al-damook, and Fahad Saleh Alkasmoul, "Heat transfer and air flow characteristics enhancement of compact plate-pin fins heat sinks–are view", received 5May 2017, Accepted 20 September 2017, Available online 11 June 2018.
        [http://dx.doi.org/10.1016/j.jppr.2018.05.003]

[7]   B. Matthew, "Optimization of The Geometry of a Heat Sink", Advisor: Hossein Haj-Hariri University of Virginia, Charlottesville, ", *VA ,* p. 22904, .

CHAPTER 13

A Comparative Study on Various Machine Learning Techniques for Brain Tumor Detection Using MRI

Samia Mushtaq[1,*], Apash Roy[1] and Tawseef Ahmed Teli[2]

[1] School of Computer Applications, Lovely Professional University, Punjab, India

[2] Department of Computer Applications, Amar Singh College, J&K, India

Abstract: The brain is the central controlling system in the human body. If the structure of the brain changes due to the enlargement of the brain cells, it is diagnosed as a brain tumor, which can lead to fatality. The techniques such as medical imaging provide evidence of whether a patient has a brain tumor or not. This paper discusses various machine learning techniques for brain tumor detection using MRI and provides a performance analysis of such methods based on the state-of-the-art. A comparison of ANN and CNN-based models have also been given after implementing the techniques in Tensorflow to understand the potential benefits of using deep learning-based techniques.

Keywords: ANN, Brain Tumor, CNN, Deep Learning, Machine learning, MRI.

INTRODUCTION

Images obtained from MRI-Magnetic resonance imaging are utilized to obtain the complete information of internal brain tissues. During the process of brain tumor [1] investigation, detection of the tumor core location is the key task to get the size and shape of the brain tumor. Segmentation techniques are vital in the detection process of tumors in the brain. To obtain complete knowledge of brain tissues like gray matter, cerebral spinal fluid, white matter, *etc.*, segmentation methods are applied. Segmentation can be done in two ways either automatically or manually. Manual segmentation takes more time and requires expert knowledge and experience but brings down computational efficiency while Automatic segmentation works with histograms based on pixel intensities. The assessment of lesions , using MRI image segmentation, in the brain of a patient

---

[*] **Corresponding author Samia Mushtaq:** School of Computer Applications, Lovely Professional University, Punjab, India; E-mail: samiamushtaq808@gmail.com

Dharam Buddhi, Rajesh Singh and Anita Gehlot (Eds.)

helps in providing significant data that in turn could help provide insights into the evaluation and detection of various brain pathologies which could help in the proper planning and monitoring of treatment. The segmentation incorporated with human intervention can prove very beneficial in terms of accuracy but it is a very cumbersome and expensive task that is also a time-consuming task. So, an efficient and automated segmentation-based detection technique is the need of the hour that may provide outcomes that could be significantly used for the quantitative evaluation of tumors.

In neurological sciences, MRI segmentation of the brain is the key basic step and has applications in functional imaging, operational planning, and quantitative analysis [2]. Although the brain's internal structure is accurately described by MRI-imaging, the segmentation of medical images is a tedious task as a result of various issues like noise, weakly defined boundaries, poor or weak spatial resolution, inhomogeneity, low contrast, partial volume, variations in the shape of objects, acquisition artifacts present in the *fetch*ed data and deficiency of anatomy models that completely give details about the deformations of the internal structures [2 - 4]. These issues can be solved by the implementation of some simple techniques which are presented in [4 - 7]. As a result of issues like tumor heterogeneity, instability in shape, size, tumor recurrence it is very complicated to formulate the best efficacious segmentation technique [3].

Hybrid techniques are utilized to obtain gainful and reliable results in the segmentation process by integrating more than one method, by utilizing their pros and rejecting their cons. Parveen [8] proposed a hybrid technique in which both SVM and FCM techniques are merged to obtain the correct classification of the brain tumor region. For segmentation of affected brain core region, FCM is utilized and in the process of classification, features are extracted using Gray level run length matrix which is then utilized by SVM for the process of classification. Region-based ACMs integrated with Machine learning are applied for fruitful segmentation results by utilizing their pros and rejecting their cons. Intensity homogeneities are dealt with by machine learning while issues of misclassified and weakly defined boundaries are handled by region-based ACMs.

Many researchers have used supervised methods for brain tumor segmentation technique that includes; Support Vector Machines (SVMs), Random Forests and K-Means, *etc*. All these methods are known to use hand-crafted feature extraction techniques which are constrained computationally as compare to any deep learning-based methods. Many researchers have used different methods of segmentation for Brain tumor detection. In the following sections, an overview of different techniques of segmentation with current literature has been discussed.

LITERATURE REVIEW

In this paper [9], a tumor classification technique based on wavelet features and DNN is used. The authors have used OFPA which stands for oppositional flower pollination algorithm, for the feature selection and Deep Neural Networks have been used to classify the tumor. Finally, a possibilistic fuzzy c-means clustering (PFCM) based technique has been used to extract the tumor-affected area. In the study, the accuracy rate of 92%is achieved with a specificity of 91% and sensitivity of 86%. Nazir *et al.* [10] proposed a segmentation method for brain tumors is used that works in various stages. In the first stage, feature extraction from input brain images is done using the Discrete Wavelet Transform technique. Decomposition of an image takes place which results in multiple wavelet sub-bands, then the bands with the highest energy are selected and divided into multiple blocks. In the second stage, from every single block high variance features are chosen using discrete cosine transformation technique and for classification transferred to neural networks. In the last stage, segmentation is achieved by k-means clustering. Shivhare *et al.* [11] provide a complete model which is based on two constituents.

- For partitioning input images into various clustered images Parameter –free clustering algorithm is used.
- Hole filling operations proceeded by morphological dilation are used to get to the core region of the tumor.

Sazzad *et al.* [12] use gray-scale images obtained from the MRI technique. The filter operation method is incorporated to eliminate undesirable noise and this results in better segmentation. OTSU segmentation technique based upon thrush hold is used to replace color segmentation. Feature information provided by pathologists is finally used for the identification of the tumor core region.

Khalil *et al.* [13] proposed the segmentation method applied for brain tumor identification involves three levels. Pre-processing for the quality improvement of the image by eliminating undesirable noise like discrepancies and inhomogeneity from the image. The technique based on two-step clustering is applied to obtain the exact basic contour points for the level set segmentation method. Dragonfly algorithm is integrated with k-means in the process of clustering. Level set segmentation helps in the speedy extraction of tumor edges

In this paper [14], the core area of the tumor is detected using four major modules. For pre-processing FNLM technique is applied for enhancement. For segmentation, the OTSU technique is applied. For extraction of various features like HOG, GEO and LBP are incorporated into one feature vector which is further used for prediction. In this study [15], a DL-based technique is used for segmentation in MR images. The patch-based technique is incorporated in the convolutional neural network. Output label is predicted while considering both contextual and local information. A two-phase training mechanism is used to solve the problem of data imbalance. Preprocessing stage involves image normalization and usage of the N4ITK technique for the correction of the bias field. After the processing stage, To enhance segmentation results small false positives are removed and Dilation proceeded by erosion is done in the same stage. The authors in [16] use tool-reading. The method includes three modules: MR images are used as input in preprocessing stage where they are converted into greyscale images followed by removal of noise using median cleanout, grey degree coincidence matrix help feature extraction, and lastly, label image is obtained using watershed segmentation. In this study [17], the researchers use FCM (fuzzy c-means) with super-resolution is used. Pretrained (CNN) convolutional neural network-based architecture – SqueezeNet is applied for extraction of features while (ELM) extreme learning machine is applied for classification. In this paper [18], the filtering technique-Triangular fuzzy Median(TFMED) is adopted for enhancing image quality. For the process of segmentation Fuzzy set method is used which is an unsupervised clustering technique. Extraction of Gobar features and similar texture features is done and for the process of classification, they act as an input to ELM-Extreme learning machine and RELM_LOO methods. In this study [19, 20], the techniques based on various wavelet transforms and SVM are proposed. In the preprocessing module removal of noise takes place using wavelet reforms. OTSU technique is applied for feature extraction in segmentation and CWT and methods are also applied to avoid dropping of edges. Finally, SVM techniques are applied for the process of classification and prediction. In this study [21], for the improvement of brain image NSCT technique is used and then an improvised image is utilized for the feature extraction process. For classification, ANFIS is used. Morphological functions are applied for classification. In this research work [22], the authors have described that the whole architecture works in three main modules. In the first module, Segmentation is done in two phases, superpixel zoning technique is applied in the first step followed by discriminative clustering that results in better clarity and accuracy of MR image of the brain. In the second module, Features are extracted using the HWT-Haar Wavelet transformation technique. In the third module, ADBRF (AdaBoost-ADB algorithm merged with RF is applied for classification. In this paper [23], a methodology is used which has three broad levels. At level first, preprocessing is done by applying both local smoothening and non-local means to eliminate noise as much as possible. At the second level GLRLM-gray level run length matrix, histogram-based method, and GLCM are applied for extraction of features, ELM (extreme learning machine)-LRF- Local receptive field techniques are used for the classification purpose. At the third level, a watershed segmentation algorithm is used for segmentation. The researchers in [24], use improved forms of Fuzzy-C means and Template-base--K-Means algorithms. The Template-based-K-Means technique is applied for the initialization of segmentation to select a perfect template that depends on gray-level image intensity. The improvised Fuzzy-C means clustering technique is applied for detection of tumor region, membership function is updated based on various attributes of tumor-image like energy, contract, entropy, correlation, and homogeneity. In this study [25], the MMR-Mathematical Morphological Reconstruction segmentation technique is used. Median Filter is applied for preprocessing of Images; MMR technique is utilized for proper segmentation. Textural features and statistical features are extracted and PCA is applied for selecting the lesser number of features. Finally, SVM is used for classification using the GRB kernel.

METHODS OF SEGMENTATION

Conventional Methods

Thresholding

For image segmentation, the thresholding method is considered the simplest and fastest method. In the thresholding technique, a scalar image is converted into a binary image, and based on the pixel intensity, a threshold value is determined. Comparison of the threshold value and pixel intensity value takes place. Value 1 is allotted to the pixels whose intensity value is equal or more than the value of threshold while value 0 is allotted with less value of intensities, thus partitioning the white pixels-foreground region and the black pixels-background region, respectively. In the situations, wherein the image region types are greater than two and equal intensity values are not shared by the objects of the image, then for the segmentation of an image into multiple regions manifold threshold values are utilized. For the mitigation of these scenarios, various dynamic and local thresh holding techniques are suggested in the literature [26].

Region Growing

In the area of image segmentation, Region growing is the prominent and foundational method utilized for segmenting homogeneous regions which have equal pixel intensity values. The region growing method can be easily applied to objects with variable shapes because this technique does not demand prior information about the shapes. In the implementation of this method, each pixel needs to be present in any region, connections should be there among the pixels in a particular region, certain similarity conditions must be satisfied, bifurcation of regions must be there and two distinct regions cannot have the identical characteristics [27]. The seed point is selected either automatically or manually and every region of concern is initialized with the same. One crucial factor due to which correctness of segmentation is badly affected is the partial volume effect, due to this effect intensity differentiation at the boundary of two region tissue types is diminished, thereupon numerous tissue types are represented by creating a voxel [28]. This issue can be resolved by introducing MRGM [29] utilizing gradient information and details for recognizing boundaries. In [30], a comparative survey of MRGM and region growing is presented and demonstrates that MRGM provides preferable results of brain tumor segmentation over region growing method in 3D T1-MRI images.

Supervised Methods

In the supervised learning method, the researchers establish an objective function that defines the data model and then fine-tunes the attributes of the classifier by the application of a set of known sample categories. The objective function helps to predict the value of the output variable from the set of input values. These methods include the training phase followed by the testing phase. Correctly labeled data acquired in the phase of training is utilized in the testing phase for the decision of classes of unlabeled data [31]. A prototype or model is built in the training phase in which extracted features from data points are mapped to classes or labels and for the determination of classes in the testing phase same model is utilized for the unlabeled data set. Human interference is needed in the training phase results in result variability demonstration. Classification precision is better in supervised methods as compared to unsupervised methods.

k-Nearest Neighbor

K-NN algorithm collates the fresh unlabeled data occurrences with a group of data samples that are labeled in a training data set. This algorithm is supervised learning based on memory. Two levels are followed in the K-NN classification. The closest neighbors for unlabeled data instances are recognized in the first level then those closest neighbors are utilized for the determination of label or class for data instances in the second level. The training data set, Class labels and the feature vector of the samples are stored in the training step. For the prediction of labels to the testing data points, gaps or similarities between the training data instances and fresh unlabeled testing data points are calculated. Using the training data set, k-Nearest neighbors for the unlabeled data points are identified and by utilizing their labels unknown data or records are labeled using majority voting.

The value of variable k hinges on the total number of cases and the number of features extracted [32]. The distance metric utilized hinges on the problem domain [33]. Steenwijk *et al.* [34] perceive a notable advancement in segmenting white matter lesions.

Support Vector Machines

SVM transforms data from input space to multidimensional feature space and then searches for a separating hyper-plane using a kernel function. Various kernel functions that are commonly used in SVM are linear polynomial, PUK, and radial basis functions. For the better classification of data in SVM, the image is divided into two categories by locating a hyperplane [35]. For example, the hyperplane having the highest distance or gap from the nearby data point from every side is taken [36].Classification of data points in SVM hinges on the closeness with segregating hyperplane taking extended time for computations for solving quadratic and linear problems so created. Wu *et al.* [37] laid more focus on dividing the image into multiple superpixels instead of voxels processing. An effort has been put in for the reduction of errors in classification utilizing SVM as portrayed in [38]. By utilizing second and first-order statistical features extraction tumor grades like II, III and IV are classified. Whereas, superpixel-based perspective or manual approach is included in ROI segmentation. For brain tumor grading using MRI images, an intensified version is portrayed [39].

Random Forests

It is a classification technique, which is supervised, straightforward, and uses an ensemble learning strategy while working effectively on big datasets. Without deleting a variable it operates on enormous input variables and estimates key features for the process of classification. Random Forests are comparatively sturdy to noise and outliers.

Artificial Neural Networks

To predict the results of disease diagnosis a machine learning technique known as Artificial Neural Network can be used. ANN is one of the most powerful tools used in the analysis of complex clinical datasets. ANNs use known data in the training process and then identify the complex relation between x and y variables, acting as input and out variables respectively, followed by the prediction of output values from a given set of input values. In animals, the nervous system is the structure of interconnected neurons due to which animals react to the surrounding environment. Artificial neural networks react to the environment and attain the optimal state by emulating the deportment of neurons. To get pertinent output at the last layer of artificial neurons, input signals are processed and then transmitted through many in-between interconnected hidden layers. The complications of this method increase with the increasing network size and thus become more time-consuming. In addition to this, a large count of images is required for effective training. Nevertheless, it has a pragmatic advantage of modeling any random non-trivial distribution. At present various kinds of neural networks are used [40, 41].

Unsupervised Methods

Enormous data is produced by the medical industry, but some data doesn't possess corresponding labels. As a result of which the use of supervised learning techniques is not suitable. In such cases, unsupervised learning can be useful to discover the relation between hidden structures of unlabeled data. It is suited for more complex scenarios where the segmentation is done based on the intensity, gradient, or texture. There is n number of segments created by applying the algorithms on features of images.

Clustering

The clustering techniques can be used for searching for useful patterns in unorganized datasets. Clustering group pixels of an image into n groups called clusters wherein each group is disjoint. There are two variants of clustering namely hard clustering and soft clustering. Hard clustering like K-clustering results in the plot of a pixel either in one of the clusters or none. Soft clustering like Fuzzy C-Means clustering results in a pixel fitting in more than one cluster, resulting in the overlapping of the clusters.

k-means Clustering

This technique is a vital method in methods that are based on pixels [42]. There is a k means defined as x_1, x_1 ... X_k for k clusters of an image. After each iteration, observations are formed and means are calculated again and again until there is no significant change in the new values.

Fuzzy C-Means Clustering

Fuzzy-based techniques are famous for representing uncertain data [43]. FCM clustering algorithm was proposed by Bezdek in 1993 [44]. This technique allocates membership values by calculating the Euclidean distance from the center to the cluster.

COMPARISON OF DIFFERENT TECHNIQUES

Table **1** given below gives the comparison in terms of accuracy for different techniques using different datasets based on the literature review done.

Table 1. Performance Comparison.

S no	Authors	Technique	Dataset	Accuracy
01	[9]	Wavelet Texture Features	BRATS 2013	92%
02	[11]	Parameter-Free Clustering		75.06%
03	[15]	MR Images Using DL	BRATS 2013	91.0%
04	[18]	extreme learning	BRATS 2015	96.5% (ELM) 87.5% (RELM-LOO)
05	[20]	Wavelet Transforms and Support Vector Machines		92.0%
06	[21]	ANFIS Classification		96.4%
07	[25]	Mathematical Morphological Reconstruction		92.0%
08	[41]	(KMANN)	RS	96.6%

EXPERIMENT AND RESULTS

Artificial Neural Network and Convolutional Neural Network were implemented using Tensorflow to ascertain the accuracy comparison and get an insight into the possible improvements using Deep Learning models. The dataset used for this experiment was taken from BRATS which included a total of 2020 images with 1060 tumorous images and 960 non-tumorous images.

After testing the models, the accuracy of ANN was calculated as 68.76% while the accuracy of CNN was 88.70%. The confusion matrix for ANN is given in Table **2**.

Table 2. Confusion matrix ANN.

-	Tumor	Tumorless
Tumor	TN=173	FP=41
Tumorless	FN=88	TP=111

The F1 score for CNN is given in Table **3**.

Table 3. F1 score CNN.

	F1 score
Testing Data	0.88
Validation Data	0.91

The accuracy of CNN may not be what one might expect but it was implemented in the most basic way without considering pre-processing or balancing of the dataset. It is quite evident from the results that CNN performs way better than ANN and looks very promising in improving the accuracy profoundly.

CONCLUSION

Machine learning techniques have been used by researchers all around the world to have an automated system for Brain Tumor Detection. In this research work, some machine learning techniques were deliberated upon and a comparative analysis was put forth on the current literature. To understand the benefits of using Deep Learning approaches like Convolutional Neural Networks, a comparison of performances between ANN and CNN was given. The performance difference is a strong endorsement of how well CNN performs even on an imbalanced dataset without any proper consideration for pre-processing. Future work should focus on using Deep Learning methods, standalone or hybrid, to achieve better results in brain tumor detection using MRI.

CONSENT FOR PUBLICATION

Not applicable.

CONFLICT OF INTEREST

The authors declare no conflict of interest, financial or otherwise.

ACKNOWLEDGEMENT

Declare none.

REFERENCES

[1]   P.B. Kanade, and P. Gumaste, *Brain tumor detection using MRI images.* vol. Vol. 3. Brain, 2015.

[2]   A. Ben Rabeh, F. Benzarti, and H. Amiri, "Segmentation of brain MRI using active contour model", *Int. J. Imaging Syst. Technol.,* vol. 27, no. 1, pp. 3-11, 2017.
      [http://dx.doi.org/10.1002/ima.22205]

[3]   K. Kamnitsas, C. Ledig, V.F.J. Newcombe, J.P. Simpson, A.D. Kane, D.K. Menon, D. Rueckert, and B. Glocker, "Efficient multi-scale 3D CNN with fully connected CRF for accurate brain lesion

segmentation", *Med. Image Anal.,* vol. 36, pp. 61-78, 2017.
[http://dx.doi.org/10.1016/j.media.2016.10.004] [PMID: 27865153]

[4]   K.B. Vaishnave, and K. Amshakala, "An automated MRI brain image segmentation and tumor detection using SOM-clustering and proximal support vector machine classifier", *IEEE Int.Conf. Engineering and Technology (ICETECH),* 2015
[http://dx.doi.org/10.1109/ICETECH.2015.7275030]

[5]   N.J. Tustison, B.B. Avants, P.A. Cook, Y. Zheng, A. Egan, P.A. Yushkevich, and J.C. Gee, "N4ITK: improved N3 bias correction", *IEEE Trans. Med. Imaging,* vol. 29, no. 6, pp. 1310-1320, 2010.
[http://dx.doi.org/10.1109/TMI.2010.2046908] [PMID: 20378467]

[6]   K. Siddiqi, Y.B. Lauzière, A. Tannenbaum, and S.W. Zucker, "Area and length minimizing flows for shape segmentation", *IEEE Trans. Image Process.,* vol. 7, no. 3, pp. 433-443, 1998.
[http://dx.doi.org/10.1109/83.661193] [PMID: 18276263]

[7]   J.G. Sled, A.P. Zijdenbos, and A.C. Evans, "A nonparametric method for automatic correction of intensity nonuniformity in MRI data", *IEEE Trans. Med. Imaging,* vol. 17, no. 1, pp. 87-97, 1998.
[http://dx.doi.org/10.1109/42.668698] [PMID: 9617910]

[8]   A. Parveen Singh, "Detection of brain tumor in MRI images, using combination of fuzzy C-means and SVM", *2nd Int. Conf. Signal Processing and Integrated Networks (SPIN),* 2015 pp. 98-102
[http://dx.doi.org/10.1109/SPIN.2015.7095308]

[9]   S. Preethi, and P. Aishwarya, "Combining Wavelet Texture Features and Deep Neural Network for Tumor Detection and Segmentation over MRI", *J. Intell. Syst.,* vol. 28, no. 4, pp. 571-588, 2017.
[http://dx.doi.org/10.1515/jisys-2017-0090]

[10]   M. Nazir, M.A. Khan, T. Saba, and A. Rehman, "Brain tumor detection from MRI images using multi-level wavelets", *2019 Int. Conf. Comput. Inf. Sci. ICCIS 2019,* 2019 pp. 1-5
[http://dx.doi.org/10.1109/ICCISci.2019.8716413]

[11]   S.N. Shivhare, S. Sharma, and N. Singh, *An efficient brain tumor detection and segmentation in MRI using parameter-free clustering.* vol. Vol. 748. Springer Singapore, 2019.
[http://dx.doi.org/10.1007/978-981-13-0923-6_42]

[12]   T.M. Shahriar Sazzad, K.M. Tanzibul Ahmmed, M.U. Hoque, and M. Rahman, "Development of Automated Brain Tumor Identification Using MRI Images", *2nd Int. Conf. Electr. Comput. Commun. Eng. ECCE 2019,* 2019 pp. 1-4
[http://dx.doi.org/10.1109/ECACE.2019.8679240]

[13]   H.A. Khalil, S. Darwish, Y.M. Ibrahim, and O.F. Hassan, "3D-MRI brain tumor detection model using modified version of level set segmentation based on dragonfly algorithm", *Symmetry (Basel),* vol. 12, no. 8, 2020.
[http://dx.doi.org/10.3390/sym12081256]

[14]   M. Sharif, J. Amin, M.W. Nisar, M.A. Anjum, N. Muhammad, and S. Ali Shad, "A unified patch based method for brain tumor detection using features fusion", *Cogn. Syst. Res.,* vol. 59, pp. 273-286, 2020.
[http://dx.doi.org/10.1016/j.cogsys.2019.10.001]

[15]   S. Sajid, S. Hussain, and A. Sarwar, "Brain Tumor Detection and Segmentation in MR Images Using Deep Learning", *Arab. J. Sci. Eng.,* vol. 44, no. 11, pp. 9249-9261, 2019.
[http://dx.doi.org/10.1007/s13369-019-03967-8]

[16]   A.N. Rakhonde, and P. Chippalkatti, *Mri Image Based Brain Tumor,* no. 3672, pp. 1-10, 2020.

[17]   F. Özyurt, E. Sert, and D. Avcı, "An expert system for brain tumor detection: Fuzzy C-means with super resolution and convolutional neural network with extreme learning machine", *Med. Hypotheses,* vol. 134, no. October, p. 109433, 2020.
[http://dx.doi.org/10.1016/j.mehy.2019.109433] [PMID: 31634769]

[18]   M. Sharif, J. Amin, M. Raza, M.A. Anjum, H. Afzal, and S.A. Shad, "Brain tumor detection based on

extreme learning", *Neural Comput. Appl.*, vol. 32, no. 20, pp. 15975-15987, 2020.
[http://dx.doi.org/10.1007/s00521-019-04679-8]

[19]   S. Reshmi, R.D. Salagar, and S.S. Veni, "a Survey on Brain Tumor Detection and Segmentation From Mri Images", *Int. J. Comput. Sci. Eng.*, vol. 6, no. 10, pp. 776-778, 2018.
[http://dx.doi.org/10.26438/ijcse/v6i10.776778]

[20]   M. Gurbina, M. Lascu, and D. Lascu, "Tumor detection and classification of MRI brain image using different wavelet transforms and support vector machines", *2019 42nd Int. Conf. Telecommun. Signal Process.*, 2019 pp. 505-508
[http://dx.doi.org/10.1109/TSP.2019.8769040]

[21]   A. Selvapandian, and K. Manivannan, "Fusion based Glioma brain tumor detection and segmentation using ANFIS classification", *Comput. Methods Programs Biomed.*, vol. 166, pp. 33-38, 2018.
[http://dx.doi.org/10.1016/j.cmpb.2018.09.006] [PMID: 30415716]

[22]   A. Panda, T.K. Mishra, V.G. Phaniharam, and M.R.I. Automated Brain Tumor Detection Using Discriminative Clustering Based, *Segmentation*. vol. Vol. 851. Springer Singapore, 2019.

[23]   A. Ari, and D. Hanbay, "Deep learning based brain tumor classification and detection system", *Turk. J. Electr. Eng. Comput. Sci.*, vol. 26, no. 5, pp. 2275-2286, 2018.
[http://dx.doi.org/10.3906/elk-1801-8]

[24]   M.S. Alam, "Automatic human brain tumor detection in mri image using template-based k means and improved fuzzy c means clustering algorithm", *Big Data Cogn. Comput.*, vol. 3, no. 2, pp. 1-18, 2019.
[http://dx.doi.org/10.3390/bdcc3020027]

[25]   B. Devkota, A. Alsadoon, P.W.C. Prasad, A.K. Singh, and A. Elchouemi, *Image Segmentation for Early Stage Brain Tumor Detection using Mathematical Morphological Reconstruction*, 2018.
[http://dx.doi.org/10.1016/j.procs.2017.12.017]

[26]   D.Y. Kim, and J.W. Park, "Connectivity-based local adaptive thresholding for carotid artery segmentation using MRA images", *Image Vis. Comput.*, vol. 23, no. 14, pp. 1277-1287, 2005.
[http://dx.doi.org/10.1016/j.imavis.2005.09.005]

[27]   R. Adams, and L. Bischof, "Seeded region growing", *IEEE Trans. Pattern Anal. Mach. Intell.*, vol. 16, no. 6, pp. 641-647, 1994.
[http://dx.doi.org/10.1109/34.295913]

[28]   M. Sato, S. Lakare, M. Wan, A. Kaufman, and M. Nakajima, "A gradient magnitude based region growing algorithm for accurate segmentation", *Proc. Int. Conf. Image Processing.*, vol. 3, 2000 pp. 448-51
[http://dx.doi.org/10.1109/ICIP.2000.899432]

[29]   S. Lakare, and A. Kaufman, *3D Segmentation Techniques for Medical Volumes.* Center for Visual Computing, Department of Computer Science, State University of New York, 2000, pp. 59-68.

[30]   N.J. Salman, "Modified technique for volumetric brain tumor measurements", *J. Biomed. Sci. Eng.*, vol. 02, pp. 16-19, 2009.
[http://dx.doi.org/10.4236/jbise.2009.21003]

[31]   N. Gordillo, E. Montseny, and P. Sobrevilla, "State of the art survey on MRI brain tumor segmentation", *Magn. Reson. Imaging*, vol. 31, no. 8, pp. 1426-1438, 2013.
[http://dx.doi.org/10.1016/j.mri.2013.05.002] [PMID: 23790354]

[32]   N.E.A. Khalid, S. Ibrahim, and P. Haniff, "MRI brain abnormalities segmentation using k-nearest neighbors (k-NN)", *Int. J. Comput. Sci. Eng.*, vol. 3, no. 2, pp. 980-990, 2011.

[33]   S.K. Warfield, M. Kaus, F.A. Jolesz, and R. Kikinis, "Adaptive, template moderated, spatially varying statistical classification", *Med. Image Anal.*, vol. 4, no. 1, pp. 43-55, 2000.
[http://dx.doi.org/10.1016/S1361-8415(00)00003-7] [PMID: 10972320]

[34]   M.D. Steenwijk, P.J. Pouwels, M. Daams, J.W. van Dalen, M.W. Caan, E. Richard, F. Barkhof, and H.

Vrenken, "Accurate white matter lesion segmentation by k nearest neighbor classification with tissue type priors (kNN-TTPs)", *Neuroimage Clin.,* vol. 3, pp. 462-469, 2013.
[http://dx.doi.org/10.1016/j.nicl.2013.10.003] [PMID: 24273728]

[35]   T.S. Kumar, K. Rashmi, S. Ramadoss, L.K. Sandhya, and T.J. Sangeetha, "Brain tumor detection using SVM classifier", *3rd Int. Conf. Sensing, Signal Processing and Security (ICSSS),* 2017, pp. 318-23

[36]   R. Ayachi, and N. Ben Amor, "Brain tumor segmentation using support vector machines", *European Conf. Symbolic and Quantitative Approaches to Reasoning and Uncertainty,* 2009, pp. 736-47
[http://dx.doi.org/10.1007/978-3-642-02906-6_63]

[37]   W. Wu, A.Y.C. Chen, L. Zhao, and J.J. Corso, "Brain tumor detection and segmentation in a CRF (conditional random fields) framework with pixel-pairwise affinity and superpixel-level features", *Int. J. CARS,* vol. 9, no. 2, pp. 241-253, 2014.
[http://dx.doi.org/10.1007/s11548-013-0922-7] [PMID: 23860630]

[38]   M. Soltaninejad, X. Ye, G. Yang, N. Allinson, and T. Lambrou, *Brain tumour grading in different MRI protocols using SVM on statistical features.* Med Image Underst Anal, 2014, pp. 259-264.

[39]   M. Soltaninejad, X. Ye, G. Yang, N. Allinson, and T. Lambrou, "An image analysis approach to mri brain tumour grading", *Oncol News,* vol. 9, no. 6, pp. 204-207, 2015.

[40]   A. Shenbagarajan, V. Ramalingam, C. Balasubramanian, and S. Palanivel, "Tumor diagnosis in MRI brain image using ACM segmentation and ANN-LM classification techniques", *Indian J. Sci. Technol.,* vol. 9, no. 1, 2016.
[http://dx.doi.org/10.17485/ijst/2016/v9i1/78766]

[41]   M. Sharma, G.N. Purohit, and S. Mukherjee, *Information retrieves from brain MRI images for tumor detection using hybrid technique K-means and artificial neural network (KMANN).* Networking Communication and Data Knowledge Engineering, 2018, pp. 145-157.
[http://dx.doi.org/10.1007/978-981-10-4600-1_14]

[42]   J. Vijay, and J. Subhashini, "An efficient brain tumor detection methodology using Kmeans clustering algorithm", *Int. Conf. Communication and Signal Processing,* 2013, pp. 653-7

[43]   T.A. Teli, and A. Wani, "Fuzzy Logic And Medicine With Focus On Cardiovascular Disease Diagnosis", *5th IEEE International Conference on Computing for Sustainable Global Development,* 2018.

[44]   J.C. Bezdek, L.O. Hall, and L.P. Clarke, "Review of MR image segmentation techniques using pattern recognition", *Med. Phys.,* vol. 20, no. 4, pp. 1033-1048, 1993.
[http://dx.doi.org/10.1118/1.597000] [PMID: 8413011]

CHAPTER 14

A Review on Health Monitoring of Electronic Passive Components

Raghav Gupta[1], Cherry Bhargava[1,*] and Amit Sachdeva[1,*]

[1] *Lovely Professional University, VLSI Design, SEEE, Phagwara, Punjab, India*

Abstract: The development of Technology in the field of electronic devices is going very rapidly. Factors such as cost, performance, complexity, and portability are contemplated over and over again. Nowadays low cost and better performance captivate the interest of a wide range of public. To meet these demands of the public various components are integrated onto a single chip. As this integration increases, the complexity of the devices increases which leads to increased chances of faults and failures in a device. It is not that today's urban human is using the electronics keeping hi-tech gadgets in hand, but today's mechanical transport means are also driven by daily changing and improved electronics devices. So many times, there are recalls of sold components or devices by leading electronics giants due to post prediction of their failure. This happens because the companies are using traditional techniques for condition monitoring and reliability testing. Even big automobile giants have to recall their cars for defects occurring in the later stage. One of the examples of recalling of TATA public transport buses sold to DTC in Delhi, India, as those automobiles are getting caught in the fire in many cases. Traditionally, to analyze electronic components reliability and condition monitoring, three techniques are used, viz., using empirical methods including standard handbooks MILHDBK-217, BELLCORE, and PRISM; analyzing using life testing experiments; collect maintenance and operating data and perform statistical analysis.

Keywords: Failure Prediction, Failure Rate, Faults, Reliability.

INTRODUCTION

Manufacturers across the globe are focusing on improving the reliability of their existing components. This is because most of the sold components received bad feedback because of low quality or fault which eventually leads to big losses even to successful companies. After some careful analysis it is concluded that the cause of this situation is ignoring the need for Reliability Calculation and fault analysis using a traditional approach in calculating RUL remaining useful life is consider-

* **Corresponding authors Cherry Bhargava and Amit Sachdeva:** Cherry Bhargava, Lovely Professional University, VLSI Design, SEEE, Phagwara, Punjab, India; E-mails: cherry.bhargava@lpu.co.in; amit.16701@lpu.co.in

Dharam Buddhi, Rajesh Singh and Anita Gehlot (Eds.)

ed a cause now the struggle of the manufacturer's is to increase component performance at a normal cost this eventually reduces their time to market the component the automobile industry is no less many big automobile manufacturers had to call back their products [1 - 18].

PREDICTION OF RELIABILITY AND FAILURE

Reliability is defined as the consistent performance of any component in electronic devices. It can help solve issues related to the failure. It helps to prevent failures. Fig. (1) shows the Bath-tub curve. Reliability plays a key role in measuring failure rates as it defines the probability of occurrence of the failures in the system with preference. Some major reasons these failures occur are:

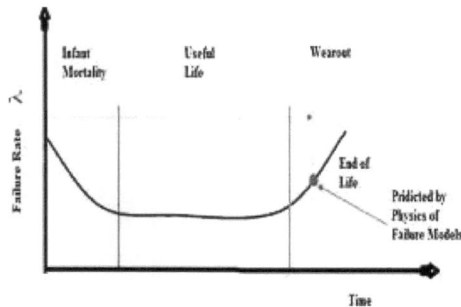

Fig. (1). Bath-tub curve [3].

1) Using components above their stress level.

2) Quality of product and weakness inherited.

3) Degradation of the component with its age and usage.

4) Failure of a component affecting others

A Bath-tub curve is used to analyze the rate of electronic devices along with time. The failures shown in infant mortality are the failures due to manufacturing. These show up at the initial stages of component usage. So, the failure rate gradually decreases as age increases. The failure rate in the useful life stage remains constant as failures occur like a randomly occurring event. Due to exposure of components to environmental stress and overloaded electrical parameters the failure rate increases as age increases. This is called the Wear out stage.

LITERATURE REVIEW

1. Zhao *et al.* (2018) [18], given the characteristics as current operational age and corresponding degradation state, conditional reliability can be estimated by using assessment method for devices that are subject to condition monitoring(CM). The degradation process is specified by a continuous-time Markov chain. Cox's proportional hazards model is used to determine the hazard rate failure time in this process. The majority of Degradation states and the transition of generation deteriorating mechanism are two main that are encountered in the health assessment. They can be properly addressed by a proposed method.

2. Vasan *et al.* (2018), we can observe degradation in a component with the change in an initial value of an electrical parameter of a component. Low performance of circuits and failure are the results of using such components. Detecting the component level errors in the circuit by deviation calculation of every parameter in the present-day process. A prognostic method that exploits the features extracted from the responses of the circuit, which consists of components that have parametric called as discussed to solve the issue.

3. Bhargava *et al.* (2018) [3], ocuses on a parameter thermal stress or high temperature the main reason for the failure of the capacitor under the high-temperature value of ESR and C change rapidly as electrolyte inside the capacitor start to evaporate. This eventually reduces the capacitor size. The constant charging and discharging of the capacitor lead to capacitor breakdown. The paper also analyses the humidity factor that enhances the capacitor's degradation cycle. Summing up the accuracy in predicting the life of capacitor increases from 55% to 90%.

4.Yao (2017) [17], an online non-invasive monitoring system was advised to inspect the dependability and life of Electrolytic capacitors as they were crucially accountable for the collapse of power electronic devices. Using capacitance and continuous analysis of Equivalent series resistance (ESR), this monitoring system is capable of yielding the electrolytic capacitors output by using an additional trigger that acts as an amplifier and amplifies the output and to analyze output voltages switching through continuous monitoring of capacitors. Offline monitoring, online monitoring are the types of monitoring. The device needed to be switched off before analyzing which was a crucial pitfall and online monitoring and quasi online monitoring played a vital role in overcoming it.

5. Chigurupati *et al.* [7], Machine learning made it easier to predict the component failure. The past behavior is used to learn and analyze *i.e.*, to train the algorithm. Then we can predict the future behavior of components more accurately. Machine learning helps in predicting the behavior before failure occurs. In this paper, the ability of machine learning in predicting future behavior using past life was explored. This technique can help in saving the whole circuit before failure occurs.

6. Bhargava *et al.* (2014) [2], the capacitor is a component that stores energy as an electrical charge. Due to lower prices and better capacitance, electrolytic capacitors are commonly used. Lakhs of components are integrated on a single chipset and among all these components the capacitor plays a vital role, as the reliability of these chipsets majorly depends on the capacitor. Different parameters affect the performance of the chipset such as environmental stress; electrical enhances downgrading in the electrolytic capacitors, *etc.* Normally electrolytic capacitors are majorly used in power electronic circuits which are highly exposed to high temperatures and abundant power is dissipated under the impact of ripple current and high voltages. Generally Downgrading in the electrolytic capacitors reflects in lower reliability of the electronic component. Fig. (**2**) shows the Mechanism to find the reliability of electrolytic capacitors.

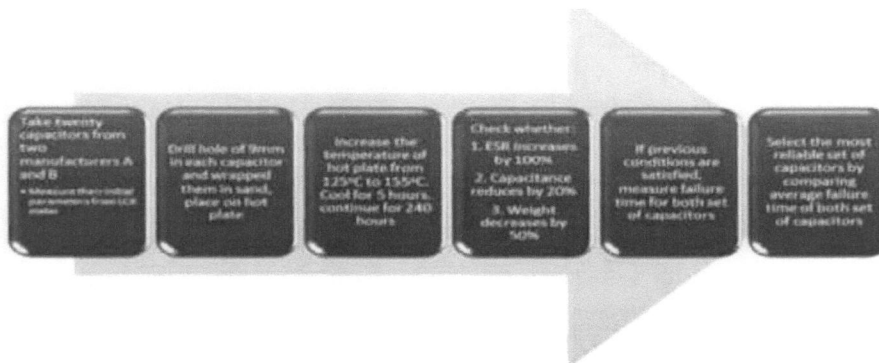

Fig. (2). Mechanism to find the reliability of electrolytic capacitors [2].

7. Challa *et al.* (2013) [4], in these contemporary times where electronics play a vital role in development, it's crucial a component that meets our specific requirements. VLSI industries are investing a lot of wealth, time, and effort in developing such components with high efficiency and comparability. Perceiving the reliability of the component under adverse conditions in virtue of qualification is a much-needed step. Qualification is a component that satisfies the specified requirements under various physical conditions. Qualification is necessary before the yielding starts as the failure of any component could cost us in times of rework and redesign of them, which could potentially increase the time to market it. Requalification is a must if any changes are made in the component to know if it is dependable to be introduced to the market. There are different stages such as contract manufacturers, component manufactures, computer manufactures, assembly companies, *etc.*, so for ideal results, it is preferable to run the process at all these stages. Above all, the real-life of a component might not be the same as that predicted by qualification and it is not necessary.

8. Chauhan *et al.* (2012) [6], solders play a vital role in electronic device functioning and monitoring of these solders is very important so that they do not result in compromising the electronic device functioning Rate of failure in the devices can be found by monitoring the solders. The paper will discuss temperature as a health assessment model in examining the health of electronic devices. By increasing the temperature and maintaining the current at 5 amperes the downgrading of solder connects can be analyzed. Temperature plays a vital role in speculating the solders as the temperature is proportionate to the fault in the device. Table **1** shows Solder Connects Reliability.

Table 1. Solder connects reliability [6].

Solder type	Sample size	Thermal cycles			
SN100C	3	750	1500	3000	4500
SnPb	3	750	1500	3000	4500

CONCLUSION

Different observations on reliability, failure rate, and remaining useful life of components by different authors have been reviewed in this paper. Multiple parameters were considered in various papers to find the defective component. This paper concludes that the failure rate has been increasing in emerging technologies even though many fault detection methods used to include the basic techniques.

CONSENT FOR PUBLICATION

Not applicable.

CONFLICT OF INTEREST

The authors declare no conflict of interest, financial or otherwise.

ACKNOWLEDGEMENT

Declared none.

REFERENCES

[1]   A. Arora, N.K. Medora, and J. Swart, "Failures of Electrical/Electronic Components: Selected Case Studies", *IEEE Symposium on Product Compliance Engineering,* 2007,pp. 1-6
[http://dx.doi.org/10.1109/PSES.2007.4378474]

[2]   C.B. Bhargava, "Failure prediction and health prognostics ofelectronic components: A review.In Engineering and Computational Sciences (RAECS)", In: *Recent Advances,* 2014, pp. 1-5.

[3]   C.B. Bhargava, "An intelligent prognostic model for electrolytic capacitors health monitoring: A design of experiments approach", *Adv. Mech. Eng.,* vol. 10, no. 10, 2018.
[http://dx.doi.org/10.1177/1687814018781170]

[4]   V.R. Challa, "Challenges in the qualification of electronic components and systems", *IEEE Trans. Device Mater. Reliab.,* vol. 1, no. 13, pp. 26-35, 2013.
[http://dx.doi.org/10.1109/TDMR.2011.2173801]

[5]   C.B.Z. Chaochao, "February). Prediction of machine health condition using neuro fuzzy and", *IEEE Trans. Instrum. Meas.,* vol. 61, no. 2, pp. 297-306, 2012.
[http://dx.doi.org/10.1109/TIM.2011.2169182]

[6]   P.O. Chauhan, "Use of temperature as a health monitoring tool for solder interconnects degradation in electronics in Prognostics and System Health Management (PHM)", *2012 IEEE Conference,* 2012, pp. 1-4

[7]   A.T. Chigurupati, "Predicting hardware failure using machine learning", *Annual Reliability and Maintainability Symposium (RAMS),* 2016, pp. 1-6

[8]   D. S. Campbell, "Reliability Behaviour of Electronic Components as a Function of Time", *Qual. Reliab. Eng. Int.,* vol. 8, pp. 161-166, 1992.
[http://dx.doi.org/10.1002/qre.4680080303]

[9]   S.K. Halgamuge, and M. Glesner, "Neural networks in designing fuzzy systems for real world", *Fuzzy Sets Syst.,* vol. 65, no. 1, pp. 1-12, 1992.

[10] M. Jaiswal, "Computer viruses: principles of exertion, occurrence and awareness", *International Journal of Creative Research Thoughts,* vol. 5, no. 4, pp. 648-651, 2017. .http://doi.one/10.1729/Journal.23273 [IJCRT].

[11] M. Jaiswal, "Big Data concept and imposts in business", *Int. J. Adv. Innov. Res.,* vol. 7, no. 4, 2018. .http://ijairjournal.in/Ijair_T18.pdf [IJAIR].

[12] M. Jaiswal, "SOFTWARE QUALITY TESTING. International Journal of Informative & Futuristic Research (IJIFR), 6(2), 114-119. Retrieved from Arora, A., Medora, N. K., & Swart, J", *IEEE Symposium on Product Compliance Engineering,* 2018, pp. 1-6

[13] N Kanari, "End-of-life vehicle recycling in the European Union", *J. Miner. Met. Mater. Soc.,* vol. 55, no. 5, pp. 15-19, 2003.
[http://dx.doi.org/10.1007/s11837-003-0098-7]

[14] F. Rugrungruang, *"An Integrated Methodology for Assessing Physical & Technological",* University of New South Wales: Thesis., 2008.

[15] A. S. Vasan, "Health and Remaining Useful Life Estimation of Electronic Circuits", *Prognostics and Health Management of Electronics: Fundamentals, Machine Learning, and the Internet of Things,* pp. 279-327, 2018.

[16] L. Wu, *Prognostics and System Health Management Conference,* pp. 1-6, 2018.

[17] K.C. Yao, "Noninvasive Online Condition Monitoring of Output Capacitor's ESR and C for a Flyback Converter", *IEEE Trans. Instrum. Meas.,* vol. 66, no. 12, pp. 3190-3199, 2017.
[http://dx.doi.org/10.1109/TIM.2017.2749838]

[18] S.E. Zhao, "Health Assessment Method for Electronic Components Subject to Condition Monitoring

and Hard Failure", *IEEE Trans. Instrum. Meas.,* vol. 68, no. 1, pp. 138-150, 2018.
[http://dx.doi.org/10.1109/TIM.2018.2839938]

CHAPTER 15

Edge Gateway and Zigbee Based SHM of Bridge Using AI

Enjeti Amareswar[1,*], **Anita Gehlot**[1] and **Rajesh Singh**[1]

[1] *SEEE, Lovely Professional University, Punjab, India*

Abstract: SHM of the bridge is the major issue for finding the condition and life span of the bridge. Recently many bridges have collapsed due to floods and environmental conditions. The up and coming age of extension SHM innovation needs to persistently screen conditions also, issue early alerts before an exorbitant fix or disastrous disappointments. With the assistance of the Internet of Things (IoT), the SHM can be transformed into a real-time monitoring system and the monitoring can be processed from any remote location through the internet. At present cloud computing is integrated with IoT for storing, monitoring, and performing analytics on the data received from the sensor node. In this study, we have proposed the Edge gateway and ZigBeecommunication-based SHM of the bridge. In the case of a bridge, the decision needs to be taken immediately for avoiding the demolition of the bridge. For immediate decisions, we have employed edge computing at the gateway node that performs analytics immediately and gives the necessary action to be undertaken. In this study, we have embedded the Zigbee module in the sensor mote for sensing the distinct parameters of the bridge and communicate them to the edge gateway. Edge gateway performs analytics by applying artificial intelligence techniques for predicting the damage and condition of the bridge and sends the alerts to the cloud server *via* a Wi-Fi module. From the cloud server, the authorities can access through the web application and mobile application.

Keywords: AI techniques, Edge gateway, IoT, Sensor mote Zigbee module.

INTRODUCTION

For better transportation in the world, bridges play a major role in interconnecting the roads between rivers, hills, canals, *etc*. In the USA about 66,405 bridge infrastructures *i.e.* more than 11 percent of bridges are in bad conditions to know the status of the bridge the SHM of the bridge is necessary [1]. Structural Health Monitoring (SHM) is characterized as a method that aims to obtain information over time on a structure's condition and behavior [2]. The stability of vital infra-

* **Corresponding author Samia Mushtaq:** SEEE, Lovely Professional University, Punjab, India;
E-mail: amar.20590@gmail.com

Dharam Buddhi, Rajesh Singh and Anita Gehlot (Eds.)

structures (VIs) such as energy grids, oil and gas systems, nuclear power plants, or transport networks is an important technique for monitoring [3]. The process of monitoring consists of the continuous compilation of the most representative parameters that indicate a structure's state and the choice of these parameters depends on a variety of factors, such as the form of structure, its function, the materials used for construction, and environmental conditions [4]. Based on the normal method, parameters are in different forms like stress, displacement, deformation, temperature, humidity temperature, pH, and oxidation of metal [5]. Three important components are included in a regular, wired SHM system: a sensor system, data collection, and transmission system, and a health evaluation system from the sensory information [6]. Here IoT is an emerging technology that is capable of real-time monitoring the physical things from any remote location *via* Internet Protocol (IP) [7]. The integration of cloud computing and IoT establishes a real-time platform for storing sensory data and applying analytics for getting understandable data. However, in the case of SHM of the bridge, the response time for detecting the damage of the bridge should be high as the bridge locates in the location where the connectivity issues also arise. With the advantage of edge computing technology, we are proposing edge gateway and Zigbee communication-based architecture for monitoring and prediction of the damage using AI techniques. Zigbee-based sensor mote is deployed to the bridge at a different point for sensing the dynamic parameters of the bridge. Zigbee communication transmits the sensory data to the IoT-based edge gateway and as the gateway is based on edge computing it performs the analytics using AI. The gateway can give understandable metadata of the health of the bridge. The gateway is also embedded with the wi-fi module, so it can transmit the data to the cloud server.

RELATED WORKS

The present approach in SHM of the bridge depends on visual assessments, which are tedious and emotional and are impossible to analyze regularly for all bridges.SHM of the bridge scour is very precise for evaluating bridges under scouring dangers, deterministic models dependent on designing decisions have been actualized throughout the long term utilizing subjective appraisal techniques [8]. Sensor-based approaches used for checking frameworks have been proposed to computerize and enhance the visual examination measure. One methodology is to introduce a variety of various sensors, for example, strain gauges and accelerometers are mounted on the bridge. The disadvantage is that such sensors require a refined and costly electronic framework with the establishment, upkeep and force uphold [9]. In the remote structural health monitoring of the bridge, GSM/GPRS wireless protocol is used for transmitting the data to a long range

[10]. Wi-Fi module is also used for transmitting and receiving the data from sensor nodes for SHM of the bridge which is an IEEE 802.11b/g/n standard with a 2.4 GHz band [11]. Wireless Global Bridge Evaluation and Monitoring System are one of the wireless communication networks which have a 902 to 928 MHz frequency band which is used in sensor nodes for collecting and processing the data [12]. Due to high power consumption and prepaid network, the GSM/GPRS should be replaced free licensed network. Zigbee communication module works on the ISM (industrial, scientific & medical) band and it is open access licensed band where the transmission of data can be sent with low power consumption a low cost based sensor and NB-IoT communication-based system is proposed for displacement monitoring of the bridge and cloud server is integrated for storing the displacement data [13]. An IoT-based flexible platform is implemented with accelerometer sensors for remote health monitoring of the bridge [14]. Applying machine learning (ML) technique from sensory data of SHM of the bridge which includes a training algorithm for characterizing the structural condition of the bridge. Based on the data analysis through algorithms, we can predict the damage and preventive measures would be taken [15].

PROPOSED ARCHITECTURE

Edge computing technology where data collected from IoT devices is computed directly at edge networks instead of transmitting to the local server [16]. By using this technology data is analyzed at the edge devices and also provides a quick response for any emergency when compared to cloud computing. In this study, an edge gateway and Zigbee-based architecture are proposed for SHM of the bridge and it is shown in Fig. (**1**). Zigbee-based sensor mote will be deployed at the distinct points of the bridge for monitoring the condition of the bridge. In the sensor mote, Acoustic Emission, ultrasonic sensors, and Accelerometer sensors are used for sensing the dynamic parameters of the bridge.

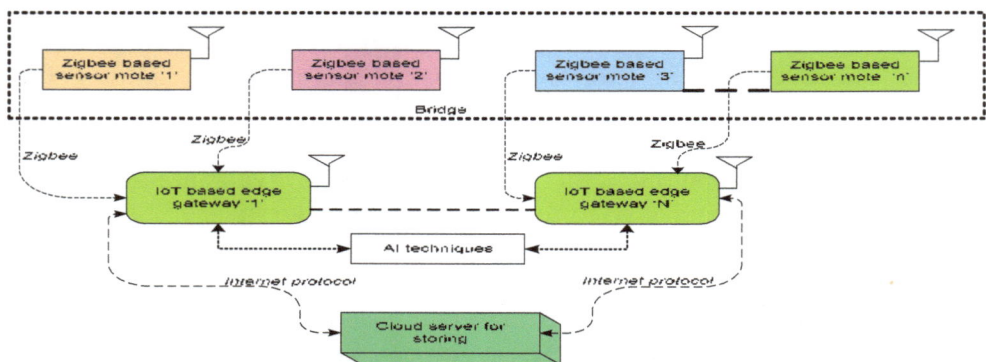

Fig. (1). Edge gateway-based architecture for SHM monitoring of bridge.

Generally, in many studies, GSM/GPRS communication module is integrated into the sensor mote for transmitting the sensory data. However, in IoT, the devices are energy-constrained so we need a low power consumption communication module to transmit the data. Here we have embedded IEEE 802.15.4 Zigbee module in the sensor mote that transmits the sensory data to the nearby IoT-based edge gateway. Usually, in IoT, the data from the sensor mote are stored in the cloud server for performing analytics to get understandable metadata. However, in the case of a bridge, the decision and analytics need to be made immediately for reducing the impact of collapse and losing lives due to the bad condition of the bridge. To overcome it, we are implementing edge computing technology in the gateway node. The edge gateway node receives the sensory data and the computing unit inside the gateway performs the analytics using AI techniques for predicting the damage. Machine learning and deep learning are evolved from the Artificial Intelligence domain. Machine learning constructs the models based on the input parameters mapped from the sensor data and generates the output which represents damage and location. In case of emergence, the gateway immediately triggers the authorities to take immediate action and it communicates the data to the cloud server *via* a Wi-Fi module. The data stored in the cloud server can be accessed *via* web applications and mobile applications. The GPS module provides the location of the based on the assessment the preventive measures would be taken.

Zigbee-based sensor mote is the integration of AE sensor, ultrasonic sensor, Accelerometer, Zigbee module, GPS, and Controller unit are shown in the Fig. (**2**). AE sensor senses the condition of the bridge based on the stress applied on the bridge. The accelerometer sensor measures the dynamic characteristic of the bridge due to vibrations caused by passing vehicles. An ultrasonic sensor is used to measure the river bed-based sonic waves reflection the scour of the bridge is identified. GPS is used to find the damage location on the bridge based on the analysis of structural parameters. Controller unit monitors the parameters of the bridge based on the condition necessary action will be taken. solar power one of the power sources is given to mote. By using Zigbee WSN the data is communicated from the transmitter to the gateway.

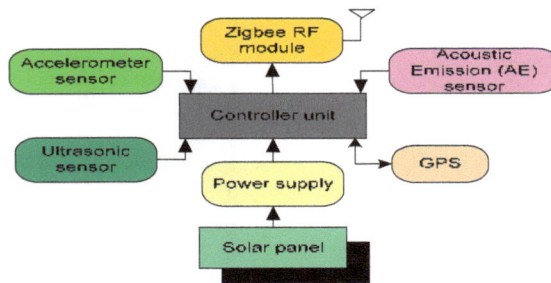

Fig. (2). Zigbee based sensor mote.

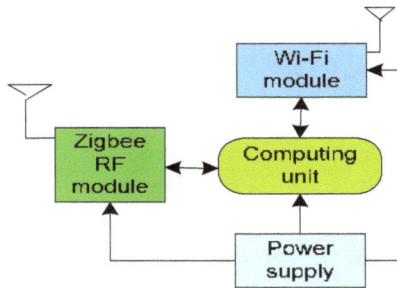

Fig. (3). IoT based Edge gateway.

An IoT-based edge gateway is the integration of an edge computing unit, Zigbee module, and Wi-Fi modem, that act as a bridge between the sensor mote and cloud server. Fig. (**3**) illustrates the components present in the IoT-based edge gateway. Zigbee is a transceiver unit that receives sensory data. The edge computing unit analyzes the received data with AI techniques for the prediction of damage. The controller unit activates the Wi-Fi modem to transmit the sensory and visual data to the cloud server over internet protocol.

HARDWARE DESCRIPTION

Acoustic Emission

AE sensor is used to monitor the damage to the structure of the bridge. AE sensors detect the damage of the bridge using the AE technique which converts mechanical parameters to electrical parameters. The relaxation ratio proportion of more than 1 shows serious damage. AE technique acquires the elastic waves generated due to a change in material stress by an external force. AE sensor converts elastic waves into electrical signals from data acquisition is called AE testing. The AE sensor peak frequency range is between f start ¼ 0 kHz and f end

¼ 1200 kHz. Signals generated by AE are very low so to increase signal strength amplifiers are used. AE sensors are mounted on the bridge when a vehicle passes on the bridge the stress is applied so the elastic waves of time domain or frequency are generated inside the material [17].

Ultrasonic Sensor

Bridge Scour detection is one of the major issues for finding bridge damage during floods. Scour monitoring devices are essential for SHM Bridge which identifies the damage and takes preventive measures. An ultrasonic sensor emits sonar rays underwater when the signal reaches the river bed the signals are reflected. Parameters of ultrasonic sensors are frequency 40kHz, range 25 to 450cm, operating voltage 0 to 5v. Based on the signal travel with time from the transmitter to receiver and wave velocity the river bed is estimated. The ultrasonic sensor is placed in the tube and is mounted on the bridge. By monitoring the waveform characteristics of the signal the depth sour can be measured based on the displacement of the transducer from the reference point [18].

Accelerometer

The SHM of the bridge is monitored by measuring the dynamic parameter of the bridge. An accelerometer is fixed to the bridge to measure the dynamic parameters of the bridge. MEMS-based accelerometer sensor measures the displacement due to vibrations caused during vehicle passage. It is a low cost, low power consumption, and 2 axis accelerometer. Vibrations of the bridge are collected and processed through time series and frequency domain. By analyzing the structural parameters of the bridge where damage is detected [19].

GPS (Global Positioning System)

There are a few strategies that can be applied in bridge damage detection and location which have been talked about by the researchers. In this GPS is used for finding the damaged location of the bridge. The GPS module is mounted on the bridge in which the data is transformed in location by using two components one is X transverse component and Y longitudinal component. GPS module finds damage location by analyzing the change in displacement with temperature measurement [20]. The received data from the GPS receiver is processed through different techniques for finding the desired location. PPP is one of the techniques which give the precise damage location based on a change in displacement [21].

Controller Unit

A microcontroller is the heart of the Network which is embedded with desired sensors. Controllers belong to the Arduino family that is Arduino mega 2560 is used for collecting and processing the data through WSN. It consists of 54 digital I/O pins, 16 analog pins, 8kB of SRAM,8kB of Flash memory, and 4 UART ports. Arduino Mega 2560 uses Arduino IDE for programming the data; it is an open-source software [22].

Zigbee Rf Module

Zigbee is one of the RF technologies used to transfer the data through WSN. Zigbee is a WSN which is a short-range communication technology with IEEE standards 802.15.4 with frequency 2.4 GHz, 900MHz, 868 MHz [23]. Zigbee WSN is a low-cost, less power consumption module, it is a 128 bit AES encryption standard for data security. It acts as both transmitter and receiver for data communication from the sender to receiver.

Wi-fi Module

Wi-Fi is an RF communication used for data transmitting and receiving through WSN. ESP8266 is a Wi-Fi module that is interfaced with a microcontroller and which supports TCP/IP protocol stack for data communication through the internet. It is an 802.11 b/g/n IEEE standard, consumes less power, baud rate for communication is 115200 bits/sec [24].

CONCLUSION

The SHM of the bridge plays a precise role in monitoring the bridge condition based on different methods. The traditional approach in SHM of the bridge is high cost and time-consuming. With the assistance of IoT, the SHM can be transformed into a real-time monitoring system and the monitoring can be processed from any remote location through the internet. In this study, we have proposed Edge gateway and Zigbee communication-based SHM of the bridge. At present cloud computing is integrated with IoT for storing, monitoring, and performing analytics on the data received from the sensor node. In the case of a bridge, the decision needs to be taken immediately from avoiding the demolition of the bridge. For immediate decisions, we have employed edge computing at the gateway node that performs analytics immediately and gives the necessary action to be undertaken. In this study, we have embedded the Zigbee module in the sensor mote for sensing the distinct parameters of the bridge and communicate them to the edge gateway. Edge gateway performs analytics by applying artificial intelligence techniques for predicting the damage and condition of the bridge and send the alerts to the cloud server *via* a Wi-Fi module. From the cloud server, the authorities can access through the web application and mobile application.

CONSENT FOR PUBLICATION

Not applicable.

CONFLICT OF INTEREST

The authors declare no conflict of interest, financial or otherwise.

ACKNOWLEDGEMENT

Declare none.

REFERENCES

[1]　S.L. Davis, and D. Goldberg, *"The fix we're in for: The state of our nation's bridges,"* 2013. Transportation for America: Washington, 2013.

[2]　C.R. Farrar, and K. Worden, "An introduction to structural health monitoring", *Philos. Trans.- Royal Soc., Math. Phys. Eng. Sci.,* vol. 365, no. 1851, pp. 303-315, 2007.
[http://dx.doi.org/10.1098/rsta.2006.1928] [PMID: 17255041]

[3]　M.A. Mahmud, K. Bates, T. Wood, A. Abdelgawad, and K. Yelamarthi, "A complete Internet of Things (IoT) platform for Structural Health Monitoring (SHM)", *IEEE 4th World Forum on Internet of Things (WF-IoT),* 2018 pp. 275-279
[http://dx.doi.org/10.1109/WF-IoT.2018.8355094]

[4]　S. Jr, "Billie F., Manuel Ruiz-Sandoval, and Narito Kurata. "Smart sensing technology for structural health monitoring", *Proc. of 13th World Conf. on Earthquake Engineering,* 2004, p. 1791

[5]　A. Abdelgawad, and K. Yelamarthi, "Internet of things (IoT) platform for structure health monitoring", *Wirel. Commun. Mob. Comput.,* 2017.
[http://dx.doi.org/10.1155/2017/6560797]

[6]　A. Myers, M.A. Mahmud, A. Abdelgawad, and K. Yelamarthi, "Toward integrating Structural Health Monitoring with Internet of Things (IoT)", *2016 IEEE International Conference on Electro Information Technology (EIT),,* 2016pp. 0438-0441
[http://dx.doi.org/10.1109/EIT.2016.7535280]

[7]　L. Alonso, J. Barbaran, J. Chen, M. Díaz, L. Llopis, and B. Rubio, "Middleware and communication technologies for structural health monitoring of critical infrastructures: A survey", *Comput. Stand. Interfaces,* vol. 56, pp. 83-100, 2018.
[http://dx.doi.org/10.1016/j.csi.2017.09.007]

[8]　S. Chen, F. Cerda, P. Rizzo, J. Bielak, J.H. Garrett, and J. Kovacevic, "Semi-supervised multiresolution classification using adaptive graph filtering with application to indirect bridge structural health monitoring", *IEEE Trans. Signal Process.,* vol. 62, no. 11, pp. 2879-2893, 2014.
[http://dx.doi.org/10.1109/TSP.2014.2313528]

[9]   https://www.dot.ny.gov/divisions/engineering/structures/manuals/hydraulics

[10]   S. Hou, and G Wu, "A low-cost IoT-based wireless sensor system for bridge displacement monitoring", *Smart*.

[11]   A. Abdelgawad, and K. Yelamarthi, "Internet of Things (IoT) Platform for Structure Health Monitoring", *Wirel. Commun. Mob. Comput.,* pp. 1-10, 2017.
[http://dx.doi.org/10.1155/2017/6560797]

[12]   C. Scuro, P.F. Sciammarella, F. Lamonaca, R.S. Olivito, and D.L. Carni, "IoT for structural health monitoring", *IEEE Trans. Instrum. Meas. Mag,* vol. 21, no. 6, pp. 4-14, 2018.
[http://dx.doi.org/10.1109/MIM.2018.8573586]

[13]   S. Panthati, and A.A. Kashyap, "Design and implementation of structural health monitoring based on IoT using lab VIEW", *Int.J.Mag. Eng. Technol. Manag. Res.,* vol. 3, no. 2, pp. 77-82, 2016.

[14]   S. Jeong, R. Hou, J.P. Lynch, H. Sohn, and K.H. Law, "A scalable cloud-based cyber infrastructure platform for bridge monitoring", *Struct. Infrastruct. Eng.,* vol. 1, pp. 82-102, 2019.
[http://dx.doi.org/10.1080/15732479.2018.1500617]

[15]   E.K. Chalouhi, I. Gonzalez, C. Gentile, and R. Karoumi, "Damage detection in railway bridges using Machine Learning: application to a historic structure", *Procedia Eng.,* vol. 199, pp. 1931-1936, 2017.
[http://dx.doi.org/10.1016/j.proeng.2017.09.287]

[16]   Y. Ai, M. Peng, and K. Zhang, "Edge computing technologies for Internet of Things: a primer2018", *Digital Communications and Networks,* vol. 4, pp. 77-86, 2018.
[http://dx.doi.org/10.1016/j.dcan.2017.07.001]

[17]   Md. Khan, and I. Tawhidul, "Structural Health Monitoring by Acoustic Emission", *Structural Health Monitoring from Sensing to Processing,* vol. 23, 2018.

[18]   B. Wu, W. Chen, and H. Li, *Real-time monitoring of bridge scouring using ultrasonic sensing technology.* Sensors and Smart Structures Technologies for Civil, Mechanical, and Aerospace Systems, 2012.
[http://dx.doi.org/10.1117/12.914994]

[19]   A. Sabato, "Pedestrian bridge vibration monitoring using a wireless MEMS accelerometer board", *IEEE 19th International Conference on Computer Supported Cooperative Work in Design (CSCWD),* 2015pp. 437-442
[http://dx.doi.org/10.1109/CSCWD.2015.7230999]

[20]   N. Manzini, "Structural Health Monitoring using a GPS sensor network", *Proc. 9th European Workshop on Structural Health Monitoring Series (EWSHM), Manchester, UK,* vol. 8, 2019.

[21]   M.R. Kaloop, "Recent advances of structures monitoring and evaluation using GPS-time series monitoring systems: A review", *ISPRS Int. J. Geoinf.,* vol. 6, no. 12, p. 382, 2017.
[http://dx.doi.org/10.3390/ijgi6120382]

[22]   https://www.arduino.cc/en/Main/arduinoBoardMega2560

[23]   X. Jiang, Y. Tang, and Y. Lei, "Wireless sensor networks in Structural Health Monitoring based on ZigBee Technology", *Conf. on Anti-Counterfeiting, Security, and Identification in Communication,* 2009
[http://dx.doi.org/10.1109/ICASID.2009.5276977]

[24]   R. Singh, P. Tyagi, and A. Gehlot, *Getting Started for Internet of Things with Launch Pad and ESP8266.* River Publishers, 2019.

CHAPTER 16

Deep Learning Based Edge Device for Diabetic Retinopathy Detection

A. Shiva Prasad[1,*], **Anita Gehlot**[1] and **Rajesh Singh**[1]

[1] *School of Electronics and Electrical Engineering, Lovely Professional University, Punjab, India*

Abstract: Diabetic Retinopathy is the major problem in Diabetic affected people, as it causes blindness to the people who are affected by this particular disease. Recently many people are being affected by Diabetes due to changes in food habits and the quality of food people take. Early and continuous testing of the eye in regular intervals leads to the identification of this disease by which the blindness problem can be overcome, but due to lack of availability of resources like optholomists and the machinery for monitoring the eyes the early detection is not that easy. The early machines and methods adopted will not provide good results in any cases because of many constraints, As the technology is advancing, with the help of neural networks and Edge Devices regular monitoring the diabetic people can be done easily and effectively in any part of the globe with much ease. At present image processing techniques are being used. In this study, we propose Deep Learning algorithms that process the date very accurately in very less time because of which we could good results in performance evaluation. In this study we have also proposed Edge Device which are end term devices, where these devices perform analytics on images of the eye and predict the condition of the eye and also it gives the information of stages of Diabetic Retinopathy, we can also store the data which is processed by edge device on cloud servers *via* wifi, From the cloud server the concerned people can obtain the records of that person who is affected with Diabetic Retinopathy.

Keywords: Convolution Neural Networks(CNN), Diabetic Retinopathy, Edge Device, Neural Networks.

1. INTRODUCTION

Diabetic Retinopathy (DR) is a disease of the eye [1] which is caused because of increased glucose levels in the body, as the body cannot produce enough insulin. This increased glucose level affects the blood vessels which are present in the eye will be swelling which in turn leads to leakage in the retina. A person who is Diabetic is subjected to failure of the Kidney, Blindness, Bleeding in teeth, nerve

---

* **Corresponding author A. Shiva Prasad:** School of Electronics and Electrical Engineering, Lovely Professional University, Punjab, India; E-mail: shivaprasadallenkis@gmail.com

Dharam Buddhi, Rajesh Singh and Anita Gehlot (Eds.)

failures, and so on. The maximum increase in glucose levels in blood leads to oozing of blood from the eye, because of which the human visual system will be damaged.

i. Diabetic Retinopathy can be classified into two types namely

ii. Non-Proliferative Diabetic Retinopathy (NPDR)

iii. Proliferative Diabetic Retinopathy (PDR)

The classification of DR is shown in Fig. **(1)**.

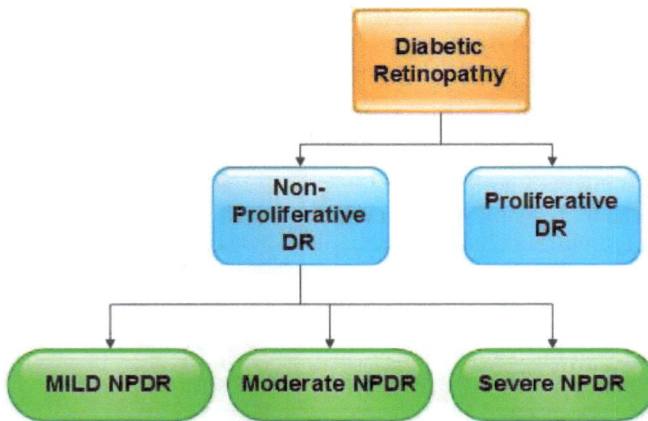

Fig. (1). Classification of Diabetic Retinopathy.

The images of NPDR and PDR are shown in Fig. **(2)**.

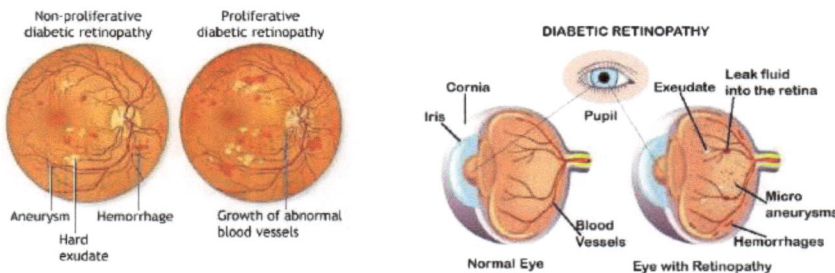

Fig. (2). Images of NPDR, PDR, and difference between normal eye and eye with retinopathy.

The aim to obtain information on various elements of the eye retina is important to identify the DR problem. The process of monitoring the eye retina consists of the continuous examination of Micro aneurysms (MA) [2], Exudates (EX), Hemorrhages (HM) which are internal elements of the eye whose properties are going to vary if it is affected. The difference between normal eye to DR is shown in Fig. (**2**).

MA represents the internal enlargements of the retinal capillaries. EX represents Internal Bleeding. HM represents inflammation in Blood vessels because of which blood flow is restricted and causes tissue damage. Parameters like MA, EX, HM are to regularly monitor and should be identified if there are any changes but with the help of existing methods that use image processing techniques will not produce a better result. The major drawback is time consumption and Performance evaluation.

The adoption of Neural Networks in Medical applications for various monitoring applications will improve the medical examination outputs concerning time consumption, ease of access. We can adopt the latest like Deep Learning techniques which could provide good results [3 - 12].

The integration of Neural Network algorithms with high-end devices like Edge Device would provide a great opportunity to provide Health Monitoring in a real-time platform by acquiring the retinal data (fundus images) and applying the analytics for getting understandable data. However, with the advent of edge computing technology, computing can perform at edge devices. With the advantage of the edge computing technology, we are proposing Edge Device and Deep Learning Algorithms which gives good results.

2. RELATED WORKS

The present approach in DR detection depends on manual assessments of patients, which is highly not possible because of an insufficient number of Optholomigists and machinery needed.

Jiawei *et al.*, (2018) proposed a method that involves two different forms from the view of MAs turnover and pathological risk factors. It is used to diagnose the DR. To get MA's turnover there are 7 other methods and pathological features. Using statistical analysis and technique of pattern classification an unchanged and new MA's can be resolved [11].

Pedro *et al.*, (2018), have proposed a method that is joint optimization of the instance encoding and the image classification stages. To get pathological images mid-level representation (MLR) is used [13].

Xianglong *et al.* (2019), have proposed a novel CNN model with Siamese-like architecture. Using a transfer learning method, the training is learned., The recommended method accepts inputs which are binocular fundus images and determines their correlation to decide a prediction [19].

YunleiSun (2019) had Proposed the CNN model is mixed with the layer to save the dispersion of the gradient, accelerate the schooling speed and enhance the accuracy of the model. The experiments show that this approach can acquire an education accuracy of 99.85 percent and a checking-out accuracy of 97. 56 percent, which is extra than 2 percent better than that of the use of logistic regression. However also for the investigation of other illnesses [20].

Sehrish *et al.*, (2019) have used the public to have a retina dataset of Kaggle pix to teach a combination of 5 deep CNN models to encode the features that are rich and increase the category for different levels of dr. The empirical effects show that the suggested version detects all of the levels of DR not like the present strategies and plays higher as in comparison to state-of-the-art strategies at the same Kaggle dataset [22].

Jordi *et al.*, (2020) have proposed a DR deep learning classifier that classifies images into different levels of severity and assigns scores, with which visual maps are generated which are very useful in DR finding [25].

3. PROPOSED ARCHITECTURE

Edge Device computes the data directly at the edge terminal itself instead of sending it to remote servers which are going to be time-consuming. By using this technology data is analyzed at the edge devices and also provides a quick response for any emergency when compared to cloud computing. In this study, Edge Device-based architecture is proposed for the Detection of Diabetic retinopathy and it is shown in Fig. (**3**).

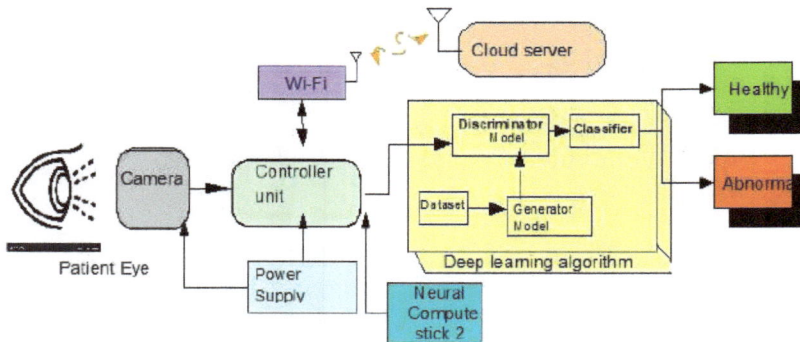

Fig. (3). Edge Device-based architecture for Diabetic retinopathy detection.

The camera is going to act as an input device that captures the image of the retina of patients [14], while taking the picture of the Retina a good and fully illuminated area will be considered for taking pictures of the Retina. Which will be called Fundus Images [15 - 18]? Where these images are processed to identify the various parameters, which gives the information of the health condition of Eye, which indicates that whether it is Normal eye or Retinopathy affected eye.

Generally, in many studies, the Camera is integrated into the controller unit for transmitting the image data. However, in IoT, the devices are energy-constrained so we need a low power consumption communication module to transmit the data. Usually, in IoT, the data from the sensor node are stored in the cloud server for performing analytics to get understandable Metadata. However, in the case of detection of DR, the decision and analytics need to be made immediately for reducing the impact on eyes and losing sight due to the bad condition of the retina using fundus images [21]. To overcome the delay in the computation of images a new technology known as Edge Computing Technology is implemented. By using Artificial Intelligence Techniques for early detection. Machine learning and deep learning are evolved from the Artificial Intelligence domain connected with edge devices to detect DR [23, 24]. These advanced techniques evolved tremendously in the present-day scenario. There are a wide variety of algorithms and modeling tools for processing huge data and data mining.

Deep learning appears to be something new concerning the term, but it is a term to describe different types of Neural Networks and related algorithms which act upon the raw input data. The Deep Learning algorithm is loaded onto the controller where the actual processing of images takes place for the identification of various elements to decide that the eye is affected or not [25]. An IoT-based edge gateway is the integration of an edge device with a computing unit [26 - 29], a Wi-Fi modem, that acts as a terminal that analyzes the data of images and gives the result. The controller unit activates the Wi-Fi modem to send the sensory and visual data to the cloud server over internet protocol.

Camera

In Retina examination normally doctors use an ophthalmoscope, 20D lens, and Fundus camera, but these are of very high cost, large size. The camera captures a field of view up to 20 degrees for a single fundus image at a distance of 1 cm from the patient eye. Cameras of the Mobile phone are used for this fundus photography.

Controller Unit

Controllers belong to the Arduino family that is Arduino mega 2560 is used for collecting and processing the data through WSN. It consists of 54 digital input and output pins, analog pins of 16, 8kB of SRAM,8kB of Flash memory, and 4 UART ports. Arduino Mega 2560 uses Arduino IDE for programming the data; it is open-source software.

Wi-Fi Module

Wi-Fi is an RF communication used for data transmitting and receiving through WSN. ESP8266 is a Wi-Fi module that is interfaced with a microcontroller and which supports TCP/IP protocol

Stack for data communication through the internet. It is an 802.11 b/g/n IEEE standard, consumes less power, baud rate for communication is 115200 bits/sec.

CONCLUSION

The regular checkup of the eyes plays a very important role in monitoring the condition of the eye based on various methods. The earlier approach in SHM of the bridge is high cost and time-consuming. With the assistance of IoT, the SHM can be transformed into a real-time monitoring system and the monitoring can be processed from any remote location through the internet. In this study, we have proposed Edge Device and Deep Learning method for examining the eye to detect DR. In the present-day world Edge Computing is used for monitoring, and performing analytics on the data received from the input node, we have considered edge computing at the gateway node that performs analytics quickly. Edge gateway performs analytics by applying artificial intelligence techniques for predicting the damage and condition of the eye.

CONSENT FOR PUBLICATION

Not Applicable.

CONFLICT OF INTEREST

The authors declare no conflict of interest, financial or otherwise.

ACKNOWLEDGEMENT

Declared none.

REFERENCES

[1]   B.E. Klein, "Overview of epidemiologic studies of diabetic retinopathy", *Ophthalmic Epidemiol.,* vol. 14, no. 4, pp. 179-183, 2007.
[http://dx.doi.org/10.1080/09286580701396720] [PMID: 17896294]

[2]   Kauppi T., Kalesnykiene Valentina, and Kamarainen Joni-Kristian, "The diaretdb1 diabetic retinopathy database and evaluation protocol", *British Conference on Machine Vision,* 2007.

[3]   B. Al Diri, A. Hunter, D. Steel, M. Habib, T. Hudaib, and S. Berry, "A Reference Data Set for Retinal Vessel Profiles", *International Conference of the IEEE Engineering in Medicine and Biology Society.,* 2008, pp. 2262-2265
[http://dx.doi.org/10.1109/IEMBS.2008.4649647]

[4]   B.B. Bruce, C. Lamirel, V. Biousse, A. Ward, K.L. Heilpern, N.J. Newman, and D.W. Wright, "Feasibility of nonmydriatic ocular fundus photography in the emergency department: Phase I of the FOTO-ED study", *Acad. Emerg. Med.,* vol. 18, no. 9, pp. 928-933, 2011.
[http://dx.doi.org/10.1111/j.1553-2712.2011.01147.x] [PMID: 21906202]

[5]   Y. Zheng, M. He, and N. Congdon, "The worldwide epidemic of diabetic retinopathy", *Indian J. Ophthalmol.,* vol. 60, no. 5, pp. 428-431, 2012.
[http://dx.doi.org/10.4103/0301-4738.100542] [PMID: 22944754]

[6]   E. Decenciere, X. Zhang, and G. Cazuguel, "Feedback on a Publicly Distributed Image Database: the Messidor Database", *Image Anal. Stereol.,* vol. 33, pp. 231-234, 2014.
[http://dx.doi.org/10.5566/ias.1155]

[7]   M.E. Ryan, R. Rajalakshmi, V. Prathiba, R.M. Anjana, H. Ranjani, K.M. Narayan, T.W. Olsen, V. Mohan, L.A. Ward, M.J. Lynn, and A.M. Hendrick, "Comparison Among Methods of Retinopathy Assessment (CAMRA) study: smartphone, nonmydriatic, and mydriatic photography", *Ophthalmology,* vol. 122, no. 10, pp. 2038-2043, 2015.
[http://dx.doi.org/10.1016/j.ophtha.2015.06.011] [PMID: 26189190]

[8]   B. Wu, W. Zhu, F. Shi, S. Zhu, and X. Chen, "Automatic detection of microaneurysms in retinal fundus images", *Comput. Med. Imaging Graph.,* vol. 55, pp. 106-112, 2017.
[http://dx.doi.org/10.1016/j.compmedimag.2016.08.001] [PMID: 27595214]

[9]   A. Dasgupta, and S. Singh, "A Fully Convolution Neural Network based Structured Prediction Approach towards the Retinal Vessel Segmentation", *International Symposium on Biomedical Imaging,* 2017, pp. 248-251

[10]   N.I. Hossain, and S. Reza, "Blood Vessel Detection from Fundus Image using Markov Random Field based Image Segmentation", *IEEE International Conference on Advances in Electrical Engineering,* 2018, pp. 123-127

[11]   Jiawei Xu, Xiaoqin Zhang, Huiling Chen, Jing Li, Jin Zhang, Ling Shao, and Gang Wang, "Automatic Analysis of Micro aneurysms Turnover to Diagnose the Progression of Diabetic Retinopathy", *IEEE Access. Machine learning for medical diagnosis,* vol. 6, pp. 9632-9642, 2018.

[12]   N. Brancati, M. Frucci, D. Gragnaniello, and D. Riccio, "Retinal Vessels Segmentation Based on a Convolution Neural Network", *Iberoamerican Congress on Pattern Recognition.,* no. Feb, pp. 119-126, 2018.

[13]   "Pedro Costa, Adrian Galdran, Asim Smailagic and AuréLio Campilho, "A Weakly- Supervised Framework for Interpretable Diabetic Retinopathy Detection on Retinal Images," IEEE Trans", *Advanced Signal Processing Methods in Medical Imaging.,* vol. 7, pp. 18747-18758, 2018.

[14]   J. Mo, L. Zhang, and Y. Feng, "Exudate-Based Diabetic Macular Edema Recognition in Retinal Images using Cascaded Deep Residual Networks", *Neurocomputing,* vol. 290, pp. 161-171, 2018.
[http://dx.doi.org/10.1016/j.neucom.2018.02.035]

[15]   W. Cao, N. Czarnek, J. Shan, and L. Li, "Microaneurysm Detection Using Principal Component Analysis and Machine Learning Methods", *IEEE Trans. Nanobioscience,* vol. 17, no. 3, pp. 191-198, 2018.
[http://dx.doi.org/10.1109/TNB.2018.2840084] [PMID: 29994317]

[16]   M. Li, Q. Yin, and M. Lu, "Retinal Blood Vessel Segmentation based on Multi-Scale Deep Learning", *Federated Conference on Computer Science and Information Systems,* 2018, pp. 1-7

[http://dx.doi.org/10.15439/2018F127]

[17] J. Latif, C. Xiao, A. Imran, and S. Tu, "Medical Imaging using Machine Learning and Deep Learning Algorithms: A Review", *2nd International Conference on Computing, Mathematics and Engineering Technologies IEEE,* 2019.
[http://dx.doi.org/10.1109/ICOMET.2019.8673502]

[18] P. Khojasteh, L.A. Passos Júnior, T. Carvalho, E. Rezende, B. Aliahmad, J.P. Papa, and D.K. Kumar, "Exudate detection in fundus images using deeply-learnable features", *Comput. Biol. Med.,* vol. 104, pp. 62-69, 2019.
[http://dx.doi.org/10.1016/j.compbiomed.2018.10.031] [PMID: 30439600]

[19] X. Zeng, H. Chen, Y. Luo, and W. Ye, ""Automated Diabetic Retinopathy Detection Based on Binocular Siamese-Like Convolution Neural Network," IEEE Access", *Deep Learning for Computer-aided Medical Diagnosis.,* vol. 7, pp. 30744-30756, 2019.

[20] Y. Sun, ""The Neural Network of One-Dimensional Convolution-An Example of the Diagnosis of Diabetic," IEEE Access", *Healthcare Information Technology for the Extreme and Remote Environments.,* vol. 7, pp. 69657-69666, 2019.

[21] N. Asiri, M. Hussain, F. Al Adel, and N. Alzaidi, "Deep learning based computer-aided diagnosis systems for diabetic retinopathy: A survey", *Artif. Intell. Med.,* vol. 99, p. 101701, 2019.
[http://dx.doi.org/10.1016/j.artmed.2019.07.009] [PMID: 31606116]

[22] S. Qummar, F.G. Khan, S. Shah, A. Khan, S. Shamshirband, Z.U. Rehman, I.A. Khan, and W. Jadoon, "A Deep Learning Ensemble Approach for Diabetic Retinopathy Detection", *IEEE Access,* vol. 7, pp. 150530-150539, 2019.
[http://dx.doi.org/10.1109/ACCESS.2019.2947484]

[23] M. Pekala, N. Joshi, T.Y.A. Liu, N.M. Bressler, D.C. DeBuc, and P. Burlina, "Deep learning based retinal OCT segmentation", *Comput. Biol. Med.,* vol. 114, pp. 103-125, 2019.
[http://dx.doi.org/10.1016/j.compbiomed.2019.103445] [PMID: 31561100]

[24] Sehrish Qummar, Fiaz Khan, Sajid Shah, Ahmad Khan, and Ahmad Din,

[25] "Jordi de la Torre, Aida Valls and Domenee puig, "A Deep Learning Interpretable Classifier for Diabetic Retinopathy Disease Grading," Elsevier", *Neurocomputing,* vol. 396, pp. 465-476, 2020.
[http://dx.doi.org/10.1016/j.neucom.2018.07.102]

[26] Pranav Sachdeva, Sankalp Rhythm, and Sangeeta Arjun,

[27] D. Das, S. Biswas, and S. Bandyopadhyay, "Laskar and Rabul, "Deep Learning Techniques for Early Detection of Diabetic Retinopathy", *5th International Conference on Computing, Communication and Security (ICCCS),* 2020, pp. 1-7

[28] A.Z. Alam, A.A. Heidari, M. Habib, H. Faris, I. Aljarah, and M.A. Hassonah, "Salp Chain-Based Optimization of Support Vector Machines and Feature Weighting for Medical Diagnostic Information Systems", *Evolutionary Machine Learning Techniques.,* no. Nov, pp. 11-34, 2020.

[29] Abdelouahab Attia, Zahid Akhtar, Samir Akrouf, and Sofiane Maza, "A survey on machine and deep learning for detection of diabetic retinopathy", *ITACT Journal On Image And Video Processing,* vol. 11, no. 2, 2020.

CHAPTER 17

Plant Disease Detection: A Survey

I. Vartika Bhadana[1] and **Pooja Asterisk Pathak**[1,*]

[1] *GLA University, Mathura, India*

Abstract: The actual framework will improve seed or plantation growth by increasing their production, efficiency, and economic benefit. It also allows one to serve nature by overseeing plant growth by balancing the climate. Many methods have shown an important role in the variety of uses, such as medical, security, *etc.* Farming, remote sensing, market research, *etc.* The use of automated simulation tools to mimic human visual capacities has proved a dynamic function of smart or precision agriculture. This principle has allowed for the automated control and observation of seeds, planting, disease control, water conservation, *etc.* to improve seed production and efficiency. In this paper, we reviewed several publications that follow the principle of machine learning, deep learning, soft computing and digital image processing (DIP) approaches for the detection and classification of plant diseases.

Keywords: Deep learning, Digital image processing, Machine learning, Plant diseases, Soft computing.

INTRODUCTION

One of the big problems for farmers has always been planting and seed disorders. It may pose a substantial danger to the production potential of agriculture. However, diagnosing the true cause of the issue with the proper or reliable diagnosis can be a large benefit in the field of agriculture [1].

It is well said that it is very straightforward to recall and evaluate the items or objects humans imagine. The eyes act as a capturing system that absorbs and converts light into a signal then transmitted to the brain. The brain then covers this to greater learning. The idea of technology for visually analyzing and recognizing real-time details using computer systems or computers. This device facsimiles to some degree the understanding capacity of living things, like humans [2]. With the assistance of a camera serving as an eye for the device, data in the information of signal and photographs are developed and handled with the help of a digital

---

[*] **Corresponding author Vartika Bhadana and Pooja Pathak:** GLA University, Mathura, India; E-mails: Vartika.bhadana_mtcs19@gla.ac.in and pooja.pathak@gla.ac.in

Dharam Buddhi, Rajesh Singh and Anita Gehlot (Eds.)

system to extract any valuable information from it. It is understood that DIP is the technique of transferring a picture from a signal to a digital form. Since it is not easy to understand details obtained in the signal type, it has been translated into 0's and 1's to well understand the results.

Many artificial intelligence techniques are commonly used for the classification and detection of diseases in the plant. The popular techniques are decision tree, logistic regression, SVM, KNN, and Deep CNN. To optimize attribute extraction, these techniques are paired with separate image pre-processing processes.

K-NN is a basic classification algorithm based on memory that takes data based on a calculation of similarity.

METHODOLOGY

The farmers are concerned about the plant which are been destroy by many diseases. Plant pathology is the process that recognizes the development of the plant. Sometimes it is referred to as Phytopathology, originating from the Greek phrases where Phyto means plant, diseases mean path, and information means emblem. The study is about plant disease and disorder affected by variables such as biological factors, nutritional shortages, temperature, and microorganisms. The classification of pathogens is shown in Fig. (**1**).

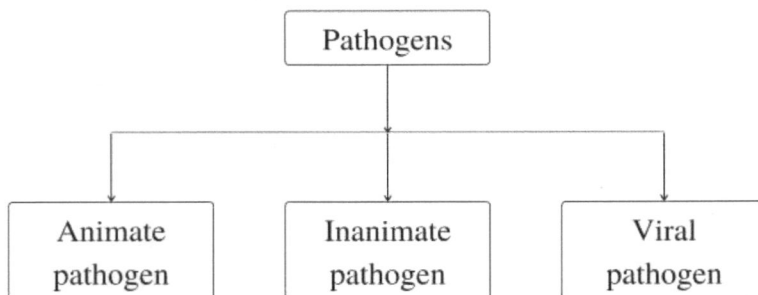

Fig. (1). Classification of pathogen.

Two types of plant diseases are there:

• Non-infectious diseases – These types of diseases aren't always connected with any animated or infectious pathogen, so they cannot be spread from an infectious agent to a healthy plant. It is mainly caused by low/high temperatures, inappropriate oxygen levels, uncertain water levels, hail, storm, air quality toxicity, *etc.*

• Infectious diseases – These are an infection caused by infectious species of bacteria in a series of environmental conditions. Microorganisms, fungus, viruses, microbes, and more bacterial infections will damage the plant.

The methodology is based on machine learning technologies to identify infectious and non-infectious diseases. The flow diagram of it is shown in Fig. (**2**), where the dataset is taken then the data augmentation process is done to add more new data and to reduce the overfitting of models, then the classification of diseases is been done by applying models like SVM, k-NN. After this, different results is been predicted for the diseases.

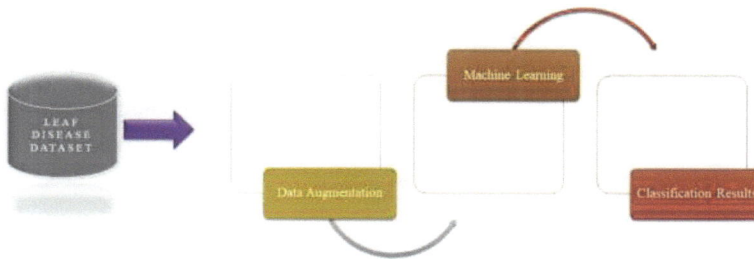

Fig. (2). Flow diagrams of classification of plant disease.

Plant diseases may be infectious or non-infectious agents, such as viruses or viroid, that cause diseases or disorders. Abiotic variables are environmental factors, such as temperature, precipitation, humidity, nutritional deficiency, *etc.* that cause sickness or disease. Table **1** shows the commonly occurring diseases and some are listed below:

Table 1. Commonly occurring diseases.

Disease's name	Caused from	Symptoms	Plant harmed
Rusts	Fungal	Orange-red, reddish, or yellow	Woody and herbaceous plant
Gall	insects	Knobs and lumps	Seen on the oak tree
Molds	Fungal	Small white grey color patches on the leaf	Mostly found on tomato plant
Curl of leaf	Virus	The leaves become thick that cause wrinkles	Mostly on peach tree
Virus	Red dwarf, mosaic, the plague of buds	Twisting, leaf discoloration	Tobacco, potatoes, soybeans, cucumbers.

Related Work

The work is divided into different categories, the authors have done the work on different technologies.

Digital Image Processing

Is the translation into digital data (0's and 1's) of optical picture data by using machines to interpret and comprehend it? Table **2** shows some of the paper related to DIP:

Table 2. Some paper on Digital Image Processing.

References	Plant	Diseases	Features
F. J. Molina *et al.* [3]	Tomato leaf	Early blight	Colors descriptor
A. Camargo *et al.* [4]	Banana, soya	Fungal	The different intensity distribution of the histogram
Qinghai He *et al.* [5]	Cotton	Fungal and bacterial	Ratio

Machine Learning

Machine learning algorithms have recently been used to identify plant diseases easily and effectively. Machine learning algorithms aim to diagnose diseases with better accuracy. The different models are used for detecting the diseases like SVM, K means, and many more. Table **3** shows the papers on machine learning.

Table 3. Some paper of machine learning.

References	Model	Purpose	Accuracy
Pal T. *et al.* [6]	SVM	Predict the sequence of plant proteins based on amino acids.	91.11%
J. Sperschneider *et al.* [7]	Naïve bayes	Prediction by fungi	AUC of 0.85
Saunders D.G. [8]	Markov clustering	Identify effectors	N/A

Deep Learning

Deep learning algorithms nowadays play a significant role in identifying plant diseases very fast and ineffective manner. Different models are been used for detecting the diseases of plant-like CNN, deep learning gives better accuracy. Table **4** shows the paper on deep learning.

Table 4. Some paper on deep learning.

References	Model	Diseases	Dataset
R. Aakanksha *et al.* [9]	ANN and k-means	Spot on leaf and scorch on leaf	28 images
S. Sanjeev *et al.* [10]	k-means and FFNN	Downy powdery mildew	33 images and self
S. Mohanty *et al.* [11]	CNN	Leaf Spot	54,306 healthy and infected images

The different authors have done the work in different technologies we have mentioned some of the authors work below:

Cope *et al.* [12] offered a study of the different methods of image processing that have been used to assess plant health and their organisms. The topics relating to the philosophy of botany were also discussed by the writer.

Barbedo *et al.* [13] published a survey on image detection techniques that can identify the species leaf and stem diseases. In the three categories given, the proposed work is then classified by identification, measurement of severity, and description. The study suggested by Huang *et al.* [14] provides a summary of the different soft computing techniques used for improved farming and material science.

Chouhan has proposed RBF neural network (NN) for detection and classifies the 6 fungal infections leaf from the photos of the plant. It was carried out using the optimization of bacterial foraging [15].

A leaf classification method called the Fuzzy Relevance Vector Machine (FRVM) was proposed by Vijaya Lakshmi and Mohan [16] based on features such as color, texture, and form. By using Particle swarm optimization, the optimal functions have been extracted.

Clement *et al.* [17] have introduced an image detection method for calculating foliar drying which appears attributable to the sycamore lace virus. The images were then split using the fuzzy c-mean algorithm, then NN was used to characterize the disorder [18].

For the segmentation purposes of the images, Tetila *et al.* [19] used a technique called Basic Linear Iterative Clustering. The leaf images were categorized with the aid of characteristics such as color, a form of texture, *etc.*

The suggested system dynamically tests very accurately the number of leaves and sizes of leaves [20]. Hamuda's *et al.*'s [21] paper provides a survey of the different methods of segmentation of plant images. Segmentation based on a color list, segmentation based on thresholds, and optimization based on learning describes the segmentation process. The segmentation based on the color index is one of them.

New sensors such as differential mobility·spectrometers and lateral flow devices, biomaterials focused on phage display and biophotonics, or remote sensing data coupled with spectroscopic approaches leading to organic agriculture have been illuminated by the authors [22]. Table **5** shows a summary of the literature review.

Table 5. Summary of literature review.

References	Models	Plants	Dataset	Diseases	Future Work
Yogesh D. and Radha K. *et al.* [23]	SIFT and Support vector machine	Soybean	120 images/self	-	Work on different fruits, plants.
Chaitali G.D. and K.H. wanjale *et al.* [24]	Color/clustering and support vector machine	Many	120 images/self-application	Sunburn, yellow-mosaic and Hopper on grass	Build real-time
D. Mrunmayee and A.B. Ingole *et al.* [25]	k-means, GLCM and Multilayer perceptron + BPA	Pomegranates	500 images/self	Spot on fruit, blight by bacteria	-
Diaz C. *et al.* [26]	Methods of Vegetation index	Soybean	48 images/self	Rust	Working on real-time
K. Ehsan and Tofik M. *et al.* [27]	Classification of Fuzzy logic algorithm	Strawberries	60 for iron and 50 for fungai/Self	Iron deficiency, fungal	-
Peter *et al.* [28]	CART	Citrus	Self	Fruit Spot	Work on more images

CONCLUSION

This paper presents the different principles and hypotheses related to all technologies for the detection and classification of diseases from plant pictures. Nearly all of the modern methodologies related to all technologies of plant pathology have been addressed in this survey article. Our goal is to make researchers and academics interested in developing and applying new technologies in this area.

CONSENT FOR PUBLICATION

Not applicable.

CONFLICT OF INTEREST

The authors declare no conflict of interest, financial or otherwise.

ACKNOWLEDGEMENT

Declare none.

REFERENCES

[1]   M. Faye, C. Bingcai, and K.A. Sada, "Plant Disease Detection with Deep Learning and Feature Extraction Using Plant Village", *Journal of Computer and Communications,* vol. 8, no. 6, pp. 10-22, 2020.
[http://dx.doi.org/10.4236/jcc.2020.86002]

[2]   M-L. Qi, "K. He, Z. N Huang, R. S. Yassar, G. Y. Xiao, Y. P. Lu, and T. Shokuhfar. "Hydroxyapatite fibers: a review of synthesis methods", *JOM,* vol. 69, no. 8, pp. 1354-1360, 2017.
[http://dx.doi.org/10.1007/s11837-017-2427-2]

[3]   M. Juan F., *2014 XIX Symposium on Image, Signal Processing and Artificial Vision.,* 2014pp. 1-5

[4]   A.C. da Silva, J.A. Ferro, F.C. Reinach, C.S. Farah, L.R. Furlan, R.B. Quaggio, C.B. Monteiro-Vitorello, M.A. Van Sluys, N.F. Almeida, L.M. Alves, A.M. do Amaral, M.C. Bertolini, L.E. Camargo, G. Camarotte, F. Cannavan, J. Cardozo, F. Chambergo, L.P. Ciapina, R.M. Cicarelli, L.L. Coutinho, J.R. Cursino-Santos, H. El-Dorry, J.B. Faria, A.J. Ferreira, R.C. Ferreira, M.I. Ferro, E.F. Formighieri, M.C. Franco, C.C. Greggio, A. Gruber, A.M. Katsuyama, L.T. Kishi, R.P. Leite, E.G. Lemos, M.V. Lemos, E.C. Locali, M.A. Machado, A.M. Madeira, N.M. Martinez-Rossi, E.C. Martins, J. Meidanis, C.F. Menck, C.Y. Miyaki, D.H. Moon, L.M. Moreira, M.T. Novo, V.K. Okura, M.C. Oliveira, V.R. Oliveira, H.A. Pereira, A. Rossi, J.A. Sena, C. Silva, R.F. de Souza, L.A. Spinola, M.A. Takita, R.E. Tamura, E.C. Teixeira, R.I. Tezza, M. Trindade dos Santos, D. Truffi, S.M. Tsai, F.F. White, J.C. Setubal, and J.P. Kitajima, "Comparison of the genomes of two Xanthomonas pathogens with differing host specificities", *Nature,* vol. 417, no. 6887, pp. 459-463, 2002.
[http://dx.doi.org/10.1038/417459a] [PMID: 12024217]

[5]   Z.H. Yan, J.Q. Guo, C.J. Zhang, L.D. Sun, X.D. Zhang, J.J. Lin, Y.H. Wang, F. Feng, M. Peng-Li, L. Cai-Hong, and L. Yan-Chun, "Climate change impacts and adaptation strategies in northwest China", *Adv. Clim. Chang. Res.,* vol. 5, no. 1, pp. 7-16, 2014.
[http://dx.doi.org/10.3724/SP.J.1248.2014.007]

[6]   T. Pal, V. Jaiswal, and R.S. Chauhan, "DRPPP: A machine learning based tool for prediction of disease resistance proteins in plants", *Comput. Biol. Med.,* vol. 78, pp. 42-48, 2016.
[http://dx.doi.org/10.1016/j.compbiomed.2016.09.008] [PMID: 27658260]

[7]   J. Sperschneider, D.M. Gardiner, P.N. Dodds, F. Tini, L. Covarelli, K.B. Singh, J.M. Manners, and J.M. Taylor, "EffectorP: predicting fungal effector proteins from secretomes using machine learning", *New Phytol.,* vol. 210, no. 2, pp. 743-761, 2016.
[http://dx.doi.org/10.1111/nph.13794] [PMID: 26680733]

[8]   D.G. Saunders, J. Win, L.M. Cano, L.J. Szabo, S. Kamoun, and S. Raffaele, "Using hierarchical clustering of secreted protein families to classify and rank candidate effectors of rust fungi", *PLoS One,* vol. 7, no. 1, 2012.e29847
[http://dx.doi.org/10.1371/journal.pone.0029847] [PMID: 22238666]

[9]   R. Aakanksha, R. Arora, and S. Sharma, "Leaf disease detection and grading using computer vision technology & fuzzy logic. In 2015 2nd international conference on signal processing and integrated networks (SPIN)", pp. 500-505. IEEE, 2015.

[10]   S.S. Sannakki, *2013 Fourth International Conference on Computing, Communications and Networking Technologies (ICCCNT),* 2013, pp. 1-5

[11]   S.P. Mohanty, D.P. Hughes, and M. Salathé, "Using deep learning for image-based plant disease detection", *Front. Plant Sci.,* vol. 7, p. 1419, 2016.
       [http://dx.doi.org/10.3389/fpls.2016.01419] [PMID: 27713752]

[12]   D.S. Ferreira, A.D. Matte Freitas, G.G. da Silva, H. Pistori, and M.T. Folhes, "Weed detection in soybean crops using ConvNets", *Comput. Electron. Agric.,* vol. 143, pp. 314-324, 2017.
       [http://dx.doi.org/10.1016/j.compag.2017.10.027]

[13]   J. S. Cope, D. Corney, J. Y. Clark, P. Remagnino, and P. Wilkin, *Plant species identification using digital morphometrics: A review." Expert Systems with Applications,* vol. 39, no. 8, pp. 7562-7573, 2012.
       [http://dx.doi.org/10.1016/j.eswa.2012.01.073]

[14]   J. Barbedo, and G. Arnal, "A review on the main challenges in automatic plant disease identification based on visible range images", *Biosyst. Eng.,* vol. 144, pp. 52-60, 2016.
       [http://dx.doi.org/10.1016/j.biosystemseng.2016.01.017]

[15]   R. Anand, S. Veni, and J. Aravinth, An application of image processing techniques for detection of diseases on brinjal leaves using k-means clustering method." In 2016 international conference on recent trends in information technology (ICRTIT) IEEE, pp. 1-6, 2016.
       [http://dx.doi.org/10.1109/ICRTIT.2016.7569531]

[16]   Y. Dandawate, and R. Kokare, "An automated approach for classification of plant diseases towards development of futuristic Decision Support System in Indian perspective", *2015 International conference on advances in computing, communications and informatics (ICACCI),* 2015, pp. 794-799
       [http://dx.doi.org/10.1109/ICACCI.2015.7275707]

[17]   A. Singh, B. Ganapathysubramanian, A.K. Singh, and S. Sarkar, "Machine learning for high-throughput stress phenotyping in plants", *Trends Plant Sci.,* vol. 21, no. 2, pp. 110-124, 2016.
       [http://dx.doi.org/10.1016/j.tplants.2015.10.015] [PMID: 26651918]

[18]   V. Balasubramanian, and M. Vasudev, "Kernel-based PSO and FRVM: An automatic plant leaf type detection using texture, shape, and color features", *Comput. Electron. Agric.,* vol. 125, pp. 99-112, 2016.
       [http://dx.doi.org/10.1016/j.compag.2016.04.033]

[19]   N.J. Vickers, "Animal communication: when i'm calling you, will you answer too", *Curr. Biol.,* vol. 27, no. 14, pp. R713-R715, 2017.
       [http://dx.doi.org/10.1016/j.cub.2017.05.064] [PMID: 28743020]

[20]   T.E. Castelão, B.B. Machado, N.A. Belete, D. Augusto Guimarães, and H. Pistori, "Identification of soybean foliar diseases using unmanned aerial vehicle images", *IEEE Geosci. Remote Sens. Lett.,* vol. 14, no. 12, pp. 2190-2194, 2017.
       [http://dx.doi.org/10.1109/LGRS.2017.2743715]

[21]   D. Mrunmayee, and A.B. Ingole, "Diagnosis of pomegranate plant diseases using neural network." In 2015 fifth national conference on computer vision, pattern recognition, image processing and graphics (NCVPRIPG)", IEEE, pp. 1-4, 2015.

[22]   D. Williams, A. Britten, S. McCallum, H. Jones, M. Aitkenhead, A. Karley, K. Loades, A. Prashar, and J. Graham, "A method for automatic segmentation and splitting of hyperspectral images of raspberry plants collected in field conditions", *Plant Methods,* vol. 13, no. 1, p. 74, 2017.
       [http://dx.doi.org/10.1186/s13007-017-0226-y] [PMID: 29118819]

[23]   A.E. Erdal, A. Abramov, F. Wörgötter, H. Scharr, A. Fischbach, and B. Dellen, "Modeling leaf growth

of rosette plants using infrared stereo image sequences", *Comput. Electron. Agric.,* vol. 110, pp. 78-90, 2015.
[http://dx.doi.org/10.1016/j.compag.2014.10.020]

[24]   H. Esmael, M. Glavin, and E. Jones, "A survey of image processing techniques for plant extraction and segmentation in the field", *Comput. Electron. Agric.,* vol. 125, pp. 184-199, 2016.
[http://dx.doi.org/10.1016/j.compag.2016.04.024]

[25]   P. Grajzl, and S. Silwal, "The functioning of courts in a developing economy: evidence from Nepal", *Eur. J. Law Econ.,* vol. 49, no. 1, pp. 101-129, 2020.
[http://dx.doi.org/10.1007/s10657-017-9570-7]

[26]   T. Beatrice, G. B. Liponi, V. Meucci, L. Casini, C. Dall'Asta, L. Intorre, and D. Gatta, *Aflatoxins M1 and M2 in the milk of donkeys fed with naturally contaminated diet." Dairy Science & Technology,* vol. 96, no. 4, pp. 513-523, 2016.

[27]   J. Francis, and B.K. Anoop, "Identification of leaf diseases in pepper plants using soft computing techniques", In: *In 2016 conference on emerging devices and smart systems (ICEDSS),* 2016.
[http://dx.doi.org/10.1109/ICEDSS.2016.7587787]

[28]   P. Ganesan, G. Sajiv, and L. M. Leo, "CIELuv color space for identification and segmentation of disease affected plant leaves using fuzzy based approach", In: *In 2017 Third International Conference on Science Technology Engineering & Management (ICONSTEM),* 2017, pp. 889-894. IEEE,

CHAPTER 18

Automated Indoor Farming System with Remote Monitoring

Dushyant Kumar Singh[1,*], **N. Ram Gopal**[2], **K. Pranay Raju**[2], **M. Jagannadha Varma**[2], **P. Balachandra**[2] and **L.B. Lokeshwar Reddy**[2]

[1] *Embedded Sytems, Dept. of ECE, Lovely Professional University, Jalandhar, Punjab, India*

[2] *Dept. of ECE, Lovely Professional University, Jalandhar, Punjab, India*

Abstract: India is one of the largest agricultural producing countries in the world where about 58% of its population's livelihood is dependent on it. The traditional method of agriculture requires a lot of human effort but then also we cannot assure the maximum yield from the crops due to various environmental factors. With the help of the Internet of Things - IoT we can modernize the traditional methods of farming and control the environmental factors. This not only helps the plants to get the maximum yield but also reduces a lot of human effort. This proposed system mainly depends on the information which is retrieved from the sensors as an automatic watering system waters the plant according to the data retrieved from the temperature and moisture sensor of the system and can be helped in reducing the water wastage.

Keywords: Agriculture Automation, Blynk, Farming, Indoor Farming, IoT, Remote Monitoring.

INTRODUCTION

For the past few years, the demand for building automation systems soars up, especially in offices and households. The term automation which is referred to the automation system that can mix household activities which include sensors to read the input condition and centralized the control of electrical appliances.

LITERATURE REVIEW

Arul Jai Singh explained the close relationship between innovations in embedded systems and how they help to farm, it doesn't have remote monitoring [1].

---

* **Corresponding author Dushyant Kumar Singh:** Embedded Sytems, Dept. of ECE, Lovely Professional University, Jalandhar, Punjab, India; E-mail: dushyant.kumar@lpu.co.in

Dharam Buddhi, Rajesh Singh and Anita Gehlot (Eds.)

Chin and Audah implemented cloud agriculture monitoring and they also included a wide range of sensors, the drawback here is it doesn't have humidity monitoring and an automatic water controlling system [2].

Saraswathi developed software to use in an Android mobile, it requires a Wi-Fi connection to centralize the server. But the drawback here is the system lacks cloud Services [3].

Latha developed a "cooperative communication-based Wireless SensorNetwork", she monitored all the parameters like Temperature, Humidity, CO_2. But the system was only limited to smart agriculture, not for zone-wise farming [4].

Ayaz *et al.*, made a perfect system which was named "Making Fields Talk" which had all the sensors related to Temperature, Humidity, Soilmoisture, *etc*. But the system was unable to determine air inflow and cannot control light for the plants [5].

Balachander *et al.*, proposed some advanced technologies to improve the efficiency in agriculture and some vast improvements to be made in the system, But there was no chance to implement the IoT in agriculture [6].

Sambath *et al.*, made a system that measures the temperature, soil moisture and controls the water inflow to the plants. It doesn't have humidity monitoring and air inflow control [7].

Sahu and Verma made an automated watering system by taking measures like the availability of water in some regions. This system mainly projects the automatic sprinkler system in agriculture [8].

METHODOLOGY

A quantitative approach was chosen for the design of the system by analyzing farming techniques and projects. Firstly, related websites, research papers, ideas, and podcasts on this topic from the last 15 years were reviewed. This gave many views on how the system can be designed. Next, the hardware used on these techniques has been assessed, and the cost and the scalability of the hardware have been reviewed.

EXPERIMENTATION

Raspberry Pi

Raspberry Pi, is often the center of the project, Collecting data from all the sensors and analyzing it in line with the program given to that and transmitting the info to the farmer. All the sensors and peripherals used here are connected through GPIO pins on the board and programmed using python, which advances because of the fundamental artificial language for Raspberry Pi. Fig. (**1**) shows the Raspberry pi 3b+ which we used in this project.

Fig. (1). Raspberry Pi 3b+.

Soil Moisture Sensor and Water Dispenser

Almost 70 percent of the freshwater existing today is employed for agriculture. Fig. (**2**) is A soil moisture sensor is employed to detect the moisture level of the soil and therefore the water dispenser drips just barely enough water for the crop.

Fig. (2). Soil moisture sensor.

Lights

Lighting is one of the most issues in Indoor farming. a lightweight device is employed to calculate the intensity of the sunshine the plant is receiving and is monitored exploitation Raspberry pi. Fig. (**3**) showing how lighting is provided to plants.

Fig. (3). Lighting for indoor plants.

Temperature and Humidity

Fig. (**4**) is a DHT11 sensor that is used to calculate the temperature and humidity inside the system. This helps in controlling these parameters with the requirements of the crop.

Fig. (4). DHT11.

Camera

Fig. (**5**) is This camera delivers high-definition photos and might additionally shoot videos. This helps in predicting the expansion time and build information points that may be used for following the crop cycle.

Fig. (5). Raspberry Pi camera.

GSM Module

Fig. (**6**) is SIM900A GSM module is employed during this for warning and causing time to time updates regarding the crop once the user isn't close.

Fig. (6). SIM900.

CO_2 Sensor

Fig. (**7**) is MG811 sensor is used to detect CO_2 concentration in the air and send it to the Raspberry pi which controls the inlet and outlet accordingly.

Fig. (7). 7 MG811 CO_2 sensor.

Circuit Diagram

Fig. (**8**) shows the circuit of the project.

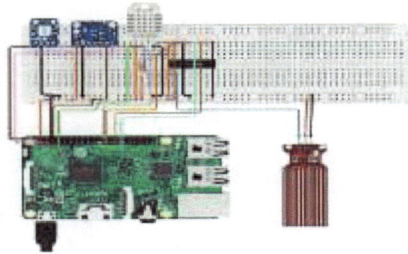

Fig. (8). Circuit.

Fig. (**9**) showing the flow chart of how this system works.

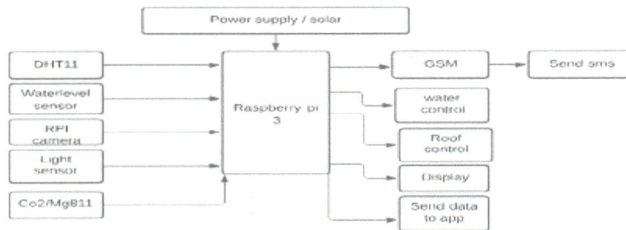

Fig. (9). Flow chart.

RESULTS AND DISCUSSIONS

The figure mentioned below shows the message being received by a user throughout the day, and Fig. (**10**) shows the working of the mobile application. It displays the data from all the sensors to the mobile application and sends an SMS to the user as a status report. Fig. (**11**) shows the user interface of the mobile application.

Fig. (10). SMS.

Fig. (11). Blynk app.

CONCLUSION

The current agricultural methods are not robust, and farmers have to work too hard to achieve a good yield in the crop. This project helps the farmers to agriculture in an efficient way and achieve good yields with less stress. This paper helps the farmers to monitor their farms from anywhere. In the coming days, this project should be made into production and helps a lot of farmers.

CONSENT FOR PUBLICATION

Not applicable.

CONFLICT OF INTEREST

The authors declare no conflict of interest, financial or otherwise.

ACKNOWLEDGEMENT

Declared none.

REFERENCES

[1]　"Embedded based greenhouse monitoring system using PIC microcontroller", *IEEE Trans Syst, Man, Cybern . A, Systems and Humans,* 2011.

[2]　Y. Chin, and L. Audah, "Vertical farming monitoring system using the internet of things (IoT)", *AIP Conf. Proc.,* vol. 1883, p. 020021, 2017.
[http://dx.doi.org/10.1063/1.5002039]

[3]　D. Saraswathi, "Automation of Hydroponics Green House Farming using IOT", *2018 IEEE International Conference on System, Computation, Automation and Networking (ICSCA),* 2018, pp. 1-4
[http://dx.doi.org/10.1109/ICSCAN.2018.8541251]

[4]　M.N.S. Latha, S.V. Girish, and A.B. Ganesh, "Cooperative communication enabled wireless sensor network for monitoring green house", *2016 International Conference on Advanced Communication Control and Computing Technologies (ICACCCT), Ramanathapuram, India,,* 2016 pp. 86-91
[http://dx.doi.org/10.1109/ICACCCT.2016.7831606]

[5]　M. Ayaz, M. Ammad-Uddin, Z. Sharif, A. Mansour, and E-H.M. Aggoune, "Internet-of-Things (IoT)-based smart agriculture: toward making the fields talk", *IEEE Access,* vol. 7, pp. 129551-129583, 2019.
[http://dx.doi.org/10.1109/ACCESS.2019.2932609]

[6]　C. Balachandar, "Vertical farm buildings", *Sameeksha,* 2014. CEvolution

[7]　M. Sambath, *J. Phys. Conf. Ser.,* vol. 1362, p. 012069, 2019.
[http://dx.doi.org/10.1088/1742-6596/1362/1/012069]

[8]　"TanuShahu, and Ashok", *Int. J. Comput. Appl.,* vol. 172, no. 6, 2017.

CHAPTER 19

An Improved Method for Diabetes Prediction through the Application of Neural Network

Roshi Saxena[1,*], **Sanjay Kumar Sharma**[1] and **Manali Gupta**[1]

[1] Department of Computer Science, School of Information and Communication Technology, Gautam Buddha University, Greater Noida, Uttar Pradesh, India

Abstract: A common disease that is affecting the whole world is diabetes, and is also known as a silent killer. Diabetes affects the nervous system, retina of the eyes, kidney, heart and affects the entire system of the body. Diabetes boosts the other diseases also whenever patients suffer from them otherwise it remains in sleep mode. The foremost reason for boosting diabetes is the lifestyle of today's generation. In this paper, we have tried to predict diabetes at the primary stage by making use of a neural network *i.e.*, multilayer perceptron. We have made use of the PIMA Indians diabetes dataset in our article and the experiments were performed using the Weka tool. After applying the proposed algorithm, the experimental result shows an increase in accuracy by 3% which is far better for predicting diabetes.

Keywords: Accuracy, Diabetes, Neural Network, Performance.

INTRODUCTION

One of the prolonged illnesses that research has shown in medical science is diabetes. Diabetes is a continuing ailment and is instigated due to a large amount of sugar present in the human blood. It is also called a silent killer. It generally remains in dormant mode but boosts other diseases such as kidney disease, cardiac problems, retina problems. When the amount of insulin is deficient in the human body, diabetes occurs. When there is improper digestion of blood glucose in the body *i.e.*, it does not get metabolized properly. The main source of energy in the human body is blood glucose. Insulin is the hormone that is created by the pancreas to help body cells in absorbing glucose that human beings get from food. If enough insulin is not produced, glucose remains in the blood and causes diabetes. Diabetic people have to take care of the whole of their lifespan by maintaining a healthy lifestyle and can remain healthy throughout their whole life.

*[*] **Corresponding author Roshi Saxena:** Amity School of Engineering & Technology, Amity University, Uttar Pradesh, Noida, India; E-mail: saxena.roshi@gmail.com

Dharam Buddhi, Rajesh Singh and Anita Gehlot (Eds.)

Diabetes is of three types: In type 1 diabetes, the pancreas is being attacked by body cells with antibodies that damage the organ, and enough insulin is not produced and the blood glucose remains unabsorbed. It is also known as insulin-dependent diabetes. In type 2 diabetes, an insufficient amount of insulin is being produced by the pancreas, thus increasing the sugar levels in the blood. This type of diabetes is also known as insulin-independent diabetes. Type 3 diabetes is gestational diabetes that occurs in expecting mothers. There is a condition known as a pre-diabetic condition where the amount of blood sugar is on the borderline and the absence of physical exercise and excess weight gain could be the reason for the pre-diabetic condition.

In this research article, the neural network is used to make predictions for diabetes at an early stage. A neural network is one of the machine learning techniques which learns from the data and builds the algorithm around the data. It is also known as multilayer perceptron.

RELATED WORK

Many researchers in the past have given strategies to detect diabetes using multilayer perceptron. Kaur *et al.* [1] have used an artificial neural network, linear support vector machines, kernel-based radial function and have classified the dataset into diabetic and non-diabetic people. Perveen *et al.* [2] discuss the ensemble of Adaboost and bagging by making use of J48 decision trees for diabetes classification after performing extensive experiments, Adaboost machine learning outperforms as compared to bagging as well as J48 technique. Robustness was enlarged by advancing techniques in diabetes prediction. K-nearest neighbor and Logistic Regression were used by the Selvakumar *et al.* [3] for detection of diabetes and classification of patients. By Maniruzzaman *et al.* [4], gaussian process-based classification technique is used by making use of linear, polynomial, and radial-basis kernel and a comparison was drawn against linear discriminant analysis, quadratic discriminant analysis, and naïve Bayes. Rigorous tests were supported out to find the best working cross-validation protocol. Their experiments revealed that Gaussian process-based classifier along with a 10-fold cross-validation protocol is the superlative classification technique for predicting diabetes. Mohapatra *et al.* [5] have made use of a Neural network and done testing on the divided dataset. Division of the dataset has been done into training and testing dataset and has proved that test data gives the classification accuracy of 77.5% when being divided.

The organization of the remaining paper is as follows: Materials and Methods section represents materials and methods, including dataset description, tool description, prediction algorithms, and classifiers evaluation. Results Section represents the consequences of all classifiers applied before feature selection and enhancement in the exactness of the classifiers after feature selection. The conclusion discusses the summary of current work and future work.

MATERIALS AND METHODS

Dataset and Tools

In this research paper, we have made use of the PIMA Indians diabetes dataset which contains 8 feature attributes and 1 class attribute. Eight attributes are several times a woman is pregnant, diastolic blood pressure, diabetes pedigree function, plasma glucose concentration, two-hour serum insulin, age of the patient, triceps skinfold thickness, and last but not the least body mass index. The description of the dataset is shown in Table 1. We have made use of weka to categorize the expecting ladies into diabetic and non-diabetic one.

Table 1. Description of PIMA Indian diabetes Dataset.

Attributes Mean	Standard Deviation	Min/Max Value
No. of times pregnant 3.8	3.4	1/17
Plasma glucose concentration 120.9	32	56/197
Diastolic Blood Pressure 69.1	19.4	24/110
Triceps skinfold thickness(mm) 20.5	16	7/52
2-hour serum insulin 79.8	115.2	15/846
Body mass index(kg/m2) 32	7.9	18.2/57.3
Diabetes pedigree function 0.5	0.3	0.0850/2.32
Age 33.2	11.8	21/81
Class	Tested Positive:	Diabetic
	Tested Negative:	Non-Diabetic

METHODOLOGY

Multilayer Perceptron: Multilayer perceptron is a feed-forward neural network [6] and it consisted of three layers. The name of the three layers is the input layer from which input is fed forward to the network, the output layer and few layers are present between the input layer and output layer which is known as the hidden layer. Input to the next coat is produced by an input layer with a linear activation function. Linearly separable problems can be solved with the help of a single-layer perceptron network [7], but the intricate complications that are linearly inseparable can't be solved with a single-layer perceptron network, to solve these problems we need a multilayer perceptron network *i.e.* neural network. The main application of neural networks is the reorganization of patterns and classification purposes.

The architecture of the neural network is shown in Fig. (**1**).

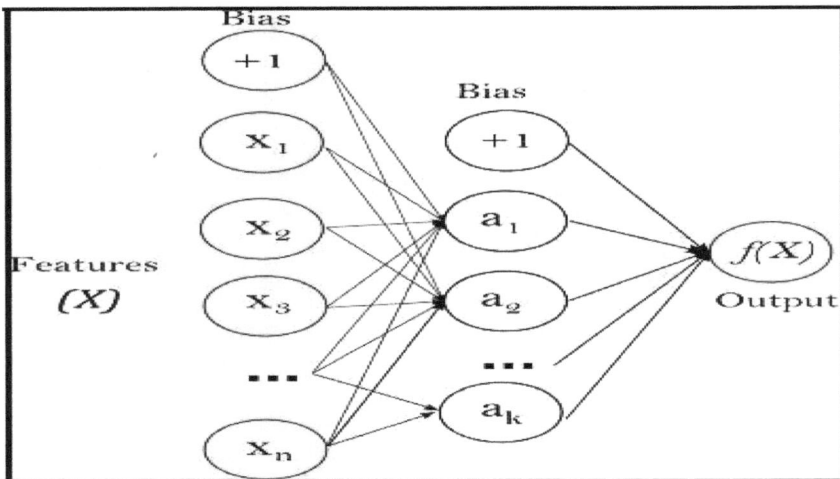

Fig. (1). Multilayer perceptron architecture.

Steps of Proposed Algorithm

- Load the dataset
- Check if the dataset is in processed form or not, if not pre-process the database.
- Find out significant features using feature selection methods.
- Discover the constraints for tuning the dataset.
- Run the multilayer perceptron classifier.
- After applying the above algorithm, a comparison of accuracy has been measured.

RESULTS

Evaluation Parameters: Predictions are made on the following evaluation parameters:

Precision: Number of correctly classified instances.

Recall: Number of people who are correctly identified with the disease.

Accuracy: How accurate is classification [8] result in predicting diabetic and non-diabetic patients.

We have applied the steps of the above-said algorithm and the confusion matrix before and after the application is shown in Tables **2** and **3** respectively.

Table 2. Confusion Matrix of Neural Network algorithm.

-	Diabetic	Non-diabetic
Diabetic	164	104
Non-diabetic	87	413

Table 3. Confusion Matrix of Neural Network algorithm.

-	Diabetic	Non-diabetic
Diabetic	139	103
Non-diabetic	54	423

Table **4** shows the accuracy before and after the application of the proposed algorithm.

Table 4. Results before and after the proposed algorithm.

-	Before Proposed Algorithm	After Proposed Algorithm
Precision	0.748	0.776
Recall	0.751	0.782
Accuracy	75.10	**78.16**

The neural network after the application of the proposed method is shown in Fig. (**2**). A graphical Comparison of evaluation parameters is shown in Fig. (**3**).

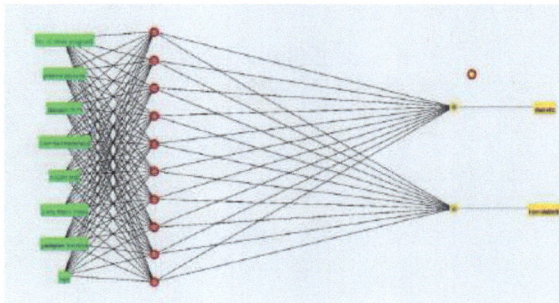

Fig. (2). Neural network model for proposed algorithm.

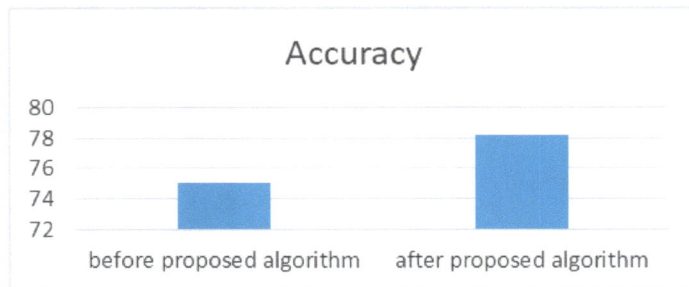

Fig. (3). Accuracy comparison.

CONCLUSION

Nowadays, diabetes can occur to anybody having an unhealthy lifestyle, excessive body weight. We have tried to predict diabetes at the preliminary stage using a neural network [9, 10] model. In this paper, we have run the proposed algorithm with multilayer perceptron on weka 3.9.4 and we have shown that accuracy has increased by 3%. Future work focuses on predicting diabetes with more accuracy using python.

CONSENT FOR PUBLICATION

Not applicable.

CONFLICT OF INTEREST

The authors declare no conflict of interest, financial or otherwise.

ACKNOWLEDGEMENT

Declared none.

REFERENCES

[1]   H. Kaur, and V Kumari, "Predictive modelling and analytics for diabetes using a machine learning approach", *Applied Computing and Informatics,* 2018.
[http://dx.doi.org/10.1016/j.aci.2018.12.004]

[2]   S. Perveen, M. Shahbaz, A. Guergachi, and K. Keshavjee, "Performance analysis of data mining classification techniques to predict diabetes", *Procedia Comput. Sci.,* vol. 82, pp. 115-121, 2016.
[http://dx.doi.org/10.1016/j.procs.2016.04.016]

[3]   S. Selvakumar, K.S. Kannan, and S. Gothai Nachiyar, "Prediction of Diabetes Diagnosis Using Classification Based Data Mining Techniques", *International Journal of Statistics and Systems,* vol. 12, no. 2, pp. 183-188, 2017.

[4]   M. Maniruzzaman, N. Kumar, M. Menhazul Abedin, M. Shaykhul Islam, H.S. Suri, A.S. El-Baz, and J.S. Suri, "Comparative approaches for classification of diabetes mellitus data: Machine learning

paradigm", *Comput. Methods Programs Biomed.,* vol. 152, pp. 23-34, 2017.
[http://dx.doi.org/10.1016/j.cmpb.2017.09.004] [PMID: 29054258]

[5]   S.K. Mohapatra, J.K. Swain, and M.N. Mohanty, "Detection of diabetes using multilayer Perceptron"
International Conference on intelligent computing and applications", pp. 109-116, 2019.

[6]   V.S. Kamadi, "R P Varma, Allam Appa Rao, T.Sitamahalakshmi, and PV Nageshwara Rao " A
Computational Intelligence approach for a better diagnosis of diabetic patients", *Journal of Computers
and Electrical Engineering,* vol. 40, no. 5, pp. 1758-1765, 2014.
[http://dx.doi.org/10.1016/j.compeleceng.2013.07.003]

[7]   T.M. Cover, "Geometrical and statistical properties of systems of linear inequalities with applications
in pattern recognition", *IEEE Trans. Electron. Comput.,* pp. 326-334, 1965.
[http://dx.doi.org/10.1109/PGEC.1965.264137]

[8]   N Sneha, "Analysis of diabetes mellitus for early prediction using optimal feature selection", *Journal
of Big data,* 2019.

[9]   A. Reinhardt, and T. Hubbard, "Using neural networks for prediction of the subcellular location of
proteins", *Nucleic Acids Res.,* vol. 26, no. 9, pp. 2230-2236, 1998.
[http://dx.doi.org/10.1093/nar/26.9.2230] [PMID: 9547285]

[10]   D. Sisodia, and D.S. Sisodia, "Prediction of diabetes using classification algorithms", *Procedia
Comput. Sci.,* vol. 132, pp. 1578-1585, 2018.
[http://dx.doi.org/10.1016/j.procs.2018.05.122]

CHAPTER 20

Enhanced Study on Security Major Issues and Challenges in the Vehicular Area Network

I. Manoj Sindhwani[1], **Charanjeet Singh**[1,*] and **Rajeshwar Singh**[1]

[1] Lovely Professional University Phagwara, India

Abstract: Ad-hoc network is a kind of temporary network used to establish connections for a temporary purpose. The ad hoc network is a kind of wireless network. This is not an infrastructure-based network. Mobile ad-hoc networks and Vehicular ad-hoc networks are two popular areas of ad-hoc networks. VANET is the most popular network nowadays. VANET is the network on wheels *i.e.* Vehicle is considered as a node; there is no issue of power management in this network, unlike MANET networks. Further to increasing the efficiency at such high speed, various issues and challenges occur in the path of this self-organized network. VANETs present the largest real-time application but lack security, scalability, efficient routing, and clustering protocols. The main aim of the study is to focus on various issues and challenges related to security requirements in VANET, which will allow researchers to work in this field for increasing the reliability of the network. Also, various attacks are focused which show the threats in the vehicular networks because of their dynamic topology.

Keywords: Intelligent Transport Network (ITN), MANET, VANET, V2I, V2V.

INTRODUCTION

Over the most recent couple of years Vehicular specially appointed systems (VANETs) have been an intriguing issue for analysts. VANETs [1, 2] draw in such a great amount of consideration of both the scholarly world and industry because of their one-of-a-kind attributes, for example, high unique topology and unsurprising versatility [1, 2]. Vehicular ad-hoc networks (VANET) are a special case of ad-hoc networks, where the vehicle's speed is dynamic and generally considered to be a fast topology varying network so sometimes it's difficult to maintain the communication between the vehicular nodes. So VANET protocols are defined which are not similar to Mobile ad-hoc networks (MANET). VANET is different from MANET in various cases like power limitation, topology changes or protocol implementation, *etc.* So VANET is a popular technique that

---

* **Corresponding author Charanjeet Singh:** Lovely Professional University Phagwara, India; E-mail: rcharanjeet@gmail.com

Dharam Buddhi, Rajesh Singh and Anita Gehlot (Eds.)

will change the future of Car communication, as to date, the vehicle mode is more mechanical, which needs to be improvised more towards software credibility. This network works more towards the communication of the vehicles and takes all the advantage of wireless communication [3, 4] between vehicle to vehicle (V2V) and Vehicle to Infrastructure (V2I). Roadside Units (RSU) also help vehicles to forward the information to another vehicle or node.

So, VANET protocols are defined which are not similar to Mobile ad-hoc networks (MANET). VANET is different from MANET in various cases like power limitation, topology changes or protocol implementation, *etc.* So VANET is a popular technique that will change the future of Car communication, as to date the vehicle mode is more mechanical, which needs to be improvised more towards software credibility. This network works more towards the communication of the vehicles and takes all the advantage of wireless communication [3, 4] between vehicle to vehicle (V2V) and Vehicle to Infrastructure (V2I). Roadside Units (RSU) also help vehicles to forward the information to another vehicle or node. The information can be security-related also. It is equipped for giving information correspondence between the portable vehicles [5 - 8]. It centers on security-related applications and the Internet getting to related applications. The Vehicular network arrangement is shown in Fig. (**1**).

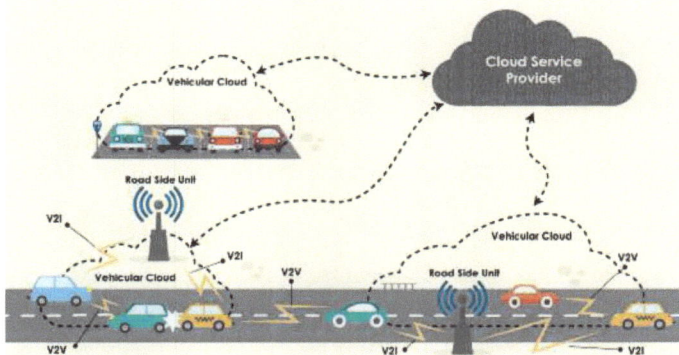

Fig. (1). VANET architecture.

Vehicular correspondence is given by the VANET which satisfies the necessities of the wellbeing as it is as creating system. The principle goal of this system is to improve well-being by trading messages between the vehicles by using remote correspondences. The state of streets, activity and diverse viewpoints which are urgent to security is an alarm by utilizing this system for the direction of vehicle stream. The transmission of data is happening in an open-access condition. The driver choice is affected by the aggressor by sending false movement cautioning messages. This assault prompts wastage of time of the driver, vehicles fuel and even prompts car crashes.

MAJOR ISSUES IN VANET

Issues are termed as the drawbacks of the VANET network but the drawbacks are useful for further research which opens the new schemes. Some issues are related to VANET. These issues are as follows:

High Mobility

Due to high mobility [6, 7] all the nodes in the network are not able to communicate properly with each other because of the rapid movement of the vehicles. Due to mobility vehicle changes its topology and the vehicle which is selected for communication purposes will leave the network. The link between the vehicle and RSU will break and the whole network is not able to communicate under high mobility. It also decreases the efficiency of the system. In the DSMC treatment, boundary interactions are treated as collisions of either purely specular reflection type, purely diffuse reflection type, or a combination of specular and diffuse reflections, with a specific description of the energy and momentum transfer using accommodation coefficients of the reflection behavior along the tangential and the normal directions to the walls. Fig. (**2**) shows mobility in VANET.

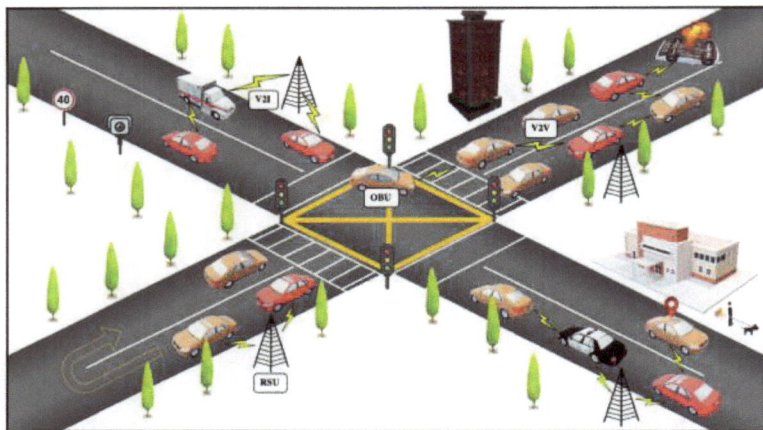

Fig. (2). Mobility in VANET [8].

Time Constraint

VANET is used for safety message delivery in case of road accidents, alternate routes, congestion, and much more. So the exact delivery time for the transmission of the information from one node to another node is difficult under

VANET. For example, suppose a safety message takes 200ns to reach the destination but due to real-time constraints it takes 300ns, so it is a drawback of the network [10].

Security

Security includes authentication, confidentiality, and integrity [9]. It is important to the vehicle that it must be authenticated by the network administrator so that it can participate in the network configuration. The message sends by the vehicles will also be authenticated. Therefore, a system wants to be introduced to such type of schemes which enables the message to known and also recognized by central authorities in the cases of accidents.

Geographical Tracking

For the actual location of the vehicle GPS tracking is required which helps in the communication over the wireless network. If there is no proper system for location identification, the delay factor will involve in the network automatically [11].

Delay

In VANET delay is a major issue that destroys the whole network under-maintained condition also. The delay should maintain under a high congestion network. The delay is increased by congestion, link breakage, overhead, packet loss, *etc.* Collision leads to the jamming problem in the network, so to overcome this problem delay should be minimum [10].

SECURITY REQUIREMENTS FOR VANET

Security in VANET is a major concern because if an intruder will come and attack the system then the life of a human can be affected. Vehicular Ad-hoc arranges are unique instances of specially appointed systems that, other than lacking foundation, imparting substances move with different increasing speeds. As needs are, this obstructs setting up solid end-to-end correspondence ways and having effective information exchange. In this manner, there are distinctive system concerns and security challenges in VANETs to get the accessibility of pervasive availability, secure correspondences, and notoriety administration frameworks which influence the trust in participation and arrangement between versatile systems administration substances. All things considered, there are different mechanical issues related to driver fewer autos and these impacts fundamentally make issues people in general. Subsequently, for the selection of these driver-less autos, it is important to have legitimate safety efforts as it can make hurt society and is considered as one of the biggest hindering variables. Security requirements include confidentiality [12], integrity, availability, privacy, access control, *etc.* Some of them are discussed here.

• **Authentication**: In the security of VANET, authenticity is a major challenge. Before accessing all the available services, it is necessary to authenticate by all the active stations in the network. The process of authentication damages the whole network due to a violation or attack within the network. The main objective of authentication is to protect nodes from outside or inside attacks in a vehicular network [13]. The authorization levels of vehicles are controlled by the process of authentication. Sybil attacks are prevented by the authentication in VANETs by specifying the identity of each vehicle.

• **Message Integrity**: Message integrity is very important as it ensures that there is no transformation in the message between the moment it was sent and received. Hence, the transferred message should match with the received message.

• **Message Non-Repudiation**: It is easy to identify the sender's location and its identity which is approved by a special authority. With the help of authenticated messages, it sends, it is easy to identify the vehicle.

• **Access control**: Vehicles haves to perform only those tasks for which they got authorization and must be as per rules. If the working of nodes is according to specified authorization, then access control is ensured and generates messages [7].

• **Message confidentiality**: To maintain privacy in the system, it is necessary to have confidential property. This privacy is enforced by the law enforcement authority only to create privacy between the communicating nodes.

• **Privacy**: It is the system that provides assured that the information is not leaked to unauthorized people. Third parties can't track the movement of vehicles as it is a violation of personal privacy. Privacy of location is very much important as it hides the past location of vehicles so that no unauthorized person can access it.

• **Real-time guarantees**: Most of the safety applications are depended upon the strict time guarantees as it plays an essential role in VANET. Hence, this feature has been utilized for the avoidance of collisions in time-sensitive road safety applications. Fig. (**3**) shows security issues in VANET.

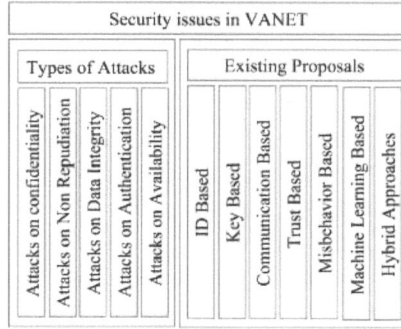

Fig. (3). Security issues in VANET [9].

ATTACKS IN VANET

The unique characteristics of VANETs have been utilized in the networks for design decisions. Hence, it is necessary to address the issues related to inter-vehicular communications which are widely deployed in the network.

DoS Attack

The network is bringing down due to a DoS attack. A node cannot deny whether it transfers the message or not due to the property of non-repudiation. It is difficult to determine the correct sequence. For the elimination of false messaging, it is required to have regular verification.

Black Hole

Dark gap assault is the assault in which hubs don't take an interest in the system. The dark opening is framed when a setup hub drops out. All the activity in this is diverted towards the particular hub which doesn't present because of which information is lost. Assailant makes a false picture because of which wastage of information happens.

Masquerade Attack

In this attack, the intruder pretends to be an authorized user of the system. It will provide authority to an intruder for using the system privileges. This attack can be made by stealing login IDs, passwords.

Replay Attack

It is also known as a playback attack because the legally correct data is illegally repeated and transmitted in the system. This is a lower-tier version of the man-i--the-middle attack [14].

CONCLUSION

VANETs have different applications in wellbeing as it gives applications in impact cautioning, location of forwarding snag and it's shirking, dispersal of crisis message, evasion of parkway/rail crash and can help with left/right turn. It is an application with the assistance of which vehicle can stop itself without the requirement for driver mediation, consequently, there is no prerequisite of human ability for this work. In this paper a hypothetical examination of various security methods of Vehicular Ad-hoc Networks has been done which think about various plans at various levels like equipment, confirmation, protection, and affirmation systems. Consequently, in VANET security is considered as a most vital issue. False activity data and message alteration are the real difficulties. For the anticipation of any conceivable pernicious assault, a computerized signature method and message confirmation calculation have been used. Thus, it is considered as the best procedure for VANET which helps in picking up validation and message legitimacy.

CONSENT FOR PUBLICATION

Not applicable.

CONFLICT OF INTEREST

The authors declare no conflict of interest, financial or otherwise.

ACKNOWLEDGEMENT

Declare none.

REFERENCES

[1]　J. Kennedy, and R. Eberhart, "Particle swarm optimization", *Proceedings of ICNN'95-international conference on neural networks,* vol. 4, 1995 pp. 1942-1948,1995
[http://dx.doi.org/10.1109/ICNN.1995.488968]

[2]　H.R. Dhasian, and P. Balasubramanian, "Survey of data aggregation techniques using soft computing in wireless sensor networks", *IET Inf. Secur.,* vol. 7, no. 4, pp. 336-342, 2013.
[http://dx.doi.org/10.1049/iet-ifs.2012.0292]

[3]　L. Morissette, and S. Chartier, "The k-means clustering technique: General considerations and implementation in Mathematica", *Tutor. Quant. Methods Psychol.,* vol. 9, no. 1, pp. 15-24, 2013.
[http://dx.doi.org/10.20982/tqmp.09.1.p015]

[4]　P. Yang, J. Wang, Y. Zhang, Z. Tang, and S. Song, "Clustering algorithm in VANETs: A survey",
[http://dx.doi.org/10.1109/ICASID.2015.7405685]

[5]　C. Cooper, D. Franklin, M. Ros, F. Safaei, and M. Abolhasan, "A comparative survey of VANET clustering techniques", *IEEE Comm. Surv. and Tutor.,* vol. 19, no. 1, pp. 657-681, 2016.
[http://dx.doi.org/10.1109/COMST.2016.2611524]

[6]   R.S. Hande, and A. Muddana, "Comprehensive survey on clustering-based efficient data dissemination algorithms for VANET", *2016 International Conference on Signal Processing, Communication, Power and Embedded System (SCOPES),* 2016pp. 629-632
[http://dx.doi.org/10.1109/SCOPES.2016.7955516]

[7]   I.T. Abdel-Halim, and H.M.A. Fahmy, "Prediction-based protocols for vehicular Ad Hoc Networks: Survey and taxonomy", *Comput. Netw.,* vol. 130, pp. 34-50, 2018.
[http://dx.doi.org/10.1016/j.comnet.2017.10.009]

[8]   Z. Lu, G. Qu, and Z. Liu, "A survey on recent advances in vehicular network security, trust, and privacy", *IEEE Trans. Intell. Transp. Syst.,* vol. 20, no. 2, pp. 760-776, 2018.
[http://dx.doi.org/10.1109/TITS.2018.2818888]

[9]   H. Zhou, H. Wang, X. Chen, X. Li, and S.J.I.A. Xu, "Data offloading techniques through vehicular ad hoc networks", *Survey (Lond.),* vol. 6, pp. 65250-65259, 2018.
[http://dx.doi.org/10.1109/ACCESS.2018.2878552]

[10]   C. Tripp-Barba, A. Zaldívar-Colado, L. Urquiza-Aguiar, and J. A. Aguilar-Calderón, *Survey on routing protocols for vehicular ad hoc networks based on multimetrics, .*
[http://dx.doi.org/10.3390/electronics8101177]

[11]   X. Bao, H. Li, G. Zhao, L. Chang, J. Zhou, and Y. J. M. Li, *Efficient clustering V2V routing based on PSO in VANETs, .*
[http://dx.doi.org/10.1016/j.measurement.2019.107306]

[12]   M. Elhoseny, and K. Shankar, Energy efficient optimal routing for communication in VANETs via clustering model.*Emerging Technologies for Connected Internet of Vehicles and Intelligent Transportation System Networks.* Springer, 2020, pp. 1-14.
[http://dx.doi.org/10.1007/978-3-030-22773-9_1]

[13]   M.M. Hamdi, L. Audah, S.A. Rashid, A.H. Mohammed, S. Alani, and A.S. Mustafa, "A review of applications, characteristics and challenges in vehicular ad hoc networks (VANETs)", *2020 International Congress on Human-Computer Interaction, Optimization and Robotic Applications (HORA),* 2020pp. 1-7

[14]   A. Katiyar, D. Singh, and R.S. Yadav, "State-of-the-art approach to clustering protocols in vanet: a survey", *Wirel. Netw.,* vol. 26, no. 7, pp. 5307-5336, 2020.
[http://dx.doi.org/10.1007/s11276-020-02392-2]

CHAPTER 21

Face Recognition Using Deep Neural Network

Namra Samin[1,*]**, Warsha Jagati**[1]**, Shailza** [1] **and Yogesh** [1]

[1] *Amity School of Engineering & Technology, Amity University, Uttar Pradesh, Noida, India*

Abstract: Abstract: There is rapid development in the field of image processing. Now, there are a lot of models for face recognition in the whole world. In the field of image recognition, the arrival of deep learning theory has evolved drastically. In today's world, biometric is used everywhere and the expectations from the system are that it provides positive results in any kind of situation as the quality, alignment, facial expression, and reflection of the picture make it difficult for the machine to validate it. Thus, there is an alternative model to the traditional neural network model which is Convolutional Neural Network (CNN) models which are deep learning models. This model includes the process in which at first the machine is trained with the data set and then the validation is done.

Keywords: Convolutional Neural Network, Deep Learning, Face Recognition, Machine Learning, Microsoft Azure.

INTRODUCTION

Image Processing has a significant stage named Image Segmentation [1]. As we know today's scenario, there are various projects/guides developed on face recognition using Artificial Intelligence and deep learning. Hence, Deep Learning is based widely on training and database collection [2]. Face Recognition is a difficult issue in the field of vision and biometric recognition because of different changes such as present varieties and outward appearances in face recognition the key issue is face extraction [3]. Also, it is quite unpredictable but deep learning makes it easier. Deep Learning makes facial information much accurate and improves the technology. So, a better result is achieved [3]. Also, CNN (Convolutional Neural Network) in the era of Deep Learning has been improved compared to the traditional ones based on precision, capacity, and speed of identification, clarification. To identify the face properly one first needs to be" clear the picture" in a sense to remove the blurriness. Hence, the equation to the motion blurring problem is generated [4]. CNN concentrates on highlight data of

---

* **Corresponding author Namra Samin:** Amity School of Engineering & Technology, Amity University, Uttar Pradesh, Noida, India, E-mail: namrashamim09@gmail.com

Dharam Buddhi, Rajesh Singh and Anita Gehlot (Eds.)

the image, which can successfully lessen the element of image information, and don't have to diminish the element image information independently [5]. Computer vision-based deep learning techniques are used to recognize human facial expressions [6]. An enormous technique of figuring out the gender classification, the iris recognition, the lip, the mouth, *etc*. have been examined. Therefore, the difference between the male and female is generated through the researches [7]. In the following paper, we have discussed the following methods using Microsoft Azure, whereas the first step is collecting a database whereas for example for a single human face we collected pictures in a way that every angle and feature is visible and then the next step is the model preparation using Microsoft Azure Machine Learning software and the accuracy of the model is observed. Feature extraction is one of the major factors that is considered as a major factor as the model accuracy depends on them. Also, a histogram is prepared to check and stick to the accuracy of the extracted feature of the human faces.

LITERATURE REVIEW

The work based on CNN was used and the realization on the face was divided into two parts which are network training and implementation. Whereas the recognition speed and accuracy are specified by 99.4% [2]. The program and implementation of each feature of the face are taken by the algorithm that is prepared, the database, and the implementation done by the common face database, but the result is not always improving [3]. DNN training system, which takes a favorable position of both the SoftMax misfortune and triplet misfortune capacities, has been proposed for productive face recognition. The viability of the proposed DNN preparing structure on the LFW dataset and four distinctive face datasets [4]. Also, the designs of both DNN and CNN were compared based on the performance and the quality in terms of the recognition rates. The face is being recognized in a way that it was taken in the real-time facial expression with emotions *etc*. and hence the recognition rate was specified [6]. The classifier determines the Euclidean distance between datasets using the K-Nearest Neighbor algorithm [8]. The local features are used to recognize the objects [9].

METHODOLOGY

Azure Machine Learning Studio (AMLS) is an Azure-based software. Azure is Microsoft's Cloud service. In 2015 Microsoft introduced AMLS, which is too new a technology. With AMLS, machine authorization models are exploited, verified, and used. Some identical technologies are Google Cloud Machine Learning Engine and Amazon Sage Maker. Getting started with the use of AMLS, one needs to have a Microsoft Account. You can create and analyze all types of non-Azure models subscriptions.

Model Description

Creating what is called a prepared data is a tedious process. Specific modules of preprocessing data are added to the raw data. It takes a long time to create prepared data from raw data. It may take a long time to select the raw data. The algorithm of the machine to run is selected by the data scientist. It also decides on the data aspects being prepared for use, and ultimately examines the outcome. The ultimate aim is to evaluate the composition of the algorithm of machine learning and trained data that provide the most valuable results. Fig. (**1**) represents the process of Model building and execution.

Fig. (1). Process of model building and execution.

For the creation of the model. Following steps must be taken and are shown in Fig. (**2**).

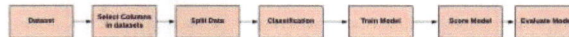

Fig. (2). Block diagram of the process.

Import Data

The feature excel sheet that is generated is provided to the model as input. In the paper, the features are taken from the image analyzer tool in MATLAB.

Selection of Columns in the Dataset

The features that should be taken from the excel sheet is sorted out and are then processed further.

Split Data

The data and the features that are selected from the above set are then split for the training and testing of the model. Further, the datasets are divided into train, test, and validate the model.

Algorithm of Machine Learning

According to the expected accuracy, the algorithm is hence developed.

Train Model

Make your data available to the configured model to learn from the patterns and generate statistics that is be used for predictions. The Studio "Train Model" module is intended to form a classification or regression model. The training takes place as soon as we defined a model and its parameters and need marked data. The model is created to generate an existing model with new data. To use the trained model to predict new values, connect it to the specified model module with the new input data.

Score Model

The evaluation model produces an anticipated incentive for the class and the likelihood of the anticipated worth. Include a prepared model and a record with new information. The information must be in an organization that is perfect with the kind of prepared model. The construction of the information record ought to for the most part additionally coordinate the mapping of the information used to assemble the model. For picture grouping models, the rating is the article class in the picture or a Boolean worth that shows whether a specific component is found.

Evaluate Model

Measure the accuracy of a trained model or compare several models.

The algorithms used in the Machine learning for calculation of the accuracy and get the results accordingly:

Multiclass Decision Forest

A decision tree is a fixed model that quickly creates a series of decision trees and learns from the marked data. The algorithm creates multiple decision trees and then selects the most popular output category. Such histograms are obtained by the averaging process and the result is transformed to determine the "possibilities" for each mark. Trees with a high degree of predictive credibility are more appropriate for the final set decision.

Multiclass Logistic Regression

It is used to predict many values. It is used to forecast the possibility of an outcome, which is especially important for scheduling tasks. The algorithm estimate that an event occurs by adapting the data to an accounting function. It also predicts several outcomes.

Multiclass Decision Jungle

This module analyzes the model and its parameters, and then connects a set of labeled training data to train the model using one of the training modules. By merging tree branches, a decision DAG generally has less memory and better generalization performance than a decision tree, but at the expense of a slightly longer training.

Multiclass Neural Network

Create a neural network model that predicts a multiple-value target. Classification through neural networks is a supervised learning method and therefore requires data entry marked with a label column. Train the model by providing the model and the log entry for the educational model. One can add many hidden layers between the input and output layers. Most anticipation work is simplified easily with a single layer or with a few hidden layers. All nodes in one layer are connected to nodes in the next layer with weighted edges. The value is determined by calculating the weighted sum of the node values from the previous level. Fig. (3) represents the model based on a Multiclass Decision Forest for face recognition.

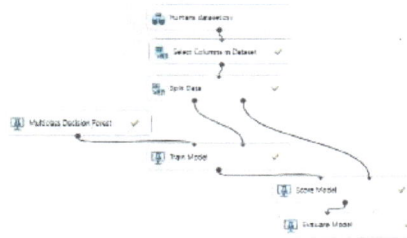

Fig. (3). Evaluating a multiclass decision forest model in microsoft azure training experiment.

The predictive experiment of the multiclass model is shown in Fig. (4).

Fig. (4). Evaluating a multiclass model in microsoft azure in predictive experiment.

ANALYSIS AND RESULT

The result obtained by running an ML algorithm for prepared data is a model. The first model created is not the most suitable. Data scientists continue to test the best machine learning algorithm and have prepared data combinations that make up the best model. Once an effective model has been obtained, the next step is to implement that model. Implementation is important because the application is used for the algorithm generated by the model. Deployment helps in recognizing patterns that help in resolving the issue much effectively than it could have been done manually. Fig. (5) shows the parameters and prediction of sample data. Thus, machine learning generates an effective and time-saving solution. The confusion matrix is shown in Fig. (6). The histogram of the model is shown in Fig. (7).

Fig. (5). Parameters and prediction of sample data.

Fig. (6). Classification model of the confusion matrix.

Metrics

Inspecting evaluation results of Fig. (**8**) after running the experiment and observed the performance of the model. The available estimation parameters for the regression models are as follows: mean absolute error, absolute base error, relative absolute error, quadratic relative error, and decision coefficient. Here the word "under" reflects the difference between the foretold value and the actual value. In general, the absolute value or square of this difference is determined to represent the entire extent of the error in all cases, since in some cases the difference between the expected value and the real value are negative. Error calculations calculate the efficiency of a regression model as the average deviation from the real values of its predictions. Each row shows the instances of the true or real class in its record, and each column represents the instances of the class that the model predicted. The importance of using different methods in the Microsoft azure Learning studio is to ensure the best result with the most appropriate accuracy is obtained. Hence both the Training experiment and the Predictive Experiment helps to get the appropriate result. The maximum accuracy obtained in the model is through the multiclass decision forest algorithm.

Fig. (7). Histogram of model.

| rows | columns |
| 1489 | 24 |

ties	Scored Probabilities for Class "mayank"	Scored Probabilities for Class "pratyush"	Scored Probabilities for Class "pujit"	Scored Probabilities for Class "rishika"	Scored Probabilities for Class "shaiba"	Scored Probabilities for Class "shambhavi"	Scored Probabilities for Class "warsha"	Scored Labels
	0.010947	0.012578	0.000459	0.024727	0.726273	0.001595	0.124128	shailza
	0.010947	0.012578	0.000459	0.024727	0.726273	0.001595	0.124128	shailza
	0.001257	0.003437	0	0.014098	0.814199	0	0.107234	shailza
	0	0	0	0	0.375	0	0	apoorva
	0	0	0	0	0.125	0	0.25	harsh
	0	0	0	0	1	0	0	shailza
	0	0	0	0	0.25	0	0.5	warsha
	0.010947	0.012578	0.000459	0.024727	0.726273	0.001595	0.124128	shailza
	0.010947	0.012578	0.000459	0.024727	0.726273	0.001595	0.124128	shailza
	0	0	0	0	0.625	0	0.25	shailza

Fig. (8). Multiclass Decision Forest Evaluation.

SUMMARY AND CONCLUSIONS

The working and the best result could be achieved by Deep Learning. Hence the face was recognized and the most accurate algorithm that is observed by the accuracy is Multiclass Decision Forest. In the future, transfer learning based on the deep neural network can be applied for face recognition for better performance with a very large amount of datasets. Model parameters optimization and standardization can also improve the model accuracy and minimize the cost function.

CONSENT FOR PUBLICATION

Not applicable.

CONFLICT OF INTEREST

The authors declare no conflict of interest, financial or otherwise.

ACKNOWLEDGEMENT

Declared none.

REFERENCES

[1]   P. Yogesh, "A comparative approach for image segmentation to identify the defected portion of apple", *6th International Conference on Reliability, Infocom Technologies and Optimization (Trends and Future Directions) (ICRITO)*, 2017, pp. 601-608

[2]   X. Qu, T. Wei, C. Peng, and P. Du, "A Fast Face Recognition System Based on Deep Learning", *11th International Symposium on Computational Intelligence and Design (ISCID)*, 2018, pp. 289-292 [http://dx.doi.org/10.1109/ISCID.2018.00072]

[3]   Z. Zhang, J. Li, and R. Zhu, "Deep neural network for face recognition based on sparse autoencoder", *2015 8th International Congress on Image and Signal Processing (CISP)*, 2015, pp. 594-598 [http://dx.doi.org/10.1109/CISP.2015.7407948]

[4]   F.Z. Zhou, G.C. Wan, Y.K. Kuang, and M.S. Tong, "An Efficient Face Recognition Algorithm Based on Deep Learning for Unmanned Supermarket", *2018 Progress in Electromagnetics Research Symposium (PIERS-Toyama)*, 2018, pp. 715-718
[http://dx.doi.org/10.23919/PIERS.2018.8597988]

[5]   G. Yue, and L. Lu, "Face Recognition Based on Histogram Equalization and Convolution Neural Network", *2018 10ᵗʰ International Conference on Intelligent Human-Machine Systems and Cybernetics (IHMSC)*, 2018, pp. 336-339
[http://dx.doi.org/10.1109/IHMSC.2018.00084]

[6]   H. Jung, "Development of deep learning-based facial expression recognition system", *2015 21ˢᵗ Korea-Japan Joint Workshop on Frontiers of Computer Vision (FCV)*, 2015, pp. 1-2
[http://dx.doi.org/10.1109/FCV.2015.7103729]

[7]   S.M. Deokar, S.S. Patankar, and J.V. Kulkarni, "Prominent Face Region Based Gender Classification Using Deep Learning", *2018 4ᵗʰ International Conference on Computing Communication Control and Automation (ICCUBEA)*, 2018, pp. 1-4 Pune, India
[http://dx.doi.org/10.1109/ICCUBEA.2018.8697761]

[8]   T. Makkar, Y. Kumar, A.K. Dubey, Á. Rocha, and A. Goyal, "Analogizing time complexity of KNN and CNN in recognizing handwritten digits", *2017 4ᵗʰ International Conference on Image Information Processing (ICIIP)*, 2017, pp. 1-6
[http://dx.doi.org/10.1109/ICIIP.2017.8313707]

[9]   Yogesh and A. K. Dubey, "Fruit defect detection based on speeded up robust feature technique", *2016 5ᵗʰ International Conference on Reliability, Infocom Technologies and Optimization (Trends and Future Directions)*, 2016.
[http://dx.doi.org/10.1109/ICRITO.2016.7785023]

CHAPTER 22

A Review on Implementation of Cloud Security in the Aadhar Card Project

Megha Malhotra[1,*] and **Yogesh Kumar**[1]

[1] *Computer Science and Engineering, Electronics and Communication Engineering, Amity School of Engineering and Technology, Amity University, Uttar Pradesh, Noida, India*

Abstract: Demands of Cloud computing have been increasing these days and becoming more prevalent in the field of Information Technology. It has been able to provide solutions to various markets in the IT sector. In this paper, we will discuss a cloud computing case study on Aadhar cards that were to be provided to all the citizens of India. This major project included cloud computing technology and due to handling of confidential data, the security needed to play a vital role. Though the cloud is a secure platform there are still some challenges that need to be encountered for the data to be saved from external and unethical attacks. So this paper discusses some solutions to the given problem of handling security on the cloud and how to overcome challenges and issues faced on the cloud.

Keywords: Aadhar Card, Cloud Security, Cryptography, Data Security.

INTRODUCTION

Cloud computing is a very fast-growing technology and will be covering around the maximum of the growing market. As cloud computing is becoming the demand of the market, security follows automatically. The only question that comes with the cloud is whether the data is secure on the cloud or not. Cloud security is a major concern of all and therefore some security measures are being adopted these days. Cloud is basically of three types namely public, private, hybrid [1]. All three types of clouds have different security issues and challenges. Further, the cloud is known to provide three types of services namely IaaS, PaaS, SaaS [1]. These three types of the environment being provided by the cloud service providers have different security issues. IaaS is known to have multi-tenancy issues, SaaS is known to have password issues, and PaaS is known to have encryption issues. Further, the major issues faced by cloud computing tech-

* **Corresponding author Megha Malhotra:** Computer Science and Engineering, Electronics and Communication Engineering, Amity School of Engineering and Technology, Amity University, Uttar Pradesh, Noida, India; E-mail: meghamalhotra1617@gmail.com

nology these days are confidentiality, integrity, availability, and Privacy. Confidentiality is maintaining privacy. Privacy in the sense of how secure the data is from attacks, breaches, or in a simple way, no unauthorized person gets access to the data in any way, let it be by any internal or external to the organization. Integrity demonstrates there should be no change in the data like if one sends the message ABC then it should not be received as ABX. This is how integrity gets affected. Availability shows whenever the data is required it should be available in a proper format as it is. Like if I need any data kept on server A then also if server A does not respond, the data should be available and as you know the cloud is known for its backup and recovery due to large data centers situated worldwide, it is also known for high availability. Many centralized data centers are known to store data but it is costly and sometimes if it fails the data becomes unavailable and this may pose a threat to security. This is solved by the high availability feature provided by cloud computing [2]. Multi-Tenancy issues are sharing the same space/infra by different customers [3]. For example- there is a flat system, in a flat, every family is given a different home but the security, water, electricity is provided by the same source. Privacy is very different from confidentiality, as any data may be personal and can't be shared with others. This is privacy in which one needs to keep all the information private.

LITERATURE REVIEW

Aadhar card in India has been a major project that requires a large amount of data to be handled. Aadhar card is known to give the complete details about the citizen, so it contains highly confidential data [4]. Moreover providing Aadhar cards to all the citizens of India has been a major and challenging task. So cloud computing has always been a solution to such a wide task. Due to the linking of bank accounts of people(Number of adult Indians with bank accounts rises to 80% of the population known to have bank accounts in India) [5] and other applications, the task has become more challenging. This will have also led to the opening of more bank accounts in rural India. For managing all the tasks, there has been a requirement of cloud computing. Due to the data being confidential, cloud security is of utmost importance. Along with the implementation of cloud-based environments for storage and security, a lot of tasks become easier. For example. counting of votes [6], linking of common data with the help of the Aadhar card-based approach. The cloud environment is also known to have some challenges. Due to these challenges the cloud security may not be reliable. Though the cloud itself is known to be a highly reliable technology there may be some challenges that may alter its features and the clients might not find the data on the cloud to be secure.

Fig. (**1**) depicts the three deployment models -Public cloud, private cloud, hybrid cloud, and community-based cloud. Along with the deployment models, service models are also depicted- Software as a Service (SaaS), Platform as a Service (PaaS), and Infrastructure as a Service (IaaS). Above the layer of service models, it shows the essential characteristics of cloud computing like Broad Network Access, Rapid Elasticity, Measured Service, On-Demand Self-Service and Resource Pooling.

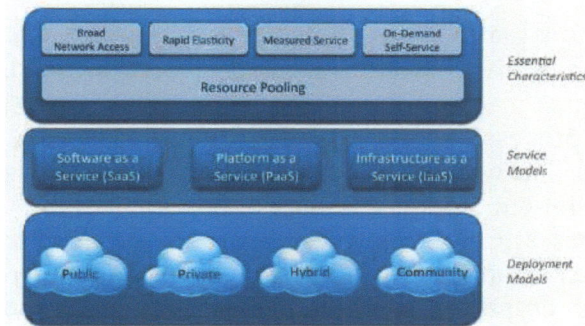

Fig. (1). Visual model of cloud computing platforms with services and characteristics [7].

Isolation failure is also one of the concerns wherein sharing of resources in a multi-tenancy cloud environment is a problem while implementing cloud for highly confidential data. Other issues may be related to guest hopping attacks [3].

Aadhar is known to link all the details from your bank account, your pan card, your fingerprints, your retina prints, so it contains a lot of private data where maintaining confidentiality is very important.

Data privacy and data protection have been a right of every citizen of the country provided by law [8]. Therefore maintaining data privacy is very important in the case of Aadhar.

METHODOLOGY

There can be many solutions to the issues discussed above, but some of them are mentioned below. The most important solution of all is to always choose a trusted cloud provider. Trust is a very important aspect where one should never compromise. Backup should always be kept for ensuring the availability of data. Whenever data may not be available there should be a backup plan to recover all the data. At the end of everything, data is of utmost priority to any organization as well as any person who is an Aadhar cardholder. The data should always be stored and transferred in encrypted form. These days many cryptographic algorithms are being developed to ensure the privacy of data. PKI (public key infrastructure) is a very common technology that uses a concept of a key pair. The pair are known to have one public key that is made available to all or in other words it is common for all and is used for encryption purposes and the other is a private key that is only available with the authorized person. It can't be shared openly. These days even the private keys are being encrypted to ensure security on the cloud. The concept of cryptographic keys has been able to be a good measure for enabling securities in various cloud-based environments [9].

Fig. (**2**) explains how plain text is encrypted using an encryption algorithm and then sent as ciphertext over the transmission or network media. Then using the decryption algorithms it is then again converted back to plain text at the receiver ends. Here in this process, there is a concept of keys also used at the encryption and decryption end. The concept of keys is public and private keys. These are two types of keys used for encryption and decryption purposes. Some similar concepts of digital signatures are being implemented with the aadhar card application. Also, there has been a concept of symmetric key cryptography being used for the implementation of cryptography for data security [2]. The next solution is monitoring and auditing data regularly. There are big companies that only do the work of auditing rather they are known just for auditing purposes. One of such companies is Ernst Young known to be a part of the big four. So the data should be properly monitored. Logs should be maintained and kept for a long time. This may help in maintaining the integrity of the whole data. Another solution to securing the data is signing the agreement. An agreement is any legal document that needs to be signed and is meant to be kept. So maintaining a service level agreement may serve as a better solution for securing data on the cloud. Based on the above solutions one can follow some very basic steps to identify suitable cloud practices for Aadhar card applications. These are as follows:

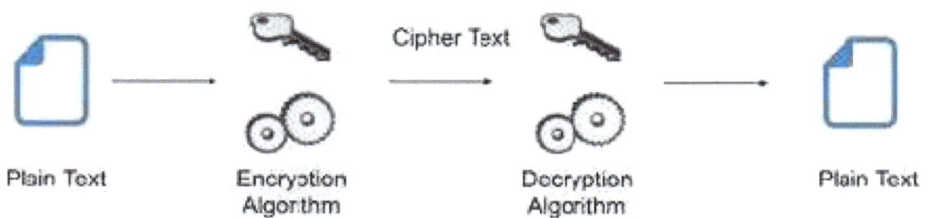

Fig. (2). Basic encryption process [3].

Step 1: Identify the requirements based on the type of data that is stored under the aadhar card application. Accordingly, check for the type of data security it requires.

Step 2: After finding the type of security that each field in an Aadhar card requires, review various cloud provider security trends, or else one can opt for setting up a private cloud with their own needs and accordingly implement security algorithms that best suit your needs.

Step 3: Regularly map cloud security and application security to perform fit analysis and performance analysis [10].

The other well-known practice that the firms have started implementing in government sectors is master data management. This can help in managing data in a well-organized way and it can be further integrated with the cloud to implement security along with other features of cloud-like reliability, availability.

Fig. (3) shows the three pie charts, the first one depicts how very large enterprises are willing to spend time, money, and skills on cloud services. On a similar background, there are around 83% of large enterprises willing to spend time, money, and skills on cloud services are being shown by the second pie chart. And the third pie chart shows similar interests in medium enterprises. A similar implementation is being done with the help of a DigiLocker that helps in providing the online aadhar card with the help of digital signatures [12]. The large data are optimized by a neural network [13]. The finest security schemes need to be implemented to prevent security threats [14].

Key CIO priorities in India over next three years

Willing to invest time, money and skills on cloud services are likely to invest in the Private cloud

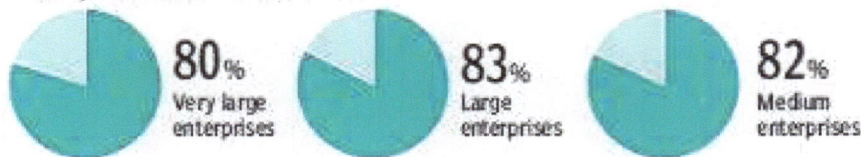

| 80% | 83% | 82% |
| Very large enterprises | Large enterprises | Medium enterprises |

Fig. (3). Statistics showing willingness for cloud usage in coming years [11].

DISCUSSION & CONCLUDING

Cloud computing is in increasing demand in the upcoming generation. Cloud security is the main reason for opting cloud as a platform for providing services and using it as a platform on the business front. It is important to maintain confidentiality, integrity, availability, and privacy in the cloud. Only the authorized and authenticated user should get access to information. It has always been a good decision to choose the cloud as a platform for major projects like Aadhar distribution to all the citizens of India. Cloud is known to be much more secure and makes computing easier. As cybercrimes are increasing these days, the cloud is a trustworthy technology adopted to decrease such crimes. But criminals are still trying to find any bug in the cloud so that cloud security can be hampered. So maintaining security in the cloud is also very important. It is known to provide services at low prices and thus being an efficient solution for maintaining security at a large scale. Various organizations have started to work on the concept of master data management to get a clearer view of their requirements and the organizational data. In the future, Master Data Management can also be combined with cloud computing practices to give a better visualization of data storage and help in securing the data more effectively. Through this paper, we have been able to discuss some real-life challenges that have been found. Also, some solutions have been discussed to tackle such issues using cloud computing and cloud security.

CONSENT FOR PUBLICATION

Not applicable.

CONFLICT OF INTEREST

The authors declare no conflict of interest, financial or otherwise.

ACKNOWLEDGEMENT

Declare none.

REFERENCES

[1]   M. Prasad, "Cloud Computing: Research Issues and Implications", *International Journal of Cloud Computing and Services Science.,* vol. 2, pp. 134-140, 2013.
[http://dx.doi.org/10.11591/closer.v2i2.1963]

[2]   A.K.B. Halvi, and S. Soma, "A robust and secured cloud based distributed biometric system using symmetric key cryptography and microsoft cognitive API", *2017 International Conference on Computing Methodologies and Communication (ICCMC), Erode,* 2017 pp. 225-229
[http://dx.doi.org/10.1109/ICCMC.2017.8282681]

[3]   Ahmed Albugmi, Madini Alassafi, Robert Walters, and Gary Wills, "Data Security in Cloud Computing", *Fifth International Conference on Future Generation Communication Technologies (FGCT),* 2016.
[http://dx.doi.org/10.1109/FGCT.2016.7605062]

[4]   Satyam P., and Kusum L., "AAnalytical Study on Aadhaar Card (UIDAI) and its inclusion into Public Services Delivery", vol. 4, pp. 64-68, 2018.

[5]   https://economictimes.indiatimes.com/industry/banking/finance/banking/number-of-adult-indi-ns-with-bank-accounts-rises-to-80/articleshow/63838930.cms?from=mdr

[6]   T. Illakiya, S. Karthikeyan, U.M. Velayutham, and N.T.R. Devan, "E-voting system using biometric testament and cloud storage", *2017 Third International Conference on Science Technology Engineering & Management (ICONSTEM), Chennai,* 2017pp. 336-341
[http://dx.doi.org/10.1109/ICONSTEM.2017.8261305]

[7]   H. Kharche, and D. Chouhan, "Building Trust In Cloud Using Public Key Infrastructure A step towards cloud trust", *Int. J. Adv. Comput. Sci. Appl.,* vol. 3, 2012.
[http://dx.doi.org/10.14569/IJACSA.2012.030305]

[8]   R.S. Raju, S Singh, and K. Khatter, "Aadhaar Card: Challenges and Impact on Digital Transformation. ", *arXiv preprint arXiv:1708.05117,* 2017.

[9]   S. Agrawal, S. Banerjee, and S. Sharma, "Privacy and security of aadhaar: A computer science perspective", *Econ. Polit. Wkly.,* vol. 52, pp. 93-102, 2017.

[10]   S. Sengupta, V. Kaulgud, and V.S. Sharma, " Cloud Computing Security--Trends and Research Directions. Proceedings", *2011 IEEE World Congress on Services, SERVICES,* 2011
[http://dx.doi.org/10.1109/SERVICES.2011.20]

[11]   https://www.financialexpress.com/industry/technology/protect-data-on-the-cloud/586271/

[12]   D. Locker, https://digilocker.gov.in/about.php

[13]   Y. Jain, "A Survey on Railway Assets: A Potential Domain for Big Data", *International Conference on Issues and Challenges in Intelligent Computing Techniques (ICICT),* pp. 1-6, 2019.
[http://dx.doi.org/10.1109/ICICT46931.2019.8977714]

[14]   A. Sharma, "A Review on Data Flow Risk based on IoT", *International Conference on Issues and Challenges in Intelligent Computing Techniques (ICICT),* pp. 1-6, 2019.
[http://dx.doi.org/10.1109/ICICT46931.2019.8977646]

CHAPTER 23

Early Detection and Classification of Breast Cancer Using Mammograms by Machine Learning

Bharath Chandra B.[1,*] and **Yogesh Kumar**[1]

[1] Department of Electronics & Communication, Amity University Uttar Pradesh, Noida, India

Abstract: Machine learning-based classification of breast cancer and its detection is possible without toxic therapy by a well-trained model. The machine learning model detects features and patterns form the data sets that used when training model which is useful for detecting tumor and classify whether it is a benign or malignant and this process simplifies the cancer detection and gives results accurately at a faster rate when compared to the other traditional methods like Magnetic resonance imaging (MRI), Coronary artery disease (CAD), Modalities using ultrasound, *etc*. Here I am proposing a new technique through which breast cancer can be easily detected by a proper training model with the help of few classifying algorithms in this research a good set of data is used for training classifier machine algorithms in Microsoft azure by comparing all those five algorithms accuracy and working these are the five algorithm models are 2-class Support vector machine, 2-class Neural Networks, 2-class Boosting Tree, 2-Class Logistic Regression, 2-Class Bayes Point and acquired better results which can lead and helpful for detecting cancer in future by using machine learning and deep learning techniques.

Keywords: Benign, Deep Neural Networks, Malignant, Modality, Simple Neural Networks, Tumor.

INTRODUCTION

A machine learning model learns by identifying patterns which is a very important and crucial phase while creating a model by which the system is faster and accurate. The selection of features and detecting patterns is done by the known training dataset. In this paper, the model is trained by various machine learning algorithms to compare their accuracy and proposing the best algorithms which are used for predicting tumors for better results. The model is capable of predicting breast cancer at an early stage by using mammograms deep neural networks, simple neural networks, automating breast cancer detection, *etc*. The

* **Corresponding author Bharath Chandra B.:** Department of Electronics & Communication, Amity University Uttar Pradesh, Noida, India; E-mail: bharath1074127@gmail.com

research is focused on improving breast cancer screening by using the best techniques from the pool of the comparative analysis to find an efficient way to increase the accuracy rate of breast cancer detection keeping in mind that only limited datasets are available.

PREVIOUS WORK

The mammography is the essential test used for screening and early recognition, and its assurance and preparing the advisers for improve bosom disease discovery. In 5593 females age thirty to sixty-nine years observed from 1995 through 2005, 896 event invasive breast cancers were recognized. 50 years is the average age at diagnosis [1]. Based on Relative Ricks from the Black Women Health Study model, the predicted risks for the calculation of fair risks are 0.73 of women who are around the age group of forty-nine a long time and 0.70 for ladies age 50 to 69 years [1]. Additionally serves as an effective examination of computerized mammograms reliant on surface division for the analysis of beginning time tumors, There changed into desirable arrangement among anticipated and decided number of bosom tumors fundamental (anticipated to-found proportion, 0.96; 95% CI, zero.88 to one.05) and in most peril segment classes. Oppressive exactness turns out to be better for females whose age is 50 years (zone beneath the bend [AUC], 0.62; 95% CI, 0.58 to0.65) than for women age $\geq$ 50 years (AUC, 0.56; ninety-five% CI, 0.53 to 0.59). Utilizing a 5-yr expected threat of 1. 66% or extra as a cut point, 2.8% of ladies younger than 50 years of age and 32.2% of women $\geq$ 50 years of age were sorted as being at raised danger of obtrusive bosom most cancers [2]. Bhupendra Gupta and Anuj Kumar Singh displayed the recognition stage pursued by the division of the tumor locale in a mammogram picture by using fundamental picture preparing strategies, for example, averaging and edge techniques [3]. The proposed strategy is basic and quick as a result of utilizing not many picture preparing strategies and is additionally be useful in other clinical imaging applications, design coordinating, highlight extraction [4] in comparison to past investigations that utilized SVM characterization that utilization basic first-request measurements highlights. Then again, with Neural Network-based classifiers, past examinations showed that computationally more costly highlights give similar outcomes to what we arrived at. Future adding the miniature calcifications (MCs) and utilizing the modern classifiers, like ANN and RBFNN, a Max-Mean and Least-Variance method for tumor recognition is moreover given by [5]. The Euclidean distance between datasets is computed by using KNN [6]. The features extracted from the image involve object recognition, registration, and recognizing parameters [7].

METHODOLOGY

To upgrade bosom disease deduction and endurance, early location is fundamental, which incorporates two early discovery methodologies for bosom malignancy: early finding and screening. Early analysis centers around giving ideal admittance to disease treatment by identifying mass (benevolent or threatening) through screening instruments like mammography. There are various markers of bosom malignant growth, like twisting of the state of the bosom disease, yet these are less demonstrative. There are two head groupings of mass. One is alluded to as benevolent (gentle) and the other as dangerous (compromising). A kind-hearted tumor is a tumor that doesn't assault it's including tissue or spread around the body, it is round or oval. A dangerous tumor is a tumor that may assault its enveloping tissue or spread around the body; it has irregular boundaries and is partially round in shape. Also, malignant tumor emerges whiter than neighboring tissues. It is the 2nd most driven reason for deaths among women, globally. It is a kind of cancer that forms in the breast tissue with the main observable symptom that is a lump that feels irregular from the remaining tissues.

In Fig. (1) there is a pie chat which shows the ratio between the deaths which are caused by all other cancers *i.e.*87% of deaths are caused by other cancers whereas 13% of deaths are caused by Breast Cancer.

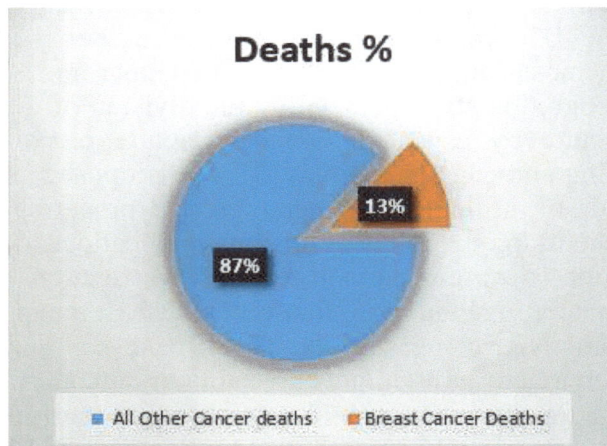

Fig. (1). The deaths ratio due to other cancer and breast cancer.

In the past years, research in breast cancer diagnosis has chiefly centered on the advancement of the computer-aided system (CAD) to help radiologists in the analysis. Conventionally, mammography depended closely on hand-engineered components of CAD system which indicated bounded accuracy in complex situations. In the recent past, with the arrival of deep learning, CAD frameworks adapt automatically with which features are highly important to be utilized to perform a diagnosis, improving the performance of the system. The advancement of deep learning, a branch of machine learning has aroused interest in its application to fuel the accuracy of cancer screening of medical imaging problems. Research reveals that most practiced physicians can determine malignancy with 77 percent exactness while 92 percent correct diagnosis is accomplished using machine learning techniques.

K- Means Clustering: K-means is one of the unsupervised learning calculations for clustering. Clustering the Picture is gathering the pixels as indicated by certain characteristics. This technique divides the image into clusters having pixels with homogeneous characteristics. This is an algorithm that recognizes the groups in the data, with the variable K be the number of groups. The algorithm takes care of every point to one among the category based on feature similarity. The main objective is to clarify the K number of centroids. This procedure keeps up till every pixel is allocated to the proximate centroids.

Feature Classification: It is a technique used for the classification of data into categories based on features extracted from an image.

Machine Learning-Based Approach: AI for most parts covers programmed registering procedures reliant on paired activities that take in an undertaking from an assortment of models. These characterization results are equipped for introducing the most tuff issue given adequate information. Machine Learning plans to create classifying expressions easy and can be easily understood by the people, also must imitate human reasoning adequately to give the understanding to provide a path into the decision process.

Classification Techniques: Arrangement is the piece of these techniques that are used to notice the having information, consider every case, and assign it to a particular class with the end goal that mistake will be least. It is utilized to extract models that precisely characterize significant information classes inside the given dataset.

Support Vector Machines (SVM) Algorithm: SVMs are commonly utilized for learning order, relapse, or positioning capacity. SVM depends on factual leaming hypothesis and operational danger diminishing rule and expects to decide the circumstance of choice limits otherwise called hyperplane that produce the ideal partition of classes. Productivity of SVM-based gathering doesn't straight forwardly rely upon the component of arranged substances. By this may be an excellent classification technique in getting the best accuracies.

Naive Bayes Algorithm: The Naive Bayes Classifier approach depends on the Bayesian hypothesis and is particularly utilized while the dimensionality of the information sources is exorbitant. The Bayesian Classifier can do Ascertaining the most extreme conceivable yield based at the center. It's moreover possible to highlight new uncooked data at runtime and have a far superior probabilistic classifier.

RESULTS

In Fig. (2) it is a Two-Class Support Vector Machine Learning model which is simulated online by using Microsoft Azure. Table **1** represents the comparative analysis of two class Machine Learning algorithms of experimental datasets.

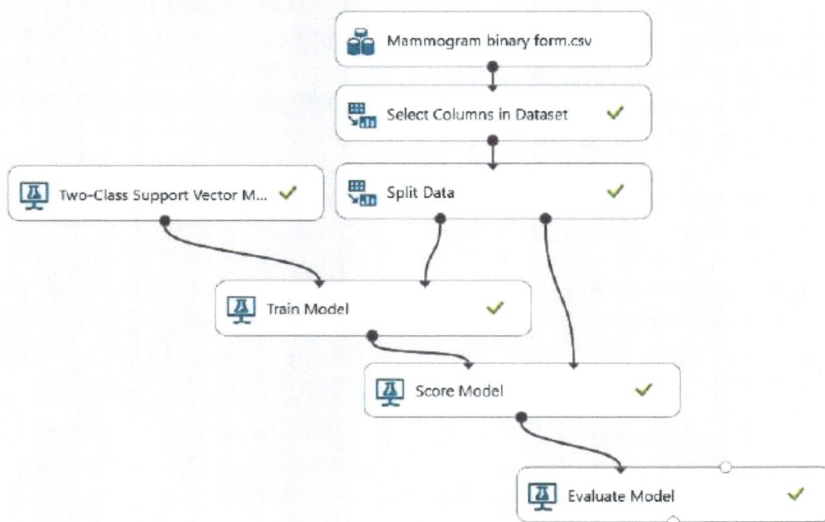

Fig. (2). Training Model of Two-Class SVM.

Table 1. Experimental Comparative Analysis of two class Machine Learning Algorithms.

Attributes	SVM (2 Class)	2-Class Neural Network	2-Class Boosting Decision Tree	2-Class Logistic Regression	2-Class Bayes Point
True +ve	30687	30687	16277	30687	30687
True -ve	0	0	0	0	0
False +ve	1	1	1	1	1
False -ve	0	0	14410	0	0
Positive Label	-7.443672	-7.443672	-7.443672	-7.443672	-7.443672
Negative Label	-7.8228898	-7.8228898	-7.8228898	-7.8228898	-7.8228898
Accuracy	1.000(100%)	1.000(100%)	0.530(53%)	1.000(100%)	1.000(100%)

(Table 1) cont.....

Attributes	SVM (2 Class)	2-Class Neural Network	2-Class Boosting Decision Tree	2-Class Logistic Regression	2-Class Bayes Point
Precision	1.000(100%)	1.000(100%)	1.000(100%)	1.000(100%)	1.000(100%)

The average accuracy is observed to be 90%. Here we can see that the comparison of all the five algorithms *i.e.* 2-class support c-vector machine, 2-class neural networks, 2-class boosting decision tree, 2-class logistic regression, and 2-class Bayes point is given and compared all the aspects like their true negatives and positives their accuracy and precision, the least accuracy acquired is 53% by the 2-class boosting decision tree and that says that it not good for these kinds of detections and the remaining models can be used for getting even more best results in detection tumor. The average accuracy acquired by all five models is 90%.

CONCLUSION

After training and testing all the models it is observed that the accuracy is increased even more by training with some more features and with more datasets. Future work is to train the model by adding a few more features and tuning data for better results. Few more algorithms will be tested for better accuracy by adding even more data with new features for better efficiency. The average accuracy can be further improved by introducing deep neural models.

CONSENT FOR PUBLICATION

Not applicable.

CONFLICT OF INTEREST

The authors declare no conflict of interest, financial or otherwise.

ACKNOWLEDGEMENT

Declare none.

REFERENCES

[1]   D.A. Boggs, L. Rosenberg, L.L. Adams-Campbell, and J.R. Palmer, "Prospective approach to breast cancer risk prediction in African American women: the black women's health study model", *J. Clin. Oncol.,* vol. 33, no. 9, pp. 1038-1044, 2015.
[http://dx.doi.org/10.1200/JCO.2014.57.2750] [PMID: 25624428]

[2]   K. Shwetha, M. Spoorthi, S.S. Sindhu, and D. Chaithra, *"Breast Cancer Detection Using Deep Learning Technique", International Journal Of Engineering Research & Technology (IJERT).* NCESC, 2018.

[3]   A.K. Singh, and B. Gupta, "A Novel Approach for Breast Cancer Detection and Segmentation in a

Mammogram", *Procedia Science,* vol. 54, pp. 676-682, 2015.
[http://dx.doi.org/10.1016/j.procs.2015.06.079]

[4]   Y. Osman, and U. Algasemi, "Breast Cancer Computer-Aided Detection System based on Simple Statistical Features and SVM Classification", *Int. J. Adv. Comput. Sci. Appl.,* vol. 11, no. 1, 2020.
[http://dx.doi.org/10.14569/IJACSA.2020.0110153]

[5]   M. Md. Islam, R. Md. Haque, H. Iqbal, M. Md. Hasan, and M. Hasan, "Breast Cancer Prediction: A Comparative Study Using Machine Learning Techniques", *SN Computer Science,* vol. 1, no. 15, 2020.

[6]   T. Makkar, Y. Kumar, A.K. Dubey, Á. Rocha, and A. Goyal, "Analogizing time complexity of KNN and CNN in recognizing handwritten digits", *2017 Fourth International Conference on Image Information Processing (ICIIP), Shimla, India,* 2017pp. 1-6 Shimla, India
[http://dx.doi.org/10.1109/ICIIP.2017.8313707]

[7]   Yogesh, and A.K. Dubey, "Fruit defect detection based on speeded up robust feature technique", *2016 5th International Conference on Reliability, Infocom Technologies and Optimization (Trends and Future Directions) (ICRITO), Noida, India,* 2016pp. 590-594
[http://dx.doi.org/10.1109/ICRITO.2016.7785023]

CHAPTER 24

Face Emotion Recognition by Machine Learning

Sarthak Patra[1,*], **Kushagra Singh Yadav**[1] and **Yogesh Kumar**[1]

[1] *Department of Electronics & Communication, Amity University Uttar Pradesh, Noida, India*

Abstract: Detection of Facial expressions and emotions is always an easy task for humans but to achieve the same task using different computer-based algorithms is a challenging task. It is possible to detect emotions from images using various machine learning algorithms as there is a huge advancement in computer vision and machine learning over the years. Programmed face appearance acknowledgment is an effectively arising research in Emotion Recognition. In this paper, the Convolutional Neural Network (CNN) which is a subset of AI is rehearsed as a way to deal with outward appearance acknowledgment tasks. Thus, the proposed method is found to be more effective than other methods and has an accuracy of 92. Face appearances are the vital qualities of non-verbal correspondence. Non-verbal explanations are imparted through outward appearances. Face looks are the delicate indications of the greater correspondence. Nonverbal correspondence implies correspondence among people and animals through the eye to eye association, signals, outward appearances, non-verbal correspondence, and paralanguage. Human facial expressions can be recognized by using deep learning.

Keywords: Convolutional Neural Networks, Deep learning, Z Face Emotion, Machine Learning, Recognition.

INTRODUCTION

Facial inclination affirmation is the route toward distinguishing human sentiments from outward appearances. The human cerebrum sees sentiments normally, and programming has now been developed that can see emotions. Facial feeling acknowledgment is an assignment in artificial intelligence (AI). This issue has grown as a fascinating subject of exploration concerning late occasions. Facial feelings are a sort of non-verbal method of correspondence which passes on the temperament of any individual. There are numerous kinds of facial feelings like cheerful, dismal, outrage, disappointment, dreadful. It has numerous applications in different fields, for example, advanced mechanics, drugs, driving help frameworks, lie finder tests. Face discovery and Recognition can be utilized to

---

* **Corresponding author Sarthak Patra:** Department of Electronics & Communication, Amity University Uttar Pradesh, Noida, India; E-mail: sarthakpatra1911@gmail.com

Dharam Buddhi, Rajesh Singh and Anita Gehlot (Eds.)
All rights reserved-© 2021 Bentham Science Publishers

improve access and security. The application of Convolutional Neural Networks (CNN) for facial feeling acknowledgment. CNN is generally utilized for picture grouping and gives the best result among all the strategies.

PREVIOUS WORK

The Facial Action Coding Systems (FACS) being among the generally utilized techniques). This technique assembles one classifier for every conceivable gathering of impediments [1]. Entropy-based element determination technique applied to 3D facial element separations is introduced for an outward appearance acknowledgment framework characterizing the looks into 6 essential classes dependent on 3-Dimensional (3D) face calculation [2]. This facial express acknowledgment framework is fabricated utilizing the Facial Landmark to get the demeanor highlight [3]. A proposed technique that breaks down the neighborhood highlights of outward appearance removed by Gabor wavelet change and plays out the different dimensional decrease on the issue of highlight excess, which adjusts the element measurement and the commitment pace of the component [4]. To begin with, in the base level, the facial element focuses on every facial segment, *i.e.*, eyebrow, mouth, and so on, catch the nitty-gritty face shape data. Second, in the middle level, facial action units, portrayed in the facial activity coding structure, address the pressing factor of a particular approach of facial muscles, *i.e.*, top tightener, eyebrow-raiser, and so on. Finally, at the undeniable level, six prototypical outward appearances address the overall facial muscle improvement and are usually used to depict the human inclination states [5]. Programmed examination of human facial expressions is one of the difficult issues in insightful frameworks and social sign handling. The calculation plots a face model chart dependent on outward appearance muscles in each edge and concentrates highlights by estimating facial diagram edges' size and point varieties [6]. For outward appearance acknowledgment, the local binary point highlight is a significant method of the surface component, yet ordinarily, the entire picture is taken as removing territory, overlooking to separate the key territories of outward appearance [7]. The local features also play an important role during image recognition [8]. It is also observed that a convolutional neural network is better in terms of producing better accuracy [9].

METHODOLOGY

The first and the premier advance in the undertaking is to gather the example pictures to prepare the model. The dataset comprised pictures of the apparent multitude of seven feelings that are glad, miserable, irate, impartial, energized, appall, and dread. The dataset being utilized is Yale Face Database B that contains 16128 pictures of 28 human subjects under 9 postures and 64 brightening conditions. The information configuration of this information base is equivalent to the Yale Face Dataset B. The features are extracted from the images using MATLAB. The samples are transformed from RGB to grayscale and from grayscale to binary during feature extraction. The seven unique features are stored for further processing using Microsoft Azure. This paper proposes a visual interface-based AI model for the prediction of human outward appearances acknowledgment. The preparation dataset comprises the zone, significant pivot length, minor hub length, border, unconventionality, Euler number, direction, and equal breadth. Most importantly, the preparation datasets are transferred to the model. The data which were missing in arranged datasets is substituted with the mean worth. The datasets are essential for two segments: one for the preparation and the second for the testing. For preparing, 70% of datasets are used, and the leftover 30% were utilized for the approval and testing of the model. Fig. **(1)** represents the proposed technique. It includes essentially three significant advances: creating a new model, model testing, and built-up for the new model

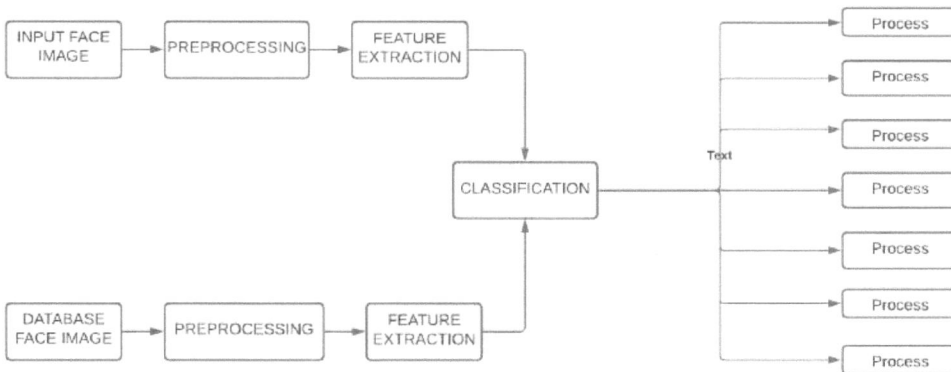

Fig. (1). Proposed method.

From providing our sample data to our output extraction there are four steps in between which are shown in Fig. **(2)**. The data is preprocessed then features are extracted from it then the machine learning algorithm and azure machine learning API work on it and then we finally get our output. shows a sample image taken from our dataset. The image is of angry emotion. Similar images of different types and emotions are used to train our model. The advancement of deep learning, a branch of machine learning has aroused interest in its application to fuel the accuracy of cancer screening of medical imaging problems. Fig. **(3)** shows the sample images of experiments from datasets. Research reveals that most practiced physicians can determine malignancy with 77 percent exactness while 92 percent correct diagnosis is accomplished using machine learning techniques. Table **1** represents the experimental comparison analysis of Face Emotion Recognition.

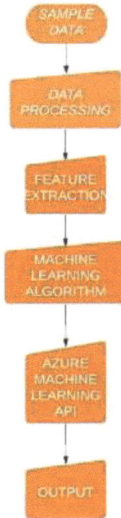

Fig. (2). Automated machine learning process.

Happy

Sad

Fig. (3). Sample Images of experiments from datasets.

Table 1. Experimental comparison analysis of face emotion recognition.

Attributes	Multiclass Neural Network	Multiclass Logistic Regression	Multiclass Decision Forest	Multiclass Decision Jungle
True +ve	38045	37512	22587	20514
True –ve	0	0	0	0
False +ve	1	1	1	1
False -ve	0	0	0	0
Positive Label	-5.45	-5.45	-5.45	-5.45
Negative Label	-5.98	-5.98	-5.98	-5.98
Accuracy	0.92(92%)	0.90(90%)	0.88(88%)	0.82(82%)
Precision	0.92(92%)	0.90(90%)	0.88(88%)	0.82(82%)

RESULTS

The main objective of this task is to utilize AI calculations for the training of human facial expression datasets. The accuracy of the model can be increased by using different machine learning algorithms. We used different available algorithms on azure which were multiclass neural network, multiclass logistic regression, multiclass decision forest, and multiclass decision jungle. The highest accuracy was given by the multiclass neural network which was 92%. Hence the proposed method for our project is a multiclass neural network. Table 1 shows an Experimental Comparison analysis of Face Emotion Recognition.

CONCLUSION

Modernized Machine learning improves the task with significant learning applications. In the future new district ought to be investigated in AI for adding relevant information to the framework for better exactness and datasets will be minimized that will save the design asset. Manual confirmation of highlights devours bunches of time and degenerates the exactness level. A more mind-blowing approach to manage lets the framework pick the huge part and AI assessment for lively dealing with and a more basic level of guess precision. For future references, we are planning to implement a full-fledged working model for human facial recognition. Different coding languages can be used for the final implementation but as of now, we are planning to use MATLAB to do the useful. We are also planning to design our working model in such a way that it plays music related to the recognized emotion for example for happy expression it will play happy music for sad expression it will play sad music and so on.

CONSENT FOR PUBLICATION ˙

Not Applicable.

CONFLICT OF INTEREST

The authors declare no conflict of interest, financial or otherwise.

ACKNOWLEDGEMENT

Declared none.

REFERENCES

[1]   J. Yang, F. Zhang, B. Chen, and S.U. Khan, *Facial Expression Recognition Based on Facial Action Unit," 2019 Tenth International Green and Sustainable Computing Conference (IGSC), Alexandria, VA, USA,,* 2019pp. 1-6 Alexandria, VA, USA
        [http://dx.doi.org/10.1109/IGSC48788.2019.8957163]

[2]   K. Yurtkan, H. Soyel, and H. Demirel, *Entropy driven feature selection for facial expression recognition based on 3-D facial feature distances," 2015 23nd Signal Processing and Communications Applications Conference(SIU), Malatya,* 2015, pp. 2322-2325
        [http://dx.doi.org/10.1109/SIU.2015.7130344]

[3]   L. Fatjriyati Anas, N. Ramadijanti, and A. Basuki, "Implementation of Facial Expression Recognition System for Selecting Fashion Item Based on Like and Dislike Expression", *2018 International Electronics Symposium on Knowledge Creation and Intelligent Computing (IES-KCIC) Bali, Indonesia,* 2018, pp. 74-78
        [http://dx.doi.org/10.1109/KCIC.2018.8628516]

[4]   L. Pang, N. Li, L. Zhao, W. Shi, and Y. Du, ""Facial expression recognition based on Gabor feature and neural network"", *2018 International Conference on Security, Pattern Analysis, and Cybernetics (SPAC) Jinan, China,,* 2018, pp. 489-493
        [http://dx.doi.org/10.1109/SPAC46244.2018.8965443]

[5]   Y. Li, S. Wang, Y. Zhao, and Q. Ji, "Simultaneous facial feature tracking and facial expression recognition", *IEEE Trans. Image Process.,* vol. 22, no. 7, pp. 2559-2573, 2013.
        [http://dx.doi.org/10.1109/TIP.2013.2253477] [PMID: 23529088]

[6]   S. Mohseni, N. Zarei, and S. Ramazani, "Facial expression recognition using anatomy based facial graph", In: *2014 IEEE International Conference on Systems, Man, and Cybernetics (SMC),San Diego,* 2014, pp. 3715-3719.
        [http://dx.doi.org/10.1109/SMC.2014.6974508]

[7]   H. Jun, C. Jian-feng, F. Ling-zhi, and H. Zhong-wen, "A method of facial expression recognition based on LBP fusion of key expressions areas", *The 27th Chinese Control and Decision Conference (2015 CCDC), Qingdao,,* 2015, pp. 4200-4204
        [http://dx.doi.org/10.1109/CCDC.2015.7162668]

[8]   Yogesh, and A.K. Dubey, "Fruit defect detection based on speeded up robust feature technique", *2016 5th International Conference on Reliability, Infocom Technologies and Optimization (Trends and Future Directions) (ICRITO), Noida, India,,* 2016, pp. 590-594
        [http://dx.doi.org/10.1109/ICRITO.2016.7785023]

[9]   T. Makkar, Y. Kumar, A.K. Dubey, Á. Rocha, and A. Goyal, "Analogizing time complexity of KNN and CNN in recognizing handwritten digits", *2017 Fourth International Conference on Image Information Processing (ICIIP), Shimla, India,,* 2017, pp. 1-6
        [http://dx.doi.org/10.1109/ICIIP.2017.8313707]

CHAPTER 25

A Review on Nanogrid Technology and Prospects

Amita Mane[1,2,*], **Shamik Chatterjee**[2] and **Amol Kalage**[3]

[1] *First Year Engineering Department, PCCOER, Pune, India*

[2] *School of Electronics and Electrical Engineering, Lovely Professional University, Phagwara, India*

[3] *Electrical Engineering Department, Sinhgad Institute of Technology, Lonavala, Pune, India*

Abstract: Global energy demand is expected to climb about 25% by 2040. This will be a challenge for the national power grid. Distance between generating stations and consumers is also another concern that may lead to more line losses and reducing the efficiency of the power system. This increasing demand leads to complexities in a national grid with increased demand for reliability, security, and environmental concern. The solution to the above challenges and requirements is the new structure of the power system known as the "Nano grid". A nanogrid can be considered as an electrical generation and distribution system, which is a building block of small loads, two or more distributed generations (DGs), and the ability to connect or disconnect from the utility grid. In this article, an overview of Nano grid technology with its advancement and the prospect has been presented.

Keywords: Distributed Generation (DG), Nano Grid, Power Grid, Renewable Energy (RE).

1. INTRODUCTION

Due to exponentially increasing demand from consumers, the national power grid is facing several challenges. This is not a single challenge in front of the centralized power grid. Distance between generating stations and consumers is also another concern that may lead to more line losses and reducing the efficiency of the power system. Environmental conditions like heavy rain, wind, *etc* may lead to a power system failure and hence the power outages. Centralized grid generating stations are dependent on fossil fuels which are responsible for the degradation of the atmosphere. In addition to the above, power supply to remote areas is one of the big tasks. To respond to 21st centuries electricity demand, it is

---

[*] **Corresponding author Amita Mane:** First Year Engineering Department, PCCOER, Pune, India; and School of Electronics and Electrical Engineering, Lovely Professional University, Phagwara, India; E-mail: maneamita4@gmail.com

Dharam Buddhi, Rajesh Singh and Anita Gehlot (Eds.)

required to design a new structure of power system which will reduce the burden on the national power grid. Distributed generation is one of the solutions to the above problems, which can generate power close to its point of consumers [1]. Photovoltaic, wind power these renewable energies (RE) sources which are omnipresent with the advantage of low carbon footprints and wide distribution comes under the DGs. One of the problems associated with renewable energy sources is their intermittent nature [2]. Output power developed by the photovoltaic module or the wind turbine is dependent on the availability of sunlight and wind respectively. Hence output power will be fluctuating and the power system has to deal with uncontrollable generation and demand. Consumers demands uninterrupted power supply and due to this intermittent nature consumers have less attention in investing into RE. Another problem associated with RE sources are high installation cost and the different energy policies related with RE [3]. To overcome these inadequacies and to bring consumers' attention towards investment in RE, a control structure is to be developed for balancing between the production and consumers' demand. In recent years such a new structure in the field of research is microgrid (MG). It is a low-voltage distribution power system, which is composed of small generating units, which may be renewable energy sources, other distributed generators, and battery energy storage systems are connected to fulfill the load demand [4]. MG can be integrated with the main grid or can be operated in off-grid mode. Another structure that is further developed and small version of the microgrid is known as the "Nano grid". Hence Nano grid is lower power and has fewer complexes than the microgrid. A Nano grid can be considered as an electrical generation and distribution system, which is a building block of small loads, two or more distributed generations (DGs), and the potential to connect or disconnect from the national power grid [5]. Nano grid is the important cornerstone of the future smart grid.

Nano grid is a recent topic in the field of research and research is going on in various areas like nanogrid control, nanogrid hardware, power quality, future development of nanogrid, *etc*. This paper will discuss an overview of the Nano grid concept, types of Nano grid Technologies, and their structure. It will also discuss new technologies and research in the Nano grid and the future scope of research.

2. NANO GRID CONCEPTS

It is difficult to define the electrical power system due to changes in opinions and structures. According to [6 - 8] Nano grid is similar to that of the microgrid. So we will start our discussion with similarities between nanogrid and microgrid. Not only microgrid but also nanogrid is a power distribution network. Another analogous point is both can operate as AC, DC, or in hybrid mode. A microgrid can operate in off-grid mode and grid-connected mode which is again true in the case of the Nano grid. To differentiate nanogrid from microgrid one should focus on the load which is to be connected. The concept of nanogrid is limited to a single house or building while microgrid extends to interconnections of multiple nano grids.

2.1. Nano Grid

By taking the reference of information given above nanogrid can be defined as A nanogrid is the distribution network of the power system for a single house or a building for load up to 20kW, with the potential to connect and disconnect from the national grid [5].

2.2. Structure of Nano Grid

The fundamental structure of nanogrid is shown in Fig. (1). It includes the following building blocks:

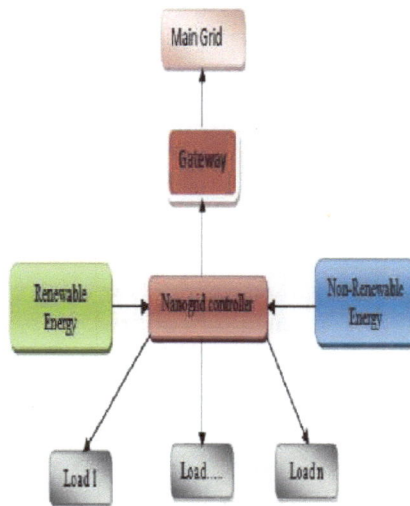

Fig. (1). Nano grid block diagram.

2.3. Distributed Generators

In the case of the nanogrid it is possible to generate the power close to the point of consumers which is known as Distributed Generation (DG) [9, 10]. It consists of both renewable and nonrenewable energy sources. Renewable energy sources are solar, wind and on the other side fuel cell, diesel generator comes under the nonrenewable DGs [11, 12].

2.4. Load

These are the household appliances like the water heater, refrigerator, microwave, television, *etc* [11, 13].

2.5. Nano grid Controller

It is known as the brain of the Nano grid, which controls the operation of the Nano grid. For Nano, grid controller may be for supply-side management or demand-side management. A properly implemented controller is a sign of healthy nanogrid [14, 15].

2.6. Gateway

A two-way connection between nanogrids, national grids, or microgrids is known as a gateway.

3. NEW TECHNOLOGIES AND RESEARCH

Research is going on various areas of nanogrid and it is implemented in various applications like household, street lighting, *etc*. This section focuses on a Review of different technologies and research of nanogrid.

- Configuration and analysis of nanogrid systems for poor villages have been presented by the author in [9]. It is not economical to extend the national grid into rural areas. Different configurations of energy are presented such as PV nanogrid, PV/wind nanogrid, solar/diesel nanogrid, *etc*. A 5 –14.5 kW PV-based nanogrid has been simulated for the demand of 12.5 – 36.5 kWh/d. The power output of different configurations is also presented.
- A hybrid nanogrid system for future community houses is presented by the author in [16]. The proposed structure of nanogrid is a combination of multiple types of AC and DC sources to supply the AC and DC loads of community houses.
- Home appliance is the main load in DC nano grid. The efficiency of nanogrid is mainly dependant on the design of these appliances. Focus on design parameters of home appliances and their simulation results are presented in [11]. An induction heating range is considered as a case for design example.
- Nano grid and Microgrid are the better solutions to meet the growing electricity demand. Solar Photovoltaic based nanogrid assessed for solar photovoltaic

charging is presented in [12]. Here, the Nano grid testbed contains solar energy as a powers supply, near about twenty electrical vehicle charging stations, smart inverter, energy storage system which are connected to the primary feeder on the university of California microgrid. For an energy management system, four different control algorithms are invented and are successfully used.

- The concept of an open energy system and its component are described in [17]. The aim of this research is a layered control scheme. The author has presented simulation results as well as an overview of the laboratory prototype of the proposed system.

- Very few publications are in the domain of the power quality of nanogrid. Analysis of voltage quality data of nanogrid in islanded and grid-connected mode has been done in [18]. Voltage quality data is compared with standard limits.

- To overcome the electricity challenge on the national grid, a hybrid AC/DC nano grid is proposed in [19]. A modified SRF controller is used to balance the supply currents and the PCC, regardless of loading condition. A model of the proposed structure is simulated in MATLAB software and output is presented for different loading conditions.

- Implementation of road lightning system integrated with nanogrid is presented by the author [20]. The feasibility of the proposed system is verified from experimental results.

Nano grid as a solution for countryside electrification has been proved in [21]. The advances in Nano grid converter technology and its implementation is also

4. FUTURE SCOPE OF RESEARCH

Nano grid is one of the healthy solutions, to overcome challenges on the national power grid. Another aim is to reduce carbon footprints by increasing the use of renewable energy sources integrated into nanogrid. The intermittent nature of renewable energy sources is responsible for the power quality issue and hence its decrease inefficiency. To overcome this it is required to improve the power quality of the Nano grid by using different compensators. The compensator can be a single-phase or three-phase type depending upon the loading condition. Very few papers are published on the power quality issue of nanogrid and its compensation. Hence further research on power quality and its compensation can be carried out also demand and supply-side management issues can be solved by different techniques.

CONCLUSION

Nano grid is a recent topic in the field of research and research is going on in various areas like nanogrid control, nanogrid hardware, future development of nanogrid, *etc*. Few papers are published on power quality issues in nanogrid; hence it is required to analyze the power quality of nanogrid and to overcome that issue by using a proper compensator. This paper mainly discusses the concept of nanogrid, different technologies and research going on the nanogrid, and finally future scope of research in the field of nanogrid.

CONSENT FOR PUBLICATION

Not applicable.

CONFLICT OF INTEREST

The authors declare no conflict of interest, financial or otherwise.

ACKNOWLEDGEMENT

Declared none.

REFERENCES

[1]   C. Rae, and F. Bradley, "Energy autonomy in sustainable communities - A review of key issues", *Renew. Sustain. Energy Rev.,* vol. 16, no. 9, pp. 6497-6506, 2012.
[http://dx.doi.org/10.1016/j.rser.2012.08.002]

[2]   C. Lupangu, and R.C. Bansal, ""Retraction Notice to " A Review of Technical Issues on the Development of Solar Photovoltaic Systems " [Renew. Sustain. Energy Rev., 73 (2017)", *Renew. Sustain. Energy Rev.,* vol. 94, p. 1234, 2018.
[http://dx.doi.org/10.1016/j.rser.2018.08.016]

[3]   E.K. Stigka, J.A. Paravantis, and G.K. Mihalakakou, "Social acceptance of renewable energy sources : A review of contingent valuation applications", *Renew. Sustain. Energy Rev.,* vol. 32, pp. 100-106, 2014.
[http://dx.doi.org/10.1016/j.rser.2013.12.026]

[4]   L. Ortiz, R. Orizondo, A. Aguila, J. W. Gonz, I. Isaac, and J. L. Gabriel, *Heliyon Hybrid AC / DC microgrid test system simulation : grid-connected mode.,* vol. 5, 2019.
[http://dx.doi.org/10.1016/j.heliyon.2019.e02862]

[5]   D. Burmester, R. Rayudu, W. Seah, and D. Akinyele, "A review of nanogrid topologies and technologies", *Renew. Sustain. Energy Rev.,* vol. 67, pp. 760-775, 2017.
[http://dx.doi.org/10.1016/j.rser.2016.09.073]

[6]   M. Di Somma, *Multi-Energy Nanogrids for Residential Applications,* 2020.

[7]   A. Burgio, D. Menniti, N. Sorrentino, A. Pinnarelli, and M. Motta, "A compact nanogrid for home applications with a behaviour-tree-based central controller", *Appl. Energy,* vol. 225, no. April, pp. 14-26, 2018.
[http://dx.doi.org/10.1016/j.apenergy.2018.04.082]

[8]   S. Javaid, Y. Kurose, T. Kato, and T. Matsuyama, "Cooperative Distributed Control Implementation of

the Power Flow Coloring over a Nano-Grid with Fluctuating Power Loads", *IEEE Trans. Smart Grid,* vol. 8, no. 1, pp. 342-352, 2017.
[http://dx.doi.org/10.1109/TSG.2015.2509002]

[9]   D. Akinyele, "Techno-economic design and performance analysis of nanogrid systems for households in energy-poor villages", *Sustain Cities Soc.,* vol. 34, pp. 335-357, 2017.
[http://dx.doi.org/10.1016/j.scs.2017.07.004]

[10] P. Burgess, M. Shahidehpour, M. Ganji, and D. Connors, "Remote power units for off-grid lighting and urban resilience", *Electr. J.,* vol. 30, no. 4, pp. 16-26, 2017.
[http://dx.doi.org/10.1016/j.tej.2017.03.012]

[11] Ó. Lucía, I. Cvetkovic, H. Sarnago, D. Boroyevich, P. Mattavelli, and F.C. Lee, "Design of home appliances for a DC-based nanogrid system: An induction range study case", *IEEE J. Emerg. Sel. Top. Power Electron.,* vol. 1, no. 4, pp. 315-326, 2013.
[http://dx.doi.org/10.1109/JESTPE.2013.2283224]

[12] L. Novoa, and J. Brouwer, "Dynamics of an integrated solar photovoltaic and battery storage nanogrid for electric vehicle charging", *J. Power Sources,* vol. 399, no. March, pp. 166-178, 2018.
[http://dx.doi.org/10.1016/j.jpowsour.2018.07.092]

[13] M.S. Pilehvar, M.B. Shadmand, and B. Mirafzal, "Analysis of Smart Loads in Nanogrids", *IEEE Access,* vol. 7, pp. 548-562, 2019.
[http://dx.doi.org/10.1109/ACCESS.2018.2885557]

[14] E. González-Romera, "Demand and storage management in a prosumer nanogrid based on energy forecasting", *Electron.,* vol. 9, no. 2, 2020.
[http://dx.doi.org/10.3390/electronics9020363]

[15] M. Rafiee Sandgani, and S. Sirouspour, "Energy Management in a Network of Grid-Connected Microgrids/Nanogrids Using Compromise Programming", *IEEE Trans. Smart Grid,* vol. 9, no. 3, pp. 2180-2191, 2018.
[http://dx.doi.org/10.1109/TSG.2016.2608281]

[16] R.P.S. Chandrasena, F. Shahnia, S. Rajakaruna, and A. Ghosh, *Dynamic operation and control of a hybrid nanogrid system for future community houses.* vol. 9. , 2015, pp. 1168-1178.
[http://dx.doi.org/10.1049/iet-gtd.2014.0462]

[17] A. Werth, N. Kitamura, and K. Tanaka, "Conceptual Study for Open Energy Systems: Distributed Energy Network Using Interconnected DC Nanogrids", *IEEE Trans. Smart Grid,* vol. 6, no. 4, pp. 1621-1630, 2015.
[http://dx.doi.org/10.1109/TSG.2015.2408603]

[18] J. Nömm, S.K. Rönnberg, and M.H.J. Bollen, "An analysis of voltage quality in a nanogrid during islanded operation", *Energies,* vol. 12, no. 4, 2019.
[http://dx.doi.org/10.3390/en12040614]

[19] A.F. Ebrahim, T.A. Youssef, and O.A. Mohammed, "Power Quality Improvements for Integration of Hybrid AC/DC Nanogrids to Power Systems", *IEEE Green Technol. Conf.,* 2017,pp. 171-176
[http://dx.doi.org/10.1109/GreenTech.2017.31]

[20] S. Yoomak, and A. Ngaopitakkul, "Investigation and Feasibility Evaluation of Using Nanogrid Technology Integrated into Road Lighting System", *IEEE Access,* vol. 8, pp. 56739-56754, 2020.
[http://dx.doi.org/10.1109/ACCESS.2020.2978897]

[21] S. Mishra, and O. Ray, "Advances in nanogrid technology and its integration into rural electrification in India", *Int. Power Electron. Conf. IPEC-Hiroshima - ECCE Asia ,* 2014, pp. 2707-2713
[http://dx.doi.org/10.1109/IPEC.2014.6869973]

CHAPTER 26

Dual Factor Authentication Bank Locker Security System

Rohit J. Sahni[1], Gourav Singh[1], Kusu Veda Krishna Uday[1] and **Suresh Kumar Sudabattula[1,*]**

[1] *School of Electronics and Electrical Engineering, Lovely Professional University, Phagwara, Punjab, India 144411*

Abstract: The foundation aim of these Projects is to increase the storage security of bank lockers based on fingerprint Sensors and Shocking mechanisms with the help of a One-time password. These systems may help the bank to attract more customers by the use of high security and smart locker system. In this framework, there is a requirement to verify individuals to recuperate the reports or cash from the storage spaces. In this security framework, unique finger impression and OTP are utilized. The first individual's fingerprint and mobile number are registered with the system. After that whenever a user will scan their finger with the system, it matches the stored print with the print received after scanning. In case the print matches at that point four-digit code will be sent on approved individual versatile to open. So biometric and OTP security are points of interest than other frameworks. This system will have an attached vibration sensor, to detect in case someone tries to hammer or break the locker. On detection of any such activity alert, there will auto-generated messages will send to high authority persons on individual's number. Also, the system goes to freeze mode and is activated only when a valid finger is scanned.

Keywords: Finger Print Sensor, GSM, Microcontroller, Motor Driver, Vibration Sensor.

INTRODUCTION

In the present scenario, well-being and safety have turned into a fundamental issue for the vast majority of individuals, particularly in the provincial and metropolitan territories. A minority of people attempt to betray or take over the property that can imperil the protection of the bank's cash, house, and office. To overcome the safety threat, the majority of individuals will propose a bundle of locks or caution framework. A variety of alert systems are accessible from the market which uses diverse sensors. These sensors can determine the changes occ-

---
* **Corresponding author Suresh Kumar Sudabattula:** Department of Electronics & Communication, Amity University Uttar Pradesh, Noida, India; E-mail: suresh.21628@lpu.co.in

urring nearby and the development is controlled by an alarm according to pre-set worth. Over time, this framework might not be used continuously. In this research article, we emerged security of cash in bank's locker, house or office by employing biometric and GSM technology which is much cautious than other systems. In Ref. [1] biometric frameworks utilize an individual's actual qualities (like fingerprints, irises, or veins), or social qualities (like voice, penmanship, or composing musicality) to decide their character or to affirm that they are who they guarantee to be. In Ref. [2] biometric information is exceptionally novel to every person. Talking about fingers prints they are one of the numerous extraordinary biometric marks which we can use to distinguish individuals precisely. In Ref. [3] However, holding someone's hand and gazing at their fingers aren't useful as we are bad at this. Be that as it may, PCs are acceptable at perceiving and coordinating examples exceptionally quickly and precisely. Before we can handle a unique finger impression design with a PC, we should "catch" it. There exist numerous strategies to digitize fingerprints; from legal techniques to ultrasound examining. R307 is an image-based mark scanner element from the R30X arrangement delivered by Hangzhou Grow Technology Co. Ltd, a Chinese merchant. Multiple types of sensors in this arrangement are capacitive sensors *viz.,* R300, R301T, R302, R303, R303T, R305, R306, R308, and R311.

Background Work

In these, we briefly introduce the previous works carried out by various researchers. Few papers in the market already worked on smart lockers Systems. S.V Tejasvi work in the smart security system by the use of fingerprint OTP by using AT89S52 microcontroller in our work we have used ATMEGA2560 and also add to more security like we are also adding one shocking mechanism where if anyone tries to harm the main door the vibration sensor senses the vibration and shocking mechanism get enable with these 220v voltages will directly give to the door so if anyone tries to touch the door they get shocked and also we add one freeze system with these if anyone tries to breach our security system our system get freeze [3]. For example, Some non-User try to enter the locker without permission the system will freeze by itself and the system will get unfreeze by the only person it can be the security head or Branch manager. These things will make add more sacristy to our bank security system. After System gets unfreeze the shock mechanism will get low command and the Shock mechanism will get disable. Our system will get back to the normal smart security locker.

Proposed Work and Methodology

The Dual factor authentication bank locker security System consists of database enrollment, data scan, data Delete, and verification. First, the client will enlist his client Name and his portable number in the framework data set through framework programming then the individual will put a finger on the unique mark module finger impression will be output and store with a finger id. In this manner, the client enrolment cycle will be finished [4, 5]. At that point client will perform login activity during login activity client will initially scan his impression and in case it, then OTP will be sent on a versatile number of the client which entered during enrolment through GSM. Then the client will punch the code through the keypad if the code gets coordinated, the storage will open and LCD will show access conceded. In case the fingerprint doesn't match an alert message will be sent to the client's number saying "someone trying to open the locker", thus alerting the client. Also, the vibration sensor attached will sense any kind of hammering or attempt to break the locker and notify it to nearby police station and client. Also, it will freeze the system functionality until a valid finger is scanned.

Enrolling: In these first, the user will get enrolled in our system by adding their fingerprint and contact number. All the details and data are saved by our system which will future use when the user will coming to open their locker.

Delete Our old users which will discontinuous our services these will help to erase their data like fingerprint and contact number from our system for their security purpose. And also help to increase the efficiency of the system so in future they don't try to breach our security.

Scan: This is the very first step toward the locker system where the user will put their finger in the figure print scanner and our system will scan the user figure if the user is a registered user then they will get access to the second step which is OTP system. If the user will not register our system will show that you are a non-user Please get enrolled your id in our system to get access the locker.

Related Work

In this section related work is discussed below and it is shown in Fig. (**1**), first, the client will enlist his client Name and his portable number in the framework data set through framework programming then the individual will put his finger on the unique mark module finger impression will be output and store with finger id. In this manner, the client enrolment cycle will be finished. At that point client will perform login activity during login activity client will initially scan his impression and in case it, then OTP will be sent on a versatile number of the client which entered during enrolment through GSM [6]. Then the client will punch the code through the keypad if the code gets coordinated, the storage will open and LCD will show access conceded. In case the fingerprint doesn't match an alert message will be sent to the client's number saying "someone trying to open the locker", thus alerting the client. Also, the vibration sensor attached will sense any kind of hammering or attempt to break the locker and notify it to nearby police station and client. Also, it will freeze the system functionality until a valid finger is scanned.

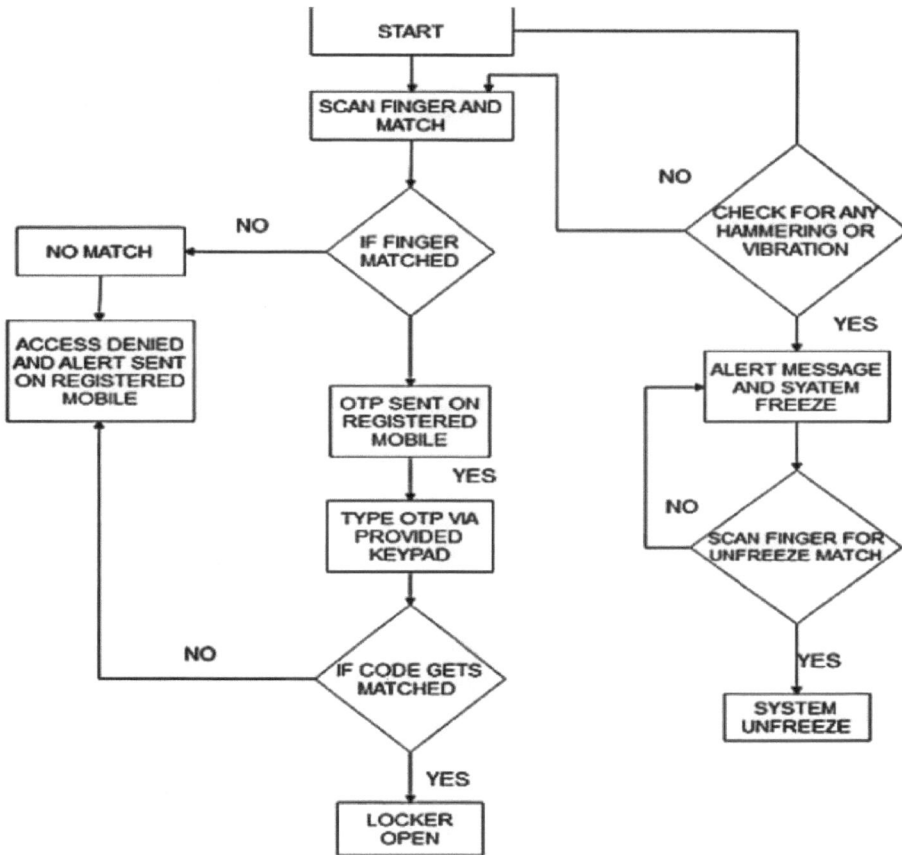

Fig. (1). Enrollment and verification stages.

Block Diagram

The overall block diagram illustrated in Fig. **(2)**, consists of a microcontroller, GSM, LCD, Finger Print Sensor, Vibration Sensor, once the client will log in by the fingerprint sensor, they will get an OTP with the help of GSM in their registered mobile. The moment the client filled OTP and it is verified by the system and if OTP is correct system gives the access of locker to the client otherwise request for locker access will be denied by the system in case of the wrong OTP.

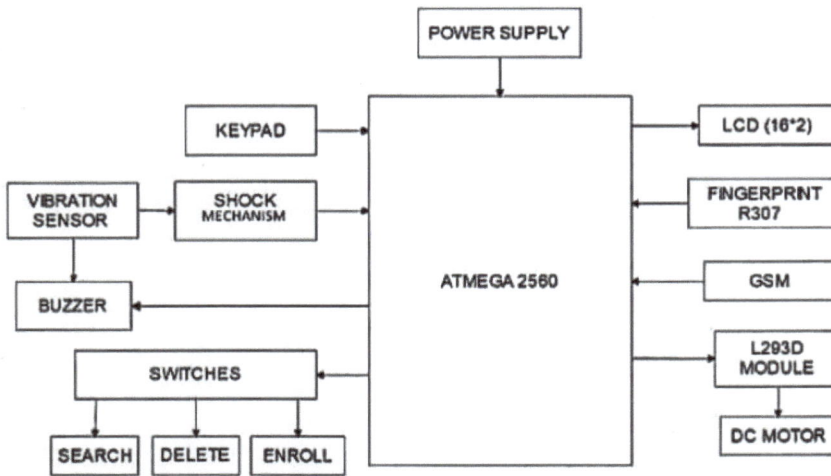

Figure 2: Overall block diagram.

Fig. (2). Overall block diagram.

Software and Hardware Implementation

Microcontroller: The ATMEGA2560 is shown in Fig. **(3)**. It is an 8-bit microcontroller with 256k bytes in a system programmable flash. The operating frequency is 16mhz which is higher as compared to the other microcontroller and low power consumption with a fast start-up. It has 54 digital input/output pins (of which 15 can be used as PWM outputs), 16 analog inputs, 4 UARTs (hardware serial ports), a USB connection, a power jack, an ICSP header, and a reset button [2].

Fig. (3). Arduino mega.

Fingerprint: The fingerprint module used is R307. It has 4 pins out of which 2 pins for the supply that is V_{cc} and ground other two pins are for serial communication. It works on serial communication. The baud rate for this module is 57600 bps. Baud rate should match so they can work in synchronization [1].

GSM: The full name of GSM is Global System for Mobile Communication it's a primary data network. Gsm module having quadrigeminal different network range frequency. We use the Sim900 GSM modem these modems support quad-band GSM850, EGSM900. This GSM modem work in 900mhz and 2G Sim can be connected with any global GSM network. With help of these GSM modems we send and receive one-time-password to customers and high authorities.

Motor Module: The driver used is L293D. it is used where we have to protect microcontroller pins from back emf. We have dc motor or geared motor. They are inductive and they have a coil inside it which produces back emf due to which all the pins can be damaged due to this motor driver is used. We can connect the DVD loader but to protect from back emf motor driver is used. It acts as isolation between motor and microcontroller, to protect the micro-controller pins. They are having 16 pins. It takes a 5v signal from Arduino and in turn provides a 12v signal for operating the dc motor.

LCD Interfacing: LCD stands for liquid crystal display and the same as shown in Fig. (**4**), the LCD used is 16*2 display which represents 16 columns and 2 rows. Its operating voltage is 4.7V-5.3V. the utilization of current is 1mA with no backlight. A 16*2 has two registers like the data register and command register. The RS (register select) is mainly used to change from one register to another. when the register set is '0', then it is known as command register. Similarly, when the register is set '1', then it is known as the data register.

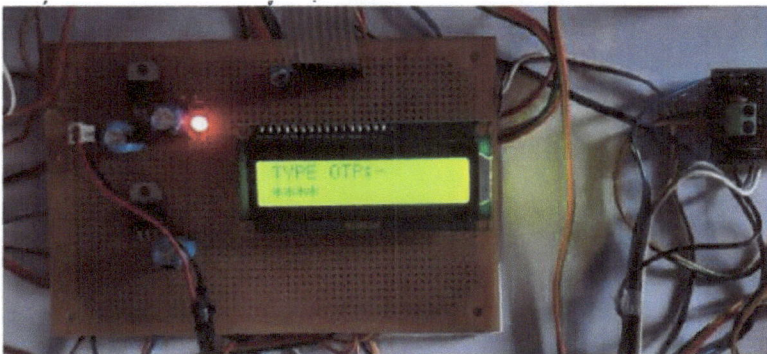

Fig. (4). LCD.

Vibration Sensor: The sensor used is SW-420. the main function of vibration sensor is that when a person is trying to break the locker a vibration sound will be produced due to which sensor will activate and send the signal to the microcontroller which further send the text messages to the Authority head and after the vibration sensor is activated the system is automatically frozen and only authorized person of the bank can unfreeze the system.

Software Program Testing

1. Installing drivers for Arduino mega.
   i. Plugin Arduino mega board in any one of the USB and want for windows to start its auto driver installation process.
   ii. In case windows fail to do that then navigate to the control panel from the start menu.
   iii. In the control panel go to the device manager option and look for COM & LPT option under the ports section.
   iv. Choose the COM option associated with Arduino mega & click on update drivers.
   v. After that choose "browse" my computer for drivers' options & navigate to file names "Arduino mega.inf located in the driver folder of the Arduino setup that you downloaded.
   vi. Windows will successfully finish the installation from there.
   vii. After that open compiler *i.e.* Arduino IDE & install the required libraries.
   viii. Libraries for fingerprint module R307 & keypad are required to be installed.
   ix. Both libraries are available on GitHub. Extract downloads files and paste them into the folder name 'Arduino' in the documents section of the system.

CONCLUSION AND FUTURE WORK

We can be inferred from the project that the implementation on hardware is successfully in running condition without any fluctuation. The stepwise procedure is done to implement the design of a microcontroller-based security system, the main aim of this project is to reduce the robberies in the bank as well as to reduce the burden of bankers which they are facing. The amount of time is significantly maintained by the dual-factor authentication system. The limitation for this project is that if a person does any mistake unknowingly then the system will automatically get freezes and if an authorized person is not present at that time in the bank then it is difficult to unfreeze the system because only the authorized person can unfreeze it. So, to avoid this limitation, the presence of an authorized person is a must. The future scope for this project is that we can add a voice-controlled setup and using the cage system outside the locker.

CONSENT FOR PUBLICATION

Not Applicable.

CONFLICT OF INTEREST

The authors declare no conflict of interest, financial or otherwise.

ACKNOWLEDGEMENTS

The study was supported by the School of Electronics and Electrical, Lovely Professional University, Phagwara, Punjab, India. The authors are highly grateful to the department for providing all necessary materials to carry out the present study.

REFERENCES

[1]   D. Hejtmánková, and R. Dvorák, "M. Drahanský And F. Orság, "A new method of finger veins detection," Int", *J. Bio-Science Bio-Technology,* vol. 1, no. 1, pp. 11-16, 2009.

[2]   S. Sridharan, "Authenticated secure bio-metric based access to the bank safety lockers", *IEEE International Conference on Information Communication and Embedded Systems,* 2014, pp. 1-7 [http://dx.doi.org/10.1109/ICICES.2014.7034063]

[3]   J. Baidya, T. Saha, R. Moyashir, and R. Patil, "Design and implementation of a fingerprint-based lock system for shared access", *In: IEEE 7th Annual Computing and Communication Workshop and Conference (CCWC,),* 2017pp. 1-6 [http://dx.doi.org/10.1109/CCWC.2017.7868448]

[4]   U.J. Ogri, D.E. Okwong, and A. Etim, "Design and construction of door locking security system using GSM", *International Journal of Engineering And Computer Science,* vol. 2, no. 7, pp. 2235-2257, 2013.

[5]   H.M. Benazir, N.S. Kalpathri, and C. Sunagar, "Fingerprint Authentication Smart Locking System Using OTP", *International Journal of Advance Research in Engineering, Science & Technology,* vol. 4, no. 6, 2017.

[6]   H.S. Detroja, P.J. Vasoya, D.D. Kotadiya, and P.C. Bambhroliya, "GSM Based Bank Locker Security System using RFID, Password and Fingerprint Technology", *IJIRST–International Journal for Innovative Research in Science & Technology,* vol. 2, no. 11, pp. 110-115, 2016.

CHAPTER 27

In Silico Identification, Analysis, and Prediction Algorithm for Plant Gene Cluster

Himanshu Singh[1], C. Vineeth[1], Bhupender Thakur[1], Atul Kumar Upadhyay[2] and Vikas Kaushik[1,*]

[1] *School of Bioengineering and Biosciences, Lovely Professional University, Punjab, India*

[2] *Department of Biotechnology, Thapar University, Punjab, India*

Abstract: The concept/phenomenon of operons, which are organized genes that work in a coordinated way in microbes, is well established. Recent developments in genetics, biochemistry, and bioinformatics have unraveled similar gene arrangements in plants. Here we aim to develop an algorithm/tool which would help us detect and identify biosynthetic gene clusters (BGCs) from any input plant genome. Through this tool, we intend to match or supersede the performance of pre-existing sting tools for BGC prediction, like the popular plantiSMASH. The predictions models were developed using the machine learning tool WEKA using the physicochemical properties as data set to classify between terpene synthases and non-terpene synthases. A set of ten physicochemical properties were selected and their values were predicted for each of the 159 proteins (terpene synthases and non-terpene synthases) Employing the random forest and SMO classifiers, we were able to obtain significantly promising accuracy of over 90 percent with 66 percent percentage split testing. Accurate prediction of BGCs in the plants, especially the major food crops like rice, wheat, and corn revolutionize farming and nutrition for the better.

Keywords: Algorithm, BGC, Mining, PlantiSMASH, Random forest, SMO WEKA.

INTRODUCTION

Metabolite gene cluster discovery techniques are improving at exponential rates which have opened up avenues of endless possibilities in the field of plant biology and natural product discovery. Improved farming (allelopathic interactions), drug discovery, better nutrition, and synthetic biology are only a few of the promising areas [1]. With over 20 wild varieties cultivated around the world, rice(Oryza *Sativa)*, a member of the family Poaceae and genus Oryza, is one of the most

---

* **Corresponding author Vikas Kaushik:** School of Bioengineering and Biosciences, Lovely Professional University, Punjab, India; E-mail: vikas31bt@gmail.com

popular and staple food crops in the world. *Oryza sativa* and *Oryzaglaberrima*are the most cultivated out of all the rice varieties. Sativa variety is a global favorite while glabberima has been around for over 3500 years, originating in West Africa. Rice species could be diploid or triploid with n=12 and *Oryza sativa* or *Oryzaglaberrima* L. are diploid species (2n = 24). Complete sequencing of the Asian cultivated rice genome has been performed and it was the first food crop to be the whole genome sequenced [2, 3]. A group of genes in the DNA of a particular organism is called a gene cluster when they collectively work in the production of protein or enzyme [4]. These genes are usually regulated by the same promoter region. BGCs produce two types of enzymes in general, signature enzymes produced by the signature genes of the BGC and tailoring enzymes produced by the tailoring genes of the BGC. Tailoring enzymes get produced first which in turn enter a cascade of reactions accelerating and catalyzing the production of the signature enzymes, which is the bigger, complex. In some gene clusters where there is the formation of enzymes. There are steps in which the formation of enzyme takes place. First, there is the synthesis of enzymes known as tailoring enzymes which helps in catalyzing as well as accelerating the processes to form the main enzyme which is a bigger and more complex molecule known as the Signature enzyme [5]. In this project, we are aiming to develop a universal cluster prediction algorithm using terpene synthase gene clusters as the reference and classifying genes into Terpene synthases and non-terpene synthases based on the distribution of selected unique physicochemical properties. Random forest and Sequential minimal optimization (SMO) [6, 7] classifiers were employed for the development of the machine learning model in the WEKA tool.

METHODOLOGY

Collection of Data

We have collected FASTA [8] sequences of Terpene synthases and non-terpene synthases (negative control), across the plant kingdom. UniProtKB database was used for the retrieval of the relevant data. Around 80 terpene synthases and non-terpene synthases each were gathered for this study. Fig. (**1**) shows the flow chart.

Selection and Prediction of Features

A total of 10 physicochemical properties were selected. They were Length, Molecular weight, Isoelectric point, Instability index, Aliphatic index, Hydropathicity Charge at pH 7, TMindex, Solubility, Extinction coefficient. Following web servers were utilized for the prediction of these features where the FASTA sequences of the proteins were fed as input.

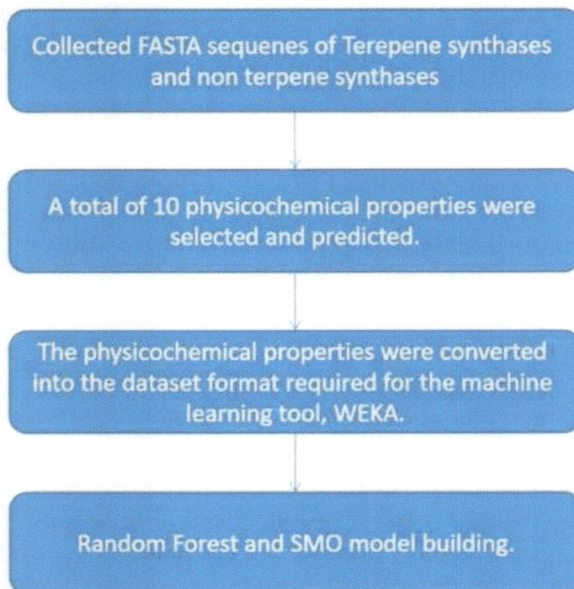

Fig. (1). Flow chart.

- ProtParam by Expasy: Chain length, molecular weight, PI, instability index, aliphatic index, GRAVY (Hydropathicity) https://web.expasy.org/protparam/ [9].
- PROTEIN CALCULATOR v3.4: Charge at pH7 http://protcalc.sourceforge.net/
- TM predictor: TMindex http://tm. life.nthu.edu.tw/
- Protein-sol: Solubility https://protein-sol.manchester.ac.uk/ [10].
- PepCalc.com-Peptide property calculator by INNOVAGEN: Extinction coefficient https://pepcalc.com/

Preparation of Data Set

The physicochemical properties were converted into the dataset format required for the machine learning tool, WEKA [11], which was used in this project, where classes were terpene or non terpene. The data set was saved in .arff format and explored through the WEKA tool for model building.

Model Building

Loaded with several algorithms such as Bayesian Network, SVMLib, Artificial Neural Network (ANN), Nearest Neighbor (IBk), Random Forest, *etc*, Weka is a convenient machine learning tool. For this study, we developed Random forest and SMO models. Fig. (**2**) shows a Display of the attribute distribution across terpene synthases and non-terpene synthases and Fig. (**3**) shows the J48 decision tree representation of attributes.

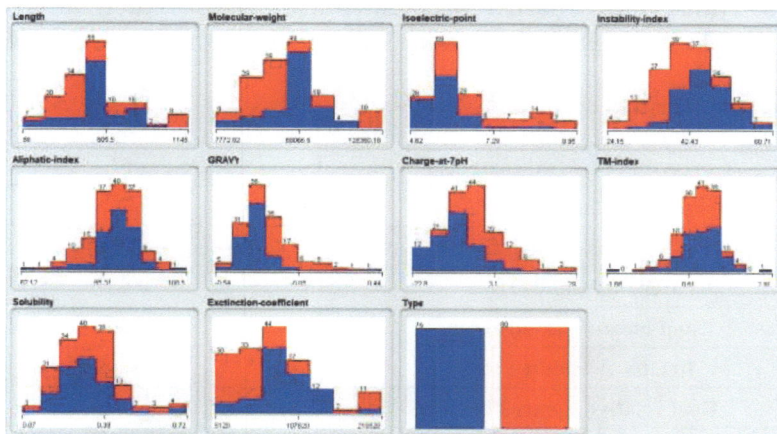

Fig. (2). Display of the attribute distribution across terpene synthases and non-terpene synthases [12].

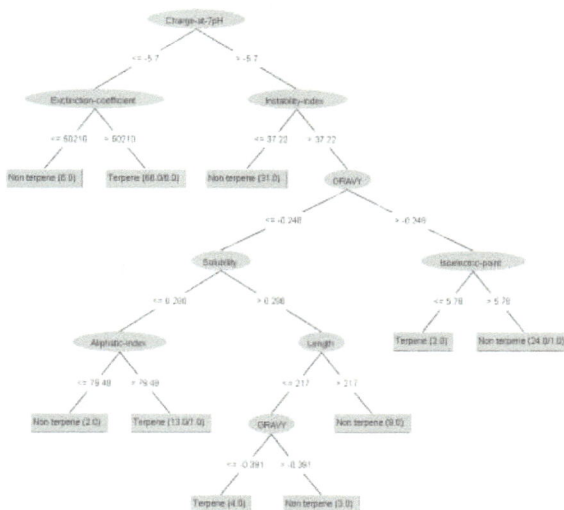

Fig. (3). J48 decision tree representation of attributes [11].

RESULT AND DISCUSSION

Physiochemical values were formatted as per the WEKA tool input requirements. The same ten physiochemical properties were considered for terpene synthases and non-terpene synthases. In this study, we have used 66% percentage split testing on the data set. We calculated the overall accuracy, true positive rate (TP), false positive rate (FP), precision, recall, Mathew's correlation coefficient(MCC), and receiver operating characteristic (ROC) along with the confusion matrices of the attributes, threshold curves and attribute vise visualization of the classification. Table **1** shows a Summary of the SMO model developed

Table 1. Summary of SMO model developed [12, 13].

Correctly Classified Instances	**52**	**96.29%**
Incorrectly Classified Instances	**2**	**3.70%**
Kappa statistic	**0.92**	
Mean absolute error	**0.037**	
Root mean squared error	**0.19**	
Relative absolute error	**7.40%**	
Root relative squared error	**38.47%**	
Total Number of Instances	**54**	

SMO model developed shows an appreciable 96.2963% accuracy. This means that there is a 96.2963% chance that the model accurately predicts the input to be a terpene synthase or a non-terpene synthase. Table **2** shows the summary of the Random forest model.

Table 2. Summary of Random forest model developed [12, 13].

Correctly Classified Instances	**51**	**94.44%**
Incorrectly Classified Instances	**3**	**5.55%**
Kappa statistic	**0.89**	
Mean absolute error	**0.16**	
Root mean squared error	**0.23**	
Relative absolute error	**32.91%**	
Root relative squared error	**45.79%**	
Total Number of Instances	**54**	

The random forest model gives an accuracy of 94.4444% percent for appropriate classification with 51 correctly classified instances and 3 incorrectly classified ones out of a total of 54. Table **3** shows the Performance of the SMO model developed by the class and Table **4** shows the performance of the Random forest developed by the class.

Table 3. Performance of SMO model developed by class [12, 13].

	TP Rate	FP Rate	Precision	Recall	MCC	ROC Area	Class
	1	0.071	0.929	1	0.929	0.964	Terpene
	0.929	0	1	0.929	0.929	0.964	Non terpene
Weighted Avg.	0.963	0.034	0.966	0.963	0.929	0.964	

Table 4. Performance of the Random forest developed by class [12, 13].

	TP Rate	FP Rate	Precision	Recall	MCC	ROC Area	Class
	0.962	0.071	0.926	0.962	0.889	0.992	Terpene
	0.929	0.038	0.963	0.929	0.889	0.992	Non terpene
Weighted Avg.	0.944	0.054	0.945	0.944	0.889	0.992	

The confusion matrix summarizes the accuracy by illustrating how many entries were classified under the appropriate class and how many were not. Table **5** shows the Confusion matrix of the SMO model and Table **6** shows the confusion matrix of the random forest model.

Table 5. Confusion matrix of the SMO model [12, 13].

Classified as→	Terpene	Non-terpene
Terpene	26	0
Non-terpene	2	26

Table 6. Confusion matrix of the random forest model [12, 13].

Classified as →	Terpene	Non-terpene
Terpene	25	1
Non-terpene	2	26

CONCLUSION

The models developed with the help of both classifiers (RF and SMO) as a part of this study to predict BGCs in plants, mainly based on the distribution of physicochemical features of the respective protein products of the said BGCs (terpene synthases and non-terpene synthases) showed significantly positive and accurate results of classification. Machine learning approaches, classifiers (SMO and RF models) have been developed to predict domain swapping at the genome level from mere protein sequence information. An accuracy of 96% and 94% were achieved for the two methods, respectively. Research in the field of plant secondary metabolic gene clusters has seen exponential growth over the past few years. It is only poised to extend farther as more discoveries of and about gene clusters emerge at an accelerated rate, thanks to high throughput screening methods. Combining systematic genome mining and functional analysis of candidate clusters along with artificial intelligence and machine learning discovers new pathways, enzymes, and chemistries well within the realm of possibility. The impact that this accelerated development can have on the way we approach everything from farming to drug discovery to nutrition will be unprecedented. Application of such technologies into staple, major food crop industry could mean a significant quality of life change for the world as a whole through better farming and nutrition.

CONSENT FOR PUBLICATION

Not applicable.

CONFLICT OF INTEREST

The authors declare no conflict of interest, financial or otherwise.

ACKNOWLEDGEMENTS

All the authors conducted research and write the MS and thoroughly studied, and are thankful towards the school of bioengineering and biosciences, LPU, Phagwara, Punjab for providing the finest computational facility for conduction of research. The authors read and approved the final manuscript. The authors hereby declare that they have no conflict of interest.

REFERENCES

[1]　H-W. Nützmann, and A. Osbourn, "Gene clustering in plant specialized metabolism", *Curr. Opin. Biotechnol.,* vol. 26, pp. 91-99, 2014.
　　　[http://dx.doi.org/10.1016/j.copbio.2013.10.009] [PMID: 24679264]

[2]　G. Yi, S-H. Sze, and M.R. Thon, "Identifying clusters of functionally related genes in genomes", *Bioinformatics,* vol. 23, no. 9, pp. 1053-1060, 2007.
　　　[http://dx.doi.org/10.1093/bioinformatics/btl673] [PMID: 17237058]

[3]　H. Singh, N. Siddique, and A.K. Upadhyay, "Genome-wide Identification and Annotation of metabolite producing Gene Clusters in Rice Genome", *Research Journal of Pharmacy and Technology,* vol. 13, no. 4, p. 1744, 2020.
　　　[http://dx.doi.org/10.5958/0974-360X.2020.00314.5]

[4]　F-J. Qin, Q-W. Sun, L-M. Huang, X-S. Chen, and D-X. Zhou, "Rice SUVH histone methyltransferase genes display specific functions in chromatin modification and retrotransposon repression", *Mol. Plant,* vol. 3, no. 4, pp. 773-782, 2010.
　　　[http://dx.doi.org/10.1093/mp/ssq030] [PMID: 20566579]

[5]　B. Ghosh, and N. Ali Md, "Response of Rice under Salinity Stress: A Review Update", *Rice Research: Open Access,* vol. 4, 2016.
　　　[http://dx.doi.org/10.4172/2375-4338.1000167]

[6]   T.K. Ho, "The random subspace method for constructing decision forests", *IEEE Trans. Pattern Anal. Mach. Intell.,* vol. 20, no. 8, pp. 832-844, 1998.
[http://dx.doi.org/10.1109/34.709601]

[7]   Z-Q. Zeng, H-B. Yu, H-R. Xu, Y-Q. Xie, and J. Gao, "Fast training Support Vector Machines using parallel sequential minimal optimization", *2008 3rd International Conference on Intelligent System and Knowledge Engineering,,* 2008

[8]   D.J. Lipman, and W.R. Pearson, "Rapid and sensitive protein similarity searches", *Science,* vol. 227, no. 4693, pp. 1435-1441, 1985.
[http://dx.doi.org/10.1126/science.2983426] [PMID: 2983426]

[9]   E. Gasteiger, C. Hoogland, A. Gattiker, S. Duvaud, M.R. Wilkins, R.D. Appel, and A. Bairoch, *Protein Identification and Analysis Tools on the ExPASy Server. The Proteomics Protocols Handbook,* 2005, pp. 571-607.
[http://dx.doi.org/10.1385/1-59259-890-0:571]

[10]   M. Hebditch, M.A. Carballo-Amador, S. Charonis, R. Curtis, and J. Warwicker, "Protein-Sol: a web tool for predicting protein solubility from sequence", *Bioinformatics,* vol. 33, no. 19, pp. 3098-3100, 2017.
[http://dx.doi.org/10.1093/bioinformatics/btx345] [PMID: 28575391]

[11]   J.R. Quinlan, "Combining Instance-Based and Model-Based Learning", *Machine Learning Proceedings,* vol. 1993, pp. 236-243, 1993.

[12]   S. Le Crom, F. Devaux, C. Jacq, and P. Marc, "yMGV: helping biologists with yeast microarray data mining", *Nucleic Acids Res.,* vol. 30, no. 1, pp. 76-79, 2002.
[http://dx.doi.org/10.1093/nar/30.1.76] [PMID: 11752259]

[13]   A.K. Upadhyay, and R. Sowdhamini, "Genome-Wide Prediction and Analysis of 3D-Domain Swapped Proteins in the Human Genome from Sequence Information", *PLoS One,* vol. 11, no. 7, p. e0159627, 2016.
[http://dx.doi.org/10.1371/journal.pone.0159627] [PMID: 27467780]

CHAPTER 28

Papaya Seeds: Treasure of Nutrients and a Promising Preservative

Shama Kakkar[1], Runjhun Tandon [1,*] and Nitin Tandon[1]

[1] *School of Chemical Engineering and Physical Sciences, Lovely Professional University, Phagwara, Punjab, India*

Abstract: Papaya is broadly known for its taste and medical advantages yet very few individuals know about the massively helpful Papaya seeds that are by-products of fruit and are generally discarded. These small round black seeds are edible and useful for our wellbeing whenever taken in a restricted amount. Their reuse will be useful not only for the economy and environment rather; it will prove to be a ray of hope for the food industries engaged in the search for plant-based food ingredients to enhance the nutritional status of functional foods. In this review, we have discussed the composition, medicinal properties, its utilization as a nutritive agent, and a promising preservative.

Keywords: Bio-Actives, Fruit Waste, Functional Foods, Papaya Seeds, Value-Added Products.

1. INTRODUCTION

Papaya is one of the affordable, nutritionally, and medicinally important fruit which belongs to the Caricaceae family. It is the third globally cultivated plant that grows wild in the Torrid Zone and excels on almost every kind of soil in the tropical rain forest. It was originated in southern Mexico and Costa Rica but now it is cultivated in many other countries [1]. It is low in calories and loaded with phytochemicals that help in keeping our bodies healthy. It is a rich source of various sorts of enzymes. Papain, vegetable pepsin present in the fruit helps to reduce digestive problems [2]. In this way, it is nothing unexpected that the high consumption of this fruit for various health benefits generates a large number of wastes. Papaya's major byproducts are peel (PP) and seeds (PSs) and these wastes consist of approximately 20-25% of total the weight of the whole fruit. PSs are not waste as they consist of many valuable bioactive such as crude protein, fatty

---

* **Corresponding author Runjhun Tandon1:** Department of Electronics & Communication, Amity University Uttar Pradesh, Noida, India; E-mail: runjhun.19532@lpu.co.in

Dharam Buddhi, Rajesh Singh and Anita Gehlot (Eds.)

acids, crude fiber, carpaine, caricine, papaya oil, glucotropacolin, and an enzyme myrosin. Despite having so many benefits, they are dumped in garbage cans. As they are high in moisture and microbes, their improper disposal can cause serious environmental problems [3, 4]. However, these nutritionally important seeds should be reused as food additives so that their recovery could be beneficial for health, economy, and environment. Therefore, there is a need to think of processes that will empower us to reuse seeds to produce commercially successful products. In this review, we are going to discuss a portion of those researches, featuring utilization of these losses as an important food ingredient and a promising preservative.

2. PAPAYA SEEDS AS FOOD ADULTERANT

PSs have been used to add bulk in black pepper for many years. To identify and separate it from black pepper, many studies have been done, in which researchers have emerged victorious [5]. However, papaya seeds are not only food adulterant but it is also a treasure of nutritious elements in itself which we will study in the upcoming sections of this review.

3. PHYSICOCHEMICAL COMPOSITION OF PAPAYA SEEDS

To get the most out of it and make maximum use of PSs, it is vital to have total information about its chemical composition. Tang *et al*. (1978) published one paper on the composition of PSs; in this research defatted PSs were taken for observation. Results showed that PSs contain 32.97% oil content, 40% crude protein, 48.9% crude fiber, 6.86% ash content, 1.11% fatty acids and minerals such as P, K, Ca, Mg, S, Mn, Fe, and Cu. A major fatty acid present in abundance was oleic acid with 71% and some toxins were also detected. The amount of benzyl-ITC content present in seed oil and benzyl glucosinolate present in seed meal was 0.56% and 1.86% respectively [6]. In another study conducted by Marfo *et al*. (1986), both defatted and undefatted PSs were taken for observation. The result showed the approximate composition of PSs in which defatted PSs contain 44.4% of crude protein, 31.8% of crude fiber, 4.48% ash content, and undefatted PSs contain 27.8% crude protein, 28.3% lipids, 22.6% crude fiber, and 3.5% ash content. Other bioactive components were 0.94% fatty acids and minerals such as P, K, Ca, Mg, S, Mn, Fe, Cu, Ni, Co, and Na. A major fatty acid in abundance was oleic acid with 79.1% and in toxins, benzyl glucosinolate (10%) was present in the highest proportion. This study revealed that carotene and monosaccharide were also present but in trace amounts and out of all sugars, sucrose was present in the highest proportion [7]. In another study of papaya seed oil (PSO) composition done by Malacrida *et al*. (2011) revealed that PSO contains a high amount of monounsaturated fats, a low amount of Tocopherols, and a substantial amount of Carotenoids and total Phenolic. Major fatty acids present in oil were oleic, palmitic, linolenic, and stearic acid, major tocopherols present were α- and δ-tocopherol, and β-cryptoxanthin were major carotenoids [8]. Thus, based on these researches, it would not be wrong to say that PSs are a potential source of crude protein, carbohydrates, fatty acids, lipids, fibers, calcium-phosphorus with some toxins.

4. MEDICINAL PROPERTIES OF PAPAYA SEEDS

PSs are successfully utilized in the treatment of viral infections such as dengue fever. They are anti-bacterial, anti-inflammatory, anti-amoebic, anti-ulcer, anti-diabetic, anti-obesity, and anti-parasitic agents, and wound healing agents. Its water extract is not at all toxic and even helpful for the protection against oxidative stress. Flavonoids are rich in water extract of PSs act as an anticancer agent. PSs are Nephroprotective as guarantees the smooth working of our kidneys and rich source of monounsaturated fatty acids and various antioxidants, so they maintain our cardiovascular health by lowering blood pressure and cholesterol level [9 - 12].

5. VALUE-ADDED FOOD INGREDIENT & A PROMISING PRESERVATIVE

The brief information about the purpose of adding PSs in various functional foods is given in Table **1** and year wise explanation is also mentioned below:

Table 1. Utilization of PSs to add value in various functional foods.

S. no.	Author	Year	Utilized In	Act As	Ref.
1.	Azevedo *et al.*	2014	Hamburger	Nutritive agent	[13]
2.	Sofi *et al.*	2015	Indian mackerel	Preservative	[14]
3.	Castro-Vargas *et al.*	2016	Edible oil	Preservative	[15]
4.	Kugo *et al.*	2018	Porridge	Deworming agent	[16]
5.	Veronezi and Jorge	2018	Soybean oil	Preservative	[17]
6.	Senrayan and Venkatachalam	2018	Papaya seed oil	Edible oil	[18]
7.	Bhosale and Udachan	2018	Functional Cookies	Nutritive agent	[19]
8.	Subandi and Nurowidah	2019	Coffee powder	Anti-obesity agent	[20]
9.	Cruz *et al.*	2019	Papaya jam	Anti-fungal agent	[21]
10.	Avila *et al.*	2020	Porridge	Nutritive agent and Preservative	[22]

(Table 1) cont.....

S. no.	Author	Year	Utilized In	Act As	Ref.
11.	Bakar *et al.*	2020	Fish sausages	Preservative	[23]

Azevedo *et al.* (2014) assessed the impact of the incorporation of papaya seed flour (PSF) on the nutritional and sensory attributes of hamburgers. During the production of hamburgers incorporation of PSF was done in four different proportions. Nutritional analysis revealed that hamburgers contain a high amount of crude protein and fiber. As the quantity of flour expanded, a great increase in cooking yield, moisture content, and a decrease in hamburger shrinkage was noticed. Moreover, 1% flour incorporated hamburgers were practically identical in shade and texture compared to control, and no sensory changes were seen up to 2% incorporation [13]. In another report published by Sofi *et al.* (2015) antioxidative properties of different concentrations of grape seed extract (GSE) and papaya seed extract (PSE) were studied on Indian mackerel. The outcomes showed that GSE contains higher phenolic and flavonoids than PSE. Linoleic acid model of GSE showed 67.67% and PSE showed 46.43% of micro-organism growth inhibition at 500 mg/L concentration. Apart from this, natural preservatives were compared with synthetic such as BHT and TBHQ and it was seen; restraint of lipid oxidation (LO) by BHT at concentration 200 mg/L was almost similar to GSE at concentration 500mg/L. The dip treatment of GSE expanded the time of usability of mackerel as long as 15 days, PSE by 12 and 9 days for control during ice storage. After studying this research, it would not be wrong to say that GSE has more promising attributes than PSE but PSE can be used to enhance the nutritional quality [14]. Castro-Vargas *et al.* (2016) isolated the benzyl glucosinolate (BG) from PSs and studied its effect on LO of edible oil (EO). A methanol extract of BG was isolated, purified then analyzed by LC-MS and NMR spectroscopy. It was observed that BG showed high efficiency than BHT and TBHQ against LO of EO, and the development of TBARS was restrained more than 80% [15]. Kugo *et al.* (2018) published one report of a survey conducted in three primary schools of rural Kenya among 326 students for deworming program. In this survey, each student of the first school was given PSF incorporated porridge, porridge without PSF was given to students of the second school and third school students received porridge with 400mg Albendazole (conventional deworming drug) one time every day. Furthermore, a stool test of kids was conducted for 2 months. It was found that porridge with the PSF diminishes of worm burden. Whereas, Albendazole still had higher inhibition and cure rate yet was less successful as a nutritive agent [16]. Veronezi and Jorge (2018) studied the effect of papaya and melon seed oil on the stability of soybean oil. Mixed oil (60% of soybean, 20% melon, and 20% Papaya seed oil PSO) showed a lower amount of free fatty acids than other oils after 20 hours. But with the heating process, increment in the quantities of saturated, monounsaturated acid and decrease in carotenoids except in PSO was noticed. Other than this, the thermo-oxidation method assessed that papaya and melon oils enhanced the nutritional status and stability of the soybean oil, with the maintenance of Carotenoids, Phytosterols, and Tocopherols [17]. Senrayan and Venkatachalam (2018) proposed that PSO can be equally useful as other EO commercially available in the market. A novel technique named Continuous ultrasound acoustic cavitation was discovered for the extraction of PSO. The 27.8% oil was recovered when Continuous ultrasound acoustic extraction was completed at 1:13(g/ml) sample solvent proportion, 80% amplitude level, 48°C, and time of 7 min. It was seen that the major fatty acid in PSO was oleic acid with 74.8%. This quantity was almost equal when compared with oleic acid present in commercial vegetable oils. The thermal stability of oil was also examined. The novel technique increased the recovery of oil using less solvent and time [18]. Bhosale and Udachan (2018) studied the Nutritional status, sensory and physicochemical properties of functional cookies made by incorporating PSs and PP powder. It was observed that developed cookies had more weight, thickness, diameter, and even harder but spreadability was slightly lower than control. Cookies were rich in protein, fiber content but low in calories and moisture. Tactile assessments showed that 3.82% PP powder and 1.24% PS powder were satisfactory regarding shade, taste, and aroma [19]. Subandi and Nurowidah (2019) developed coffee powder from PSs and its pancreatic lipase inhibiting effect was analyzed. Pancreatic lipase is an enzyme that is responsible for lipid absorption thus causes obesity. To test the organoleptic properties, a survey was conducted at 50 consumers and most of them liked the aroma, taste, and color of PSs coffee. Using the titrimetric method, the anti-obesity attributes of PSs were compared with a commercially available anti-obesity drug. It was seen that 1.42g of PSs powder have anti-obesity properties equivalent to 120mg of anti-obesity drug available in the market [20]. Cruz *et al*. (2019) formulated PSF-incorporated papaya jam. The analyses were done for antioxidant activities and sensory properties. The outcomes showed that with an increase in the proportion of flour, crude fiber content also increased in the jam. No development of micro-organisms during 30 days of storage in the jams was noticed; this proved that seeds can act as fungicides. On basis of sensory attributes, the overall acceptance score was near to control, which indicated the positive response of consumers [21]. Avila *et al*. (2020) investigated the nutritional status of porridge produced using PSs and maize. This result revealed that mixing cornmeal with PSs upgraded the nutritional value by enhancing phytochemicals and improves the microstructural properties of the formulated porridge by bringing down enthalpy of gelatinization. Furthermore, the study demonstrated that phenolic present in the free and bound parts of PSs essentially adds to the anti-oxidative action of the product. The fortification proved to be a creative way to improve the medicinal properties of porridge [22]. Bakar *et al.* (2020) studied the change in quality, moisture, color, pH, TBA, and total plate count (TPC) of fish sausages coated in edible sago starch-gelatin incorporated with PSE during storage at 7°C. Results revealed the coating had a significant positive effect on the shade of sausages during production but no change in the shade was noticed during storage. The best outcome for TPC was seen in a sample with 7% PSE incorporated coating. 5% PSE fortified coatings proved the best preservative for fish sausages by expanding the period of usability of the product for an extra week [23].

6. SIDE EFFECTS REPORTED

The brief information about experiments conducted to observe the side effects of PSs is summarized in Table **2**. The outcomes reported are the result of high dosage consumption of PSs.

Table 2. Side effects reported from literature.

1. ABORTIFACIENT PROPERTIES	An experiment was conducted on two groups of Female Rats. Group 1 was given 100 mg/kg and Group 2 was given an 800 mg/kg dose of PSs.	No abnormalities or death were found in fetuses. Impaired growth without mutation was noticed. A changed toxicological profile is high doses result.	[24]
2. CONTRACEPTIVE PROPERTIES	An experiment was conducted on 35 Human semen samples and treated with water extract of PSs.	The decrease in total motility and increase in DNA-fragmented results were noticed. High doses slow human sperm thus act as a natural contraceptive	[25]

7. FUTURE DIRECTION

Despite many health benefits, the literature showed that papaya seeds are still not used in the food industry as strongly as they should be. The reason may be the presence of some toxins. It was also observed that its limited consumption is not at all harmful. Still, separating and utilizing these toxins from beneficiary phytochemicals can also be an interesting approach which could be a new dawn for the researchers doing efforts in this direction.

CONCLUSION

PSs, which are said to be in vain, are full of medicinal and nutritional properties are a boon for the human body, yet dumped in the waste bin. The re-use of this nutrient-packed waste will prove to be good for the betterment of our health and environment. Therefore it can be estimated that future studies going to be done in this direction can prove to be the foundation for the progress of the economy, environment, and food industries.

CONSENT FOR PUBLICATION

Not Applicable.

CONFLICT OF INTEREST

The authors declare no conflict of interest, financial or otherwise.

ACKNOWLEDGEMENTS

The study was supported by the School of Electronics and Electrical, Lovely Professional University, Phagwara, Punjab, India. The authors are highly grateful to the department for providing all necessary materials to carry out the present study.

REFERENCES

[1]   R. Ming, and P.H. Moore, "Genetics and genomics of papaya", *Genet. Genomics Papaya,* no. 130, pp. 1-438, 2014.
       [http://dx.doi.org/10.1007/978-1-4614-8087-7]

[2]   V. Yogiraj, P. K. Goyal, C. S. Chauhan, A. Goyal, and B. Vyas, "Goyal, and B. Vyas, Carica papaya Linn: an overview", *Int. J. Herb. Med,* vol. 2, no. 5 Part A, pp. 1-8, 2014.

[3]   P.D. Pathak, S.A. Mandavgane, and B.D. Kulkarni, "Waste to Wealth: A Case Study of Papaya Peel", *Waste Biomass Valoriz.,* vol. 10, no. 6, pp. 1755-1766, 2019.
       [http://dx.doi.org/10.1007/s12649-017-0181-x]

[4]   T. Vij, and Y. Prashar, "A review on medicinal properties of Carica papaya Linn", *Asian Pac. J. Trop. Dis.,* vol. 5, no. 1, pp. 1-6, 2015.
       [http://dx.doi.org/10.1016/S2222-1808(14)60617-4]

[5]   S. Bansal, A. Singh, M. Mangal, A.K. Mangal, and S. Kumar, *Food adulteration: Sources, health risks, and detection methods*, 2017.

[6]   "R. A. HEU, C. □S TANG, E. N. OKAZAKI, and S. M. ISHIZAKI, "Composition of Papaya Seeds", *J. Food Sci.,* vol. 43, no. 1, pp. 255-256, 1978.
       [http://dx.doi.org/10.1111/j.1365-2621.1978.tb09785.x]

[7]   E.K. Marfo, O.L. Oke, and O.A. Afolabi, "Chemical composition of papaya (Carica papaya) seeds", *Food Chem.,* vol. 22, no. 4, pp. 259-266, 1986.
       [http://dx.doi.org/10.1016/0308-8146(86)90084-1]

[8]   C.R. Malacrida, M. Kimura, and N. Jorge, "Characterization of a high oleic oil extracted from papaya (Carica papaya L.) seeds", *Food Sci. Technol. (Campinas),* vol. 31, no. 4, pp. 929-934, 2011.
       [http://dx.doi.org/10.1590/S0101-20612011000400016]

[9]   "The surprising health benefits of papaya seeds: A review", *J. Pharmacogn. Phytochem.,* vol. 6, no. 1, pp. 424-429, 2017.

[10]   N. Pathak, S. Khan, A. Bhargava, G.V. Raghuram, D. Jain, H. Panwar, R.M. Samarth, S.K. Jain, K.K. Maudar, D.K. Mishra, and P.K. Mishra, "Cancer chemopreventive effects of the flavonoid-rich

fraction isolated from papaya seeds", *Nutr. Cancer,* vol. 66, no. 5, pp. 857-871, 2014.
[http://dx.doi.org/10.1080/01635581.2014.904912] [PMID: 24820939]

[11]   J.A. Olagunju, "Nephroprotective activities of the aqueous seed extract of Carica papaya Linn. In carbon tetrachloride induced renal injured wistar rats: A dose- and time-dependent study", *Biol. Med. (Aligarh),* vol. 1, no. 1, pp. 11-19, 2009.
[http://dx.doi.org/10.4172/0974-8369.1000002]

[12]   L.F. Santana, A.C. Inada, B.L.S.D. Espirito Santo, W.F.O. Filiú, A. Pott, F.M. Alves, R.C.A. Guimarães, K.C. Freitas, and P.A. Hiane, "Nutraceutical potential of carica papaya in metabolic syndrome", *Nutrients,* vol. 11, no. 7, p. E1608, 2019.
[http://dx.doi.org/10.3390/nu11071608] [PMID: 31315213]

[13]   L.A. Azevedo, and P.C.B. Campagnol, "Papaya seed flour (Carica papaya) affects the technological and sensory quality of hamburgers", *Int. Food Res. J.,* vol. 21, no. 6, pp. 2141-2145, 2014.

[14]   F.R. Sofi, C.V. Raju, I.P. Lakshmisha, and R.R. Singh, "Antioxidant and antimicrobial properties of grape and papaya seed extracts and their application on the preservation of Indian mackerel (Rastrelliger kanagurta) during ice storage", *J. Food Sci. Technol.,* vol. 53, no. 1, pp. 104-117, 2016.
[http://dx.doi.org/10.1007/s13197-015-1983-0] [PMID: 26787935]

[15]   H.I. Castro-Vargas, W. Baumann, and F. Parada-Alfonso, "Valorization of agroindustrial wastes: Identification by LC-MS and NMR of benzylglucosinolate from papaya (Carica papaya L.) seeds, a protective agent against lipid oxidation in edible oils", *Electrophoresis,* vol. 37, no. 13, pp. 1930-1937, 2016.
[http://dx.doi.org/10.1002/elps.201500499] [PMID: 26756135]

[16]   M. Kugo, L. Keter, A. Maiyo, J. Kinyua, P. Ndemwa, G. Maina, P. Otieno, and E.M. Songok, "Fortification of Carica papaya fruit seeds to school meal snacks may aid Africa mass deworming programs: a preliminary survey", *BMC Complement. Altern. Med.,* vol. 18, no. 1, p. 327, 2018.
[http://dx.doi.org/10.1186/s12906-018-2379-2] [PMID: 30526582]

[17]   C.M. Veronezi, and N. Jorge, "Effect of Carica papaya and Cucumis melo seed oils on the soybean oil stability", *Food Sci. Biotechnol.,* vol. 27, no. 4, pp. 1031-1040, 2018.
[http://dx.doi.org/10.1007/s10068-018-0325-1] [PMID: 30263832]

[18]   J. Senrayan, and S. Venkatachalam, "A short extraction time of vegetable oil from Carica papaya L. seeds using continuous ultrasound acoustic cavitation: Analysis of fatty acid profile and thermal behavior", *J. Food Process Eng.,* vol. 42, no. 1, pp. 1-9, 2019.
[http://dx.doi.org/10.1111/jfpe.12950]

[19]   B. Priyanka, ""Studies on Utilization of Papaya Peel and Seed Powder for Development of Fiber Enriched Functional Cookies," IJRAR1944444 Int", *J. Res. Anal. Rev.,* vol. 5, no. 4, pp. 459-466, 2018.

[20]   V.A. Cruz, "Manufacturing of Formosa papaya (Carica papaya L.) jam containing different concentrations of dehydrated papaya seed flour", *Int. Food Res. J.,* vol. 26, no. 3, pp. 849-857, 2019.

[21]   "The Potency of Carica papaya L. Seeds Powder as Anti- Obesity 'Coffee' Drinks", *IOP Conf. Ser. Mater. Sci. Eng.,* vol. vol. 515, 2019
[http://dx.doi.org/10.1088/1757-899X/515/1/012098]

[22]   S. Ávila, M. Kugo, P. Silveira Hornung, F.B. Apea-Bah, E.M. Songok, and T. Beta, *Carica papaya seed enhances phytochemicals and functional properties in cornmeal porridges,* 2020.
[http://dx.doi.org/10.1016/j.foodchem.2020.126808]

[23]   J. Bakar, N.S. Abdul Kadir, A.S. Ahmad Mazlan, and M.R. Ismail-Fitry, "Effect of edible coating of sago starch-gelatine incorporated with papaya seed extract on the storage stability of Malaysian fish sausage (keropok lekor)", *Int. Food Res. J.,* vol. 27, no. 4, pp. 618-624, 2020.

[24]   O. Oderinde, C. Noronha, A. Oremosu, T. Kusemiju, and O.A. Okanlawon, "Abortifacient properties of aqueous extract of Carica papaya (Linn) seeds on female Sprague-Dawley rats", *Niger. Postgrad.*

Med. J., vol. 9, no. 2, pp. 95-98, 2002.
[PMID: 12163882]

[25]   V. Ghaffarilaleh, D. Fisher, and R. Henkel, "Carica papaya seed extract slows human sperm", *J. Ethnopharmacol.,* vol. 241, no. May, p. 111972, 2019.
[http://dx.doi.org/10.1016/j.jep.2019.111972] [PMID: 31128152]

CHAPTER 29

Cryptocurrency Price Prediction Using FB Prophet Model

Kanksha[1,*] and **Harjit Singh**[1]

[1] *School of Computer Science And Engineering, Lovely Professional University, Phagwara , Punjab, India*

Abstract: A new method encountered for securing cryptocurrency *i.e* cryptographic algorithms for example Secure Hash Algorithm (SHA-2) and Message Digest (MD5). It uses Blockchain technology to make the transactions secure, transparent, traceable, and immutable. This is the reason cryptocurrencies have gained popularity in almost all sectors, especially in the financial sector. Cryptocurrency price prediction has become a trending research topic globally. Many Machine Learning algorithms have been developed such as Linear Regression, SVM, Random forest, and Facebook Prophet. Facebook Prophet is a time-series forecasting model for predicting the future price of bitcoins. In this paper, Facebook prophet Model is used, and two cryptocurrencies are considered, namely Bitcoin and Litecoin. The result depicts that FB prophet Model accurately predicts the prices of bitcoin cryptocurrencies. We considered the data from yahoofinace.com for BTC-USD and LTC-USD.

Keywords: Bitcoin, Cryptocurrency Price Prediction, Facebook Prophet Model, Hash Algorithm, Litecoin, Machine Learning, Time-series Forecasting Model.

INTRODUCTION

One of the common new financial assets is crypto-currencies. Even though their exchange rates and market capitalization have experienced many drastic ups and downs over the last decade, following the appearance of the first cryptocurrency, namely Bitcoin [1].

At the beginning of 2017, the overall market capitalization of Crypto-currencies amounted to $15.6 billion; it was nearly $230 billion at the beginning of 2020 and the maximum market cap hit almost $860 billion in mid-2018.

A contentious and debatable issue is the role and place of Crypto-currencies in the global financial market. Large fluctuations in their rates and the legal complexity

---

* **Corresponding author Kanksha:** School of Computer Science And Engineering, Lovely Professional University, Phagwara, Punjab, India; E-mail: kanksha24@gmail.com

Dharam Buddhi, Rajesh Singh and Anita Gehlot (Eds.)

of the deals made for them in the majority of countries are creating significant uncertainty and, as a consequence, a high risk of the money being spent. Prediction of prices for cryptocurrencies, but they concentrated only on small but popular cryptocurrencies such as Monero and Ethereum. But other coins can be broadly accepted by financial institutions, such as Litecoin, Bitcoin, and Stellar. In the top 10 currencies, Litecoin is located [1]. They have the potential to be broadly embraced by financial institutions. The price history of bitcoin and litecoin are shown in Fig. (**1**).

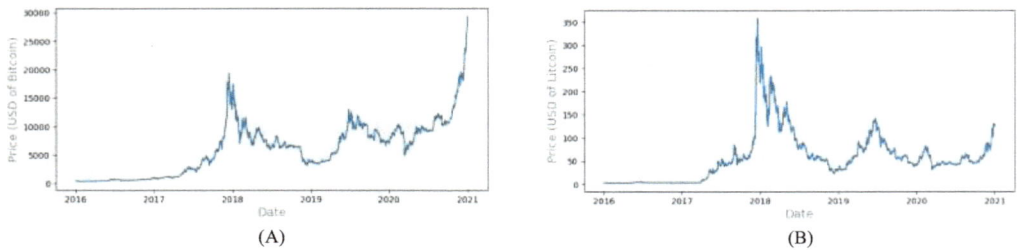

(A)　　　　　　　　　　　　　　　　　(B)

Fig. (1). Price History of (a) Bitcoin (b) Litecoin [1].

Due to its non-traceable and analytical transactions, the coin is branded as a privacy measure. Because of this property, their demand is very likely to grow in the future. This paper is designed to attack these cryptocurrencies and provide a suitable market prediction scheme [1, 2].

RELATED WORK

We prepared a comparative analysis of papers as shown in Table **1**.

Table 1. Comparative Analysis of Papers [3 - 12].

Study	Features	Finding
Applied of feed-forward neural network and FB Prophet Model for train passengers forecasting [3]	Data collected on Java Island from January 2006 to August 2019 are the data collected every month by train passengers.	Accurately predict the number of passengers
Using the Facebook model of air temperature estimates and long-term memory [4]	Five-year daily Bandung air temperature forecast	Prophet improves at the highest air temperature and LSTM improves at the lowest air temperature.
Models are used for price prediction and analysis [5]	Focused on a popular cryptocurrency for example Bitcoin	Two series of deep learning approaches and demonstrated their effectiveness in predicting bitcoin prices.

(Table 1) cont.....

Study	Features	Finding
GRU Prediction scheme [6]	Used the mean root squared error to study and compared various approaches (RMSE).	Model GRU with repeated dropout is better than established common models
Bitcoin Price Prediction using Machine Learning [7]	Forecasts 30 days ahead	Compare three separate neural networks: the FNN, the Exogenous Input Network, and the Nonlinear Autoregressive Neural Network (NARX), by obtaining the expected results of each model.
Evidence from a quantile cross-spectral approach [8]	Formulates and bond markets between 2011 and 2019	Support the idea that in some return quantiles Bitcoin will have financial diversification
Improving Stock Price Prediction with GAN-based Data Augmentation [9]	The prediction of price series stock data using GAN produced increased time-series data	24.47% of AMZN and 30.27% of lower RMSE, 15.84% of B, and 13,88% of lower RMSE and MAE.AMZN data collection
Prediction of Bitcoin prices with machine learning methods using time-series data [10]	For different window lengths filters with different weight coefficients are used.	A 10-fold cross-validation approach is used for building a model during testing.
Crypto-Currency price prediction using Decision Tree and Regression techniques [11]	Until the present date, the dataset is taken with open, high, low, and close Bitcoin value price information.	Compared the accuracy of the Bitcoin prediction with various ML algorithms.
Bitcoin Price Prediction Using Machine Learning Methods [12]	Kaggle Bitcoin Dataset 2010-2019 data set	Accuracy rates are 97.2%.

MATERIALS AND METHODS

This research will be based on the Fb prophet which is an open-source library provided by Facebook. The architecture of the Fb Prophet is shown in Fig. (2). It is used for Machine Learning Model. It is a Model which has opened for additivity, primarily in the Time series. Users use Prophet to forecast revenue and buy-back prices. The prophet is a method that has been using for Time-series Data.

It is open-source software released by Facebook's Core Data Science Team.

DATASET USED

Dataset Description and Preprocessing of Data: The data used for the analysis was collected by Yahoofinance.com. It is an online forum for analyzing and providing global financial market statistics. Data have been collected for the two cryptocurrencies Litecoin and Bitcoin. The regular opening price, highest price, lowest price, and closing price for Bitcoin are formatted in a standard format. We have removed the null columns from the CSV files.

A. Types Of Cryptocurrencies Used

B. We used Litecoin and Bitcoin as shown in Table **2**.

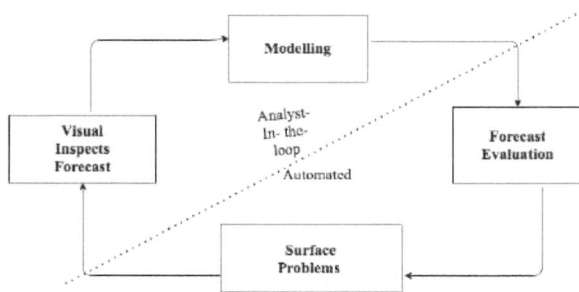

Fig. (2). Architecture of Facebook Prophet.

Table 2. Cryptocurrencies Used [13, 14].

Name	Duration	Size
Litecoin	January 1, 2016 - January 1, 2021	1829 rows × 7 columns Data points
Bitcoin	January 1, 2016 - January 1, 2021	1829 rows × 7 columns Datapoints

RESULTS AND DISCUSSION

The Studies aim to test Facebook Prophet Machine Learning Model on the python notebook platform while it is possible to forecast Bitcoin's price behavior. Our model explicitly shows the superior results of Bitcoin Currency (USD) than Litecoin as shown in Fig. (**3**).

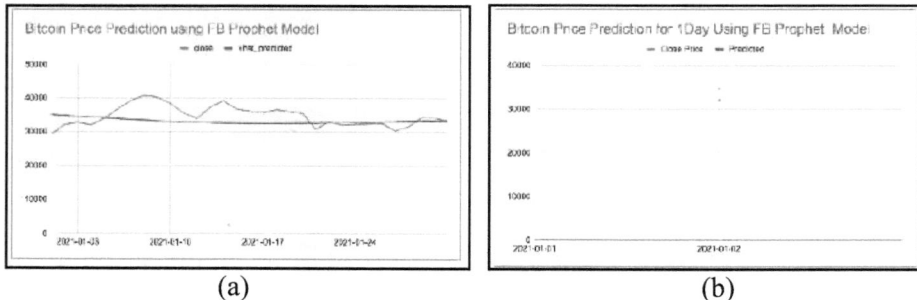

(a)   (b)

Fig. (3). Prediction of Bitcoin (a) 30 Days (b) 1 Day [13].

As shown in Fig. (**4**), the 30Day and 1Day Price forecast with the help of the variable Date and Close Price of Bitcoin in the above figures (USD). The 30 and 1Day prediction is presented in other figures.

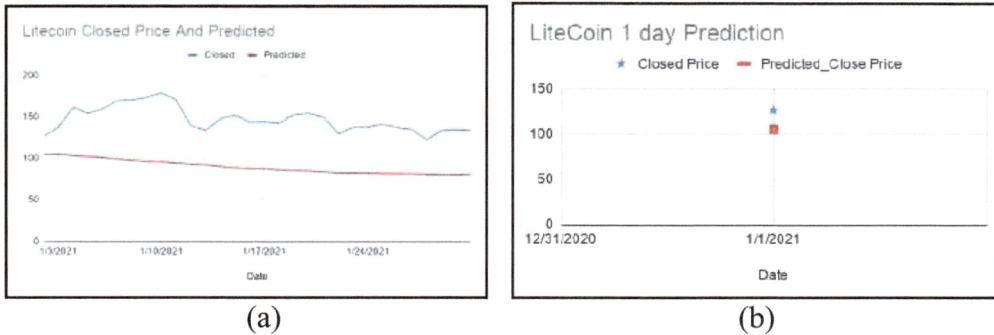

| (a) | (b) |

Fig. (4). Prediction of Litecoin (a) 30 Days (b) 1 Day [14].

CONCLUDING REMARKS

This paper describes a Bitcoin Prediction Time Series Facebook Prophet Model. For the potential cost of two main coins, *i.e.* Bitcoin and Litecoin, we use a yahoo financing data collection. The dataset consists of 7 columns, including the stock price from the year 2016 to 2021. The open, high, low, close, and close columns constitute Bitcoin's closing price. After the FB Prophet model has been developed and trained, we have downloaded a test data set and added it to our trained model and a line map shows how the real bitcoin price is expected. Compared to Litecoin prediction, Bitcoin prediction was close to the actual price when we used the Facebook prophet model for 1 day and 30-day prediction. This model can be used to predict the price of Bitcoin and Litecoin that helps the dealer to invest and earn profits.

CONSENT FOR PUBLICATION

Not applicable.

CONFLICT OF INTEREST

The authors declare no conflict of interest, financial or otherwise.

ACKNOWLEDGEMENT

Declared none.

REFERENCES

[1]   Mohil Maheshkumar Patel, Sudeep Tanwar, Rajesh Gupta, and Neeraj Kuma, "A Deep Learning-

based Cryptocurrency Price Prediction Scheme for Financial Institutions", *Journal of Information Security and Applications,* 2020.

[2]   F. Ferdiansyah, S.H. Othman, R. Zahilah Raja Md Radzi, D. Stiawan, Y. Sazaki, and U. Ependi, "A LSTM-Method for Bitcoin Price Prediction: A Case Study Yahoo Finance Stock Market", *2019 International Conference on Electrical Engineering and Computer Science (ICECOS), Batam, Indonesia,,* 2019,pp. 206-210 Batam, Indonesia
[http://dx.doi.org/10.1109/ICECOS47637.2019.8984499]

[3]   S. Resa Septiani Pontoh, *Zahroh, H R Nurahman, R I Aprillion, A Ramdani, D I Akmal,"Applied of feed-forward neural network and facebook prophet model for train passengers forecasting.* IOP, 2021.

[4]   T. Toharudin, and R.S. Pontoh, "Rezzy Eko Caraka ,Solichatus Zahroh,Youngjo Lee & Rung Ching Chen"Employing long short-term memory and Facebook prophet model in air temperature forecasting", *Commun. Stat. Simul. Comput.,* p. 17, 2020.

[5]   T. Awoke, M. Rout, L. Mohanty, and S.C. Satapathy, "Bitcoin Price Prediction and Analysis Using Deep Learning Models", In: *Communication Software and Networks. Lecture Notes in Networks and Systems,,* V. Bhateja, M. Ramakrishna Murty, N. Gia Nhu, K. Jayasri, Eds., vol. 134. , 2021.

[6]   Aniruddha Dutta, Saket Kumar, and Meheli Basu, "A Gated Recurrent Unit Approach to Bitcoin Price Prediction", *Temesgen AwokeMinakhi Rout,ipika MohantySuresh Chandra Satapathy,* 2019.*MDPL,* 2019.

[7]   E. Loh, and S. Ismail, "Emerging Trend of Transaction and Investment: Bitcoin Price Prediction using Machine Learning", *International Journal of Advanced Trends in Computer Science and Engineering.,* vol. 9, pp. 100-104, 2020.
[http://dx.doi.org/10.30534/ijatcse/2020/1591.42020]

[8]   A. Maghyereh, *Hussein Abdoh,"Tail dependence between Bitcoin and financial assets: Evidence from a quantile cross spectral approach.* Elsevier, 2020.

[9]   *Improving Stock Price Prediction with GAN-based Data Augmentation.* IJAIDM, 2021.

[10]   S. Karasu, A. Altan, Z. Saraç, and R. Hacioğlu, "Prediction of Bitcoin prices with machine learning methods using time series data", *2018 26th Signal Processing and Communications Applications Conference (SIU), Izmir, Turkey,* 2018, pp. 1-4
[http://dx.doi.org/10.1109/SIU.2018.8404760]

[11]   K. Rathan, S.V. Sai, and T.S. Manikanta, "Crypto-Currency price prediction using Decision Tree and Regression techniques", *2019 3rd International Conference on Trends in Electronics and Informatics (ICOEI), Tirunelveli, India,,* 2019, pp. 190-194
[http://dx.doi.org/10.1109/ICOEI.2019.8862585]

[12]   A. Demir, B.N. Akılotu, Z. Kadiroğlu, and A. Şengür, "Bitcoin Price Prediction Using Machine Learning Methods", *2019 1st International Informatics and Software Engineering Conference (UBMYK),Ankara, Turkey,* 2019, pp. 1-4
[http://dx.doi.org/10.1109/UBMYK48245.2019.8965445]

[13]   "Bitcoin USD (BTC-USD)", https://in.finance.yahoo.com/quote/BTC-USD/history?p=BTC-USD

[14]   "Litecoin USD (LTC-USD)", https://in.finance.yahoo.com/quote/LTC-USD?p=LTC-USD

Study on An Evolutionary Feature of Canine Circovirus Genome: Nucleotide Composition and Codon Usage Bias in Canine Circovirus

Pankaj Jain[1], Amit Joshi[1] and **Vikas Kaushik[1,*]**

[1] *Department of Bioinformatics, LPU, Punjab, India*

Abstract: Canine Circovirus (CaCV) or Dog Circovirus (DogCV) causes hemorrhagic gastroenteritis, thrombocytopenia, vasculitis, hemorrhages, and neutropenia and is usually found as co-infection with other canine infectious agents like Parvovirus and Distemper virus in dogs including wild canines (wolf, fox, badger). To know the evolutionary feature of the viruses, nucleotide composition and Codon usage bias (CUB) have been used which further ascertain their adaptability towards the suitable host. In the present study CUB and Nucleotide composition of both cap & rep gene of around 88 CaCV strains were compared. Nucleotide composition of CDSs at 3rd codon position (G3% A3% T3% C3%) together with overall AT% and GC%, GC12, and GC3 were analyzed. using codon W 1.4.2 and CAIcal server, Aromo, and Gravy values, effective no of codons and relative synonymous codon usage values were also analyzed. The results show that CaCV has an AT-rich genome and codons ending with A/T have to preference over GC ending codons. Nc-GC3 plot of CanineCV reveals that selection pressure was dominant over mutation pressure. Correlation analysis between Gravy, Aroma, and CAI indicate natural selection over mutational pressure. The RSCU values of Cap & Rep genes were analyzed to find out overrepresented codons.

Keywords: CaCV, Canine Circovirus, Codon usage bias, Dog Circovirus, RSCU.

INTRODUCTION

Canine Circovirus (CanineCV) [1] or Dog Circovirus (Dog CV) belongs to the genus Circovirus within the family Circoviridae. Canine CV has been isolated in 2012 as a part of the genetic screening of canine samples for a new virus [2]. Since then various strains of CanineCV have been isolated in different parts of the world including the USA, Italy, UK, Croatia, China, Thailand, and Argentina. Canine CV is small, naked, icosahedrons, with circular, monomeric, ambisence, covalently closed, single-stranded DNA (ssDNA) genome ≈2000bp in size. They

* **Corresponding author Vikas Kaushik:** Department of Bioinformatics, LPU, Punjab, India; Tel: +918558841359; E-mail: vikas.19644@lpu.co.in

Dharam Buddhi, Rajesh Singh and Anita Gehlot (Eds.)

are the tiniest known self-replicating animal pathogen, encoding capsid protein. The genome comprising of 2 units encoding 2 open reading frames, Rep and Cap proteins [3]. Using a double-stranded replicative form DNA intermediate, Canine CV replicates their genomes [4]. A few studies on synonymous codon usage characteristics were done earlier on PorcineCV, PigeonCV, and DuckCV [5]. This study is going to be the first effort in evaluating the codon usage pattern of CanineCV. No earlier attempt to evaluate factors (Mutational Pressure/Natural selection) responsible for the host-specific adaptation of CanineCV. There is no vaccine against CanineCV. No earlier attempt for development of vaccine using codon usage bias manipulation strategy in CanineCV.

METHODOLOGY

Codon Usage Bias Analysis

(a) *Analysis of Nucleotide Composition*

Using coding sequences of CanineCV genomes, we intend to determine: (a) GC content (b) nucleotide's frequency (A%, G%, U%, and C %) (c) Nucleotide's frequency at 3rd position of synonymous codons (U3S%, A3S%, G3S%, and C3S %) (d) Nucleotide's frequency (G + C) at 3rd synonymous codon positions (GC3S %) (e) Nucleotide's frequency (G + C) at 3rd codon position (GC3) and mean frequency of both G + C at the 1st and 2nd position (GC12). 5 codons UGG and AUG (Tryptophan and Methionine) and UAG, UAA, and UGA (termination codons) will be excluded from the analysis. Using CodonW/EMBOSS/DAMBE/BioEdit/CAIcal Server, the composition of Nucleotide will be calculated.

(b) *Relative Synonymous Codon Usage*

Distinct as the ratio of the observed frequency of **codons** to expected frequency given that all synonymous **codons** for the same amino acids are used equally. RSCU will be calculated using the software CodonW/CAIcal Server.

(c) *An Effective Number of Codons (ENC)*

The degree of codon usage bias can be determined using ENC and is calculated using CodonW.

(d) *Codon Dinucleotide Frequency Analysis*

DAMBE software is used to calculate the dinucleotide frequencies.

(e) *Gravy and Aroma Statistics*

The Gravy value indicates protein hydrophobicity on codon usage bias. The effect of aromatic hydrocarbon proteins on codon usage bias can be measured by Aroma value. Both of these statistics are calculated using CodonW.

EFFECT OF NATURAL SELECTION AND MUTATIONAL PRESSURE ON CUB

(a) Enc-plot analysis: To reveal factors driving CUB, ENC value against GC3s value will be used. Points will lie on a standard curve in case of mutational pressure.

(b) Parity rule 2 analysis: Used to measure the effect of natural selection and mutational pressure.

(c) Neutrality analysis: To determine the dominant factor affecting CUB, values of GC12s against GC3s will be used using EMBOSS CUSP /CodonW.

STUDY OF HOST-SPECIFIC ADAPTATION

(a) *Codon Adaptation Index (CAI)*: Among highly expressed genes, CAI determines the preference for codon usage of a gene. It also represents virus adaptation to the host. Its value lies between 0 and 1. Higher the CAI stronger is the adaptation to the host. CAI is calculated using CAIcal Server

(b) *Relative Codon Deoptimization Index (RCDI):* Codon deoptimization of virus to its hosts is measured by RCDI. It is calculated using RCDI/eRCDI Server

(c) *Correspondence Analysis (CoA):* Usage patterns among viruses coding sequences are determined by CoA. It is determined using CodonW.

(d) *Statistical Analysis:* SPSS22.0/SPSS26.0 software is used to determine the factors influencing synonymous codon usage patterns.

All step-wise methodology is represented in Table **1**. It represents all the In-Silico tools used in codon bias analysis.

Table 1. Methodology used.

Analysis	Parameters to be Measured	Software Used
Synonymous codon usage characteristics analysis of CanineCV	Codon usage bias analysis, *Composition of Nucleotide Analysis*, *RSCU, ENC analysis, Codon dinucleotide frequency analysis, Gravy and aroma statistics*	CodonW/EMBOSS/DAMBE/BioEdit/CAIcal Server CodonW/EMBOSS/DAMBE/BioEdit/CAIcal Server CodonW/CAIcal Server CodonW, DAMBE, CodonW
Evaluate factors such as mutational pressure and natural selection responsible for the host-specific adaptation of CanineCV	*Enc-plot analysis, Parity rule 2 (PR2) analysis, Neutrality analysis Codon adaptation index (CAI) analysis, Relative codon deoptimization index (RCDI) analysis, Correspondence Analysis (CoA), Statistical Analysis*	EMBOSSCUSP program/CodonW CAIcalServer (http://genomes.urv.cat/ CAIcal/RCDI/)RCDI/eRCDI Server (http://genomes.urv.cat/CAIcal/RCDI/), CodonW SPSS22.0/SPSS26.0

RESULTS

Some of the strains which we have used in this study are being mentioned in Table **2**.

Table 2. Canine Circovirus variants sequences were used for this study.

Ser No	Accession Number	Country	Year of Isolation	Strain	Virus Specie	Host Species	Gene/CDS
1	JQ821392	USA	2012	Strain 214	Canine Circovirus	Dog	Cap/Rep
2	KC241982	USA	2012	UCD1-1698	Canine Circovirus	Dog	Cap/Rep
3	KC241983	USA	2012	UCD3-478	Canine Circovirus	Dog	Cap/Rep
4	KC241984	USA	2012	UCD2-32162	Canine Circovirus	Dog	Cap/Rep
5	NC_020904	USA	2013	UCD1-1698	Canine Circovirus	Dog	Cap/Rep
6	MF457592	USA	2017	OH19098-1	Canine Circovirus	Dog	Cap/Rep
7	MK033608	South America	2018	UBA-Baires	Canine Circovirus	Dog	Cap/Rep
8	KJ530972	Italy	2014	Bari/411-13	Canine Circovirus	Dog	Cap/Rep
9	KT734812	Italy	2015	CB6293/1-14	Canine Circovirus	Dog	Cap/Rep
10	KT734813	Italy	2015	AZ2972-13	Canine Circovirus	Dog	Cap/Rep
11	KT734814	Italy	2015	TE4016-13	Canine Circovirus	Dog	Cap/Rep

(Table 2) cont.....

12	KT734815	Italy	2015	AZ4133/1-13	Canine Circovirus	Dog	Cap/Rep
13	KT734816	Italy	2015	AZ4438-13	Canine Circovirus	Dog	Cap/Rep
14	KT734817	Italy	2015	AZ5212/1-14	Canine Circovirus	Dog	Cap/Rep
15	KT734818	Italy	2015	AZ5212/2-14	Canine Circovirus	Dog	Cap/Rep
16	KT734819	Italy	2015	AZ5586-13	Canine Circovirus	Dog	Cap/Rep
17	KT734820	Italy	2015	AZ663/1-13	Canine Circovirus	Dog	Cap/Rep
18	KT734821	Italy	2015	TE6685/1-13	Canine Circovirus	Dog	Cap/Rep
19	KT734822	Italy	2015	TE7482-13	Canine Circovirus	Dog	Cap/Rep
20	KT734823	Italy	2015	PE8575/1-13	Canine Circovirus	Dog	Cap/Rep
21	KT734824	Italy	2015	AZ663/2-13	Canine Circovirus	Dog	Cap/Rep
22	KT734825	Italy	2015	TE6685/2-13	Canine Circovirus	Dog	Cap/Rep

Codon usage bias of *cap* & *rep* gene and Nucleotide composition of 88 CaCV strains were compared. The overall nucleotide composition of CDSs at the 3rd codon position (A3% C3% G3% T3%) along with overall GC% and AT%, GC12, and GC3 were analyzed. The Aroma and Gravy values, effective no of codons (ENC), and relative synonymous codon usage (RSCU) values were also analyzed using codon W 1.4.2 and CAIcal server. The results show that the CaCV genome is AT-rich and A/T ending codons were preferred over GC ending codons. Nc-GC3 plot (Fig. **1**) of CanineCV reveal that selection pressure was dominant over mutational pressure. The correlation analysis between CAI, Gravy, and Aroma indicate natural selection over mutational pressure. The RSCU values of *Cap* & *Rep* genes were analyzed to find out overrepresented codons. In Table **3** all RSCU values are represented.

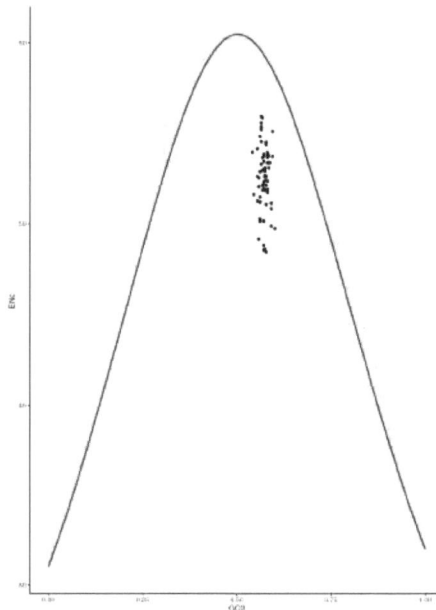

Fig (1). Nc-GC3 plot of CanineCV reveal that selection pressure was dominant over mutational pressure.

Table 3. RSCU values.

S. No.	Amino Acid	Codons	Coding Sequences (CDSs)	
			C	R
1	Phenylalanine (F)	TTT	0.6345	1.25
		TTC	1.3654	0.7494
2	Leucine (L)	TTA	0.4469	0.5103
		TTG	0.9232	1.6306
		CTT	0.5977	1.1133
		CTC	1.0821	1.0514
		CTA	1.0748	0.0706
		CTG	1.8747	1.6237
3	Isoleucine (I)	ATT	0.7359	1.215
		ATC	0.8424	1.175
		ATA	1.4214	0.6088
4	Valine (V)	GTT	0.8074	1.2879
		GTC	0.8324	1.0201
		GTA	1.2438	0.3155
		GTG	1.1158	1.3767
5	Serine (S)	TCT	1.3975	1.7173
		TCC	0.6125	0.1988
		TCA	2.0131	0.5917
		TCG	0.4743	0.8141
		AGT	0.7614	0.9998
		AGC	0.7411	1.6778
6	Proline (P)	CCT	0.7839	1.5291
		CCC	0.1906	1.1443
		CCA	2.2456	0.1546
		CCG	0.7801	1.1719
7	Threonine (T)	ACT	1.0025	0.8869
		ACC	0.5531	1.6699
		ACA	2.0829	0.2641
		ACG	0.3614	1.1789
8	Alanine (A)	GCT	1.1787	1.598
		GCC	1.0211	1.0846
		GCA	1.5243	0.3621

(Table 3) cont.....

S. No.	Amino Acid	Codons	Coding Sequences (CDSs)	
		GCG	0.2757	0.9554
9	Tyrosine (Y)	TAT	1.0886	1.0205
		TAC	0.9113	0.9794
10	Histidine (H)	CAT	1.4577	1.2806
		CAC	0.5422	0.7193
11	Glutamine (Q)	CAA	1.4125	0.725
		CAG	0.5875	1.275
12	Asparagine (N)	AAT	1.0714	0.9169
		AAC	0.9285	1.083
13	Lysine (K)	AAA	1.6914	0.816
		AAG	0.3085	1.184
14	Aspartic acid (D)	GAT	0.8274	1.2857
		GAC	1.1725	0.7142
15	Glutamic acid (E)	GAA	1.1948	0.4393
		GAG	0.8051	1.5606
16	Cysteine (C)	TGT	0.0227	0.8746
		TGC	0.0454	1.1253
17	Arginine (R)	CGT	0.2766	0.8916
		CGC	1.0568	0.9323
		CGA	0.2839	1.2023
		CGG	0.6218	2.2048
		AGA	2.3553	0.6986
		AGG	1.4054	0.07008
18	Glycine (G)	GGT	0.5603	1.2742
		GGC	0.7602	1.091
		GGA	1.99	0.62
		GGG	0.6891	1.0145
			0.967205085	0.999938644

CONCLUSION

We identified in this study the Codon usage analysis and selection pressure for evolutionary perspective. The nucleotide composition of CDSs was determined at the 3rd codon position (A3% C3% G3% T3%) along with overall GC% and AT%, GC12, and GC3. The Aroma and Gravy values, ENC values, and RSCU values were also analyzed using codon W 1.4.2 and CAIcal server.

CONSENT FOR PUBLICATION

Not Applicable.

CONFLICT OF INTEREST

The authors declare no conflict of interest, financial or otherwise.

ACKNOWLEDGEMENTS

All the authors PJ, AJ, and VK conducted this research study and also manuscript preparation under the Dept. of Bioinformatics, LPU, Punjab, India.

REFERENCES

[1]   A. Anderson, K. Hartmann, C.M. Leutenegger, A.L. Proksch, R.S. Mueller, and S. Unterer, "Role of canine circovirus in dogs with acute haemorrhagic diarrhoea", *Vet. Rec.,* vol. 180, no. 22, pp. 542-542, 2017.
        [http://dx.doi.org/10.1136/vr.103926] [PMID: 28242782]

[2]   A. Kapoor, E.J. Dubovi, J.A. Henriquez-Rivera, and W.I. Lipkin, "Complete genome sequence of the first canine circovirus", *J. Virol.,* vol. 86, no. 12, pp. 7018-7018, 2012.
        [http://dx.doi.org/10.1128/JVI.00791-12] [PMID: 22628401]

[3]   F. Kotsias, D. Bucafusco, D.A. Nuñez, L.A. Lago Borisovsky, M. Rodriguez, and A.C. Bratanich, "Genomic characterization of canine circovirus associated with fatal disease in dogs in South America", *PLoS One,* vol. 14, no. 6, p. e0218735, 2019.
        [http://dx.doi.org/10.1371/journal.pone.0218735] [PMID: 31237902]

[4]   M. Breitbart, E. Delwart, K. Rosario, J. Segalés, and A. Varsani, "ICTV Virus Taxonomy Profile: Circoviridae", *J. Gen. Virol.,* vol. 98, no. 8, pp. 1997-1998, 2017.
        [http://dx.doi.org/10.1099/jgv.0.000871] [PMID: 28786778]

[5]   P-C. Liao, K-K. Wang, S-S. Tsai, H-J. Liu, B-H. Huang, and K-P. Chuang, "Recurrent positive selection and heterogeneous codon usage bias events leading to coexistence of divergent pigeon circoviruses", *J. Gen. Virol.,* vol. 96, no. 8, pp. 2262-2273, 2015.
        [http://dx.doi.org/10.1099/vir.0.000163] [PMID: 25911731]

CHAPTER 31

Evaluation of Thoracic Mobility in Different Stages of Chronic Obstructive Pulmonary Disease: A Pilot Study

Mahvish Qaiser[1,*], Nahid Khan[2], Jyoti Ganai[2], Prem Kapur[3] and **Abhinav Jain[4]**

[1] *Department of Rehabilitation Sciences, School of Nursing Sciences and Allied Health, Jamia Hamdard, India*

[2] *Department of Rehabilitation Sciences, School of Nursing Sciences and Allied Health, Jamia Hamdard, India*

[3] *Department of Medicine, Hamdard Institute of Medical Sciences and Research associated HAHC Hospital, India*

[4] *Department of Radiology, Hamdard Institute of Medical Sciences and Research associated HAHC Hospital, India*

Abstract: Patients suffering from COPD have decreased thoracic mobility which affects the functioning of the respiratory muscles. As the literature on thoracic mobility in COPD patients is scarce the present study focuses on assessing thoracic mobility in different grades of COPD with the help of a digital inclinometer. A total of 59 subjects were included in the present study (37 COPD and 22 Controls) and subcategorized into 4 groups according to the severity (GOLD's Classification). Thoracic mobility on the left and right sides, both were found to be less in the COPD group as compared to the controls. There was a statistically significant difference for the left side thoracic mobility ($p=0.00$) between both groups with the mean value of 35.38 ± 6.92 for the COPD group and 44.49 ± 3.92 for control. There was a statistically significant difference for right side thoracic mobility between both groups ($p = 0.01$) with the mean values of 40.08 ± 9.28 in the COPD group and 44.43 ± 4.22 for the control. The right and left thoracic mobility was found to be decreasing with increasing severity but the difference was not statistically significant. There is an alteration in the respiratory mechanic and shoulder girdle kinematics which alters the thoracic mobility. In the present study, the sample size for each subgroup of COPD was less. The mild and severe COPD subgroup had a limited sample and hence more studies should be conducted to understand the changes that happen in thoracic mobility due to an increase in severity. The thoracic mobility is affected in COPD patients and with increasing severity of COPD, the thoracic mobility decreases in this population.

[*] **Corresponding author Mahvish Qaiser:** Department of Rehabilitation Sciences, School of Nursing Sciences and Allied Health, Jamia Hamdard, India; E-mail: mahvishqaiser@gmail.com

Keywords: COPD, Digital Inclinometer, Left Thoracic Mobility, Right Thoracic Mobility, Thoracic Excursion.

INTRODUCTION

Chronic respiratory disease (CRD) is a disease of the airway and other structures of the lung. Some of the most common are Chronic Obstructive Pulmonary Disease (COPD), asthma, occupational lung disease, and pulmonary hypertension. Estimates suggested that COPD and asthma are the most common [1]. The mortality rate of chronic respiratory disease increased by 18% from 1990 to 2017. In 2017, deaths due to COPD comprises 81.7% of the total number of deaths from chronic respiratory diseases. COPD related death in 2017 was 23% more than in 1990 [2]. A review of the published reports revealed 384 million cases of COPD in 2010 which is 11.7% globally [3].

COPD brings changes in chest wall configuration [4]. Patients with COPD develop hyperinflated chest which impairs their chest mobility. 12 thoracic vertebrae are connected to 12 pairs of ribs. Additionally, it provides muscle attachments and allows those muscles to function as muscles of ventilation. Therefore, it is necessary to focus on thoracic mobility. Thorax and thoracic vertebrae are part of the upper body quadrant [5]. Postural re-alignment and mobility of the upper body quadrant have been recommended as part of pulmonary rehabilitation.

To document the stability, evaluation, and progression of chest function; a physical therapist measures chest mobility. Chest mobility can be measured by measuring the chest circumference during breathing using an inch-tape [5]. It is a simple and inexpensive method in clinical practice but defined reference values are absent. With the development of time, there is more equipment available to assess chest mobility. Debrunner kyphometer, digital inclinometer, bubble inclinometer, are some of them.

Movement available in the thoracic region is rotation so by assessing thoracic rotation to left and to the right we can easily know the thoracic mobility. Our study evaluates thoracic mobility by using a digital inclinometer (DI) which is a valid and reliable method [6]. To the best of our knowledge no such study was done which assesses thoracic mobility in different stages of COPD. For better understanding, we took healthy controls to compare the data. This study found a new opportunity of using a digital inclinometer as a tool to assess the severity of COPD in different stages.

MATERIAL AND METHODS

All COPD patients were recruited from the medical department of HAHC hospital, Delhi. After explaining the study protocol, interesting subjects signed the informed consent form. A total of fifty-nine (59) subjects were enrolled in the present study of which, thirty-seven (37) were diagnosed with COPD and labeled as group A. Twenty-two (22) were non-COPD patients and labeled as group B. Present study is a quantitative descriptive cross-sectional study.

Clinically stable patients and those who don't require oxygen supplementation were included in the COPD group. History of smoking was taken and both smokers and non-smokers were recruited in both the groups. Patients with recent exacerbation in the past 3 months, unable to cooperate and understand the protocol, and with comorbidities were excluded from the study. Thoracic mobility was assessed by using a digital inclinometer. Subjects were instructed to adopt heel sit position (Fig. **1**) and time was provided to them so that they could get familiarize with the movement of thoracic rotation in the heel-sit position but no formal warm-up was done to reflect clinical practice. The Digital Inclinometer was calibrated before data collection according to the manufacturers' guidelines.

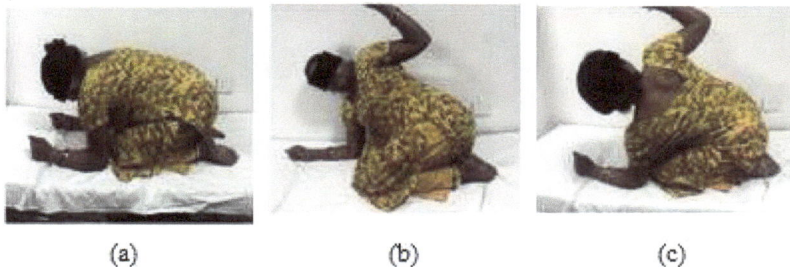

(a)   (b)   (c)

Fig. (1). (a) Starting heel sit position, (b) Left side rotation, (c) Right side rotation.

The device was placed over the C7– T1 interspinous space perpendicular to the spine, which can be located by palpating the subject's cervical spine. Subjects were instructed to place their ipsilateral upper extremity at the side of the body in full elbow flexion, keep the head aligned with the thoracic spine in the horizontal plane, and maintain the kneeling position. Compensatory patterns of movement were observed. We took three measurements by using Digital Inclinometer in the starting position (Fig. **1a**) and then at the end-range left- and right-rotation positions (Figs. **1b** and **1c**). The mean of the 3 measurements was calculated for each left- and right-rotation position [6].

Sample size was calculated by using n = [DEFF*N*p*q]/ [(d2 /Z2 1- α/2*(N-1) +p*q] formula [7, 8]. Data were analyzed by using the SPSS Statistics software package, version 16 for windows. Independent samples t-test was used to measure demographic data and thoracic mobility. One-way analysis of variance (ANOVA) was used to compare the differences among four stages of COPD. The level of significance was less than 0.05 ($p < 0.05$). To explore differences between four grades of COPD, a post hoc test was used.

Result: Characteristics of the sample population are shown in Table **1**.

Table 1. Depicting age, Body mass index (BMI) for both groups.

Variable	Mean ± SD		t	P
	Group A COPD (n = 37)	Group B Non-COPD (n = 22)		
Age	62.27 ± 8.18	55.59 ± 10.34	2.58	0.01
BMI	21.93 ± 4.37	23.62 ± 3.53	1.62	0.11

The mean and standard deviation of left and right thoracic mobility of group A (COPD) and group B (non-COPD) are shown in Table **2**. We divided the COPD group into four subgroups based on COPD stages (GOLDS' Classification) and measured thoracic mobility for the right and left side which is depicted in Table **3**.

Table 2. Depicting thoracic mobility (TM) of the left (LT) and right (RT) sides of both the groups.

Variable (in degrees)	Mean ± SD	Mean ± SD	t	P
	Group A COPD (n = 37)	Group B Non-COPD (n = 22)		
TM- RT	40.08±9.28	44.43±4.22	2.45	0.01
TM- LT	35.38±6.92	44.49±3.92	5.64	0.00

Table 3. Depicting thoracic mobility (TM) of the left (LT) and right (RT) sides of all the four stages of the groups. SD= standard deviation.

Variable (in degrees)	Stage A	Stage B	Stage C	Stage D	f	P
	Mean ± SD					
TM-RT	44±6.92	43.01±6.5	39.41±11.79	34.03±3.96	1.51	.22
TM-LT	38.66±5.03	35.75±3.28	35.54±9.73	32.6±3.34	0.54	.65

Post hoc test revealed that there is no statistically significant mean difference in left side thoracic mobility of stage D with stage B, C, and A Table **4**. Post hoc test revealed that there is a statistically significant mean difference in right side thoracic mobility of stage D with stage B ($p < 0.05$) (Table **4**).

Table 4. Mean difference of left side and Right side thoracic mobility in four grades of COPD.

Stage	Stage	Left		Right	
		Mean Difference ± Standard Deviation Error	Level of Significance	Mean Difference ± Standard Deviation Error	Level of Significance
A	B	2.91±4.55	.52	.98±5.86	.86
	C	3.12±4.44	.48	4.58±5.71	.42
	D	6.06±4.99	.23	9.96±6.42	.13
B	A	-2.91±4.55	.52	-.98±5.86	.86
	C	.20±2.69	.94	3.59±3.47	.30
	D	3.15±3.53	.37	8.98±4.54	.05
C	A	-3.12±4.44	.48	-4.58±5.71	.42
	B	-.20±2.69	.94	-3.59±3.47	.30
	D	2.94±3.38	.39	5.38±4.35	.22
D	A	-6.06±4.99	.23	-9.96±6.42	.13
	B	-3.15±3.53	.37	-8.98±4.54	.05
	C	-2.94±3.38	.39	-5.38±4.35	.22

DISCUSSION

It can be concluded from the present study that thoracic mobility is lower in COPD patients as compared to the non-COPD population. Also, it was observed that thoracic mobility decreased with an increase in the severity of COPD. Though the decrease in thoracic mobility from one subgroup of COPD to another was not found to be statistically significant. The only statistically significant difference was noted between group B and group D on the right side.

Decreased thoracic mobility in COPD is due to possible alterations seen in respiratory mechanics and shoulder girdle kinematics. Structural and functional changes in rib change are related to the biomechanical changes in thoracic structure and region. Normally, during inspiration, the bucket handle and the pump handle movement takes place to increase the transverse and anteroposterior diameter of the chest wall. Elastic recoil capacity of the lung becomes less and results in hyperinflation. These changes lead to a loss in bucket handle movement and exaggerated pump handle movement during inspiration [9]. These findings corroborate with those of Nuno Morais *et al.,* who found lesser shoulder range of motion and thoracic mobility in COPD [5]. They explained the relationship between upper quadrant muscle and thoracic mobility. Upper quadrant muscle kinematics alterations lead to decreased thoracic spine extension. Although they described, thoracic spine extension and flexion as movement assessing thoracic mobility, our research uses thoracic rotation to left and right as movement assessing thoracic mobility [5]. Thoracic rotation is the axial rotation around the thoracic spine and it purely occurs in the thoracic region while thoracic flexion and extension is the result of combined movement in the thoracic and lumbar region [5, 9]. Therefore, we focused on thoracic rotation.

Right thoracic rotation was slightly lower in a study done by Jonathan bucke *et al* [9]. In the present study, left thoracic rotation was found to be lower as compared to right rotation. The reason could be the population differences, that they included normal athletic population while our population was COPD in which pulmonary deterioration changes the normal physiology of respiratory mechanics which impacts the scapulothoracic kinematics.

The objective of our study was to compare thoracic mobility in different stages of COPD. Post hoc test revealed that there is a statistically significant difference in right thoracic mobility of stage B with stage D. This relationship was not found in left side thoracic mobility. Although mean values were found to be lower in stage D compared to other stages but were statistically not significant.

Increasing severity decreases thoracic rotation in COPD stages. Progression of COPD leads to kyphotic deformity and it changes thoraco-kyphosis angle [9]. With the disease progression, postural malalignment and deterioration in lung function have been observed in COPD patients. These changes further decrease the thoracic movement and its function. We suggest further researches in this field with large populations as it is still unclear why this relationship is not found in left thoracic mobility. Another limitation is that stage A had only 3 subjects and 6 subjects of stage D (very severe). Stage B had a total of 12 subjects and stage C had 16 subjects.

In summary, our study explored the use of a digital inclinometer in all the stages of COPD to assess thoracic mobility. Thoracic mobility assessment can be used in pulmonary rehabilitation to re-evaluate the treatment protocol. Future studies should be conducted to evaluate the changes in thoracic mobility with the progression of the disease.

CONSENT FOR PUBLICATION

Not applicable.

CONFLICT OF INTEREST

The authors declare no conflict of interest, financial or otherwise.

ACKNOWLEDGEMENT

Declared none.

REFERENCES

[1]   "World health organization", *Chronic Respiratory diseases,[online]. Available:* .https://www.who. int/health-topics/chronic-respiratory-diseases#tab=tab_3 [Accessed: 12th Jan, 2020].

[2]   X. Li, X. Cao, M. Guo, M. Xie, and X. Liu, "Trends and risk factors of mortality and disability adjusted life years for chronic respiratory diseases from 1990 to 2017: systematic analysis for the Global Burden of Disease Study 2017", *BMJ,* vol. 368, no. m234, p. m234, 2020.
[http://dx.doi.org/10.1136/bmj.m234] [PMID: 32075787]

[3]   D. Adeloye, S. Chua, C. Lee, C. Basquill, A. Papana, E. Theodoratou, H. Nair, D. Gasevic, D. Sridhar, H. Campbell, K.Y. Chan, A. Sheikh, and I. Rudan, "Global and regional estimates of COPD prevalence: Systematic review and meta-analysis", *J. Glob. Health,* vol. 5, no. 2, p. 020415, 2015.
[http://dx.doi.org/10.7189/jogh.05.020415] [PMID: 26755942]

[4]   D. Parmar, and A. Bhise, "The Immediate Effect of Chest Mobilization Technique on Chest Expansion in Patients of COPD with Restrictive Impairment", *Int. J. Sci. Res.,* vol. 4, no. 6, pp. 2413-2416, 2015.
[IJSR].

[5]   N. Morais, J. Cruz, and A. Marques, "Posture and mobility of the upper body quadrant and pulmonary function in COPD: an exploratory study", *Braz. J. Phys. Ther.,* vol. 20, no. 4, pp. 345-354, 2016.
[http://dx.doi.org/10.1590/bjpt-rbf.2014.0162] [PMID: 27556391]

[6]   J. Bucke, S. Spencer, L. Fawcett, L. Sonvico, A. Rushton, and N.R. Heneghan, "Validity of the Digital Inclinometer and iPhone When Measuring Thoracic Spine Rotation", *J. Athl. Train.,* vol. 52, no. 9, pp. 820-825, 2017.
[http://dx.doi.org/10.4085/1062-6050-52.6.05] [PMID: 28787176]

[7]   P. Rajkumar, K. Pattabi, S. Vadivoo, A. Bhome, B. Brashier, P. Bhattacharya, and S.M. Mehendale, "A cross-sectional study on prevalence of chronic obstructive pulmonary disease (COPD) in India: rationale and methods", *BMJ Open,* vol. 7, no. 5, p. e015211, 2017.
[http://dx.doi.org/10.1136/bmjopen-2016-015211] [PMID: 28554925]

[8]   A.G. Dean, K.M. Sullivan, and M.M. Soe, *OpenEpi: open-source epidemiologic statistics for public health, version.2.3.1,* 4th May 2015.www.OpenEpi.com

[9]   M.A. Gonçalves, B.E. Leal, L.G. Lisboa, M.G.S. Tavares, W.P. Yamaguti, and E. Paulin, "Comparison of diaphragmatic mobility between COPD patients with and without thoracic hyperkyphosis: a cross-sectional study", *J. Bras. Pneumol.,* vol. 44, no. 1, pp. 5-11, 2018.
[http://dx.doi.org/10.1590/s1806-37562016000000248] [PMID: 29538536]

CHAPTER 32

Wearable Antennas-An Overview

Narendra Gali[1,*] and **Narbada Prasad Gupta**[2]

[1] *Research Scholar, Lovely Professional University, Punjab/ Assistant Professor, Vignan's Inst. of Mgmt. and Tech. for Women, Hyd., India*

[2] *Professor, School of Electronics and Electrical Engineering Lovely Professional University, Punjab, India*

Abstract: The most popular antenna for portable devices in current communication technologies is the wearable antenna due to its compactness and flexibility; demand was rapidly growing and can communicate through signals with the human body and the wearable devices. The advantages of wearable antennas are flexible, hidden, low profile, and no harm to humans. The key benefit of this antenna is that it is placed on the human body or included in clothing, effortlessly transmits, and receives signals through clothes or on-body. These antennas play a vital role in the number of applications, viz. navigation (118MHz to 137MHz), medicine (750MHz to 2.6GHz), military (225MHz to 400MHz), RFID (433MHz to 5.4GHz), physical training, tracking, and health monitoring, *etc*. This paper discussed the important aspects of wearable antennas, which include materials used, substrate, and fabrication techniques. Next, discussed a clear overview of wearable antennas existing and design aspects, their advantages, and drawbacks.

Keywords: Fabrication Technique, Flexible Antennas, ISM Band, Substrate Integrated Waveguide, Textile Antennas, Wearable Antennas.

1. INTRODUCTION

It has been seen that during the last decade of years, portable devices play a proximity role in human life those are mobiles and tablets. The technology is rapidly changing year by year and the size of the device, visibility decreases. In forthcoming days, sensors are used to control human activities; further devices are used to monitor the different requirements of the human including medical

* **Corresponding author Narendra Gali:** Research Scholar, Lovely Professional University, Punjab/ Assistant Professor, Vignan's Inst. of Mgmt. and Tech. for Women, Hyd, India; E-mail: narendra.gali@gmail.com

Dharam Buddhi, Rajesh Singh and Anita Gehlot (Eds.)

conditions. These devices and sensors play a crucial role, communicate with each other and stuffs outside. In real-world communications, all these advancements are possible only because of wearable devices and antennas [1].

Generally, Wearable devices are carried by the person within the body or on-body and are capable of communicating among them *via* cellular connectivity. The devices used in these types of communications would include other components such as like sensors, antennas, and batteries. One of the most important components in the wearable device is antennas and they contribute overall efficiency of a wearable wireless link [2 - 4]. The wearable antennas must be light in weight, conformal design, low cost, easy system integrable, *etc.* and the design ought to be specified which is not deteriorated even if they are bent [5].

In our life, Wearables play a significant role and are found to be used in many portable devices used for Security and Entertainment like fitness bands, wristwatches, reality glasses, and cover a lot of medical applications and rescue operations [6]. In the field of health monitoring, wearable devices are used to monitor the patient's health conditions who are in the critical stage like, the sugar level of the patient, glucose level of the patient, Blood pressure, inner intestinal system, body temperature, *etc.* of the patient minute to minute. Wearables are also used in rescue operations as well as entertainment. The Table **1** describes the field involved in wearable antennas with their applications.

Table 1. Applications of Wearable devices in different fields [7].

Field	Applications
Health Monitoring	Glucose Level Monitoring / Oximetry/ Endoscopy / GPS tractor / Wearable Doppler unit
Entertainment	Smartwatches / LED dress / Music Jackets / Intelligent
Rescue and security	Helmet / Tractors / E-shoes / Fitness-bands / Life jackets/ Raincoat

In day-to-day life, Wearable antennas are integrated into humans wearing gadgets like shoes, jackets, helmets, lenses, cooling glasses, raincoats, *etc.*, and numerous aspects to be considered while designing wearable antennas as they need to be flexible, hidden, unobtrusive, low profile and should not be considerable degradation in proximity to the human body [8].

This paper discusses the overview of wearable antennas and mainly about on-body as well as textile-based antennas. Section II describes important aspects involved in wearable antennas and section III involved the discussion about the design of substrate integrated waveguides. Section IV is about the existing different wearable antennas and ended with concluding remarks.

2. IMPORTANT ASPECTS IN WEARABLE ANTENNAS

This section discusses important aspects of wearable antennas and includes a discussion about materials, the substrate as well as fabrication techniques.

2.1. Materials

The different kinds of dielectric, conductive materials are used to implement Wearable antennas, and these can be carefully chosen to prevent mechanical bending, mechanical wrapping at different weather conditions for the protection of EM radiation [9]. Table **2** reports the flexible conductive materials with their Thickness and conductivity.

Table 2. Thickness and conductivity of Flexible conductive materials.

Materials for Conductor	Thickness (mm)	Conductivity
Egain Liquid fillet [9]	0.08	250K
Ployleurethene nanoparticle composite sheet [10]	0.0065	1.1M
Zoflex plus copper [11]	0.175	193K
Copper coated taffeta [12]	0.15	3.4M
Sliver flakes plus fluorine rubber [13]	NA	85K
PANI/ CCo composite [14]	0.075	7.3K

2.2. Substrates

The dielectric substrate plays a very dominant role in the design of wearable antennas. The standard substrates like Rogers, Teflon, RT-Duriod, and silicon have been not preferred in the case of textile antennas due to bending, stre*tc*hing, and rotating capabilities [15 - 18]. This needs to invent flexible conductive materials and some of the materials which having loss tangent, dielectric constant as revealed in Table **3** .

Table 3. Flexible materials (Dielectric constant and loss tangent).

Conductive Material	Dielectric Constant (εr)	Loss Tangent (Tan δ)
Polydimethylsiloxane [13]	3.2	0.01
Polydimethylsiloxane -ceramic composite [14]	6.25	0.02
Ethylene Vinyl Acetate [15]	2.8	0.002

2.3. Fabrication Techniques

Fabrication methods play a very important role in the design of wearable antennas concerning the accuracy, and speed of antenna while choosing the design of an antenna. The most common wearable fabrication techniques are Screen Printing [19], Inkjet Printing [20,21], Wet *etch*ing [22,23], Embroidery methods [24].

3. SUNSTRATE INTEGRATED WAVEGUIDE

SIW is one of the families of substrate integrated circuits (SICs). SIW is like a transmission line and is invented to overcome the drawbacks that occurred in existing transmission lines like microstrip and coplanar waveguide [25]. SIW is purely invented for higher frequencies that are millimeter and centimeter wavelength frequency applications and most preferred for millimeter-wave applications. Laminated waveguides, post-wall waveguides are the other names and general structures being illustrated in Fig. (**1**). The outline is familiar; derived from a rectangular waveguide, integrated with two rows of cyclic vias or holes placed in between the two ground planes through a substrate. The planar structure is used for developing SIW based wearable devices, called System on Substrate (SoS) [26 - 28].

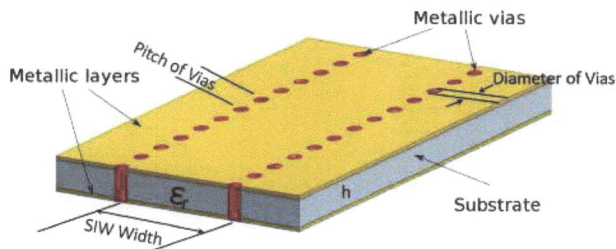

Fig. (1). Schematic representation of SIW.

3.1. Design of SIW

The Width of SIW is derived from the rectangular waveguide that is also called an air-filled waveguide, the schematic is shown in Fig. (**2a**), and generalized formulae for cutoff frequency are represented in equation 1. The dominant mode of the rectangular waveguide is TE_{10} and simplified formulae for the width of the rectangular waveguide as mentioned in equation 2 [29 - 31].

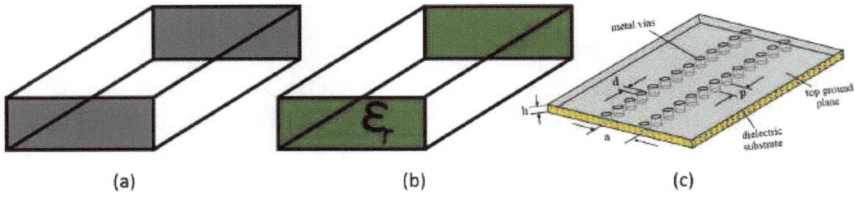

Fig. (2). (a) Air-Filled Waveguide **(b)** Dielectric Filled Waveguide **(c)** SIW.

$$f_c = \frac{C}{2\pi}\sqrt{\left(\frac{m\pi}{a}\right)^2 + \left(\frac{n\pi}{b}\right)^2} \tag{1}$$

$$a = \frac{c}{2 * f_c} \tag{2}$$

The rectangular waveguide packed with a dielectric material is called a dielectric-filled rectangular waveguide (DFW) and its schematic representation is shown in Fig. **(2b)**, the width is mentioned in equation 2 [32].

$$a_R = \frac{c}{2 * f_c * \sqrt{\varepsilon_r}} \tag{3}$$

DFW is filled with two rows of holes is called SIW and their representation is shown in Fig. **(2c)**. The standard equation is used to find the width of the SIW is represented in equation 3 [32].

$$a_{siw} = a_R + \frac{d^2}{0.95*s} \tag{4}$$

Where 'd' is a diameter of holes and 's' is spacing between holes. The diameter (d) to a width (a_R) is not included in equation 4. Sometimes, give an error value and the more appropriate equation is

$$a_{siw} = a_R + 1.08 * \frac{d^2}{s} - 0.1 * \frac{d^2}{a_{SIW}} \tag{5}$$

The condition used to maintain the loss-free radiation is [33],

$$d \leq \frac{\lambda_g}{5} \text{ and } s \leq 2d \qquad (6)$$

The Planar technology is used in the substrate integrated waveguide fabrication process. So, SIW can be extended to implement like a SIW antenna, SIW active component, and SIW passive component.

4. WEARABLE BASED ANTENNAS

The wearable antenna was implemented on eyeglasses for 4G applications [34], which contain shapes like a wire and ground plane act as a glass frame with a thickness of 0.85mm, PCB fabrication technique is used in this model. The patch-based wearable antenna was introduced for mobile phones [35], which is implemented with FR-4 material and ferrite material used to act as an antenna, which was more expensive.

The monopole-based wearable antenna [36] was introduced for 802.11ac an application that contains wire mesh sheets, PCB *etc*hing was used, facing complexity while fabricating. The textile-based wearable antenna [37] is implemented for ISM band applications. The shape of the antennas used in this design is the circular patch which is migrating soft surfaces and the material used as an implement of an antenna is a textile material. The truncated patch antenna [38] is implemented for on-body applications which are operated between 2 GHz to 10.6GHz frequency bands. The full ground reflector plane was used as mitigation techniques and fabricated manually used as felt substrate.

A Quarter Mode Substrate Integrated Waveguide-based slot antenna is proposed with microstrip line feed using wool felt substrate, a relative permittivity of $\varepsilon_r =$ 1.2 and a loss tangent of tan $\delta = 0.02$ is shown in Fig. (**3**). The shorting vias are used near the feed line to implement dual-band operation are suitable for the 3.5-GHz WiMAX band and the 5.8-GHz industrial, scientific, and medical (ISM) band [39].

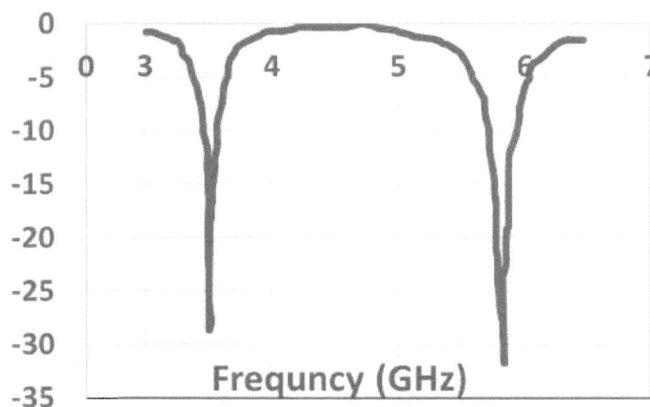

Fig. (3). Quarter mode substrate integrated waveguide-based slot antenna [39].

The unilateral patch-based antenna [40] was implanted for mobile phone applications which is feed by CPW and used radiating patch as asymmetric, and flexibility is the major drawback of the antenna. The textile-based antenna [41] is used for wearable application with a fractal-based monopole patch antenna and introduced EBG structure for SAR reductions.

The polyester cotton-based fabric used in the design of the patch and implemented for scientific, industrial, a medical band (ISM) [42] with an operating frequency of 2.4 GHz. Polyurethane material is used as a paste and printed on the cotton, but it damages the substrate for reduced sizes. In inkjet printing, post-heating is the major issue, to avoid that deposition and sintering have been introduced [43].

The mixing of two substrates was used to design an antenna for wearable tracking applications [44] and both the substrates being printed by applying the interface layer with Dimatix printer [45], avoided ink smear onto the fabric material. Inkjet printing technology [46] has been used to reduce the cost of the equipment, eliminate the process of printing layer. A conductive layer with silver nanoparticles (AgNP) is printed using an instant chemically curing process, which avoids the need for post-processing [47]. Mandal *et al.* [48], a wearable antenna was fabricated using the printing method and is designed for operating ISM band applications, PET substrate material is used in this design, and the cost of the material is low. Compare to the inkjet printing method, the cost is low and effective using surface morphological measurements. Moreover, a 3D surface propeller has been used for measuring surface roughness.

The Substrate integrated waveguide-based wearable antennas were introduced, and two antennas are designed using the same substrate [49], wearers garment material is used as substrate. Another antenna was designed with the leather substrate for multiband band applications [50]. On the top of the substrate, a copper sheet was placed to enable the operations like WiMAX, Wi-Fi. The different methods used for fabricating textile antennas are presented in [51].

The woolen substrate is added with pure copper fabric for wide-band operation [52], these wearable antennas have more radiation efficiency like 84% and fractional bandwidth is 46%. A meta-material impressed SIW based antenna has been projected on woolen felt substrate using conductive fabric [53]. This antenna was placed on the body and measured the radiation efficiency is approximately 75% with a reduction of size is 80% as compared with other structures and performance of the antennas was satisfactory, back radiation was low towards the human body.

The miniaturized dipole antenna is introduced for 2.4GHz operating a frequency application, which is designed by using paper-based substrate material [54]. The mixing of nanowire/ nano-paper high-h sliver material has reduced the size by 50% compared with ordinary low nano-paper antennas. The EBG structures are used to reduce the size of an antenna and designed a complementary split-ring resonator (CSRR) antenna to miniaturization [55] and gain of the antennas was slightly decreased.

The wearable antenna is introduced for mobile phone applications and designed by using FR-4 substrate material, ferrite sheet used as copper material [56]. The monopole-based wearable antenna is introduced for 802.11ac applications and thin wire mesh sheet used for designing, PCB *etc*hing, and the main drawback is complex fabrication [57].

The monopole-based wearable antenna is introduced for energy harvesting applications, GPS and DCS band, the adhesive sheet on zelt-fiber was used and a laser machine is used to cut the shape [58]. Agarwal *et al.* [59], the truncated patch antenna has been introduced for on-body application with an operating frequency of 2 to 10.6 GHz. The material used in this design is felt substrate and the shape was etched manually.

The antenna is with a circular disc monopole structure designed on a thin and flexible Ultralam 3850 laminate, which can support with the bandwidth of 1.74 to 100 GHz. The experimental results were suited to be used for a vast range of applications including 5G technologies [60]. The design structure is depicted in Fig. (**4a**) and results are shown in Fig. (**4b**).

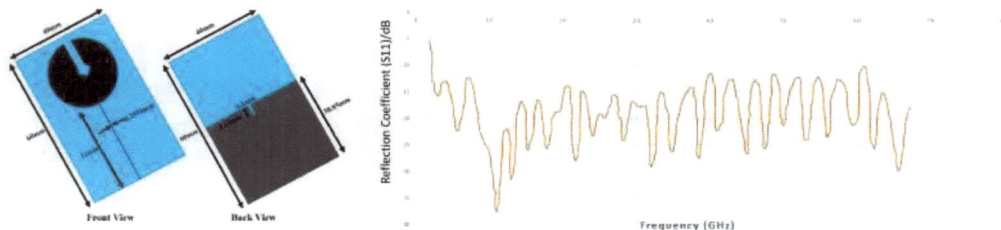

Fig. 4. (a) Disc monopole structure (front and back views), **(b)** Simulated results [60].

CONCLUSION

In the recent past, Wearable antennas attracted huge attention due to their tremendous features like being cost-effective, flexible, lightweight, used for portable communications. These antennas are generally preferred to implement on different parts of a human body, which is very attractive to implement by using flexible materials as well as low profile structure. This paper provided an overview of wearable antenna-based textile antennas as well as wearable antennas for on-body applications. The materials preferred for designing wearable antennas and their merits and demerits are exposed. The different types of wearable antennas are discussed and observed that gain, efficiency, *etc.* were less as demanded. The design of the Substrate integrated waveguide is also discussed in this review to extend this work to implement in SIW based structures.

CONSENT FOR PUBLICATION

Not applicable.

CONFLICT OF INTEREST

The authors declare no conflict of interest, financial or otherwise.

ACKNOWLEDGEMENTS

We acknowledge the encouragement provided by the School of Electronics and Electrical Engineering, Lovely Professional University Punjab, and Vignan's Institute of Management and Technology for Women, Hyderabad.

REFERENCES

[1] A. Ericsson, "Cellular networks for massive IoT-enabling low power wide area applications", *Ericsson, Stockholm, Sweden, Tech. Rep. Uen* , vol. 284 23- 3278, pp. 1-13, 2016.

[2] *Global Mobile Data Traffic Forecast Update, 2014-2019.* CISCO: San Jose, CA, USA, 2015, pp. 1-48.

[3]   S. Bhattacharyya, N. Das, D. Bhattacharjee, and A. Mukherjee, *Handbook of Research on Recent Developments in Intelligent Communication Application.* IGI Global: Philadelphia, PA, USA, 2016.

[4]   S. Bhavani, and T. Shanmuganantham, *Analysis of Different Substrate Material on Wearable Antenna for ISM Band Applications.* Springer Science and Business Media LLC, 2020.
[http://dx.doi.org/10.1007/978-981-15-3992-3_63]

[5]   N.P. Gupta, R. Maheshwari, and M. Kumar, "Advancement in Ultra-Wideband Antennas for Wearable Applications", *International Journal of Scientific & Engineering Research,,* vol. 4, no. 8, 2013. 341 ISSN 2229-5518

[6]   M. Chan, D. Estève, J.-Y. Fourniols, C. Escriba, and E. Campo, "Smart wearable systems: Current status and future challenges", *Artif. Intell. Med,* vol. 56, no. 3, pp. 137-156, 2012.

[7]   S. Seneviratne, "A survey of wearable devices and challenges", *IEEE Commun. Surveys Tuts,* vol. 19, no. 4, pp. 2573-2620, 2017. 4th Quart.

[8]   D.H. Werner, and Z.H. Jiang, *Electromagnetics of Body Area Networks: Antennas, Propagation, and RF Systems.* Wiley: Hoboken, NJ, USA, 2016.
[http://dx.doi.org/10.1002/9781119082910]

[9]   I. Locher, M. Klemm, T. Kirstein, and G. Tester, "Design and characterization of purely textile patch antennas", *IEEE Trans. Adv. Package., ,* vol. 29, no. 4, pp. 777-788, 2006.

[10]   J-H. So, J. Thelen, A. Qusba, G.J. Hayes, G. Lazzi, and M.D. Dickey, "Reversibly deformable and mechanically tunable fluidic antennas", *Adv. Funct. Mater.,* vol. 19, pp. 3632-3637, 2009.
[http://dx.doi.org/10.1002/adfm.200900604]

[11]   Y. Kim, J. Zhu, B. Yeom, M. Di Prima, X. Su, J.G. Kim, S.J. Yoo, C. Uher, and N.A. Kotov, "Stretchable nanoparticle conductors with self-organized conductive pathways", *Nature,* vol. 500, no. 7460, pp. 59-63, 2013.
[http://dx.doi.org/10.1038/nature12401] [PMID: 23863931]

[12]   A. Kumar, "A highly deformable conducting traces for printed antennas and interconnects: Silver/fluoropolymer composite amalgamated by triethanolamine", *Flexible Printed Electron,* vol. 2, no. 4, 2017.

[13]   R.A. Liyakath, A. Takshi, and G. Mumcu, "Multilayer stretchable conductors on polymer substrates for conformal and reconfigurable antennas", In: *IEEE Antennas Wireless Propag. Lett.* vol. 12. , 2013, pp. 603-606.

[14]   R.B.V.B. Simorangkir, Y. Yang, R.M. Hashmi, T. Björninen, K.P. Esselle, and L. Ukkonen, "Polydimethylsiloxane-embedded conductive fabric: Characterization and application for the realization of robust passive and active flexible wearable antennas", *IEEE Access,* vol. 6, pp. 48102-48112, 2018.
[http://dx.doi.org/10.1109/ACCESS.2018.2867696]

[15]   S. Velan, *Dual-band EBG integrated monopole antenna deploying fractal geometry for wearable applications,* vol. 14, pp. 249-252, 2015.
[http://dx.doi.org/10.1109/LAWP.2014.2360710]

[16]   Z. Hamouda, "Wojkiewicz jean-luc, Wojkiewicz, A. A. Pud, L. Kone, S. Bergheul, and T. Lasri, "Magnetodielectric nanocomposite polymer-based dual-band flexible antenna for wearable applications," IEEE Trans", *AntennasPropag.,* vol. 66, no. 7, pp. 3271-3277, 2018.

[17]   N.P. Gupta, and M. Kumar, "Radiation Performance Improvement in Wearable UWB Antenna through Slot Insertion Technique", *2015 Fifth International Conference on Communication Systems and Network Technologies, Gwalior,* 2015, pp. 83-87
[http://dx.doi.org/10.1109/CSNT.2015.41]

[18]   N.P. Gupta, R. Maheshwari, and M. Kumar, "Advancement in ultra wideband antennas for wearable applications", *Int. J. Sci. Eng. Res.,* vol. 4, no. 8, pp. 341-348, 2013.

[19]   N.P. Gupta, and M. Kumar, "Development of a Reconfigurable and Miniaturized CPW Antenna for Selective and Wideband Communication", *Wirel. Pers. Commun.,* vol. 95, pp. 2599-2608, 2017. [http://dx.doi.org/10.1007/s11277-017-3942-8]

[20]   N.P. Gupta, M. Kumar, and R. Maheshwari, "Development and performance analysis of conformal UWB wearable antenna under various bending radii", *IOP Conf. Series: Materials Science and Engineering,* vol. 594, 2019

[21]   S.B. Roshni, M.P. Jayakrishnan, P. Mohanan, and K.P. Surendran, "Design and fabrication of an E-shaped wearable textile antenna on PVBcoated hydrophobic polyester fabric", *Smart Mater. Struct.,* vol. 26, no. 10, 2017. [http://dx.doi.org/10.1088/1361-665X/aa7c40]

[22]   M.C. Tang, B. Zhou, and R.W. Ziolkowski, "Flexible uniplanar electrically small directive antenna empowered by a modified CPW-feed", *IEEE Antennas Wirel. Propag. Lett.,* vol. 15, pp. 914-917, 2015. [http://dx.doi.org/10.1109/LAWP.2015.2480706]

[23]   H.F. Abutarboush, M.F. Farooqui, and A. Shamim, "Inkjet-printed wideband antenna on the resin-coated paper substrate for curved wireless devices", *IEEE Antennas Wirel. Propag. Lett.,* vol. 15, pp. 20-23, 2016.

[24]   A. Tsolis, W.G. Whittow, A.A. Alexandridis, and J. Vardaxoglou, "Embroidery and related manufacturing techniques for wearable antennas: Challenges and opportunities", *Electronics (Basel),* vol. 3, no. 2, pp. 314-338, 2014. [http://dx.doi.org/10.3390/electronics3020314]

[25]   S. Ahmed, F.A. Tahir, A. Shamim, and H.M. Cheema, "A compact Kapton-based inkjet-printed multiband antenna for flexible wireless devices", *IEEE Antennas Wirel. Propag. Lett.,* vol. 14, pp. 1802-1805, 2015. [http://dx.doi.org/10.1109/LAWP.2015.2424681]

[26]   M. Akbari, M.W.A. Khan, M. Hasani, T. Bjorninen, L. Sydanheimo, and L. Ukkonen, ""Fabrication and characterization of graphene antenna for low-cost and environmentally friendly RFID tags," IEEE AntennasWireless Propag", *Lett.,* vol. 15, pp. 1569-1572, 2016.

[27]   Z. Stempien, E. Rybicki, A. Patykowska, T. Rybicki, and M. I. Szynkowska, *Shape-programmed inkjet-printed silver electroconductive layers on textile surfaces,* vol. 47, pp. 1321-1341, 2017.

[28]   M. Grouchko, A. Kamyshny, C. F. Mihailescu, D. F. Anghel, and S. Magdassi, *Conductive inks with a `built-in' mechanism that enables sintering at room temperature,* vol. 5, pp. 3354-3359, 2011. [http://dx.doi.org/10.1021/nn2005848]

[29]   M. Nanda Kumar, and T. Shanmuganantham, "Broadband I shaped SIW slot antenna for V-band Applications", *Applied Computational Electromagnetic Society,* vol. 34, no. 11, p. 2019, 2019. [ACES].

[30]   M. Nanda Kumar, " Broadband Substrate Integrated Waveguide Venus shaped Slot Antenna for V-band Applications", In: *Microwave and Optical Technology Letters* vol. 61. , 2019, no. 10, pp. 2342-2347.

[31]   M.N. Kumar, and T. Shanmugnantham, *Division shaped SIW slot antenna for millimeter wireless/automotive radar applications," Computer and Electrical Engineering* vol. 71. , 2018, pp. 667-675.

[32]   M. Nanda Kumar, and T. Shanmuganantham, *Broad-Band H-Spaced Head Shaped Slot with SIW Based Antenna for 60GHz Wireless Communication Applications,* 2019.

[33]   M. Nanda Kumar, and T. Shanmuganantham, *Substrate Integrated Waveguide based Slot Antenna for 60GHz Wireless Applications,* 2019.

[34]   M. Nanda Kumar, and T. Shanmugnantham, *"SIW based slot antenna fed by microstrip for 60/79GHz Applications," Journal: Lecturer Notes in Electrical Engineering.* Scopus, 2019, pp. 741-748.

[35]   M Nanda Kumar, *"Microstrip Fed SIW Venus shaped slot Antenna for Millimeter Wireless Communication Applications," International Journal of Engineering & Technology,* vol. 7, no. 2.33, pp. 878-881, 2018.

[36]   M. Nanda Kumar, and T. Shanmugnantham, *"Design of Substrate Integrated Waveguide back to back π-shaped Slot Antenna for 60GHz Applications," Journal: Lecturer Notes in Electrical Engineering (Springer).* Scopus, 2018, pp. 215-224.

[37]   N. Gunavathi, and D. Sriramkumar, *CPW-fed monopole antenna with reduced radiation hazards toward human head using metallic thin-wire mesh for 802.11ac applications,"* Microw. Opt. Technol. Lett., vol. 57, pp. 2684-2687, 2015.
[http://dx.doi.org/10.1002/mop.29411]

[38]   E. Rajo-Iglesias, I. Gallego-Gallego, L. Inclan-Sanchez, and O. Quevedo-Teruel, *Textile soft surface for back radiation reduction in bent wearable antennas,"* IEEE Trans. Antennas Propag, vol. 62, no. 7, pp. 3873-3878, 2014.
[http://dx.doi.org/10.1109/TAP.2014.2321133]

[39]   C. Loss, R. Gonçalves, C. Lopes, P. Pinho, and R. Salvado, "Smart coat with a fully-embedded textile antenna for IoT applications", *Sensors (Basel),* vol. 16, no. 6, p. 938, 2016.
[http://dx.doi.org/10.3390/s16060938] [PMID: 27338407]

[40]   L. A. Y. Poffelie, P. J. Soh, S. Yan, and G. A. E. Vandenbosch, *A high-_delity all-textile UWB antenna with low back radiation for off-body WBAN applications,* 2016.

[41]   K. Agarwal, Y.-X. Guo, and B. Salam, *Wearable AMC backed near end-fire antenna for on-body communications on latex substrate,* 2016.

[42]   X. Zhu, X. Liu, and H. Yang, "Compact Dual-Band Wearable Textile Antenna Based on Quarter-Mode Substrate Integrated Waveguide", *2020 9th Asia-Pacific Conference on Antennas and Propagation (APCAP), Xiamen, China,* 2020pp. 1-2
[http://dx.doi.org/10.1109/APCAP50217.2020.9246045]

[43]   S. Velan, *"Dual-band EBG integrated monopole antenna deploying fractal geometry for wearable applications,"* IEEE Antennas Wireless Propag. Lett, vol. 14, pp. 249-252, 2015.
[http://dx.doi.org/10.1109/LAWP.2014.2360710]

[44]   S.J. Chen, B. Chivers, R. Shepherd, and C. Fumeaux, "Bending impact on a flexible ultra-wideband conductive polymer antenna", *Proc. Int. Conf. Electromagn. Adv. Appl. (ICEAA),* 2015, p. 422
[http://dx.doi.org/10.1109/ICEAA.2015.7297148]

[45]   S.M. Salleh, M. Jusoh, A.H. Ismail, M.N.M. Yasin, M.R. Kamarudin, and R. Yahya, *Circular polarization textile antenna for GPS application,* 2015.

[46]   B. Krykpayev, M. F. Farooqui, R. M. Bilal, M. Vaseem, and A. Shamim, *A wearable tracking device inkjet-printed on textile,* 2017.
[http://dx.doi.org/10.1016/j.mejo.2017.05.010]

[47]   F. Ghanem, R. Langley, and L. Ford, "Propagation control using SIW technology", *Proc. IEEE Antennas Propag. Soc. Int. Symp.,* 2010, p. 1

[48]   B. Mandal, and S. K. Parui, *Wearable tri-band SIW based antenna on the leather substrate,* 2015.
[http://dx.doi.org/10.1049/el.2015.2559]

[49]   M. E. Lajevardi, and M. Kamyab, *A low-cost wideband quasi-yagi SIWbased textile antenna,* 2017.

[50]   M. E. Lajevardi, and M. Kamyab, *Ultraminiaturized metamaterial inspired SIW textile antenna for off-body applications,* 2017.
[http://dx.doi.org/10.1109/LAWP.2017.2766201]

[51]   M. Bozzi, S. Moscato, L. Silvestri, N. Delmonte, M. Pasian, and L. Perregrini, "Innovative SIW components on paper, textile, and 3D-printed substrates for the Internet of Things", *Proc. Asia Pacific Microw. Conf. (APMC),* 2015, p. 1

[http://dx.doi.org/10.1109/APMC.2015.7411615]

[52]   K. Black, J. Singh, D. Mehta, S. Sung, C.J. Sutcliffe, and P.R. Chalker, "Silver ink formulations for sinter-free printing of conductive films", *Sci. Rep.,* vol. 6, no. Feb, p. 20814, 2016.
[http://dx.doi.org/10.1038/srep20814] [PMID: 26857286]

[53]   K.N. Paracha, S.K.A. Rahim, H.T. Chattha, S.S. Aljaafreh, S.U. Rehman, and Y.C. Lo, "Low-cost printed flexible antenna by using an office printer for conformal applications", *Int. J. Antennas Propag.,* vol. 2018, no. Feb, 2018.3241581
[http://dx.doi.org/10.1155/2018/3241581]

[54]   T. Inui, H. Koga, M. Nogi, N. Komoda, and K. Suganuma, *A miniaturized flexible antenna printed on a high dielectric constant nano paper composite,* 2015.

[55]   M. Ramzan, and K. Topalli, "A miniaturized patch antenna by using a CSRR loading plane", *Int. J. Antennas Propag.,* vol. 2015, 2015.
[http://dx.doi.org/10.1155/2015/495629]

[56]   X. Y. Liu, Z. T. Wu, Y. Fan, and E. M. Tentzeris, *A miniaturized CSRR loaded wide-beamwidth circularly polarized implantable antenna for subcutaneous real-time glucose monitoring,* 2017.
[http://dx.doi.org/10.1109/LAWP.2016.2590477]

[57]   E. Rajo-Iglesias, I. Gallego-Gallego, L. Inclan-Sanchez, and O. Quevedo-Teruel, *Textile soft surface for back radiation reduction in bent wearable antennas,* 2014.
[http://dx.doi.org/10.1109/TAP.2014.2321133]

[58]   C. Loss, R. Gonçalves, C. Lopes, P. Pinho, and R. Salvado, *Smart coat with a fully-embedded textile antenna for IoT applications,* vol. 16, no. 6, p. 938, 2016.
[http://dx.doi.org/10.3390/s16060938]

[59]   K. Agarwal, Y.-X. Guo, and B. Salam, *Wearable AMC backed near end-fire antenna for on-body communications on latex substrate,* vol. 6, no. 3, pp. 346-358, 2016.

[60]   S. Dey, M.S. Arefin, and N.C. Karmakar, "Design and Experimental Analysis of a Novel Compact and Flexible Super Wide Band Antenna for 5G", In: *in IEEE Access* vol. 9. pp. 46698-46708.
[http://dx.doi.org/10.1109/ACCESS.2021.3068082]

CHAPTER 33

A Novel Energy-Efficient Routing Algorithm for Reduction of Data Traveling Time in Wireless Sensor Networks

Y. Venkata Lakshmi[1,*], **Parulpreet Singh**[1] and **Narbada Prasad Gupta**[1]

[1] *Department of Electronics and Communication Engineering, Lovely Professional University, Punjab, India*

Abstract: Wireless Sensor Network Lifetime improvement is very well demonstrated using sink mobility effectively in the literature. The challenge taken up in the past research was to identify the shortest route that avoids obstacles and delivers the mobile sinks at the designated nodes. In this paper, we present a cluster-based approach for collecting the data and a heuristic algorithm for a well-planned work. The routing protocols may differ from application to application, due to a large number of sensor nodes the algorithm should be studied in a novel way. In this method, the nodes known as cluster heads collect information from different cluster points and sends data to the mobile sink following this clustering process. We also propose an energy-efficient data gathering algorithm for the collection of mobile sinks. Simulations were conducted using NS2 software to verify the efficiency of the proposed algorithm.

Keywords: Clusters, Energy efficiency Algorithms, Heuristic tour plan algorithm, Mobile Sinks, Wireless sensor Networks.

INTRODUCTION

Wireless sensor networks are a source of many applications in a range of aspects, including medicine, disaster management, and environment, safety, and surveillance [1, 2]. in parallel with the Internet of things. As Power consumption is the main drawback in many applications, Wireless sensor networks designed for energy efficiency can be also avoided. However, in WSNs with limited power consumption, we use effective sensors, whereby the nodes near sinks are still responsible for the transmission of data to topless photos far away from sinks. While there is space for further energy consumption. Due to the failure of these

* **Corresponding author Y. VenkataLakshmi:** Department of Electronics and Communication Engineering, Lovely Professional University, Punjab, India; E-mail: venkatlaki@gmail.com

Dharam Buddhi, Rajesh Singh and Anita Gehlot (Eds.)

kinds of nodes at the sinks, the entire network gets disconnected and this is a situation that causes worry and keeps the lifetime of the network in doubt. Therefore, this is also a challenging task to increases the lifetime of the network by reducing the power consumption at these kinds of sensor nodes. The important limitation of this node is also the energy supplied and the bandwidth allocated. Scaling of voltage dynamically, the hardware used in the radio communication system, partitioning of the communication systems, issues that arise due to small duty cycles are some of the areas where the utilization of energy is concentrated such that it is effective in the literature.

This paper focused on using a routing algorithm that is applied exclusively in multiple paths at a cross-layered geographical node [3]. Instead of using the static nodes, this paper employed mobile nodes with the interest of increasing the network lifetime. With the use of mobile nodes installed on mobile vehicles which move across the sensing fields, the network will tend to collect data from all the static nodes. Mobile nodes can adopt single hopping or multiple hopping mechanisms.

In this work, we identify the mobile nodes as mobile sinks which can reduce the overhead on various sensor nodes, both of which are located nearer as well as farther from the mobile sinks. By processing the movement of the mobile nodes, the network lifetime can be increased.

As the physical atmosphere includes many more difficulties, the sensing field is reorganized by finding the shortest routing path which can avoid obstacles and the mobile sink can move all over the network. This paper also proposes an algorithm for finding such a routing path.

Also, the mobile sink should consider the effective performance of energy among the nodes when passing through the sensing field at the same time. This was done with a cluster-based approach [4]. The sensors in the network are known as the cluster heads and the cluster members in two categories (Sensor nodes). Cluster members collect the environmental data and thereby collect the data from them by the cluster heads and send it to the mobile sinks. The mobile sinks start this process from an initial point and return to this beginning point after collection. Fig. (**1**) shows a Cluster-based approach.

When the path of this mobile sink is designed well to avoid the obstacles, then there is every chance to utilize the power effectively and thereby contribute to the overall energy efficiency.

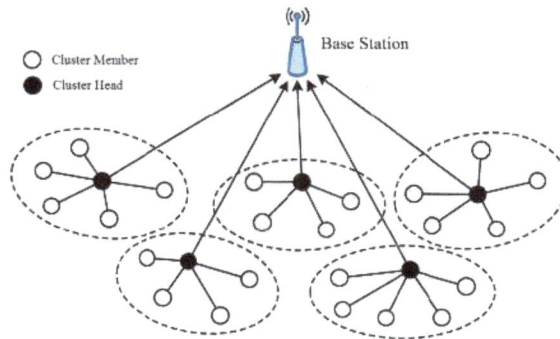

Fig. (1). Cluster-based approach (CH- Cluster Head).

When a problem is considered, a regularized grid-based approach is used to design the mobile sink network, in which the whole sensor field is described uniformly in a two-Dimensional plane, using grids of similar sizes. These grid cells now cover the entire sensing field, which may also consist of obstacles. A grid is considered an obstacle when a particular obstacle falls in a grid cell region. These grid cells are also regularised and a spanning graph is also constructed. This spanning graph makes the scheduling of the mobile sink easy by identifying the grid cells with and without obstacles which helps us in identifying the shortest route which avoids obstacles.

The rest of the paper is organized in different sections covering the different algorithms proposed followed by the simulated experimental results.

A HEURISTIC APPROACH BASED OBSTACLE AVOIDING ALGORITHM

This section presents a heuristic approach Obstacle that eliminates the shortest algorithm for the mobile sink. It involves various steps which are discussed below.

- A minimum Spanning tree is used for finding the shortest route as that of it is used in comparison to the traveling salesman problem. The spanning graph which is the set of edges is obtained by connecting the obstacle corners with the terminals. This graph reduces the infinite possible sites into finite sets.
- The spanning graph for the physical environments cannot be directly obtained as discussed above, as the shapes of the obstacles are quite irregular. Therefore, a grid-based technique [5 - 9] is used as a solution to the problem of scheduling the mobile sink path effectively. The region of the sensor is divided into units of the

same grid through the use of these grid-based techniques. Edges of boundaries connect grid cells and walls will be part of those grid cells.

Scope of Implementation

They have been suggested for many applications wherein a vast area of small and low-cost sensors are used to senses the environment, collect data, and transmit data to a sink hop-by-hop through sensor nodes themselves. Every node consumed more or less energy, which was normally provided by the limited energy sources of the node themselves concerning batteries, both the processing and transmission of information. Given that sensor deployment normally occurs for a long time once, the energy supply of the nodes is expected to last a long time. The energy of each node is important not only for the life of the network but the balance of the energy throughout the network is important for the whole region and finding a path to a sink.

Hierarchic algorithms split the nodes into sub-areas called clusters such that the control environment areas are isolated. A leader (Cluster Head) is selected from every cluster to permit contact between the clusters. Management (data aggregation, consultancy dispatch) and transmission of data obtained in the area under control are then the responsibility of the leaders. Hierarchical or cluster-based routing is typically composed of two-layer routing: one layer is concerned with the selection of the cluster head and the other with routing. The main purpose of the protocols is to save energy by engaging in multi-hop communication with the nodes.

Cluster Head Selection Algorithm

[1] *(C,R)* ← *Welzl's (n, SNS)*

[2] *CLH* ← *{Ø}*

[3] WSN Square grids Partition Gi side 2* Tr, with C as the Centre, as far as possible.

[4] Determine the grid center points cen_i for each G_i from a set of grid center points

[5] *GP* ← *{G$_0$ . . . G$_m$}*

[6] *fori = 0tom* where m is the number of grids

[7] *begin*

[8] Identify *sj* closest to each G_i in welzl_circle

[9] *CLH* ← *CLHusj* //append *sj* to list of CHs

[10] *S* ← *S-sj* //remove *sj* from SNS

[11] *endfor*

[12] End

The output of this algorithm will be a set of cluster heads.

PROPOSED SYSTEM

This section presents the detailed steps for the construction of clusters within the near location of the mobile sink.

This method adopts the division of the sensing field into clusters of unequal sizes, dividing the entire sensing field into unequal regions. As the Cluster head (CH) is located at the center of these regions, the cluster members can communicate in a single hopping method. The CH is also responsible for keeping the track of mobile sink position and also reduce the overhead of the members by contributing to the shortest path. Cluster member's job is to establish intercommunication within the cluster as well as between other clusters.

For constructing a model which suits the algorithm, the network is assumed to have the following desired characteristics:

- All cluster members are static and are located over the entire field.
- They have the same energy level and bandwidth.
- For rerouting, the cluster heads are supplemented with information related to the present position of the sinks and their time of arrival.

The cluster forming process based on the division into equal sizes of the sensing field and unparalleled clusters can be used to obtain data from the cluster heads. Based upon the LEACH Protocol, only some of the nodes can be used as CH. Generally, a node at the center will be identified as CH and after a round of data collection, the CH is again chosen with high energy.

The route adjustment is done based upon the present Area of the mobile sink inured to have accordance with the network topology. Only a set of cluster heads are employed for the maintenance of new routes.

After each round of the data collection, the energy at each CH is calculated and measured concerning a threshold. If it falls below the threshold, a new CH is identified based upon the distance from the center of the previous CH. By frequent reflection of CH, the energy consumption is avoided.

Collection Point Selection Algorithm

[1] **Start:**

[2] **for**i=0 to$m$

[3] $CLi \leftarrow \{\varnothing\}$

[4] $S' \leftarrow S\text{-}CLH$ // remove cluster heads from S

[5] *for*j=0 to$m$

[6] **begin**

[7] *for*i= to$n$

[8] **begin**

[9] **if***dist* $(si,chj) \leq Tr$

[10] $CLj \leftarrow CLJUsi$ // add node si to the cluster clj

[11] **End for**

[12] $S' \leftarrow S'\text{-}clj$ // remove nodes joined in clj from S'

[13] **End for**

[14] $L \leftarrow S'$

[15] $CLP \leftarrow CLHUL$ //final set of collection points

[16] $CLP \leftarrow lin\text{-}kernighan\ (CLP)$ //determine shortest path

[17] **End**

The output of this algorithm will be the distance d between points P_1 and P_2.

As the k-mean algorithm attempts to divide the network graph into K clusters, the distance between the node in each cluster is minimized. K number of nodes is randomly selected first as the initial center nodes.

SIMULATION RESULTS

We considered an area of 1000 X 500 square meters in this paper, with at least 25 nodes, which are spread throughout the field. During the simulation phase, all ideal cases concerning sensors are assumed. The mobile sink is supposed to locate at the top-left corner of the two-dimensional plane (50,50). The sink starts its movement avoiding the obstacles and finally returns after data collection.

The simulations presented in Fig. (2) in two different columns represent the networks with and without application of the proposed algorithms. The figures in the first column represent the network and its broadcasting phenomenon, nodes deployment, and the data movement. In comparison, the simulation in column 2 indicates the network with the data movement between the cluster members and the mobile sink by application of proposed algorithms on the network.

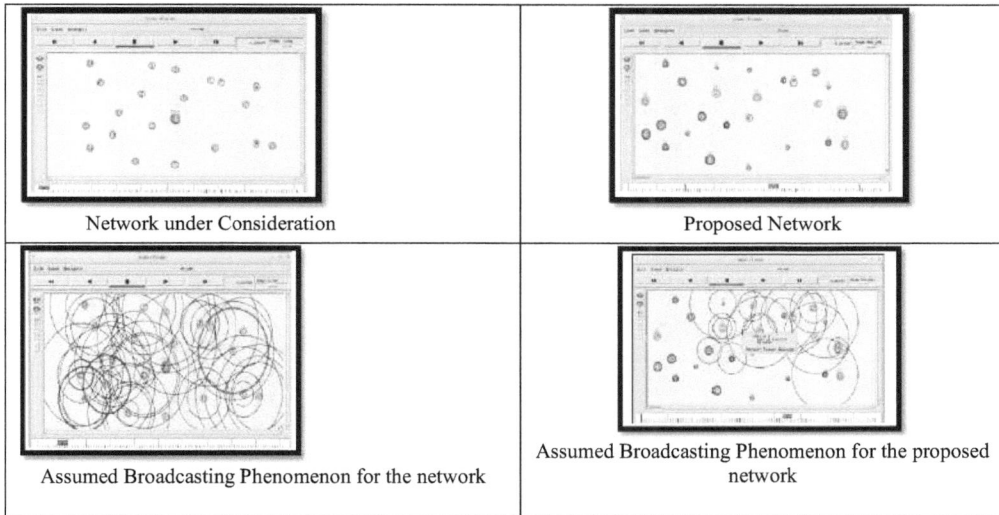

Network under Consideration	Proposed Network
Assumed Broadcasting Phenomenon for the network	Assumed Broadcasting Phenomenon for the proposed network

Fig. 2 cont.....

Nodes deployment in the Network	Data Movement from the Cluster Point 10 to the Sink Node in the Network
Data Movement from the Cluster Head to the Sink Node in the Network	Data Movement from the Cluster point 13 to the Sink Node in the Network

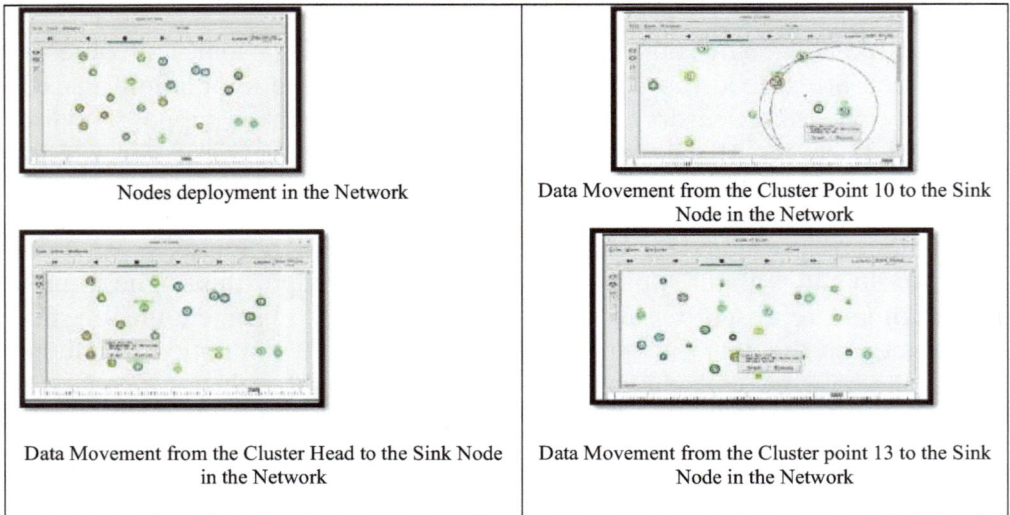

Fig. (2). Comparison of simulation results for the Network under consideration and the Proposed network.

In Fig. (**3**), various graphs related to different performance parameters such as delay, throughput, energy utilization, and overall network performance are presented in comparison concerning LEACH protocol, Heuristic Tour planning method, and Cluster-based method.

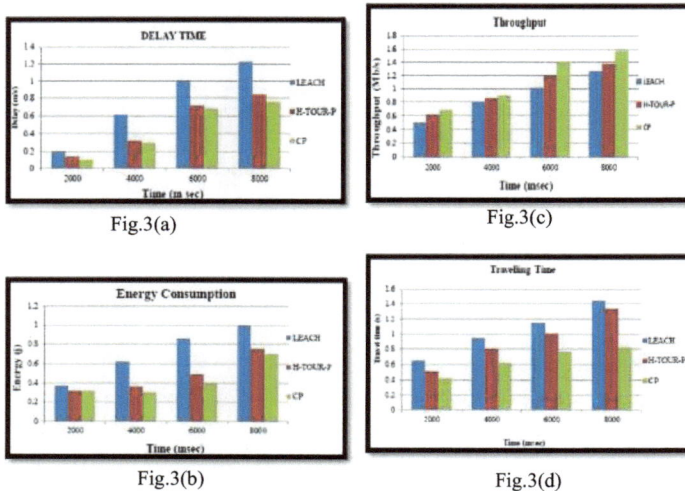

Fig.3(a)

Fig.3(c)

Fig.3(b)

Fig.3(d)

Fig. (3). a) Performance chart w.r.to Delay b) Energy level Routing c) Performance of the network d) Travelling time in routing.

CONCLUSION

Routing in sensor networks becoming a novel approach in reducing data traveling time with feasible and easier algorithm implementation did in the present work. The correlation between cluster head algorithm and collected point algorithm variation is observed in the results. This work presented an important obstacle to prevent a mobile sink with a minimum route algorithm. The grid-based approach was used to divide the regions and an advanced network is designed to find the minimum path. A heuristic obstacle avoiding algorithm was considered based on the Clustered based approach. Experimental results support that the proposed method is feasible for dispatching the mobile sink avoiding obstacles in the shortest path and how it supports extending the network lifetime. In WSNs, most of the data routing methods proposed are intended to achieve a certain objective according to the relevant application. These objectives can be classified into three main categories: energy efficiency, speedy distribution, and tolerance of fault. As every approach should take into account the energy constraint nature of the WSNs, most routing approaches focus primarily on one of the three objectives listed. In other words, a request in which one of those objectives is the main requirement was presented.

CONSENT FOR PUBLICATION

Not applicable.

CONFLICTS OF INTEREST

The authors declare no conflict of interest, financial or otherwise.

ACKNOWLEDGEMENT

Declared none.

REFERENCES

[1]   J. C. Cuevas-Martinez, J. Canada-Bago, J. A. Fernandez-Prieto, and M. A. Gadeo-Martos, "Knowledge-based duty cycle estimation in wireless sensor networks: Application for sound pressure monitoring", *Appl. Soft Comput,* vol. 13, pp. 967-980, 2013. [http://dx.doi.org/10.1016/j.asoc.2012.10.005]

[2]   H.-L. Fu, H.-C. Chen, and P. Lin, "Aps: Distributed air pollution sensing system on wireless sensor and robot networks", *Comput. Commun.,* vol. 35, pp. 1141-1150, 2012. [http://dx.doi.org/10.1016/j.comcom.2011.08.004]

[3]   G. Han, "Cross-layer optimized routing in wireless sensor networks with duty-cycle and energy harvesting", In: *Wireless Commun. Mobile Comput* vol. 15. , 2015, pp. 1957-1981. [http://dx.doi.org/10.1002/wcm.2468]

[4]   W. B. Heinzelman, A. P. Chandrakasan, and H. Balakrishnan, "An application-specific protocol architecture for wireless microsensor networks", *IEEE Trans. Wireless Commun,* vol. 1, pp. 660-670,

2002.
[http://dx.doi.org/10.1109/TWC.2002.804190]

[5]   C. Ai, L. Guo, Z. Cai, and Y. Li, "Processing area queries in wireless sensor networks", *In Proc. 5th Int. Conf. Mobile Ad-Hoc Sensor Netw.,,* 2009,pp. 1-8
[http://dx.doi.org/10.1109/MSN.2009.72]

[6]   I.-H. Peng, and Y.-W. Chen, "Energy consumption bounds analysis and it applications for grid based wireless sensor networks", *J. Netw. Comput. Appl,* vol. 36, pp. 444-451, 2013.
[http://dx.doi.org/10.1016/j.jnca.2012.04.014]

[7]   K. Xu, G. Takahara, and H. Hassanein, "On the robustness of grid-based deployment in wireless sensor networks", *Proc. Int. Conf. Wireless Commun. Mobile Comput.,* 2006,p. 1183
[http://dx.doi.org/10.1145/1143549.1143786]

[8]   Z. Zhou, J. Tang, L.-J. Zhang, K. Ning, and Q. Wang, "EGF-tree: An energy-efficient index tree for facilitating multi-region query aggregation in the Internet of Things", *Pers. Ubiquitous Comput,* vol. 18, pp. 951-66, 2014.
[http://dx.doi.org/10.1007/s00779-013-0710-y]

[9]   G. Han, A. Qian, J. Jiang, N. Sun, and L. Liu, "A grid-based joint routing and charging algorithm for industrial wireless rechargeable sensor networks", *Comput. Netw., in press.*
[http://dx.doi.org/10.1016/j.comnet.2015.12.014]

CHAPTER 34

Covid-19 X-Ray Image Classification Using Deep Learning Models

Atul Sharma[1,*] and **Gurbakash Phonsa**[1]

[1] *School of Computer Science And Engineering, Lovely Professional University, Phagwara, Punjab, India*

Abstract: Coronavirus (COVID-19) disease is spreading rapidly and is becoming increasingly common every day. This study can be helpful in the basic steps for doctors to diagnose the diseases and treat the patient accordingly. We used 1200 images of COVID-19,1300 Normal images and 1345 Viral Pneumonia images for the research work. We compared three popular deep learning methods which were CNN, VGG19, and Resnet-18 Model. We found that the Resnet-18 model with different useful parameters outperforms all the other proposed models. The accuracy we got was 97% when 50 images from each category were being tested.

Keywords: CNN, CORONAVIRUS, COVID-19 Radiography Dataset, Normal, Resnet-18, VGG19, Viral Pneumonia.

INTRODUCTION

In the field of research, image classification has played a vital role in the case of medical images classification as it reduces the efforts made by doctors to detect the disease. Due to high computation power, we can achieve many tasks in a few minutes. Researchers are always working on different algorithms and models for getting more accurate results and to make the work easier [1].

Image classification generally has two basic steps which are classification and labeling. If an image belongs to class A, it must be classified as a class A image and it must be labeled as a class A image by the model. Whenever a researcher is working on a particular algorithm, the output must be better and efficient to make the process easy [2].

There are different deep learning models which can be used to solve various image classification problems. Each model has different parameters, and based on

---

[*] **Corresponding author Atul Sharma:** School of Computer Science And Engineering, Lovely Professional University, Phagwara, Punjab, India; E-mail: atul24437@gmail.com

Dharam Buddhi, Rajesh Singh and Anita Gehlot (Eds.)

different steps one can find the actual accuracy after performing the training and testing part. Researchers are working on hybrid models so that more accurate results can be generated in a shorter span [3].

The dataset contains viral pneumonia as some symptoms of it matched with the COVID-19 symptoms [4]. The three categories are shown in Fig. (1) .

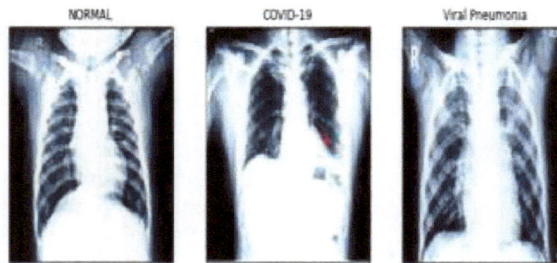

Fig. (1). The three classes used from the COVID-19 radiography database.

In the current conditions, the adverse effects of COVID-19 can be very close to viral pneumonia, as medical clinics are crowded, and not stopped to treat patients. As a result, COVID-19 would misrepresent non-COVID Viral Pneumonia, thereby postponing a cure procedure [5].

RELATED WORK

Asif *et al*. [5] used the deep learning model Inception V3 to classify the images and proposed a model that outperformed all the stated models in the paper, the results were promising and the dataset used consists of more than 3000 images.Toraman *et al*. [6] used the Capsnet model in their research work and compared the model with terms of accuracy for binary as well as multiclass classification. Das *et al*. [7] used the architecture of the extreme Inception model in their research work and did a literature review of different papers. Shaban *et al*. [8] proposed a methodology in their research work that was based on fuzzy inference engine and Deep Neural Network. They proposed a hybrid model to get more accurate results. Ly [9] compared various factors of the ANFIS (Adaptive Nuero-Fuzzy Inference System). Sahlol *et al*. [10] used the different CNN models for their research work. Singh *et al*. [11] proposed a new CNN approach that was being compared with the state of art models and it also outperformed them by various accuracy parameters. Khan *et al*. [12] proposed a CTNet model and compared it with 9 deep learning models in their research. . Apostolopoulos and Mpesiana [13] proposed a CNN model with transfer learning and compared it with other state-of-the-art deep learning models.Yadav [14] proposed SVR based model and stated the number of cases along with the number of deaths.

DATASET USED

The dataset was rendered in a Kaggle repository. It contains three classes of x-rays that have been stored: COVID-19, pneumonia, and normal [4, 5]. The comparative analysis of papers showed that CNN types were widely used in the image classification process as shown in Table **1** .

Table 1. Comparative Analysis of Different Papers [5 - 14].

Paper No.	Deep Learning Models Used, Accuracy%
[5]	Inception V3, Classification Accuracy more than 98%
[6]	Capsnet Model, More than 90% in binary and More than 80% in multiclass classification.
[7]	Inception V3, More than 95% Accuracy
[8]	Fuzzy Inference engine and Deep Neural Network (DNN), More than 95% Accuracy
[9]	ANFIS 1 to 12, Training Error, Validation Error, and Testing Error was calculated
[10]	CNN types, More than 98% Accuracy
[11]	CNN,F-measure:1.9789%
[12]	CTNet,Accuracy:99%
[13]	Transfer Learning with CNN,Accuracy:96%
[14]	Method of SVR, Compared the correlation

RESULTS AND DISCUSSION

We took all the images and labeled them to the category to which they belong for comparison purposes with our prediction as shown in Fig. (**2**). We compared three popular models which are Resnet-18, CNN, and VGG19. We used 50 images from each category for the testing part. Batch size was 10 [15 - 17]. We used 20 epochs and trained and tested all three models and plotted the accuracy using graphs. In the case of the Resnet-18 model, steps were considered [18, 19].

We took 20 epochs for the training accuracy and the models we used were Resnet18, CNN, and VGG19. When we compared these three best models we got the results as shown in Fig. (**3**) [19]. CNN is a basic model that has approximately 2Lakh parameters, whereas VGG19 has 144 Million and Resnet-18 has 11 Million Parameters [19]. We compared the models and predicted the images and labeled them according to the class they belong to as shown in Fig. (**4**) .

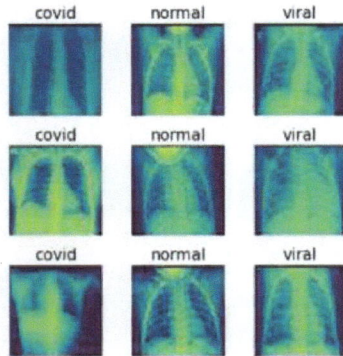

Fig. (2). Different images with their image labels [2].

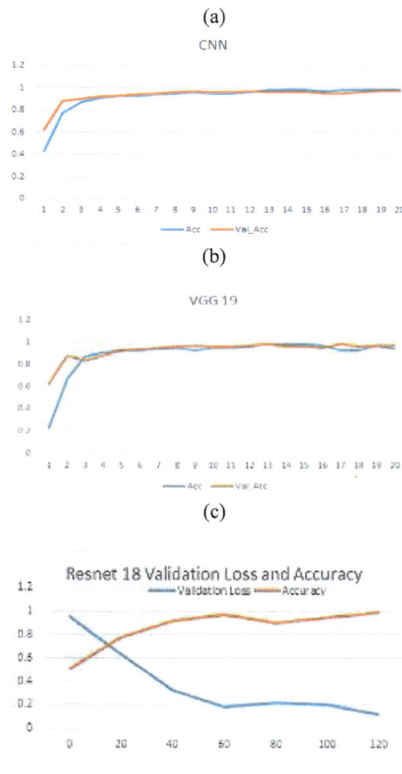

Fig. (3). Accuracy graph of **(a)** CNN Model **(b)** VGG-19 Model **(c)** Resnet18 Model.

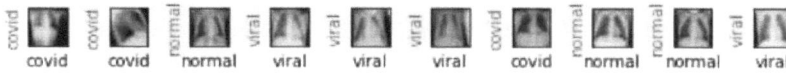

Fig. (4). 10 Images with image label and predicted class.

Table **2** shows the accuracy after testing 50 images from each category and their overall accuracy predicted by the models.

Table 2. Models used with their accuracy.

Models Used	Covid Accuracy	Normal Accuracy	Viral Pneumonia Accuracy	Overall Accuracy
Resnet-18	96%	97%	100%	97%
VGG19	90%	88%	100%	93%
CNN	88%	85%	100%	91%

CONCLUDING REMARKS

The performance of the models depends on what parameters are being used. Different deep learning models are used to classify the medical images efficiently and accurately. When tested, Resnet-18 performed well as compared to the other models when we took 50 images from each category. 145 out of 150 images were predicted and labeled correctly. It took around 12 hours to train and test all the models.

CONSENT FOR PUBLICATION

Not applicable.

CONFLICT OF INTEREST

The authors declare no conflict of interest, financial or otherwise.

ACKNOWLEDGEMENT

Declare none.

REFERENCES

[1]   https://towardsdatascience.com/10-papers-you-should-read-to-understand-image-classificat-on-in-the-deep-learning-era-4b9d792f45a7

[2]   H. Lee, and H. Kwon, "Going deeper with contextual CNN for hyperspectral image classification", *IEEE Trans. Image Process.,* vol. 26, no. 10, pp. 4843-4855, 2017.
[http://dx.doi.org/10.1109/TIP.2017.2725580] [PMID: 28708555]

[3]   S.Y. Wang, O. Wang, R. Zhang, A. Owens, and A.A. Efros, "CNN-generated images are surprisingly easy to spot... for now", *Proceedings of the IEEE/CVF Conference on Computer Vision and Pattern Recognition,* 2020, pp. 8695-8704
[http://dx.doi.org/10.1109/CVPR42600.2020.00872]

[4]   T. Rahman, "covid19-radiography-database,2021.Available at: ", https://www.kaggle.com/ tawsi furrahman/covid19-radiography-database (Accessed: 1 October 2020).

[5]   S. Asif, Y. Wenhui, H. Jin, Y. Tao, and S Jinhai, "Classification of COVID-19 from Chest X-ray images using Deep Convolutional Neural Networks", *medRxiv,* .
[http://dx.doi.org/10.1101/2020.05.01.20088211]

[6]   S. Toraman, T.B. Alakus, and I. Turkoglu, "Convolutional capsnet: A novel artificial neural network approach to detect COVID-19 disease from X-ray images using capsule networks", *Chaos Solitons Fractals,* vol. 140, 2020.
[http://dx.doi.org/10.1016/j.chaos.2020.110122] [PMID: 32834634]

[7]   N. Narayan Das, N. Kumar, M. Kaur, V. Kumar, and D. Singh, "Automated Deep Transfer Learning-Based Approach for Detection of COVID-19 Infection in Chest X-rays", *IRBM,* vol. 1, pp. 1-6, 2020.
[http://dx.doi.org/10.1016/j.irbm.2020.07.001] [PMID: 32837679]

[8]   W.M. Shaban, A.H. Rabie, A.I. Saleh, and M.A. Abo-Elsoud, "Detecting COVID-19 patients based on fuzzy inference engine and Deep Neural Network", In: *Network,"* *Appl. Soft Comput.* vol. 99. , 2021, p. 106906.
[http://dx.doi.org/10.1016/j.asoc.2020.106906]

[9]   K.T. Ly, "A COVID-19 forecasting system using adaptive neuro-fuzzy inference", *Finance Res. Lett.,* vol. 41, no. November, p. 101844, 2021.
[http://dx.doi.org/10.1016/j.frl.2020.101844] [PMID: 34131413]

[10]   A.T. Sahlol, D. Yousri, A.A. Ewees, M.A.A. Al-Qaness, R. Damasevicius, and M.A. Elaziz, "COVID-19 image classification using deep features and fractional-order marine predators algorithm", *Sci. Rep.,* vol. 10, no. 1, p. 15364, 2020.
[http://dx.doi.org/10.1038/s41598-020-71294-2] [PMID: 32958781]

[11]   D. Singh, V. Kumar, Vaishali , and M. Kaur, "Classification of COVID-19 patients from chest CT images using multi-objective differential evolution–based convolutional neural networks", *Eur. J. Clin. Microbiol. Infect. Dis.,,* pp. 1-11, .

[12]   S. H. Khan, A. Sohail, A. Khan, and Y. S. Lee, "Classification and region analysis of COVID-19 infection using lung CT images and deep convolutional neural networks", *arXiv, no. Mcc,,* 2020.

[13]   I.D. Apostolopoulos, and T.A. Mpesiana, "Covid-19: automatic detection from X-ray images utilizing transfer learning with convolutional neural networks", *Phys. Eng. Sci. Med.,* vol. 43, no. 2, pp. 635-640, 2020.
[http://dx.doi.org/10.1007/s13246-020-00865-4] [PMID: 32524445]

[14]   M. Yadav, M. Perumal, and M. Srinivas, "Analysis on novel coronavirus (COVID-19) using machine learning methods", *Chaos Solitons Fractals,* vol. 139, p. 110050, 2020.
[http://dx.doi.org/10.1016/j.chaos.2020.110050] [PMID: 32834604]

[15]   M.F. Aslan, M.F. Unlersen, K. Sabanci, and A. Durdu, "CNN-based transfer learning-BiLSTM network: A novel approach for COVID-19 infection detection", *Appl. Soft Comput.,* vol. 98, p. 106912, 2021.
[http://dx.doi.org/10.1016/j.asoc.2020.106912] [PMID: 33230395]

[16]   J. Fang, Y. Sun, Q. Zhang, Y. Li, W. Liu, and X. Wang, "Densely connected search space for more flexible neural architecture search", *Proceedings of the IEEE/CVF Conference on Computer Vision and Pattern Recognition,* 2020,pp. 10628-10637
[http://dx.doi.org/10.1109/CVPR42600.2020.01064]

[17]   M. Esmaeilpour, P. Cardinal, and A.L Koerich, "From sound representation to model robustness", *."*

arXiv preprint arXiv:2007, p. 13703.

[18]   S. Alinsaif, and J. Lang, "Histological image classification using deep features and transfer learning", *17th Conference on Computer and Robot Vision (CRV). IEEE,* 2020 [http://dx.doi.org/10.1109/CRV50864.2020.00022]

[19]   S. Ayyachamy, *"Medical image retrieval using Resnet-18." Medical Imaging 2019: Imaging Informatics for Healthcare, Research, and Applications.* vol. 10954. International Society for Optics and Photonics, 2019.

CHAPTER 35

Design and Analysis of Triangular MIMO Antenna with a Truncated Edge for IoT Devices

Narbada Prasad Gupta[1,*], **Parulpreet Singh**[1,*], **Neelesh Gupta**[2] and **Kapil Kumar**[3]

[1] *SEEE, Lovely Professional University, Punjab, India*

[2] *AKGEC, Ghaziabad , India*

[3] *BIT, Meerut, India*

Abstract: The Internet of Things (IoT) is a developing system of articles, gadgets, and types of machinery each ready to interconnect with the other utilizing a remote system to get to the Internet. These frameworks permit users to achieve further mechanization, examination, and integration inside a structure. IoT gadgets have an adaptable scope of both wired and remote availability choices. IoT conventions generally use Industrial Scientific & Medical (ISM) band 915 MHz, 2.4 GHz (Zigbee), 5GHz. Right now, a short review of IoT highlights has been given in the paper. Likewise, a triangular MIMO reception apparatus (antenna) with a truncated edge has been intended to work at the frequency of 2.4 GHz (Zigbee). The proposed reception apparatus has been structured utilizing FR4 material having a dielectric constant of 4.4 and loss tangent 0.002. The thickness of the substrate is taken 1 mm. The proposed reception apparatus is reverberating at 2.4 GHz, which is reasonable to be utilized at Zigbee. The simulated Gain of the proposed antenna is 1.18 dB, its radiation efficiency is approximately 30%. All the simulated parameters such as multiplexing efficiency (ME), envelop correlation coefficient (ECC), and mean effective gain (MEG) is fulfilling the requirement to be used as MIMO antenna for IoT applications.

Keywords: IoT, Microstrip Antenna, MIMO Antenna, Truncated Antenna, Zigbee.

INTRODUCTION

An IoT device is a remote associating device to the established network to transmit information. As the adaptable increment in the IoT devices in the current world, because of this effect, such a large number of organizations had recognized this IoT innovation and executing the miniaturized and integrated communication

---

* **Corresponding authors Narbada Prasad Gupta and Parulpreet Singh:** Lovely Professional University, Punjab, India; E-mails: narbada.24806@lpu.co.in, parulpreet.23367@lpu.co.in

Dharam Buddhi, Rajesh Singh and Anita Gehlot (Eds.)

systems and modules. They improve the range of areas and their accuracy. IoT gadgets are a piece of a situation wherein each gadget bury relate with each other gadget in a domain to the programmed world and convey an ever-increasing number of usable information to clients, specific frameworks for the information projections, and so forth [1].

IoT Technology

IoT predominantly exploits standard agreements and system administration advances. Notwithstanding, the major authorizing developments and protocols of IoT are near-field communication (NFC), low power Bluetooth, RFID, low-power wireless, LTE-A, and WiFi-Direct. The double antenna design framework is very much followed in the current innovation because the use of double antenna builds the exhibition of the framework and the transmission of information is increasingly solid [2]. The MIMO innovation is successful in the multipath moderating circumstances and afterward it is executing in the ongoing advancements, for example, various modules. This paper proposes a triangular MIMO antenna with a truncated edge, which involves two scaled-down triangular antennas that work under the ISM band for the IoT devices with the resonance frequency of 2.4 GHz as well as for the Bluetooth, Wi-Fi, and WLAN. It utilizes for the system devices, for example, WLAN and Bluetooth conventions for the information transmission [3 - 4].

IoT-Key Features

Artificial Intelligence, sensors, network, and small device usage are the absolute most significant highlights of IoT. It makes essentially anything "keen", which means it advancements each portion of presence with strength of information collection, AI algorithms, and networks.

IoT-Pros & Cons

The advantages of IoT range over each section of the way of life and occupational, it incorporates Technology Optimization Improved Customer Assignation, Concentrated Waste, Enhanced Data Collection and so forth. A portion of the key difficulties incorporates Security, Privacy, Complexity, Flexibility, and Compliance.

Hardware Utilization

The apparatus used in IoT structures integrates devices for a distant dashboard, sensors, devices for control, and servers. These devices comprise strength elements, sensing modules, and RF modules. IoT may incorporate Accelerometer

Temperature Sensor, Auditory Sensor, Gas RFID sensor Pressure Sensor, Humidity Sensor, Microflow sensor, Proximity Sensor, *etc.*

DESIGN OF MIMO ANTENNA FOR IOT DEVICES

The proposed antenna has been designed to work for Zigbee so it should resonate at 2.4 GHz. This frequency has been considered as the design frequency. Also, it has been reported earlier that an inset fed formation is the common method to advance the impedance band of microstrip patch antennas (MPA) because of its simplicity. Therefore to feed the proposed antenna inset feed is used. According to the cavity model, the resonance input impedance of an MPA is given by the following expression:

$$R_{in} = R_0 Cos^2 \left(2\pi \frac{d}{\lambda}\right) \tag{1}$$

Where R_0 is the input resistance when the patch is fed at the edge, d is the feed distance (or inset depth) from the edge, and λ is the guided wavelength at the resonance frequency. For a triangular MPA, the resonance frequency for any TM_{mn} mode is given by [4]:

$$f_r = \frac{2c}{3a\sqrt{\varepsilon_r}}\sqrt{m^2 + mn + n^2} \tag{2}$$

Where m and n are several modes, c is the speed of light, a is the side length of triangle, ε_r is the Dielectric constant of the substrate. The calculation of the patch can be made using the effective length, a_e of the patch as in the equation below:

$$f_{10} = \frac{2c}{3a_e\sqrt{\varepsilon_r}} \tag{3}$$

a_e is the effective side length of the triangle and it is given by:

$$a_e = a\left[1 + 2.199\frac{h}{a} - 12.853\frac{h}{a\sqrt{\varepsilon_r}} + 6.182\left(\frac{h}{a}\right)^2 - 9.802\frac{1}{\sqrt{\varepsilon_r}}\left(\frac{h}{a}\right)^2\right] \tag{4}$$

The technique of changing feed depth can be used efficiently to properly match antenna using microstrip line feed, whose characteristic impedance is given by [4]:

$$Z_c = \begin{cases} \dfrac{60}{\sqrt{e_{reff}}} ln\left[\dfrac{8h}{W_0} + \dfrac{W_0}{4h}\right] & \dfrac{W_0}{h} \leq 1 \\[2em] \dfrac{120\pi}{\sqrt{e_{reff}}\left[\dfrac{W_0}{4h} + 1.393 + 0.667 ln\left(\dfrac{W_0}{h} + 1.444\right)\right]} & \dfrac{W_0}{h} > 1 \end{cases} \qquad (5)$$

Where W_0 is the width of the microstrip line.

Antenna Configuration

Fig. (1) shows the configuration of the proposed triangular MIMO antenna with a truncated edge for IoT devices. Two antennas designed separately have been connected to form a MIMO antenna system. It is ensured to get good isolation between the antennas. There are various methods to improve isolation. In [5], asymmetric coplanar strip fed with an I-shaped slot in the radiator and by attaching a rectangular patch on the back, while in [5], a mushroom type electromagnetic bandgap structure is used between two antennas to increase the isolation. In this paper, isolation between the antennas has been achieved by putting them distant apart. Table 1 describes the design parameters of the proposed antenna. The antenna has been designed with a substrate size of $90\times90\times1.6$ mm^3

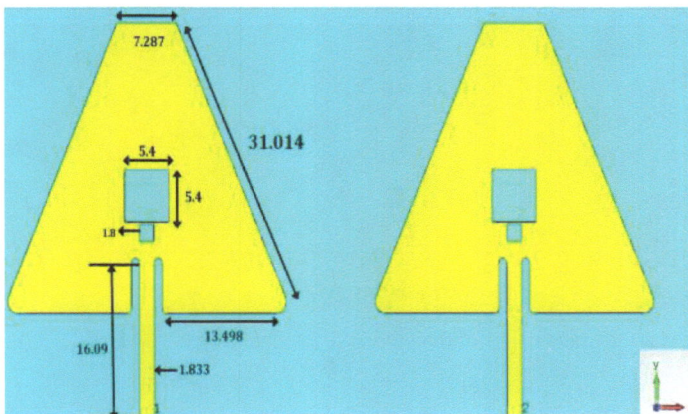

Fig. (1). MIMO Antenna configuration.

Table 1. Dimension of the antenna.

Parameter	Value (mm)	Parameter	Value (mm)
Side of the triangle	36.5 mm	Big Slot side	5.4 × 5.4 mm^2
Truncation radius	2 mm	Small slot side	1.8 × 1.8 mm^2

Results and Discussion

A triangular antenna has been designed initially using the formula given in the literature above and then its edges are truncated to get the optimized results. Fig. (**2**) shows the parametric study of the side of the triangular antenna. It has been shown here that as the side length is increasing the resonance frequency decreases. So, by a parametric study conducted in various iterations. The length of the side has been made varying from 33.13 mm and 36.57 mm. The optimized S-parameters of the designed antenna have been shown in Fig. (**3**) separately.

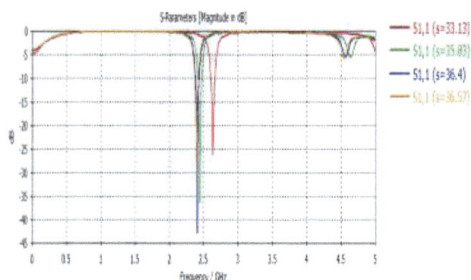

Fig. (2). Parametric study of the side length of the proposed antenna.

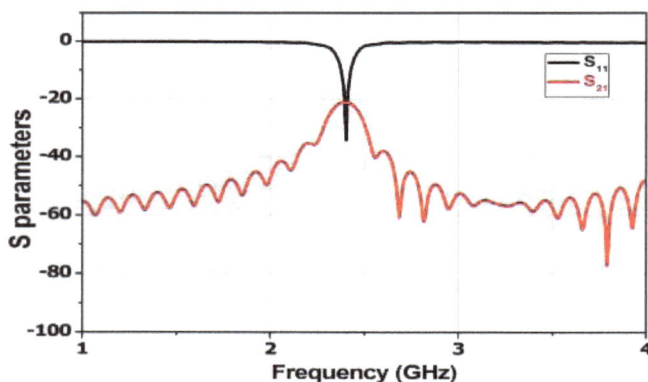

Fig. (3). Optimized S_{11} and S_{21} characteristics of the proposed antenna.

The MIMO behavior of the designed Antenna is assessed by considering the envelope correlation coefficient (ECC), multiplexing efficiency (ME), and mean effective gain (MEG). Fig. (**3**) also shows the optimized isolation between the two antennas. The two monopole elements have been placed orthogonally and symmetrically on the substrate for good isolation between the two input ports. It is observed to be well below -20 dB it is more than -23 dB at the resonant frequency, which conforms to the requirement for the MIMO antenna system. The ECC can be calculated using S-parameters [6]:

$$ECC = \frac{|S_{11}^* S_{12} + S_{21}^* S_{22}|^2}{(1 - |S_{11}|^2 - |S_{21}|^2)(1 - |S_{22}|^2 - |S_{12}|^2)} \tag{6}$$

The ECC should preferably be zero for an uncorrelated diversity Antenna but its practical limit is <0.5. Fig. (**4**) shows the envelop correlation coefficient (ECC) of the proposed antenna. Its simulated value at the resonance is 0.236. For uniform 3-D angular power spectrum and high signal-to-noise ratio, multiplexing efficiency (ME) is preferred and it is given by [6]:

$$ME = \sqrt{(1 - |\rho_c|^2)\eta_1\eta_2} \tag{7}$$

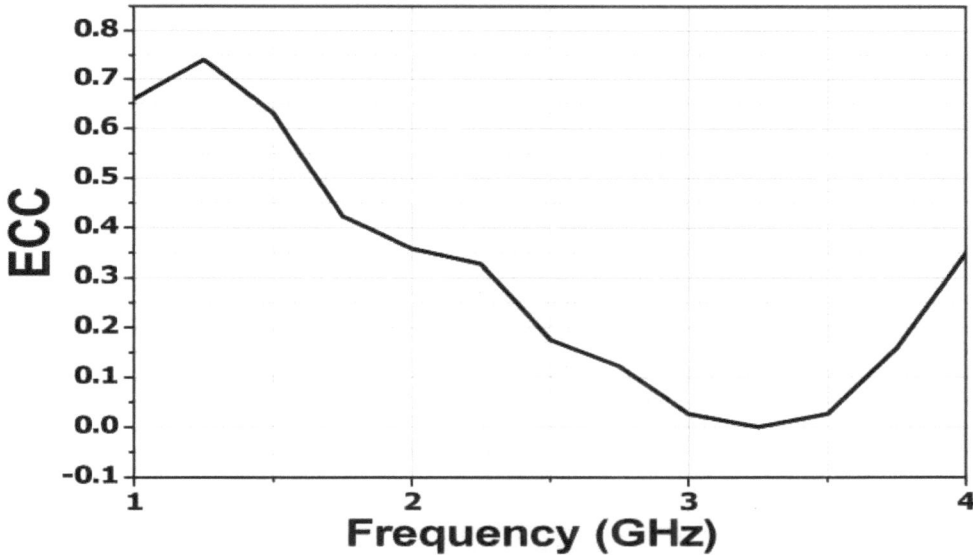

Fig. (4). ECC of the proposed antenna.

Where, ρ_c is the complex correlation coefficient between the two elements ECC is approximately equal to ρ_c^2 and η_i is the total efficiency of the ith antenna element.

Fig. (**5**) shows the multiplexing efficiency (ME) of the proposed Antenna. The simulated value of ME is -11.9 dB, which is within the limits. Fig. (**6**) depicts the radiation efficiency of the proposed antenna. It has been observed that the efficiency is close to 60% at the designed frequency.

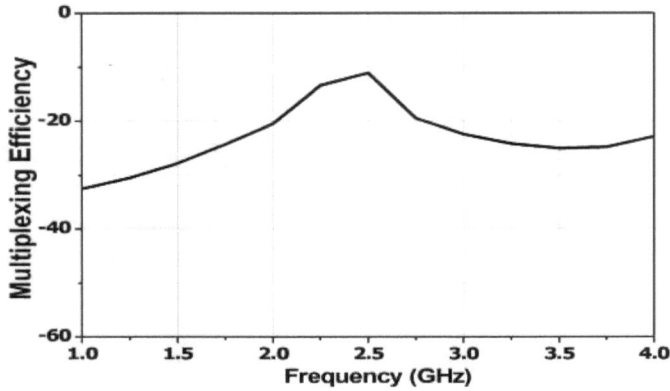

Fig. (5). Multiplexing efficiency of proposed antenna.

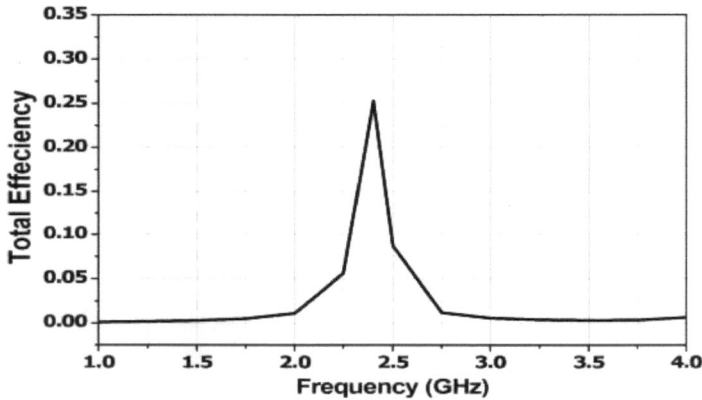

Fig. (6). Total efficiency of the proposed antenna.

The relative mean power levels between the signals delivered from each Antenna are measured using the MEG. It is one of the most significant parameters for the characterization of antennas in wireless channels. Figs. (**7** and **8**) show the MEG and the electric field of the proposed antenna respectively. It is a single parameter that describes the impact of the antenna on the link budget. Its simulated value for the proposed antenna is below -3 dB, which is expected. For a good diversity performance and channel characteristics, the ratio of MEG of the two Antenna elements should fulfill the criteria $|MEG_i/MEG_j| < \pm 3$ dB [6]. Here, i and j denote the Antenna elements 1 and 2, respectively.

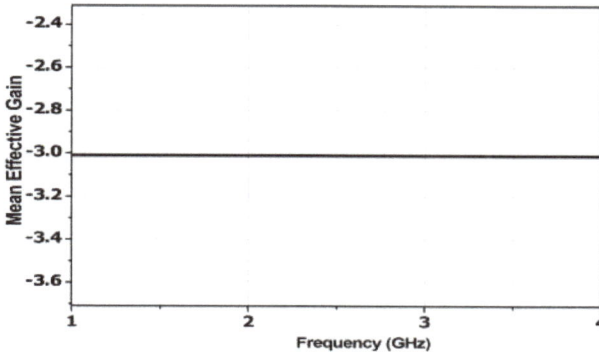

Fig. (7). Mean effective gain of the proposed antenna.

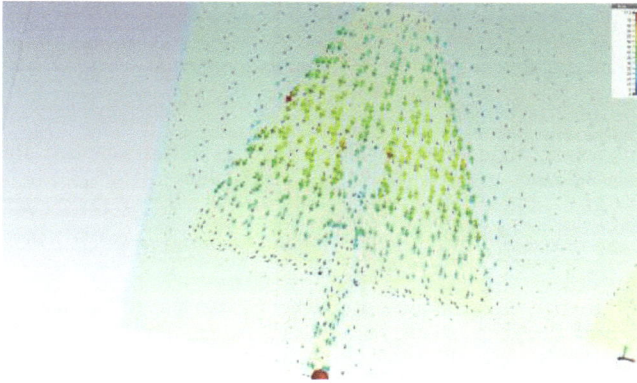

Fig. (8). Electric field of the proposed antenna.

CONCLUSION

In this paper, a brief description of IoT (Internet of Things) and its features are presented. A triangular MIMO reception apparatus (antenna) with a truncated edge has been designed to operate at a frequency of 2.4 GHz, which is a suitable frequency to be used for IoT devices. The proposed reception apparatus has been structured utilizing FR4 material having a dielectric constant of 4.4 and loss tangent 0.002. The thickness of the substrate has been chosen to be 1 mm. The proposed reception apparatus is reverberating at 2.4 GHz, which is reasonable to be utilized for IoT applications. The simulated gain of the antenna is 1.18 dB, its total efficiency is close to 60%. Transmission loss and reflection are fulfilling the demand for MIMO antenna design. It has been shown that the designed antenna is well suited to be used in IoT devices using Zigbee technology. All the simulated parameters such as multiplexing efficiency (ME), envelop correlation coefficient (ECC), and mean effective gain (MEG) is fulfilling the requirement to be used as MIMO antenna for IoT applications.

CONSENT FOR PUBLICATION

Not applicable.

CONFLICT OF INTEREST

The authors declare no conflict of interest, financial or otherwise.

ACKNOWLEDGEMENTS

The authors would like to acknowledge the encouragement provided from time to time from their respective workplaces *i.e.* SEEE, Lovely Professional University, Punjab, AKG Engg. College, Ghaziabad, and MIT, Meerut.

REFERENCES

[1]　S.R. Best, "Low Q electrically small linear and elliptical polarized spherical dipole antennas", *IEEE Trans. Antenn. Propag.,* vol. 53, no. 3, pp. 1047-1053, 2005.
[http://dx.doi.org/10.1109/TAP.2004.842600]

[2]　L. Lizzi, and F. Ferrero, "Use of ultra-narrow band miniature antennas for internet-of-things applications", *Electron. Lett.,* vol. 51, no. 24, pp. 1964-1966, 2015.
[http://dx.doi.org/10.1049/el.2015.3142]

[3]　*Prakash Bhartia, Inder Bahl, Apisak Ittipiboon, "Microstrip antenna design handbook.* Artech House, 2001.

[4]　N.P. Gupta, and M. Kumar, "Development of a Reconfigurable and Miniaturized CPW Antenna for Selective and Wideband Communication", *Wirel. Pers. Commun.,* vol. 95, pp. 2599-2608, 2017.
[http://dx.doi.org/10.1007/s11277-017-3942-8]

[5]　J-Y. Zhang, F. Zhang, W-P. Tian, and Y-L. Luo, "ACS-fed UWB-MIMO Antenna with shared radiator", *Electron. Lett.,* vol. 51, no. 17, pp. 1301-1302, 2015.
[http://dx.doi.org/10.1049/el.2015.1327]

[6]　R. Chandel, A.k. Gautam, and K. Rambabu, *Tapered fed compact UWB MIMO-diversity antenna with dual band-notched characteristics,* 2018.
[http://dx.doi.org/10.1109/TAP.2018.2803134]

CHAPTER 36

A Forest Fire Detection and Reporting System Using a Wireless Sensor Network

Mekapotula Bhuvan Sundhar Reddy[1,*], **Nelavelli Chandu**[1], **Yarra Raviteja**[1], **Rongsennungsang Jamir**[1] and **Koushik Barman**[1]

[1] *Lovely Professional University, Jalandhar, Punjab, India*

Abstract: Forest fires are a great threat to ecologically healthy grown forests and the protection of the environment. The root cause is due to the lack of a scalable network to monitor the physical conditions over vast forest areas. To overcome this, we used the NRF24L01+ module configured in mesh network mode interfaced with Arduino along with DHT22, MQ2 sensors, powered by Lithium-ion batteries which can be recharged with solar panel, this allows the node to reconnect to the network automatically in case of any damage to its parent node.

Keywords: Arduino, DHT22, MQ2, Mesh Network, NRF24L01+, Raspberry Pi, Wireless Network.

INTRODUCTION

Ever since the life of humans progresses, we have changed the land cover of the earth gradually. In Consequence, one of the most alerting issues today is the conservation of forests. Forest protection is a branch of forestry that is concerned with the preservation or improvement of a forest and prevention and control of damage to the forest by natural or man-made causes like forest fires, plant pests, and adverse climatic conditions (Global warming). Conservation of forests is the practice of planning and maintaining forested areas for the benefit and sustainability of future generations. Forest conservation involves the upkeep of the natural resources within a forest that is beneficial for both humans and the ecosystem. In these recent years, we have experienced many major incidents of forest fires in many parts of the world. Fig. (**1**) shows the number of forest fires that occurred worldwide in 13 years [1].

---

* **Corresponding author Mekapotula Bhuvan Sundhar Reddy:** Lovely Professional University, Jalandhar, Punjab, India; E-mail: bhuvanreddy54@gmail.com

Dharam Buddhi, Rajesh Singh and Anita Gehlot (Eds.)

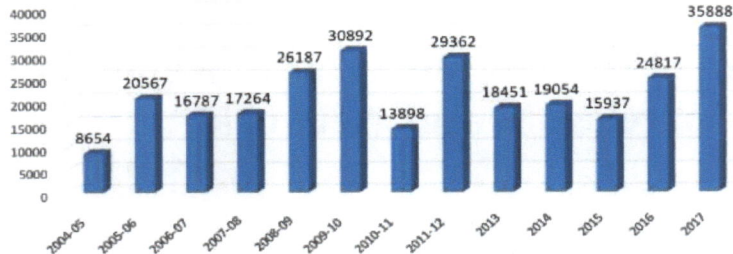

Fig. (1). Total forest fires in 13 years. Adapted from [1].

When a forest fire occurs, there are many disadvantages for both humankind and the ecosystem. The major disadvantages are air pollution, soil damage, animal habitats will be burnt which will cause unbalance in the ecosystem, *etc.*

In 2018, the annual rainfall measured in India was over 1020 millimeters. This was a decrease from 2017 where around 1127 millimeters of rainfall was recorded. The basic steps in preventing forest fires are reducing both the risk and hazard of the situation. The next question that always comes to mind is how to prevent or how to detect the fire in the forest. Because the fire is still at its origins, how do let the authorities know and reach the place on time. These days we can see a huge leap in the development of technologies from year to year but why can't we still solve this problem. One of the main reasons will be the connectivity issues that we face in the forests. The improper network is a major problem that delays the information reaching the authorities on time. Now the question is how can we face this major issue and fix the problem. The solution is to design a network that will help the transmitters to transmit signals or data out from the forests so that they can reach the authorities on time. Hence, we need devices that can send the data or the information out of the forest. we will be using NRF24L01+ as a trans-receiver in our project.

Related Work

Alexander A. Khamukhin and Silvano Bertoldo determined the type of forest fire in WSNs by analyzing the noise power spectrum of forest fires [2].

With the use of unmanned aerial vehicles (UAVs) with specialized cameras and Lora WAN sensor networks Georgi Hristov; Jordan Raychev and the rest of the authors worked on early fire detection [3].

Adnan; A. Ejah Umraeni Salam and the rest of the authors showed Forest Fire Detection using LoRa Wireless Mesh Topology [4].

Similarly, in [5,6], and [7] different methods are used but using nrf24l01+ saves cost as well as we can get more optimized location and higher data transfer rates.

System Architecture

Mesh Network

Mesh Network is a network topology where nodes connect directly and dynamically to other nodes and transmit information efficiently using different routing techniques [8]. Fig. (**2**) shows an example of a mesh network [8].

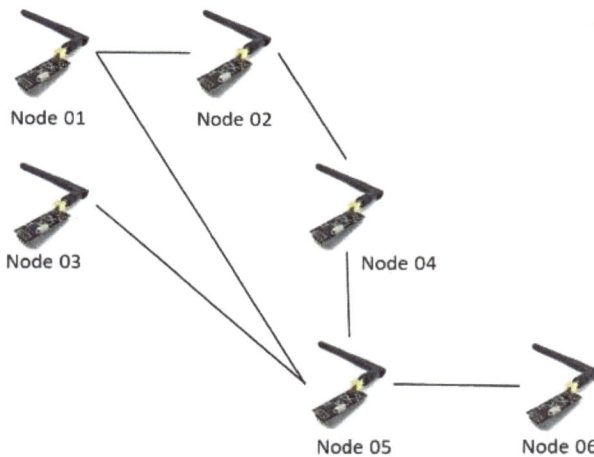

Fig. (2). Mesh network. Adapted from [8].

Sensor Node and Gateway Node

In this project, we are using Arduino Nano/Mini as a sensor node interfaced with DHT22 for temperature and humidity data and an MQ2 sensor for smoke data. The information is transmitted using low-cost NRF24l01+module. In Ref [9] other available methods are discussed. The location of each node is manually fixed in the database after placing the node in the forest so that there is no need for a GPS module which cuts down cost. Fig. (**3**) represents a sensor node with basic interfacing.

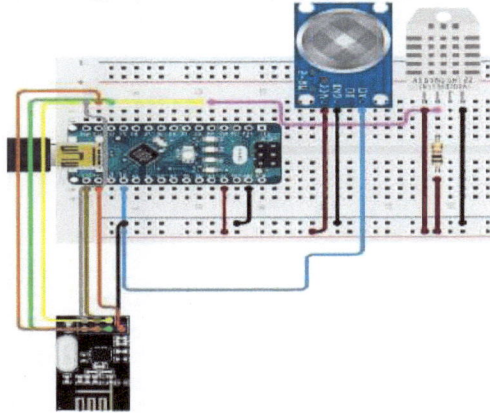

Fig. (3). Sensor Node with DHT22, MQ2, and NRF4l01.

The received information is filtered and stored in the MySQL database in form of a table for further analysis and representation. To access MySQL database from python code we need MySQL-python connector installed and MySQL server should be running. To access the data the gateway node and pc/laptop both should be on the same network. The web application takes the data from the database and represents it in form of graphs like battery health, time *vs* temperature. In case of fire, the flag in incoming data goes high which will raise an alert on the web application screen mentioning the node id and its location and its status. Fig. (**4**) represents the gateway node.

Fig. (4). Gateway Node with NRF24L01.

We are using Raspberry Pi as a gateway node. Raspberry Pi is interfaced with NRF24L01+ and receives information from sensor nodes in successive intervals. The Raspberry Pi is running Raspbian OS with python 3 environment and is accessed through SSH.

The radio network is set up and managed using the RF24 library containing different sub-modules for the standard network and Mesh network. The data of mq2 is read through A0 (analog pin 0) Arduino and dht22 data are read through digital pin 2.

The latency, payload delivery, and current consumed by NRF24L01+ observed under different configurations can be inferred from Ref [10]. The performance of NRF24L01+ can be inferred from Ref [11] where different parameters like throughput *vs* payload, mesh routing recovery time, and algorithms are discussed. However, nrf24l01 does not provide any means to calculate the signal strength received.

DESIGN

System Design

The system is designed such that even in case of failure of one node it should not affect other nodes which are connected through the failed node. We made sure of this by using a dynamic Mesh network library which ensures that in case of connection loss the node renews its address and reconnects to the mesh network. The Ref [12 - 14] are the optimized libraries RF24, RF24Mesh, and Network respectively, that we used. The sensor node monitors sensor data continuously but sends data after a certain time interval to save the battery. However, in case the fire is detected then data is sent immediately without any regard to time interval.

System Outline

The plot of the system is explained through a flowchart (Fig. **5**) where the temperature, humidity, and smoke particle concentration values will receive into the database and compared with predetermined threshold values. Those values are 45-70 ^{0}C temperature, 50% of humidity, and 2,368 ppm of smoke particle concentration. If the value reaches above threshold value then fire detection is confirmed and data is transmitted to the gateway node immediately and an alert is raised on the web application showing the location of the node and its status for rescue. If data does not meet threshold values then keep on reading the sensor data and transmit data in fixed time intervals to the gateway node.

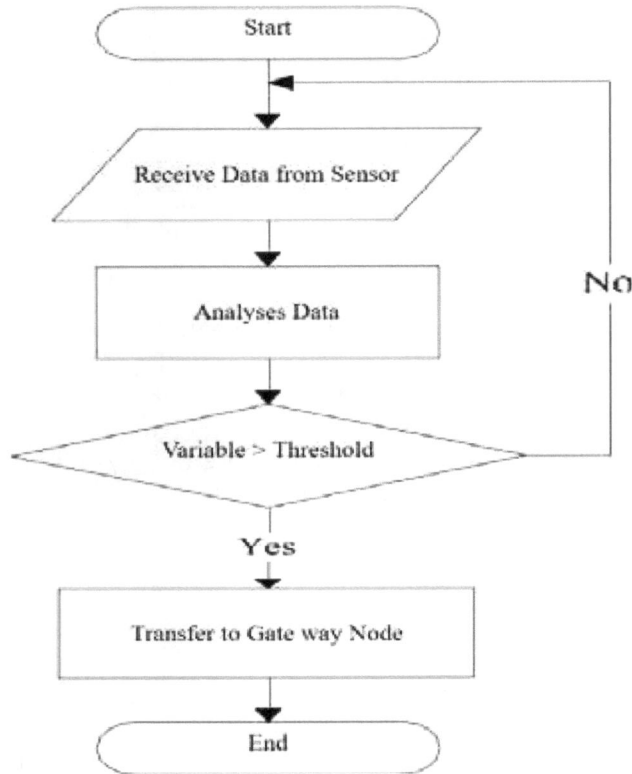

Fig. (5). Flowchart system outline.

Power

To power the nodes, we are using a 18650 Li-ion battery (in parallel configuration). The battery is rated at 3.7v nominal. We are using 2 or 3 batteries in parallel to increase the capacity. To charge the batteries we are using a 5v solar panel. The power from the solar panel is fed to a TP4056 module. TP4056 is connected to batteries that have reverse-voltage and discharge protection. Later the battery voltage (3.5-4.2) is boosted to 5v using a voltage booster which is then supplied to the microcontroller and sensors. When the battery voltage is less than 3.5v TP4056 charges the battery.

Panic Button Feature

One of the additional features we included is the Panic button. A button is attached to the device in case when anyone is in danger or lost in the forest, they can hold the panic button for 5sec and an alert will be sent to the rescue team, this is more beneficial if the forest area is prone to landslides or other natural calamities and people are lost in the forest.

EXPERIMENT AND RESULT

Table 1. Data received before the fire and moderate fire. Adapted from [4].

-	Data as per conducted experiment		
	Temperature	*Humidity*	*Smoke sensor Value*
Before fire	32°Celsius	60%	180-200
In Moderate fire	68°Celsius	50%	>250

We created a forest fire scenario in our locality on small scale and measured the temperature, humidity, and smoke data using one sensor node. Table **1** depicts the data before and after the test-fire [4]. The smoke data is a value directly proportional to the gas concentration in smoke. We can find the gas concentration in ppm by finding out Ro and Rs from the slope of values from the smoke sensor. Fig. (**6**) represents temp data Fig. (**7**) represents smoke data.

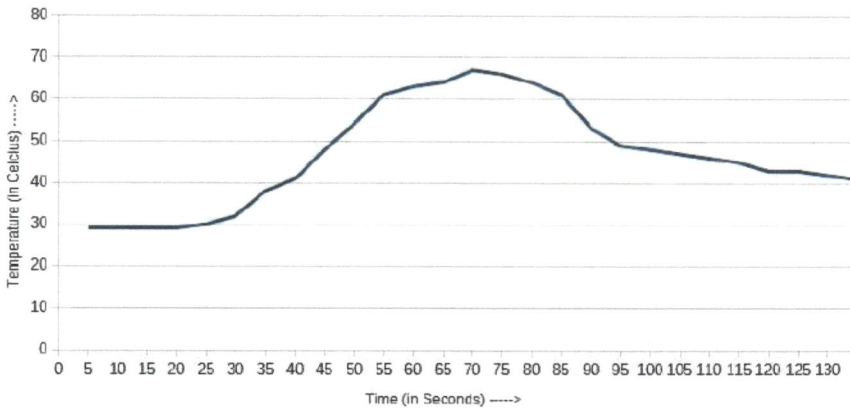

Fig. (6). Temperature data.

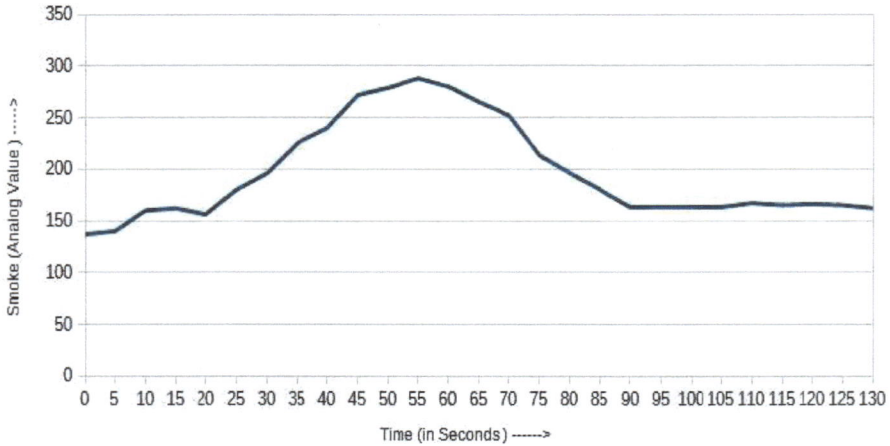

Fig. (7). Smoke data.

There is a 5sec time interval between successive data. The nrf24l01+ module is configured in channel 90, 2_MBPS data rate, and power_min mode to save battery. Data-rate can be set to minimum for long-distance connection or medium or medium-range connection. For optimal range, data-rate is set to 1Mbps and distance of 500m for power amplified version of nrf24l01. The height of the sensor to be placed depends on the type of plants growing and surroundings, Forest Fires can go up to 6 feet (1.8m) in height.

CONCLUSION

In this section, we conclude our research. The dense forest restricts the range of the nrf24l01+ module due to LOS(line of sight). The weather conditions can change at any time so the sensor nodes should be water-resistant. In case of an extended rainy or cloudy season, there will not be enough solar light to charge the batteries. In that case, the batteries should be replaced or we should use an extra 18650 battery in parallel to increase the capacity(ah). We can get extra information like a particular gas concentration in ppm from smoke data and its slope. The dataset collected can be further used in developing a machine learning model which can predict the occurrence of forest fire before it can even occur. In the future, we can add a camera or other sensing modules and can be used to track wild animals. This can also be used to measure forest environmental changes like season changes and weather shifts due to the effect of pollution and global warming.

CONSENT FOR PUBLICATION

Not applicable.

CONFLICT OF INTEREST

The authors declare no conflict of interest, financial or otherwise.

ACKNOWLEDGEMENT

Declared none.

REFERENCES

[1]   https://fsi.nic.in/uploads/documents/technical_information_series_vol1_no1.pdf

[2]   A.A. Khamukhin, and S. Bertoldo, "Spectral analysis of forest fire noise for early detection using wireless sensor networks", *2016 International Siberian Conference on Control and Communications (SIBCON)*, 2016, pp. 1-4 Moscow, Russia
[http://dx.doi.org/10.1109/SIBCON.2016.7491654]

[3]   G. Hristov, J. Raychev, D. Kinaneva, and P. Zahariev, "Emerging Methods for Early Detection of Forest Fires Using Unmanned Aerial Vehicles and Lorawan Sensor Networks", *2018 28th EAEEIE Annual Conference (EAEEIE), Hafnarfjordur, Iceland*, 2018pp. 1-9
[http://dx.doi.org/10.1109/EAEEIE.2018.8534245]

[4]   A.E.U. Adnan, "Salam, A. Arifin and M. Rizal, "Forest Fire Detection using LoRa Wireless Mesh Topology", *2018 2nd East Indonesia Conference on Computer and Information Technology (EIConCIT), Makassar, Indonesia,*, 2018, pp. 184-187

[5]   S. Wu, and L. Zhang, "Using Popular Object Detection Methods for Real Time Forest Fire Detection", *2018 11th International Symposium on Computational Intelligence and Design (ISCID), Hangzhou, China*, 2018pp. 280-284
[http://dx.doi.org/10.1109/ISCID.2018.00070]

[6]   H. Dang-Ngoc, and H. Nguyen-Trung, "Aerial Forest Fire Surveillance - Evaluation of Forest Fire Detection Model using Aerial Videos", *2019 International Conference on Advanced Technologies for Communications (ATC)*, 2019, pp. 142-148 Hanoi, Vietnam
[http://dx.doi.org/10.1109/ATC.2019.8924547]

[7]   T. Wang, J. Hu, T. Ma, and J. Song, "Forest fire detection system based on Fuzzy Kalman filter", *2020 International Conference on Urban Engineering and Management Science (ICUEMS)*, 2020, pp. 630-633 Zhuhai, China
[http://dx.doi.org/10.1109/ICUEMS50872.2020.00138]

[8]   I.F. Akyildiz, and X. Wang, "A survey on wireless mesh networks", *IEEE Commun. Mag.,* vol. 43, no. 9, pp. 23-30, 2005.
[http://dx.doi.org/10.1109/MCOM.2005.1509968]

[9]   D. Ibrahim, "Low-Power Early Forest Fire Detection and Warning System", *Indian J. Sci. Technol.,* vol. 13, no. 3, pp. 286-298, 2020.
[http://dx.doi.org/10.17485/ijst/2020/v13i03/142755]

[10]   A. Weder, "An energy model of the ultra-low-power transceiver nRF24L01 for wireless body sensor networks", *2nd Int. Conf. Comput. Intell. Commun. Syst. Networks, CICSyN 2010*, 2010, pp. 118-123
[http://dx.doi.org/10.1109/CICSyN.2010.24]

[11]   H. Saha, S. Mandal, S. Mitra, S. Banerjee, and U. Saha, "Comparative Performance Analysis between nRF24L01+ and XBEE ZB Module Based Wireless Ad-hoc Networks", *Int. J. Comput. Netw. Inf.*

Secur., vol. 9, no. 7, pp. 36-44, 2017.
[http://dx.doi.org/10.5815/ijcnis.2017.07.05]

[12]　" S. 2016 TMRH20", ―*Optimized High Speed NRF24L01+ DriverClass Documentation V1.0*‖, *"rf24,* 2016.https://github.com/nRF24/RF24 [Online].

[13]　"TMRH20,", ―*Newly Optimized RF24Network Layerv1.0.7*‖, .https://github.com/nRF24/RF24Mesh [Online].

[14]　"TMRH20,",　―*Newly　Optimized　RF24Network　Layerv1.0.7*‖,.https://　github.com/nRF24/ RF24Network [Online].

CHAPTER 37

Bioremediation of Textile Dyes for Safer Waters

Vinay Thakur[1], Daljeet Singh Dhanjal[1], Reena Singh[1], Saurabh Satija[2,3] and **Chirag Chopra[1,*]**

[1] *School of Bioengineering and Biosciences, Lovely Professional University, Jalandhar-Delhi GT Road, Punjab, India*

[2] *School of Pharmaceutical Sciences, Lovely Professional University, Phagwara, Punjab-144411, India*

[3] *Discipline of Pharmacy, Graduate School of Health, University of Technology Sydney, Ultimo, SW, Australia*

Abstract: The earth holds several natural resources in itself. Water is the most crucial resource among them. However, anthropogenic activities in the proximity of water bodies are leading to water pollution. The primary concern is the textile industry, where water consumption and environmental risk are significant. The manner the water reaches into the water body is of primary concern as no proper treatment is done in many cases. The review mainly focuses on the use of microorganisms for degrading textile dyes in water through bioremediation. Among the different methods being explored for dye degradation, bioremediation is one of the most promising. It is easy to alter and more economical when applied to a commercial scale. Bioremediation is also a sustainable solution that can be harnessed at a large scale and for several generations. However, microbes with specific new biotechnological applications such as nanobiotechnology can give better results. Still, significantly less literature is available on this subject matter. Microbes are a powerful alternative for dye degradation as several bacteria and fungi can degrade many acidic and basic dyes in short periods. The effluent obtained at the end of the process does not have a tertiary effect, making these microbes an efficient choice for dye degradation.

Keywords: Biomagnification, Bioremediation, Dyes, Microbes, Nanobiotechnology.

INTRODUCTION

The textile industry is a big player when it comes to economic aspects. After agriculture, it is the second biggest industry in terms of employment generation. Per the current statistics, this industry globally has emerged as a trillion-dollar

* **Corresponding authors Chirag Chopra**: School of Bioengineering and Biosciences, Lovely Professional University, Jalandhar, Delhi GT Road, Punjab, India; E-mail: chirag.18298@lpu.co.in

Dharam Buddhi, Rajesh Singh and Anita Gehlot (Eds.)

industry [1]. One of the WHO (World Health Organization) reports states that residues from the textile industries alone contribute nearly 18 to 20% of the water population. The reason is the use of several synthetic dyes, which are available at a lower cost [2]. However, from an environmental point of view and concern, they possess no benefit or potential as sustainable or even eco-friendly compounds. Various synthetic dyes nowadays are readily available in the market and fit in the pocket of industries. Some of them are toluidine blue, neutral red, safranine O, eosin yellowish, coomassie brilliant blue, methyl violet, malachite green, methyl green. These dyes, when released either untreated or undertreated, are a cause of great environmental concern. Such dyes are insoluble in water and tend to enter the aquatic fauna's living system as suspended particles, responsible for tumor formation [3]. The bioaccumulation of such toxic compounds in water can have potential threats and be hazardous not for aquatic species but for humans and other animals indirectly linked to water bodies. Bioremediation is one of the most suitable methods for overcoming this problem [4]. Specific improvements in this method with the incorporation of new technology can make this technique reach its optimum potential, providing relief from water pollution [4].

MICROBES USED FOR DYE DEGRADATION

There are several classes of microbes that are used for the degradation of dyes. The focus is primarily on microbes, which are economical, easy to culture, and do not produce any secondary or tertiary by-products after the reaction with the dyes. Bacteria like *Brevibacillus* are effective in degrading toluidine blue. The identity of the bacterial species is routinely confirmed using 16S rRNA sequencing in conjunction with morphological identification and biochemical characterization [5]. Studies show that the textile industrial effluents removed by microbial use have low-enough BOD levels to meet the standard criteria [6]. The BOD parameter fit for commercial and environmental concerns is generally set at 10-30mg/L. With the help of these bacteria, the target of the standard can be easily achieved. It is often seen that natural dyes can provide an alternative to harsh synthetic dyes.

BIOREMEDIATION OF TEXTILE DYES

The textile industry is a big industry with a market capitalization of billions. In our country, too, it holds a significant economic role. The use of hazardous chemicals in the form of dyes can be an actual theft to the environment. Microbes are a significant alternative to chemicals when it comes to dye degradation. As compared to the general practice where one microbe is used, the use of more than one microbe or even a consortium is a more promising protocol, especially when it comes to azo dyes degradation. For degrading congo red, bacterial strains like

Bacillus thuringiensis RUN1 are popularly used. This is a decolorizing dye bacterium. However, this bacterium can degrade some other dyes as well, namely Reactive blue 13, Reactive red 58, and reactive yellow 42 [7]. The bacteria produce necessary enzymes for the biodegradation of dyes, which include *azoreductases, laccases*, among others. Bacteria like *Providencia* sp and *Pseudomonas aeruginosa* can degrade other highly reactive dyes like the species Red HE3B [8]. Untreated effluents of textile industries are one of the major pollutants of the water. In the case of dyes, the azo dye class dominates the segment. Being the largest class, it has various compounds that are highly soluble in water and can easily escape treatment systems, which is when they cause the most problems. One of the most commonly used and effective methods is bioremediation. The use of microorganisms is economical and practical, just as it is of a drawback that minor literature is available for the same. Microbes from class bacteria, fungi, yeast can clean the effluent water. In fungus, species like *Phaerochaete chrysoporium* are competent to metabolize several xenobiotics compounds. Another widely usable microbial strain is *Pseudomonas aeruginosa ETL-1*. This bacterial species is capable of decolorizing triarylmethane dyes (Maulin, Kavita, *et al.*). For the same dye, another bacterial strain, namely *Aeromonas hydrophila,* is also effective. The operational parameters are generally the decolorization efficiency of the bacterium [9]. When it comes to azo dyes, the degradation becomes even more challenging. One of the novel fungal strains used for dye-degradation is *Aspergillus oryzae* [10]. It is observed that this fungus can decolorize and remove the toxicity of reactive textile dyes like PR-HE7B and PV-H3R. However, better productivity is achieved when the fungus is in pellet form or is having dead biomass. *Aspergillus* is capable of removing metals like copper and chromium from mixed wastewater streams. It is essential to consider metal waste as most textile industry effluents had residual metal and dye in waste effluents [11]. For metal degradation *A. lentulus* and *A, terreus* are popular. This fungus shows high productivity even in the presence of mixed pollutants. Mixed pollutants are generally known to exhibit high toxicity [11]. Several new techniques, such as bioremediation, are emerging for the current scenario. Several industrial effluents are treated with nanoparticles like Fe, Au, Sn, Ag, *etc.* Like ZnO nanoparticles effectively degrade Rhodamine B, a colorant [12]. Several sulfate-reducing bacteria (SRB) are found to be effective in azo dye degradation. They are also beneficial in the mineralization of intermediates when given anaerobic conditions [13]. The fungal strain *Trichoderma harzianum* in a semisolid medium can also degrade dyes, which is identified by decolorization. Fungal mycelium exhibits color, confirming the presence of dyes [14]. Another fungal species, *Aspergillus flavus,* is reported to degrade bromophenol blue [15].

ENZYMES IN BIOREMEDIATION OF DYES

Bioremediation is when microbes are used to treat the contaminated water or remove harmful pollutants from wastewater. It must be noted that several enzymes may facilitate the degradation of textile dyes. Several bacteria tend to degrade dyes as they can produce certain redox enzymes like reductase, laccases, and peroxidases. At the same time, they can degrade at unfavorable conditions [16]. Dye degradation using microbes is very cost-effective, as the culturing of microbes does not require much-elaborated infrastructure. Azo reductases effectively degrade dyes like direct blue-green (DBG). The enzyme catalyzes the initial reduction of azo bonds followed by cleavage in a microaerophilic environment [17]. The majority of dye degradation processes are redox reactions [18]. Enzymes also perform other crucial functions like the transfer of electrons. Enzymes named NADH dehydrogenase are essential to transfer an electron from NADH to coenzyme Q. It is seen that when *in-vivo* NADH becomes an electron donor, the azo dyes are easily degraded by reduction. Enzymes like NADH-ubiquinone reductase and NADPH quinone reductase are found to be active in the TCA cycle. The same enzymes are found actively participating in glycolysis, too, and aids in dye degradation. There are several pieces of research done on dye degradation. The azo dye makes a large group of textile dyes. It is observed that NADH-azoreductase, which is FMN dependent, is a key enzyme when it comes to azo-dye degradation [19]. Azo reductase is one of the most important enzymes. This is mainly attached to the plasma membrane. It releases the electrons when it binds with complex 1 of the electron transport chain. Azo reductase is the enzyme that reacts physically with dye molecules and transfers electrons, making it a direct mechanism (Fig. **1**).

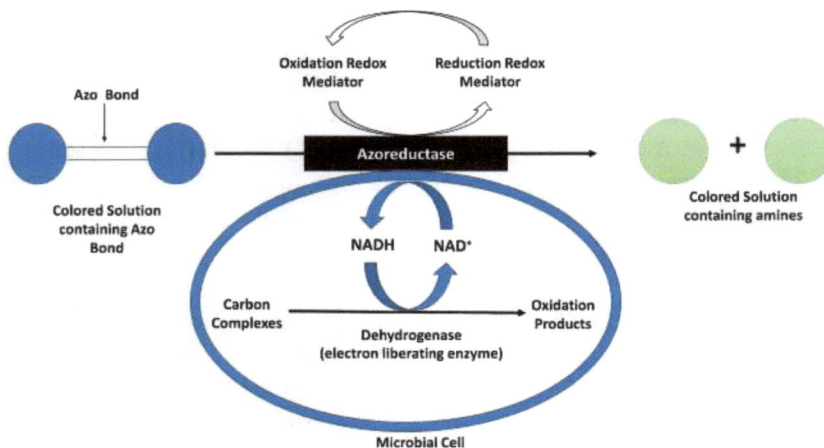

Fig. (1). Illustration of the mechanism of action of Azoreductase for dye degradation.

In some cases, it can be through an indirect mechanism where a coenzyme comes into the picture. One such coenzyme is nicotinamide adenine dinucleotide (NAD). The main enzyme donates electrons to this coenzyme which transfers it to the textile dye molecule and eventually breaks the dye bonds [18]. Enzymes also have a slight disadvantage because they are prone to inactivation by the inhibitors present in wastewater. But certain modifications and a better understanding of crucial enzymes can help improve the process and make it more effective in adverse conditions.

ENVIRONMENTAL IMPACT

The majority of industrial effluents are reported to have harmful pollutants consisting of heavy metals too. Many azo dyes have a toxic effect on the microbial population [20]. In some cases, the microbes use the carbon sources of dye molecules as an energy medium, but the absorbed dyes tend to move in the food chain, leading to biomagnification of these molecules in the ecosystem.

In the case of wool industries, several bleaching processes use oxidants such as peroxidase enzymes and hypochlorite. These chemicals give nearly 10% contribution to the total pollutant load in the end-stage [21]. The use of synthetic dyes became famous after the industrial revolution. This also led to severe pollution, harming the ecosystem. Generally, when the dye gets mixed in water, it becomes opaque, causing an imbalance in the water ecosystem.

Several benzidine-based azo dyes are observed as carcinogenic and very allergic to the skin [22]. Several textile dyes like blue HFRL have anti-estrogenic activity. It is highly disappointing that several industrial dyes are endocrine-disruptive agents [23]. Many harmful dyes are also known for promoting tumorigenesis. The toxicological analysis is critical when it comes to dye degeneration. Tests like the "Allium Test" are used for such analysis. "Mitotic index" is of great concern in such tests. The rise in the mitotic index indicates tumor formation, rapid and uncontrolled cell division [8]. Comet assay is used to test the genotoxic potential of industrial dyes.

CONCLUSION

The review focuses on microbial use for degrading the dyes, which are highly used in the textile industries. There are several species of microbes that can degrade the dyes, and their efficiency is highly appreciable. When the use of microbial load is combined with nanotechnology, the results are more effective. A good combination of both aspects may provide more efficient methods for dye degradation shortly. Some alternative dye molecules need to be explored, which are easily degradable and preferably obtained from natural resources. This is an amicable and sustainable solution to the environmental problems caused due to dye molecules.

CONSENT FOR PUBLICATION

Not applicable.

CONFLICT OF INTEREST

The authors declare no conflict of interest, financial or otherwise.

ACKNOWLEDGEMENT

Declared none.

REFERENCES

[1]   S. Madhav, A. Ahamad, P. Singh, and P.K. Mishra, "A review of textile industry: Wet processing, environmental impacts, and effluent treatment methods", *Environ. Qual. Manage.,* vol. 27, no. 3, pp. 31-41, 2018.
[http://dx.doi.org/10.1002/tqem.21538]

[2]   S. Khan, and A. Malik, "Environmental and health effects of textile industry wastewater", In: *Environmental Deterioration and Human Health: Natural and Anthropogenic Determinants* vol. 9789400778. Springer: Netherlands, 2014, pp. 55-71.
[http://dx.doi.org/10.1007/978-94-007-7890-0_4]

[3]   G. Li, "Decolorization and biodegradation of triphenylmethane dyes by a novel Rhodococcus qingshengii JB301 isolated from sawdust", *Ann. Microbiol.,* vol. 64, no. 4, pp. 1575-1586, 2014.
[http://dx.doi.org/10.1007/s13213-014-0801-7]

[4]   A. A. Juwarkar, S. K. Singh, and A. Mudhoo, "A comprehensive overview of elements in bioremediation", *Reviews in Environmental Science and Biotechnology,* vol. 9, pp. 215-288, 2010.
[http://dx.doi.org/10.1007/s11157-010-9215-6]

[5]   H.A. Alhassani, M.A. Rauf, and S.S. Ashraf, "Efficient microbial degradation of Toluidine Blue dye by Brevibacillus sp", *Dyes Pigments,* vol. 75, no. 2, pp. 395-400, 2007.
[http://dx.doi.org/10.1016/j.dyepig.2006.06.019]

[6]   P. Shanooba, P. Dhiraj, and P. Yatin, "Microbial degradation of textile industrial effluents", *Afr. J. Biotechnol.,* vol. 10, no. 59, pp. 12657-12661, 2011.
[http://dx.doi.org/10.5897/AJB11.1618]

[7]   O.D. Olukanni, A.A. Osuntoki, A.O. Awotula, D.C. Kalyani, G.O. Gbenle, and S.P. Govindwar, "Decolorization of dyehouse effluent and biodegradation of Congo red by Bacillus thuringiensis RUN1", *J. Microbiol. Biotechnol.,* vol. 23, no. 6, pp. 843-849, 2013.
[http://dx.doi.org/10.4014/jmb.1211.11077] [PMID: 23676913]

[8]   S.S. Phugare, D.C. Kalyani, A.V. Patil, and J.P. Jadhav, "Textile dye degradation by bacterial consortium and subsequent toxicological analysis of dye and dye metabolites using cytotoxicity, genotoxicity and oxidative stress studies", *J. Hazard. Mater.,* vol. 186, no. 1, pp. 713-723, 2011.
[http://dx.doi.org/10.1016/j.jhazmat.2010.11.049] [PMID: 21144656]

[9]   C.J. Ogugbue, and T. Sawidis, "Bioremediation and Detoxification of Synthetic Wastewater Containing Triarylmethane Dyes by Aeromonas hydrophila Isolated from Industrial Effluent", *Biotechnol. Res. Int.,* vol. 2011, p. 967925, 2011.
[http://dx.doi.org/10.4061/2011/967925] [PMID: 21808740]

[10]   C.R. Corso, and A.C. Maganha de Almeida, "Bioremediation of dyes in textile effluents by Aspergillus oryzae", *Microb. Ecol.,* vol. 57, no. 2, pp. 384-390, 2009.
[http://dx.doi.org/10.1007/s00248-008-9459-7] [PMID: 18989608]

[11]   A. Mishra, and A. Malik, "Novel fungal consortium for bioremediation of metals and dyes from mixed waste stream", *Bioresour. Technol.,* vol. 171, no. 1, pp. 217-226, 2014.
[http://dx.doi.org/10.1016/j.biortech.2014.08.047] [PMID: 25203229]

[12]   N.T. Nandhini, S. Rajeshkumar, and S. Mythili, "The possible mechanism of eco-friendly synthesized nanoparticles on hazardous dyes degradation", *Biocatal. Agric. Biotechnol.,* vol. 19, p. 101138, 2019.
[http://dx.doi.org/10.1016/j.bcab.2019.101138]

[13]   W. Miran, M. Nawaz, A. Kadam, S. Shin, J. Heo, J. Jang, and D.S. Lee, "Microbial community structure in a dual chamber microbial fuel cell fed with brewery waste for azo dye degradation and electricity generation", *Environ. Sci. Pollut. Res. Int.,* vol. 22, no. 17, pp. 13477-13485, 2015.
[http://dx.doi.org/10.1007/s11356-015-4582-8] [PMID: 25940481]

[14]   L. Singh, and V.P. Singh, "Microbial degradation and decolourization of dyes in semi-solid medium by the fungus-Trichoderma harzianum", *Environ. We Int. J. Sci. Tech.,* vol. 5, pp. 147-153, 2010.

[15]   L. Singh, and V.P. Singh, "Biodegradation of Textile Dyes, Bromophenol Blue and Congo red by Fungus Aspergillus Flavus", *Environ. We Int. J. Sci. Tech.,* vol. 5, pp. 235-242, 2010.

[16]   S. Kumar Garg, M. Tripathi, S.K. Singh, and J.K. Tiwari, "Biodecolorization of textile dye effluent by Pseudomonas putida SKG-1 (MTCC 10510) under the conditions optimized for monoazo dye orange II color removal in simulated minimal salt medium", *Int. Biodeterior. Biodegradation,* vol. 74, pp. 24-35, 2012.
[http://dx.doi.org/10.1016/j.ibiod.2012.07.007]

[17]   H.A. Hassan, A.B.A.Z. Eldein, and N.M. Rizk, "Cloning and Kinetic Properties of Catechol 2,3-dioxygenase from Novel Alkaliphilic BTEX-degrading Pseudomonas sp. HB01", *Life Sci. J.,* vol. 11, no. 2, pp. 376-384, 2014.

[18]   J. Guo, L. Kang, X. Wang, and J. Yang, Decolorization and degradation of azo dyes by redox mediator system with bacteria.*Biodegradation of azo dyes.* Springer, 2010, pp. 85-100.
[http://dx.doi.org/10.1007/698_2009_46]

[19]   G. Chen, X. An, H. Li, F. Lai, E. Yuan, X. Xia, and Q. Zhang, "Detoxification of azo dye Direct Black G by thermophilic Anoxybacillus sp. PDR2 and its application potential in bioremediation", *Ecotoxicol. Environ. Saf.,* vol. 214, p. 112084, 2021.
[http://dx.doi.org/10.1016/j.ecoenv.2021.112084] [PMID: 33640726]

[20]   G.B. Michaels, and D.L. Lewis, "Sorption and toxicity of azo and triphenylmethane dyes to aquatic microbial populations", *Environ. Toxicol. Chem.,* vol. 4, no. 1, pp. 45-50, 1985.
[http://dx.doi.org/10.1002/etc.5620040107]

[21]   R. Singh, and Y. Yadav, "Effluents quality of woolen industrial units and efficiency of wastewater treatment plant at Jorbir, Bikaner, Rajasthan (India)", *Orient. J. Chem.,* vol. 30, no. 1, pp. 49-56, 2014.
[http://dx.doi.org/10.13005/ojc/300106]

[22]   S. Sarkar, A. Banerjee, U. Halder, R. Biswas, and R. Bandopadhyay, "Degradation of Synthetic Azo Dyes of Textile Industry: a Sustainable Approach Using Microbial Enzymes", *Water Conserv. Sci. Eng.,* vol. 2, no. 4, pp. 121-131, 2017.
[http://dx.doi.org/10.1007/s41101-017-0031-5]

[23]   I. Bazin, A. Ibn Hadj Hassine, Y. Haj Hamouda, W. Mnif, A. Bartegi, M. Lopez-Ferber, M. De Waard, and C. Gonzalez, "Estrogenic and anti-estrogenic activity of 23 commercial textile dyes", *Ecotoxicol. Environ. Saf.,* vol. 85, pp. 131-136, 2012.
[http://dx.doi.org/10.1016/j.ecoenv.2012.08.003] [PMID: 22947508]

CHAPTER 38

Advancements of Transdermal Patches in Psychiatric Disorders

Kunwar Shahbaaz Singh Sahi[1,*] and **Anjuvan Singh[1]**

[1] *Department of Biomedical Engineering, School of Bioengineering and Biosciences, Lovely Professional University, Phagwara, India, 144001*

Abstract: Non-adherence and non-compliance to the course of psychiatric medications affect the standard treatment plan of a patient. Inability to monitor the correct time and dosage also results in lesser efficiency of the drug and its action mechanism. Factors that may cause instability and non-compliance with the treatment plan are the possible side effects of drugs or the ease of use. With technological advancements in the field of drug delivery, transdermal patches, while Being non-invasive, ensure proper dosage and delivery of drugs through the skin, minimizing the first-pass metabolism. This review examines the existing literature, working mechanism, preclinical studies, and advancements in the application of transdermal patches in psychotropic drugs and de-addiction while evaluating various psychiatric disorders and comparing their efficacy and remission rate concerning standard oral treatment. It also addresses the challenges, drawbacks, and strategies required to increase its efficiency in clinical use.

Keywords: Affective Disorders, Transdermal Drug Delivery Systems, Schizophrenia, Substance Abuse Disorders, Sumatriptan Iontophoretic patch.

INTRODUCTION

Over the past century, psychiatric care has mostly been given in form of oral medication or through brain-stimulating techniques such as electroconvulsive therapy. With recent advancements in drug delivery systems, psychiatric care has widened its sphere of influence by providing alternatives like patches that are self-administered and not patient depended. However, compliance to medication and optimization of dose can get hampered by various psychological as well as physiological factors such as dose routine, route of administration, personal beliefs, and nature of the illness [1]. To counter these, transdermal patches are non-invasive with each patch having a period of 1 to 7 days of sustained drug

---

[*] **Corresponding author Kunwar Shahbaaz Singh Sahi:** Department of Biomedical Engineering, School of Bioengineering and Biosciences, Lovely Professional University, Phagwara, India, 144001; E-mail: kunwarsahi@gmail.com

Dharam Buddhi, Rajesh Singh and Anita Gehlot (Eds.)

release and hence have paved a way for an alternative route of drug administration.

Advantages of Transdermal patches include the drug delivery through the skin avoids hepatic first-pass metabolism and gastrointestinal incompatibility without vexatious experiences of injections or rectal applications [2], and steady and sustained drug levels in the body [3]. With recent advancements, applications of Transdermal patches has widely increased in the field of Neuropsychopharmacology, specifically in psychotropic drugs which alter mood, behavior, and thoughts through drug delivered *via* skin. Examples of these include medicines for Attention Deficit Hyperactivity Disorder (ADHD) and For Major Depressive Disorders (MDD) [4] which would be discussed in the coming sections

TRANSDERMAL THERAPEUTIC SYSTEMS

Transdermal Patches, also known as transdermal therapeutic systems, consist of three structures: an adhesive, a pharmacologic agent, and an enhancing agent. Further, they are of three main types: reservoir, matrix, and iontophoretic patches. Originally, the patches were made of a liquid reservoir system where they consisted of a drug reservoir, a backing material (which functioned as adhesive and protective), and a release membrane. On the other hand, the liquid-reservoir design has been used to develop patches of testosterone, scopolamine, estrogen, *etc* [4]. An updated configuration, the adhesive matrix system consists of only three layers and the drug and the adhesive layer are combined into one, leaving only the backing layer and protective layer which is removed before applying on the skin. This design is much more popular and is used to deliver new drugs except for the ones mentioned above.

Another approved patch, the sumatriptan iontophoretic transdermal system makes use of a low voltage current to transfer the drug through the stratum corneum. Iontophoresis entails the movement of charged molecules through electrophoresis whereas the weak and uncharged molecules move through electro-osmotic force [5]. The advantage of this design is as it uses the application of electric current, the system can be controlled through a microprocessor and in this case, the patient itself.

MECHANISM OF TRANSDERMAL DRUG DELIVERY

To simplify the mechanism, drugs are transferred *via* three main pathways: through the sweat ducts, through hair follicles and sebaceous glands (collectively called shunts), or directly through the stratum corneum. In the passive reservoir or matrix configuration, the drug diffuses into the stratum corneum through intercellular lipids. Drugs administered Transdermally need to have low molecular weight and lipophilicity, which corresponds to penetration through the skin and efficient solubility. Apart from this, drugs having a low melting point and which are more volatile can be easily composed into a Transdermal patch as they tend to permeate through the skin effectively [5].

APPLICATION OF TRANSDERMAL PATCHES IN CLINICAL PSYCHOPHARMACOLOGY

With recent advancements in the field of drug delivery and especially in neuro-psychiatry, new patches have been made available for transporting psychopharmacological drugs to treat neuropsychiatric disorders such as ADHD, Depression, Dementia, Obsessive-Compulsive Disorder, and addictions such as Nicotine and Opioid. Examples of such patches are Buprenorphine, Clonidine, Fentanyl, Methylphenidate, Nicoderm, Rivastigmine, Selegiline, *etc*. However, Scopolamine, a Transdermal patch used for the treatment of motion sickness has been found to induce toxic psychosis within a short period [6]. This is one of the side effects which would be discussed in the coming sections.

Dementia

Patients receiving drug therapy for Dementia for a greater period have shown its effects of cognitive decline which has reduced patient admission into in-patient wards. This has led to increased requirement of prolonged wear Transdermal patches [7] since a complicated dosing schedule can lead to patients taking sub-therapeutic doses [8], making the treatment less effective.

Rivastigmine has evolved as the first transdermal patch for Alzheimer's and Parkinson's disease dementia and its efficacy level reaching as high as standard oral dosage, making it a sustained drug through Transdermal transport [3]. Rivastigmine, a cholinesterase inhibitor that is administered Transdermally avoids the central cholinergic gastrointestinal side effects [9] which in the titration phase, is believed to cause a rapid increase in acetylcholine levels in the brain after inhibition of targeted enzymes. This drug instead provides sustained levels, reducing fluctuations in plasma concentrations. These properties are responsible for reducing the probable side effects of oral medication [10].

In a study by Winblad and colleagues [11], they concluded that the Transdermal Rivastigmine patch of $10cm^2$ showed improvement likewise as oral administration of the drug but with lesser side effects. Additionally, they also found that patch of $20cm^2$ showed better and superior cognitive scores as compared to standard $10cm^2$.

Major Depressive Disorder

FDA approved Emsam (Selegiline), the first 24-hour antidepressant patch in 2006 which is a monoamine oxidase inhibitor. This patch has many advantages; one being it doesn't restrict the diet which is a factor when using oral MAO inhibitors since it spikes the blood pressure by tyramine hypertensive crisis. This patch also reduces sexual dysfunction, sedation, and weight gain when compared with SSRIs and SNRIs. The only drawback of the patch is a mild redness on the skin and if exposed to heat such as sunlight can result in an increased amount of drug being released into the skin [12].

Additional studies show the development of Fluoxetine, an SSRI, as a Transdermal patch. This research concluded that at 65% vol/vol ethanolic solution, the permeation of this drug was the most to deliver sufficient doses of 20-80mg [13].

Attention Deficit Hyperactivity Disorder (ADHD)

Sympathomimetic medicines such as methylphenidate have been reported to have the best treatment for ADHD. Patients with ADHD require prolonged and sustained drug delivery to cover the maximum symptoms overall. This concept led to the development of Methylphenidate Transdermal Systems (MTS). The system solves the problem of frequent dosing and increases the ease of use making it better suited for hyperactive children aged 6-12 years. Studies on this drug have shown increased tolerability and effectiveness when applied the patch [14,15]. The peak plasma level has been recorded near to 8 hours with side effects being nausea, vomiting, and decreased appetite [16].

Dextroamphetamine, a drug used to treat ADHD was delivered Transdermally by Pelham and colleagues in 2011 [16]. They concluded on the note that transdermal delivery was giving more relief through the day, therefore was sustaining, but the onset was late when compared to oral administration.

Another study by Levin and colleagues [17] found nicotine patches administered 7mg/day for 4.5 hours increased attentiveness in patients with ADHD but non-smokers.

Anxiety, Schizophrenia, and Affective Disorders

There is a lack of literature available for anti-anxiety patches except for "Go-patch" a recently developed transdermal patch that is documented to relieve stress and anxiety over 24 hours. On the other hand, FDA in 2019 approved the first Schizophrenia patch composed of antipsychotic asenapine applied once a day.

Apart from this, in a study by Evins and colleagues [18], they found varenicline, bupropion, and transdermal nicotine patch to be useful in the treatment of psychosis, anxiety, and mood disorders. In double-blinded, randomized study by Kulkarni and colleagues [19], found estradiol 100µg (patch of estrogen), useful in the treatment of psychotic symptoms. The scores on the Positive and Negative Syndrome Scale (PANSS) of patients showed significant improvement in positive and psychopathological symptoms in Schizophrenia than patients taking oral antipsychotics. Further, Estrogen patches have also shown efficacy in treating affective/mood disorders, especially in females during pre-menopausal and postpartum studies [20]. On the other hand, nicotine patches for Obsessive-Compulsive Disorder have shown a remarkable improvement and are open for further research [21].

Autism Spectrum Disorder (ASD)

Clonidine, an α-adrenergic agonist that acts on presynaptic neurons primarily in the brainstem to inhibit norepinephrine activity [22] has shown effective treatment in children with ASD by controlling their aggressive behavior, mood swings, and sleep cycle. This medicine has also shown efficacy in treating aggression in ADHD and insomnia in sleep disorders and can be used as monotherapy or adjunct therapy in the treatment of sleep disorders [23,24]. In a study by Frankhauser and colleagues [25], they found better results in Ritvo-Freeman Real Life Rating Scale and The Clinical Global Impressions scale and results were seeing in aggression and social conduct.

Apart from Clonidine, nicotine has also shown great results in managing aggression and agitation in ASD [26]. In a trial by Lewis and colleagues [27], they found significant results in Aberrant Behavior Checklist-Irritability and reduction in agitation, irritability, and sleep in patients with ASD.

Addiction and Substance Abuse Disorder

The most popular transdermal systems for addiction are nicotine and Opioid addiction. Patches for other drugs such as synthetics are still in research.

Nicotine

In the past decade, Carbon Nanotubes have been found useful in designing transdermal patches for nicotine replacement therapy (NRT) [28] therefore giving an alternative path to smokers. However, NRT does not eliminate all the withdrawal symptoms because it cannot replicate the instant high doses of arterial nicotine which is achieved during cigarette smoking. Hence, to counter these, NRT patches come in a variable dosage range (5 to 22mg), giving a wide spectrum to light and heavy smokers over 24 hours [29]. But according to contrary studies, the conventional patch of the highest transdermal dosage, *i.e.* 22mg can only replace half of the serum levels of nicotine and cotinine in smokers [30]. Therefore, >42mg NRT patches were evaluated and their efficacy was found to be considered good with a high abstinence rate in smokers [31 - 33]. Further Studies have combined passive mode of delivery from NRT patches and libitum mode from gums, inhalers, *etc.* This architecture of nicotine delivery has shown efficacy in reducing the urge to smoke and safely managing the withdrawal symptoms [29,33]. Selegiline has also shown a high abstinence rate when combined with nicotine patches with almost no additional side effects [34].

Opioids

Sustainable and Managed withdrawal, which is detoxification, is a widespread clinical issue for patients with Opioid dependence. Buprenorphine Transdermal System is the most popular rehabilitation drug for Opioid addiction in respect to transdermal patches. In a study by Lanier and colleagues [35], they administered transdermal Buprenorphine for 7 days and found its efficacy in treating opioid withdrawal syndrome with no adverse effects including no respiratory depression, no opioid intoxication, and no itching at the site of the patch.

Other Natural/Synthetic Drugs

There has been a marginalized amount of research in the field of synthetic and organic drugs in terms of transdermal therapy. Psychotropic drugs like selegiline have been used to treat cocaine dependence and be well tolerated in patients and attenuate cardiovascular, physiological, and certain subjective effects of cocaine [36]. However, in a study by Elkashef and colleagues [37], they concluded that selegiline was not very efficient when compared to placebo.

DISCUSSION

Transdermal Therapeutic Systems have proved to be a breakthrough in treating various types of disorders, primarily in psychiatric care since our patients find trouble in adhering to the dosage routine. TDS increases the ease of use as it is self-administered throughout usually 24hours. Another advantage is decreased side effects as we have seen in rivastigmine which avoids the central cholinergic gastrointestinal side effects. Similar to this, many psychotropic drugs administered Transdermally reduce the side effects when compared to oral forms. One unique and significant property of TDS is the sustained and prolonged drug release which in psychiatry is excessively needed to control the mood, behavior, and thoughts throughout the day. However, despite TDS gaining popularity, the research is still lacking behind for the treatment of Anxiety, Affective Disorders,

Schizophrenia, and Substance Use Disorders. Furthermore, a patch that Transdermally administers benzodiazepines over 12 or 24 hours can help control anxiety, whether it be OCD, Social Anxiety, Panic Disorder, *etc*. Transdermal Nicotine, on the other hand, has shown efficacy in controlling the agitation and aggressive behavior seen in ADHD and ASD, and can also be used to treat the same symptom in personality disorders as well as mood disorders combined with transdermal administration of mood stabilizers. These systems can also be used for experimental purposes of natural/synthetic drugs by micro-dosing them Transdermally in mice or Non-Human Primate models. Table **1** mentions the summary of all preclinical and clinical studies given below.

Table 1. Summary of Pre-clinical and Clinical study data available on TDS.

S.No.	Therapeutic Drug	Preclinical and Clinical Studies Design	Disorder	Inference	References
1.	Rivastigmine	A placebo-controlled, double-blind study	Alzheimer's	$10cm^2$ patch showed similar efficacy to capsules and $20cm^2$ patch showed superior cognitive scores	[10]
2.	Methylphenidate	Randomized, double-blind, placebo-controlled, laboratory classroom assessment	ADHD	Showed increased efficacy on all measures compared to placebo	[14]
		Randomized, double-blind, placebo-controlled, parallel-group study		Improvement in ADHD-Rating Scale-IV from baseline	[15]
3.	Estradiol	Randomized, double-blind	Psychosis	$100\mu g$ patch showed remarkable improvement in positive and general psychopathological symptoms.	[19]
4.	Clonidine	Double-blind, placebo-controlled	Autism	Preliminary studies showed effectiveness in reducing hyperarousal behaviors and improved social relationships	[25]
5.	Buprenorphine	Open-label evaluation	Opioid Addiction	Showed efficacy in opioid detoxification	[35]
6.	Haloperidol and Nicotine	A randomized, double-blind study	Tourette's syndrome	Nicotine patches showed greater efficacy in the treatment of dyskinetic symptoms	[38]

CONCLUDING REMARKS

Transdermal Systems have shown great efficacy in reducing withdrawal and general symptoms of various psychiatric disorders such as Major Depressive Disorders, Dementia, Autism, ADHD, and Nicotine, and Opioid Addiction same as oral medication but with more advantages including the ease of use and its property of self-administration. The only side effects of transdermal patches are skin reactions which might include rashes on the point of contact. However, it also lacks research in the development of patches for Anxiety, Affective disorders, Schizophrenia, and substance use disorders of organic and synthetic drugs.

CONSENT FOR PUBLICATION

Not applicable.

CONFLICT OF INTEREST

The authors declare no conflict of interest, financial or otherwise.

ACKNOWLEDGEMENT

Declared none.

REFERENCES

[1]   S. Griffith, "A review of the factors associated with patient compliance and the taking of prescribed medicines", *Br. J. Gen. Pract.,* vol. 40, no. 332, pp. 114-116, 1990.
        [PMID: 2112014]

[2]   "Transdermal patches: The emerging mode of drug delivery system in Psychiatry", *Article in Therapeutic Advances in Psychopharmacology,* 2012.

[3]   F. Mercier, "G. Lefèvre, H.Huang, Schmidli, H., B. Amzal, and S. Appel-Dingemanse,"Rivastigmine exposure provided by a transdermal patch versus capsules", *Curr. Med. Res. Opin.,* vol. 23, no. 3, pp. 199-3204, 2007.

[4]   R. Jonathan, "Stevens, M. Justin Coffey, Megan Fojtik, Kristina Kurtz, Theodore A. Stern, "The Use of Transdermal Therapeutic Systems in Psychiatric Care." A Primer on Patches", *Psychosomaticsvol,* vol. 56, no. 5, pp. 423-444, 2015.
        [http://dx.doi.org/10.1016/j.psym.2015.03.007]

[5]   B. Vecchia, and A. Bunge, "Evaluating the transdermal permeability of chemicals", *Transdermal Drug Delivery,* pp. 25-55, 2002.

[6]   A.A. Ziskind, and M.D. Ziskind, "Transdermal scopolamine-induced psychosis", *Postgrad. Med.,* vol. 84, no. 3, pp. 73-76, 1988.
        [http://dx.doi.org/10.1080/00325481.1988.11700397] [PMID: 2901077]

[7]   A. Harada, and A. Vanderplas, "PNL27 the effect of adherence to Alzheimer's disease treatmenton health care costs in managed care", *Value Health,* pp. A87-A88, 2006.
        [http://dx.doi.org/10.1016/S1098-3015(10)64576-7]

[8]   B. Roberto, and M.L. Pablo, "Clinical Benefits Associated with a Transdermal Patch for Dementia", *Eur. Neurol. Rev.,* vol. 3, no. 1l, pp. 10-13, 2008.

[9]   A. Di Stefano, A. Iannitelli, S. Laserra, and P. Sozio, "Drug delivery strategies for Alzheimer's disease treatment", *Expert Opin. Drug Deliv.,* vol. 8, no. 5, pp. 581-603, 2011.
[http://dx.doi.org/10.1517/17425247.2011.561311] [PMID: 21391862]

[10]   T. Darreh-Shori, and V. Jelic, "Safety and tolerability of transdermal and oral rivastigmine in Alzheimer's disease and Parkinson's disease dementia", *Expert Opin. Drug Saf.,* vol. 9, no. 1, pp. 167-176, 2010.
[http://dx.doi.org/10.1517/14740330903439717] [PMID: 20021294]

[11]   B. Winblad, G. Grossberg, L. Frölich, M. Farlow, S. Zechner, J. Nagel, and R. Lane, "IDEAL: a 6-month, double-blind, placebo-controlled study of the first skin patch for Alzheimer disease", *Neurology,* vol. 69, no. 4, suppl. Suppl. 1, pp. S14-S22, 2007.
[http://dx.doi.org/10.1212/01.wnl.0000281847.17519.e0] [PMID: 17646619]

[12]   A.M. Bied, J. Kim, and T.L. Schwartz, "A critical appraisal of the selegiline transdermal system for major depressive disorder", *Expert Rev. Clin. Pharmacol.,* vol. 8, no. 6, pp. 673-681, 2015.
[http://dx.doi.org/10.1586/17512433.2016.1093416] [PMID: 26427518]

[13]   D.K. Parikh, and T.K. Ghosh, "Feasibility of transdermal delivery of fluoxetine", *AAPS PharmSciTech,* vol. 6, no. 2, pp. E144-E149, 2005.
[http://dx.doi.org/10.1208/pt060222] [PMID: 16353971]

[14]   J.J. McGough, S.B. Wigal, H. Abikoff, J.M. Turnbow, K. Posner, and E. Moon, "A randomized, double-blind, placebo-controlled, laboratory classroom assessment of methylphenidate transdermal system in children with ADHD", *J. Atten. Disord.,* vol. 9, no. 3, pp. 476-485, 2006.
[http://dx.doi.org/10.1177/1087054705284089] [PMID: 16481664]

[15]   R.L. Findling, O.G. Bukstein, R.D. Melmed, F.A. López, F.R. Sallee, L.E. Arnold, and R.D. Pratt, "A randomized, double-blind, placebo-controlled, parallel-group study of methylphenidate transdermal system in pediatric patients with attention-deficit/hyperactivity disorder", *J. Clin. Psychiatry,* vol. 69, no. 1, pp. 149-159, 2008.
[http://dx.doi.org/10.4088/JCP.v69n0120] [PMID: 18312050]

[16]   W.E. Pelham, J.G. Waxmonsky, J. Schentag, C.H. Ballow, C.J. Panahon, E.M. Gnagy, M.T. Hoffman, L. Burrows-MacLean, D.L. Meichenbaum, G.L. Forehand, G.A. Fabiano, K.E. Tresco, A. Lopez-Williams, E.K. Coles, and M.A. González, "Efficacy of a methylphenidate transdermal system versus t.i.d. methylphenidate in a laboratory setting", *J. Atten. Disord.,* vol. 15, no. 1, pp. 28-35, 2011.
[http://dx.doi.org/10.1177/1087054709359163] [PMID: 20439487]

[17]   E.D. Levin, C.K. Conners, D. Silva, S.C. Hinton, W.H. Meck, J. March, and J.E. Rose, "Transdermal nicotine effects on attention", *Psychopharmacology (Berl.),* vol. 140, no. 2, pp. 135-141, 1998.
[http://dx.doi.org/10.1007/s002130050750] [PMID: 9860103]

[18]   A.E. Evins, N.L. Benowitz, R. West, C. Russ, T. McRae, D. Lawrence, A. Krishen, L. St Aubin, M.C. Maravic, and R.M. Anthenelli, "Neuropsychiatric Safety and Efficacy of Varenicline, Bupropion, and Nicotine Patch in Smokers With Psychotic, Anxiety, and Mood Disorders in the EAGLES Trial", *J. Clin. Psychopharmacol.,* vol. 39, no. 2, pp. 108-116, 2019.
[http://dx.doi.org/10.1097/JCP.0000000000001015] [PMID: 30811371]

[19]   J. Kulkarni, "A de Castella, P. B. Fitzgerald, C. T Gurvich, M. Bailey., C. Bartholomeusz, H. Burger. "Estrogen in Severe Mental Illness", *Arch. Gen. Psychiatry,* vol. 65, no. 8, p. 955, 2008.
[http://dx.doi.org/10.1001/archpsyc.65.8.955] [PMID: 18678800]

[20]   U. Halbreich, and L.S. Kahn, "Role of estrogen in the aetiology and treatment of mood disorders", *CNS Drugs,* vol. 15, no. 10, pp. 797-817, 2001.
[http://dx.doi.org/10.2165/00023210-200115100-00005] [PMID: 11602005]

[21]   S. Lundberg, A. Carlsson, P. Norfeldt, and M.L. Carlsson, "Nicotine treatment of obsessive-

compulsive disorder", *Prog. Neuropsychopharmacol. Biol. Psychiatry,* vol. 28, no. 7, pp. 1195-1199, 2004.
[http://dx.doi.org/10.1016/j.pnpbp.2004.06.014] [PMID: 15610934]

[22] X. Ming, E. Gordon, N. Kang, and G.C. Wagner, "Use of clonidine in children with autism spectrum disorders", *Brain Dev.,* vol. 30, no. 7, pp. 454-460, 2008.
[http://dx.doi.org/10.1016/j.braindev.2007.12.007] [PMID: 18280681]

[23] H.J. Horacek, "Extended-release clonidine for sleep disorders", *J. Am. Acad. Child Adolesc. Psychiatry,* vol. 33, no. 8, p. 1210, 1994.
[http://dx.doi.org/10.1097/00004583-199410000-00022] [PMID: 7982875]

[24] S. Rubinstein, L.B. Silver, and W.L. Licamele, "Clonidine for stimulant-related sleep problems", *J. Am. Acad. Child Adolesc. Psychiatry,* vol. 33, no. 2, pp. 281-282, 1994.
[http://dx.doi.org/10.1097/00004583-199402000-00021] [PMID: 8150803]

[25] M.P. Fankhauser, V.C. Karumanchi, M.L. German, A. Yates, and S.D. Karumanchi, "A double-blind, placebo-controlled study of the efficacy of transdermal clonidine in autism", *J. Clin. Psychiatry,* vol. 53, no. 3, pp. 77-82, 1992.
[PMID: 1548248]

[26] G.I. Van Schalkwyk, A.S. Lewis, Z. Qayyum, K. Koslosky, M.R. Picciotto, and F.R. Volkmar, "Reduction of Aggressive Episodes After Repeated Transdermal Nicotine Administration in a Hospitalized Adolescent with Autism Spectrum Disorder", *J. Autism Dev. Disord.,* vol. 45, no. 9, pp. 3061-3066, 2015.
[http://dx.doi.org/10.1007/s10803-015-2471-0] [PMID: 25982311]

[27] A.S. Lewis, G.I. van Schalkwyk, M.O. Lopez, F.R. Volkmar, M.R. Picciotto, and D.G. Sukhodolsky, "An Exploratory Trial of Transdermal Nicotine for Aggression and Irritability in Adults with Autism Spectrum Disorder", *J. Autism Dev. Disord.,* vol. 48, no. 8, pp. 2748-2757, 2018.
[http://dx.doi.org/10.1007/s10803-018-3536-7] [PMID: 29536216]

[28] C.L. Strasinger, N.N. Scheff, J. Wu, B.J. Hinds, and A.L. Stinchcomb, "Carbon Nanotube Membranes for use in the Transdermal Treatment of Nicotine Addiction and Opioid Withdrawal Symptoms", *Subst. Abuse,* vol. 3, pp. 31-39, 2009.
[http://dx.doi.org/10.4137/SART.S1050] [PMID: 20582253]

[29] L. Umesh Wadgave, "Nagesh. "Nicotine Replacement Therapy: An Overview", *J Health Sci (Qassim),* vol. 10, no. 3, pp. 425-435, 2016.

[30] R.D. Hurt, L.C. Dale, K.P. Offord, G.G. Lauger, L.B. Baskin, G.M. Lawson, N.S. Jiang, and P.J. Hauri, "Serum nicotine and cotinine levels during nicotine-patch therapy", *Clin. Pharmacol. Ther.,* vol. 54, no. 1, pp. 98-106, 1993.
[http://dx.doi.org/10.1038/clpt.1993.117] [PMID: 8330471]

[31] C. Silagy, T. Lancaster, L. Stead, D. Mant, and G. Fowler, "Nicotine replacement therapy for smoking cessation", *Cochrane Database Syst Rev,* vol. 3, 2004. CD000146
[http://dx.doi.org/10.1002/14651858.CD000146.pub2]

[32] J.R. Hughes, G.R. Lesmes, D.K. Hatsukami, R.L. Richmond, E. Lichtenstein, D.E. Jorenby, J.O. Broughton, S.P. Fortmann, S.J. Leischow, J.P. McKenna, S.I. Rennard, W.C. Wadland, and S.A. Heatley, "Are higher doses of nicotine replacement more effective for smoking cessation?", *Nicotine Tob. Res.,* vol. 1, no. 2, pp. 169-174, 1999.
[http://dx.doi.org/10.1080/14622299050011281] [PMID: 11072398]

[33] J.O. Ebbert, L.C. Dale, C.A. Patten, I.T. Croghan, D.R. Schroeder, T.P. Moyer, and R.D. Hurt, "Effect of high-dose nicotine patch therapy on tobacco withdrawal symptoms among smokeless tobacco users", *Nicotine Tob. Res.,* vol. 9, no. 1, pp. 43-52, 2007.
[http://dx.doi.org/10.1080/14622200601078285] [PMID: 17365735]

[34] R. Biberman, R. Neumann, I. Katzir, and Y. Gerber, *A randomized controlled trial of oral selegiline plus nicotine skin patch compared with placebo plus nicotine skin patch for smoking cessation.,* 2003.

[http://dx.doi.org/10.1046/j.1360-0443.2003.00524.x]

[35]   R.K. Lanier, A. Umbricht, J.A. Harrison, E.S. Nuwayser, and G.E. Bigelow, "Opioid detoxification via single 7-day application of a buprenorphine transdermal patch: an open-label evaluation", *Psychopharmacology (Berl.),* vol. 198, no. 2, pp. 149-158, 2008.
[http://dx.doi.org/10.1007/s00213-008-1105-z] [PMID: 18327673]

[36]   E.J. Houtsmuller, L.D. Notes, T. Newton, N. van Sluis, N. Chiang, A. Elkashef, and G.E. Bigelow, "Transdermal selegiline and intravenous cocaine: safety and interactions", *Psychopharmacology (Berl.),* vol. 172, no. 1, pp. 31-40, 2004.
[http://dx.doi.org/10.1007/s00213-003-1616-6] [PMID: 14605792]

[37]   A. Elkashef, P.J. Fudala, L. Gorgon, S-H. Li, R. Kahn, N. Chiang, F. Vocci, J. Collins, K. Jones, K. Boardman, and M. Sather, "Double-blind, placebo-controlled trial of selegiline transdermal system (STS) for the treatment of cocaine dependence", *Drug Alcohol Depend.,* vol. 85, no. 3, pp. 191-197, 2006.
[http://dx.doi.org/10.1016/j.drugalcdep.2006.04.010] [PMID: 16730924]

[38]   A.A. Silver, R.D. Shytle, M.K. Philipp, B.J. Wilkinson, B. McConville, and P.R. Sanberg, "Transdermal nicotine and haloperidol in Tourette's disorder: a double-blind placebo-controlled study", *J. Clin. Psychiatry,* vol. 62, no. 9, pp. 707-714, 2001.
[http://dx.doi.org/10.4088/JCP.v62n0908] [PMID: 11681767]

CHAPTER 39

Review of Intangible Urban Planning Aspects for Sustainable Brick & Mortar Retail Markets

Raminder Kaur[1,*] and **Mahendra Joshi**[1]

[1] *Lovely Professional University, Phagwara, Punjab, India*

Abstract: Undeniably, public precincts are the stage that unfolds urban life every day, they foster social & economic bonds, bringing people together. Brick & Mortar retail market along streets is one of major mode for retail shopping in a city, in addition, these places play a pivotal role for connecting people with the physical environment and this is not a recent phenomenon but B & M retail market along streets have played this role since the origin of towns and cities. But, in recent past, rather than pedestrian there have been more flow of private vehicles in these public precincts. The position is deteriorating day by day as private vehicles are growing drastically, due to which on one hand quality of life is degraded and on other hands the aspects of public spaces had undermined in several cities. Thus, there is an immediate demand to give attention to all users at these places, so that they can stroll, communicate, shop, and comply with other social acts that are important for sustaining socio-culture aspects of cities. Along with tangible, intangible characteristics also play a pivotal role in maintaining the social & culture of any area, and as retail depicts one of the important elements of urban development so these non-physical characteristics should be incorporated in retail planning. This paper will provide insights to make planning parameters that can revive footfall in B & M stores and increase user satisfaction while physical shopping. It will also help in addressing sustainable social and cultural aspects of the city and its impact on users.

Keywords: B&M Retail, Intangible Aspects, Planning, Social & Cultural Aspects, Sustainable, Tangible Elements.

1. INTRODUCTION

Undoubtedly, in process of the retail sector, like others also changed with the advancement in technology and urban development. Each change does not eliminate the previous process but if considered appropriately they can comple-

---

[*] **Corresponding authors Raminder Kaur:** Lovely Professional University, Phagwara, Punjab, India; E-mail: raminder.18258@lpu.co.in

Dharam Buddhi, Rajesh Singh and Anita Gehlot (Eds.)

ment each other. For example, Omnichannel was described by Rigby [1], as "an integrated sales experience that melds the advantages of physical stores with the implementation of rich experience of online". Rigby heads the firm global innovation practice and in the article "The future of shopping" discussed that every 50 years retailing undergo interference in the process, but these changes will not eliminate the previous process, but planners and retailers need to pick up the pace of changes and reshape the process in such way that customer expectations are fulfilled. Change in customer behavior is the main reason for organized retail growth in India, is claimed by Vij [2], "The study and the analysis: an impact of organized Retail on unorganized retail in India" recommended that organized and unorganized retail sectors in India can co-exist and flourish side by side but both sectors need to keep customer satisfaction on top. Obviously, along with advancements in m-commerce & e-commerce which are providing comfort and convenient platforms to attract customers, another reason for the decline of footfall in B&M markets is increased vehicles that are not safe, especially for pedestrian shoppers. In addition, another reason for user dissatisfaction while physical shopping in present B &M markets is missing basic amenities and infrastructure.

Thus, it is time to strictly consider both tangible and intangible elements as an integral part of planning, as they play a pivotal role regarding all B & M retail markets. In the Smart city proposal, retrofitting of retail markets is one of the area-based projects in many proposals so while planning these areas emphasis should be given to the need of all users. In addition, intangible characteristics also play a pivotal role in maintaining the social & culture of any area, and as retail depicts one of the important elements of urban development so these non-physical characteristics should be incorporated in retail planning.

2. LITERATURE REVIEW

This review is to understand and analyze the need of all users while shopping in the existing B&M retail market and the impact of tangible and intangible characteristics on user satisfaction. This review also aims to compare various retail modes over decades and how it is affecting people of the city in terms of sustainability, socio-economic aspects, social life opportunities and aspirations of people.

2.1. Tangible and Intangible Measures in a Retail Environment

Irreversibly, considering retail store image, a fusion of tangible and intangible measures, and decisive customer relationship with retailers is accepted for its success in terms of profit as well it adds value to store. Burt and Carralero-Encinas [3], "The role of store image in retail internationalization" by taking an example in the UK and Spain explained connection and contrast in customer perception of store image attribute both tangible and intangible, which must be handled carefully to have a consistent position in host market by international retail companies.

Kaltcheva [4], "When Should a Retailer Create an Exciting Store Environment?" recommended that the consumer motivational orientation controls the effect of the arousal produced by a store environment. They elaborated task and recreational motivational orientation, on one hand, task-oriented customers focus on output rather than process so high arousal decreases pleasantness & harms shopping behavior. On other hand, for recreational-oriented consumers, high arousal enhances their shopping experience & increases intentions to visit and positively make purchases.

Diallo *et al.* [5], "Factors influencing consumer behavior towards store brands: evidence from the French market" explored virtue of consumer image factors and store acquaintance on store brand purchase style. By sampling three French towns they found various factors like value awareness, store image impression and price-image have a significant impact on store brand purchase style.

Mullick [6], "A Study of Shopping Experience in Selected Retail Centres in NCR.," interprets the shopping experience in Delhi retail stores. He claims that along with shopping the ambiance should be created in such a way that customers can relish the space with their family, eventually, it will increase the reliability in customers and motivate them to visit the store again.

2.2. Urban Planning / Inclusive Approach

Gert-Jan Hospers [7], "Lynch's The Image of the City after 50 Years: City Marketing Lessons from an Urban Planning Classic" claimed that even after fifty years of publication, still the study of the perception of the image by the user in making city image in their mind, is applicable and up to date. Importantly it is not only relevant for urban planners, but psychologists, geographers, and today's city marketers can also take advantage of highlighting the city concerning five elements: - paths, edges, districts, nodes, landmarks.

Reardon [8], "The Death and Life of Great American Cities" by Jane Jacobs comments that, in her book, through the hymn, she argues in fighting for planning from street- level perspective, having mixed-use of building types is accepted, but at the same time her criticism against high rise development ignores other sets of people who are in favor of living in group towers.

Greed [9], "Planning for sustainable urban areas or everyday life and inclusion" proclaim by examining micro streel level situations that inclusive urban design applications are not considered in many designs, policies regarding sustainable urban settlements. In these policies, although enough considerations are given to the disabled but needs of older people & women are not considered and she claims that this is one of the other reasons for their discomfort as they cannot access and move comfortably in these public spaces.

Afacan [10], "Achieving Inclusion in Public Spaces: A Shopping Mall Case Study" discuss need and expectation of various users in Shopping Mall in Turkey. In the survey, it was cleared from the user's point of view that exclusive approach in public space is more considered and by this approach expectations of the diverse user is not fulfilled. Various factors of inclusive approach were elaborated which increase the quality of life by making public space more user friendly.

2.3. Impact of Urban Landscape Elements

Jain [11], "Regeneration and Renewal of old Delhi," discusses the need and evolvement for the redevelopment process to enhance life and economic sustainability in one of the most important parts of Delhi. He described the importance of the renewal of selected areas by improving infrastructure and adding urban landscape elements which will increase the interest of visitors along with our heritage could be preserved by restoration of buildings without changing the traditional character.

Thompson [12], "Activity, exercise and the planning and design of outdoor spaces" suggested a strategy through a model of action and context. In this paper is researcher elaborated that urban landscape elements play a pivotal role in aesthetics and activities performed in space, which is admired by all age groups.

Mosler [13], "Everyday heritage concept as an approach to place-making process in the urban landscape", elaborated that to engage activity and place, urban design should respond to the urban landscape of that space. The researcher concluded, by considering Lynch's approach of mental map, this paper claims that if an urban landscape/space is designed with the historic fabric of that city, it becomes a landmark and leaves a mark on tourist mental map.

Capitanio [14], "Attractive Streetscape Making Pedestrians Walk Longer Routes: The Case of Kunitachi In Tokyo" explained the pleasurable influence of streetscape features on pedestrian behavior in urban design. Simulation of the pedestrian was done and found that pedestrians prefer the attractive route, irrespective of the fact that it is longer than other optional routes which were also safe, short but had the least streetscape elements. Thus, streetscape features like shading greenery of different kinds and heights, sitting opportunities, and broad sidewalks make the route more interesting till our destination.

2.4. Economic Aspect in the Retail Sector

Doley [15], "Golden Temple heritage street: Revamps heralds Amritsar makeover" claimed that after adding urban landscape elements and restoration of façade, the existing 1km str*etc*h which earlier appeared congested but now same space appear to be huge and attractive resembling open monuments. Added features like identical signboards for the shops like the markets of Jaipur, improved services, various streetscape furniture and fountains depicting our heritage, *etc.* are having a positive impact on social as well as economical aspects.

Teulings, *et al.* [16], "The urban economics of retail" discussed these developments from a land market point of view." They tested microdata from big shopping areas in the Netherlands and found that consumers are simulated by amenities and facilities which is positive for the B&M retail industry. Along with this, they interpreted that retail profits and shop rents are affected by walking distance to the center of the shopping area.

Ganesha, *et al.* [17],"Theory of B & M Retailing in India", explained that in the Indian context the overall phenomenon of B&M retailing is truly complex, and that complexity is necessary to an adequate description of the overall retailing phenomenon, we cannot rely on theories which are practiced in developed countries. They recommended for sustainable success, the image on the customer's mind is of utmost importance, thus, retailers need to think beyond revenue and profit.

Ganesha, *et al.* [18],"Rational Organizational Structure: For Brick-and-Mortar Lifestyle Retailers in India to Overcome Diseconomies of Scale and Protect Firm's Sustainability", demonstrate that the existing belief of B & M lifestyle retailers in India which assumes economies of scale and long-term firm's sustainability as the retailer increases the store count is just a misconception and does not hold. The sustainable success of a B & M lifestyle retailer in India significantly depends on the trueness level of their image.

3. EXAMPLES OF INCLUSIVE APPROACH

Tiwari [19], "Developing a Sense of Place by Humanizing Public Pedestrian Precincts" focused on Abraham Maslow's theory for uplifting public-pedestrian zones as he believed that along with physical, psychological needs of all human beings should be taken care of. Further considering international successful examples of city cores where planners were able to make these areas fully or partially pedestrianized, it was recommended same should be implemented in contemporary cities in India to preserve the original character of the place for whom it was designed *i.e.* "The Pedestrians".

In many cities of India, still, a major mode of transport is walking, but in public precincts, absence of facilities and amenities for pedestrians and increased number of vehicles is the main reason for road accidents, thus planners along with decision-makers should fulfill a pedestrian need for safety and comfort. Parida,*et al.* [20], "Feasibility of Providing a Skywalk for Pedestrian in Chandni Chowk, Delhi" recommended that by providing grade-separated, the movement of pedestrians would be safe and comfortable.

Krishna [21], "The Catalysts for Urban Conservation in Indian Cities: Economics, Politics, and Public Advocacy in Lucknow" discusses that for preserving culture and heritage economic interests, political support, and public participation act as stimulators in the whole process. Qualitatively examination of iconic traditional market precincts *i.e.*, Hazratganj of Lucknow's was interpreted and recommended that all three mentioned catalysts were effective in uplifting the Hazratganj precinct and as stakeholders were part of the process, so they felt connected with the success of the project.

Malhotra [22],"Pedestrianisation of Commercial Area: A Case Study of Aminabad, Lucknow" has intended that for better socio-economic and environmental sustainability public spaces should retain their main purpose i. e. better accessibility and mobility for pedestrian movement which make space lively and vibrant. Through the case example of Aminabad Bazar, she emphasized the need for restoration of the market is need of the hour to bring back the color, life and make it an attractive place as it was in past.

Undeniably, for furnishing the needs of the city population, urban cores play a pivotal role, as they are considered a catalyst in the economic growth of any urban city. But in recent times these clusters are transformed into the non-livable environment due to many issues like pollution, congestion, longer travel distances and time, *etc.* Randhawa & Kumar [23] "Reviving the Urban Core: Ludhiana

City, Punjab, India" intended to identify the urban development issues in Ludhiana city core and provide appropriate strategies based on the Smart Development principles.

CONCLUSIONS

After intensive research, it is evident that there is a drastic decline in footfall in B & M retail in recent years, but at the same time impact of urban landscape elements along with other intangible planning, parameters are also visible in sustaining different urban areas. In India, concerning pedestrian interest and safety sufficient guidelines like appropriate lighting, signage, street furniture, sidewalk, *etc.* are present, but in practice, these amenities & facilities are absent in existing B & M markets; thus, pedestrians are declining at a fast pace, which is one of the major reasons of declining profit in these retail markets. Qualitative research with the context of socio-culture aspects of the city is needed to be accomplished in potential tier-II cities with a systematic methodology in these public spaces. Thus, this highlights the need to have comprehensive research to understand the pedestrian user experience and their relationship with intangible characteristics of retail market infrastructure and its impact on them by considering B & M markets in different Cities.

CONSENT FOR PUBLICATION

Not applicable.

CONFLICT OF INTEREST

The authors declare no conflict of interest, financial or otherwise.

ACKNOWLEDGEMENT

Declare none.

REFERENCES

[1]   D. K. Rigby, "The Future of Shopping. Harvard Business Review", *Innovation issue of Harvard Business Review,* 2011.

[2]   P. Vij, "The study and the analysis: an impact of organized Retail on unorganized retail in India", *International Journal of Multidisciplinary Management Studies,* vol. 3, pp. 174-184, 2013.

[3]   S.B.J. Carralero-Encinas, "The role of store image in retail internationalisation", *Int. Mark. Rev.,* vol. 17, no. 4/5, pp. 433-453, 2000.
      [http://dx.doi.org/10.1108/02651330010339941]

[4]   *J. Mark. Channels,* vol. 70, pp. 107-118, 2006.

[5]   J-L.C.G.C.J.P. Mbaye Fall Diallo, "Factors influencing consumer behaviour towards store brands", *Int.*

J. Retail Distrib. Manag., vol. 41, no. 6, pp. 422-441, 2013.
[http://dx.doi.org/10.1108/09590551311330816]

[6]   N. H. Mullick, "A Study of Shopping Experience in Selected Retail Centres in NCR", *Retail Marketing in India: Trends and Future Insights,* 2016.

[7]   G-J. Hospers, "Lynch's the Image of the City after 50 Years: City Marketing Lessons from an Urban Planning Classic", *Eur. Plann. Stud.,* pp. 2073-2081, 2010.
[http://dx.doi.org/10.1080/09654313.2010.525369]

[8]   P. T. Reardon, "Book review: The Death and Life of Great American Cities by Jane Jacobs", 2012.

[9]   C. Greed, "Planning for sustainable urban areas or everyday life and inclusion", *Urban Design and Planning,* vol. 164, no. DP2, pp. 107-119, 2010.

[10]   Y. Afacan, *Achieving Inclusion in Public Spaces: A Shopping Mall Case Study.* Designing Inclusive Systems, 2012, pp. 85-92.

[11]   A.K. Jain, *Regeneration And Renewal of Old Delhi (Shahjahanabad).* ITPI Journal, 2004, pp. 29-38.

[12]   C.W. Thompson, "Activity, exercise and the planning and design of outdoor spaces", *J. Environ. Psychol.,* vol. 34, pp. 79-96, 2013.
[http://dx.doi.org/10.1016/j.jenvp.2013.01.003]

[13]   S. Mosler, "Everyday heritage concept as an approach to place-making process in the urban landscape", *J. Urban Des.,* vol. 24, pp. 1-17, 2019.
[http://dx.doi.org/10.1080/13574809.2019.1568187]

[14]   M. Capitanio, "Attractive Streetscape Making Pedestrians Walk Longer Routes: The Case of Kunitachi In Tokyo", *J. Archit. Urban.,* vol. 14, pp. 131-137, 2019.
[http://dx.doi.org/10.3846/jau.2019.10359]

[15]   K. Doley, *Golden Temple heritage street: Revamp heralds Amritsar's makeover.* Financialexpress: New Delhi, 2017.

[16]   *J. S. Coen N. Teulings, "The urban economics of retail.* ERSA: Vancouver, 2018.

[17]   "Theory of Brick-and-Mortar Retailing in India", *Munich Personal RePEc Archive,* 2020.

[18]   "Rational Organizational Structure: For Brick-and-Mortar Lifestyle Retailers in India to Overcome Diseconomies of Scale and Protect Firm's Sustainability", *Munich Personal RePEc Archive,* 2020.

[19]   S. Tiwari, "Developing a Sense of Place by Humanizing Public Pedestrian Precincts", *International Journal of Architecture and Urban Development,* vol. 3, no. 3, 2013.

[20]   J.S.S.G. Purnima Parida, *Feasibility of Providing a Skywalk for Pedestrian in Chandni Chowk.* Indian Highways: Delhi, 2014, pp. 20-29.

[21]   A. Krishna, "The Catalysts for Urban Conservation in Indian Cities: Economics, Politics, and Public Advocacy in Lucknow", *J. Am. Plann. Assoc.,* 2016.
[http://dx.doi.org/10.1080/01944363.2015.1132390]

[22]   C. Malhotra, *Pedestrianisation of Commercial Area: A Case Study of Aminabad.* International Research Journal of Engineering and Technology: Lucknow, 2017, pp. 930-936.

[23]   A. Kumar, *Reviving the Urban Core: Ludhiana City, Punjab, India,* 2020.

CHAPTER 40

Design of Arduino and Ultrasonic Based Smart Shoe

Bankuru Gowthami[1], Teetla Anand[1], Kesu Manoj Kumar[1], Vishal Agrawal[1] and Suresh Kumar Sudabattula[1,*]

[1] School of Electronics and Electrical Engineering, Lovely Professional University, Phagwara, Punjab, India 144411

Abstract: With the increase in population, usage of vehicles is also increasing at an alarming rate. People working round the clock and earning handsome amounts usually prefer four-wheelers to ease their journey from one location to another location. But this ease has now created a lot of problems as now more traffic jams can be seen than before and traveling time is again increasing for people bound with deadlines. So, now people have moved from four-wheelers to either two-wheelers or by foot. Also, while traveling to unknown places and roads, GPS is a must nowadays. But on two-wheelers, it is tough to use GPS on mobile phones. Moreover, the shoe is the necessity of life and when it comes to a blind person shoes can help to protect from pebbles and roads but obstacles cannot be avoided. So, we are trying to make it more convenient for blind people as well, so that with the help of sensors in front of shoes he/she will be indicated with vibrations to avoid the obstacles in the path. we are trying to make it more efficient by using new technology *i.e.*, Smart shoe where information of path is retrieved from Google navigation database, and accordingly, the instructions are sent to the sensors installed in the shoes to take turns (*i.e.*, U-Turn, left, right, *etc.*). All this can be achieved by interfacing the Bluetooth sensor placed in shoes and it would give signals to 4 sensors placed in four directions in shoes and when to take any specific turn it would vibrate. Also, this   signal is provided to the Bluetooth sensor from a mobile application which will be connected to the Google navigation database.

Keywords: Bluetooth, Microcontroller, Navigation Database, Sensors, Smart Shoe, Ultrasonic Sensor, Vibration Motor.

INTRODUCTION

With the fast-moving world, technology is also growing faster to ease the life of human beings. Multiple options are already available for tracking routes but they sometimes lack fulfillment in certain circumstances. Also, when it comes

* Corresponding author Suresh Kumar Sudabattula: School of Electronics and Electrical Engineering, Lovely Professional University, Phagwara, Punjab, India 144411; E-mail: suresh.21628@lpu.co.in

Dharam Buddhi, Rajesh Singh and Anita Gehlot (Eds.)

to handicapped people there is no efficient and affordable technology available. So, we are coming up with the idea of smart shoes which will be able to track the location and accordingly give direction to the person without actually holding his/her smartphone in hand. Also, if a blind person is wearing it he/she can feel comfortable as it will be notified with the help of vibrations against any obstacles in the path.

This Mobile phone plays a major role as we are concentrating on the Bluetooth module. So, for handicapped people, this is the best advantage. Therefore, a Mobile phone with Bluetooth acts as a port between Smart Shoe and GPS. Concentrating on GPS which acronyms for Global Positioning System is also in trend. People nowadays rely on it when exploring new routes or when distance measuring. GPS has been enabled and introduced in a way to detect the exact endpoint of geographical locations by military officers for the main operations and also by the civil users. This is based on the use of satellites that transfer particulars in earth orbit which allow computing the distance between the user and the satellites. Now, in the smart shoes with the accumulation of these both technologies *i.e.* GPS and Sensors, we can make the life of humans more comfortable. Now with the use of these smart shoes if a person is holding his/her Mobile phone which should be connected to shoes then with the 4-way sensors he can move along that path even if he/she is holding the phone. And for the blind also his/her life can be much more.

BACKGROUND WORK

In this project, our work will be concentrated on making a complete design for the prototype for the shoe with all the components inside it, and to be perfectly distributed in the shoe without affecting the feet. On the other side, we have a serious challenge which is to make the circuit small and to choose the smallest components as possible as we can to fit in the shoe as a primitive prototype design made by our hands, and in the long term to be abroad manufactured.

Proposed Work and Methodology

Many other solutions are available for tracking paths which are very useful in places of pilgrims and tracking on mountains and others. The techniques used in these already available systems will be discussed in this section. Global Processing System Navigation algorithm, mainly designed and introduced for hearing and sight-impaired people [1]. It consists of a GPS device with Bluetooth which interfaces with mobile. But this technology has its limitations as it does not capture all areas of markets [2]. The main keywords and Necessary information required for GPS Navigation functionality are GPS status, GPS Provider, Speed, HDOP, VDOP, longitudinal, GPS Dispatcher as shown in Fig. (**1**) [3].

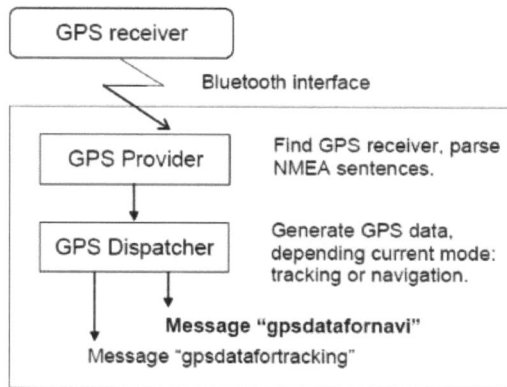

Fig. (1). Obtaining the necessary GPS data.

GPS Receiver: This communication uses GPS Provider. This mainly helps to carry through search and communication with GPS receivers by interfacing with Bluetooth. This process of connection and search is done without any human interaction [4].

GPS Provider: Find GPS receiver, parse NMEA sentences.

GPS Dispatcher: The use of class GPS Dispatcher, which helps in implementing and interfacing GPS Listener, speed modules, and filters GPS position for the availability of new data [5].

Analyzing GPGGA NMEA-0183 and CPGGA Sentences, Smooth processing of GPS information are obtained: They are Direction HDOP and VDOP, Latitude speed, Latitude, Status, Longitude approach to the GPS data, GPS Listener and Generates mainly Global Positioning System Information, depending on current mode tracking or navigation [6].

Mailbox: This is used for communication between classes. The dispatcher class generates new message "GPS data for NAVI" [7]. Whenever there is any new message, this message is generated in 1.5 to 10 seconds determined on the path of fast filtered speed.

Purpose

Our project mainly concentrating on the idea of pairing smartphones using Bluetooth with smart shoe prototypes and help the person to give necessary navigational data to those who are using it.

Related Work

The main idea we are using to design smart shoes accrues from an already designed prototype that uses another technology. Yet other advanced technologies have so many drawbacks. There is a various number of shoes available in the market named Smart shoes which are mainly used for gaming and workout purposes. The most common use is Pressure sensors and accelerometers. Also, for providing the power it uses cells that are not that reliable. Also, for Alzheimer's patients, the watches with installed GPS are essential as this will help the family members of the patient to keep track of them. Also, in hajj yatra famous in Saudi Arabia, to find out the exact location of pilgrims, wireless sensors are used which are connected to Bluetooth. Our project is kind of unique, and its work plays a prominent role in the people who are using it. We did great research if any earlier would match the work as we did, we noticed that Apple Company cooperating with Nike that has a similar idea like us, but we found it's not the same as ours. Now if we explore the usage of smart shoes in fitness then the available system can check for step count, calories burned and also can set daily reminders. Motion sensing games are also available which fetch the position of the user and gives the feeling of real-time gaming. Also, GPS-enabled shoes are much more useful for the safety of girls and children so that parents can keep track of them 24/7.

Block Diagram

The block diagram consists of a Microcontroller, Vibration Motor, Bluetooth device, Ultrasonic Sensor, Battery, and Piezoelectric Sensor. This prototype involves the programming of microcontrollers which will give instructions to sensors (*i.e.*, vibration motor and ultrasonic sensors) and receive it from a Bluetooth sensor. Depending on the important and main commands encountered, the Micro controller then enables many vibrations to lead the person using this to reach out to his/her landing place. Also, an important part of this prototype focuses on creating circuits that will power the microcontroller and also generating the power according to requirement. Fig. (**2**) shows the overall block diagram.

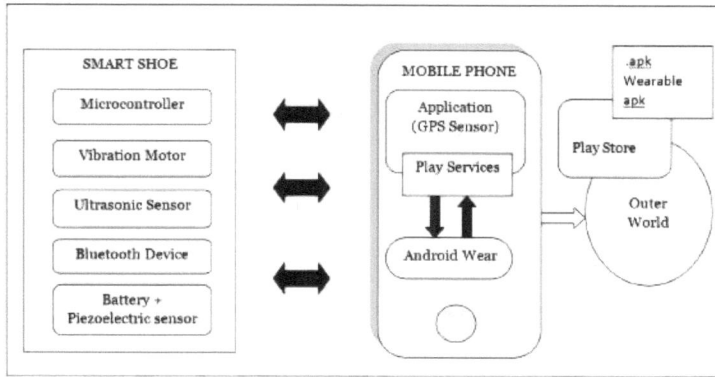

Fig. (2). Overall block diagram.

SOFTWARE AND HARDWARE IMPLEMENTATION:

Microcontroller

The microcontroller unit which comprises Arduino Pro Mini which is small in size suits our compact distributed design is used to send signals to vibration motors according to the signals received from ultrasonic sensors and android applications. It is also used to measure the voltage across the Force Sensitive Resistor, Which mainly calculates the current and pressure of the person's feet who are using it. These all information are required so that the piezoelectric sensor will be able to produce equivalent amounts of energy to charge the battery which will provide power to the Arduino Pro Mini. To function correctly Arduino Pro Mini operates at 2.7 - 5.

Fig. (**3**) shows Arduino Pro Mini and Fig. (**4**) shows the Ultrasonic sensor.

Fig. (3). Arduino Pro Mini.

Fig. (4). Ultrasonic sensor.

Ultrasonic Sensor

We are using HCSR-04 ultrasonic sensors and here we are using two ultrasonic sensors which are mainly used to detect the obstacles that occur in the pathway of the user. One sensor among two ultrasonic sensors is organized on the front end part of the shoe for obstacle detection in the front pathway of the user, whereas the second sensor among two is organized on the side part of the shoe for sideward detection of obstacles.

Coin Vibration Motor

Vibration Motor Unit consists of 4 sensors that are placed around the four sides of the shoes for the guidance of four directions. In which it provides the necessary information to guide the person, that is using in different directions based on the signals that are received from the microcontroller. As these motors are mounted inside the shoe, the user that he/ she uses the shoe must be comfortable to use them, the motors we used should be lightweight and consume a minimum amount of energy to be able to provide necessary and sufficient vibration output. Fig. (**5**) shows the coin vibrator motor and Fig. (**6**) shows the Bluetooth module.

Fig. (5). Coin Vibrator Motor.

Fig. (6). Bluetooth module.

Bluetooth Module

Here we are using HC-06 Bluetooth that communicates in serial connection and interfaces with Arduino Pro Mini. The Bluetooth unit is powered by a regulated power supply from the power unit. Bluetooth module ensures compatibility with most smartphones out in the market today. This Bluetooth module allows always us to send and receive messages or information from the android application that we have designed and organized to micro-controller. Bluetooth module needs 3.0 – 6V for better performance and it combines and syncs with another Bluetooth unit within a connective area and a distance of 5ft. Fig. (7) shows the Arduino pro mini and HC-06 schematic.

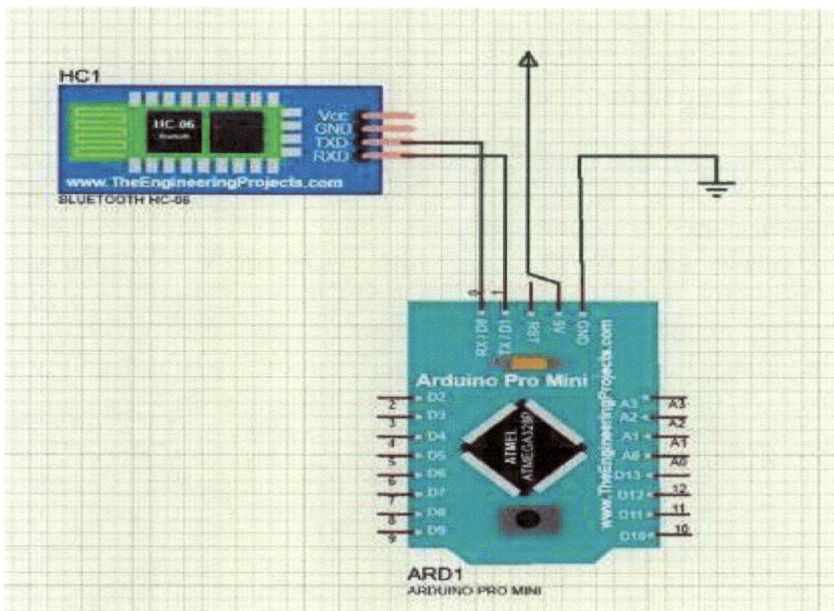

Fig. (7). Arduino pro mini and HC-06 schematic.

INTERFACING DIAGRAM

Google Mapping Database

Google has a mapping database that can be used for both outdoor and indoor setouts. This data will be set up to calculate signals by an android application unit for the vibration motors and with this data, we can send the signals to the appropriate sensors. Navigation information from the Google Mapping Database will android Application depending on his/her location. Figs. (**8a and b**) show complete prototype designing.

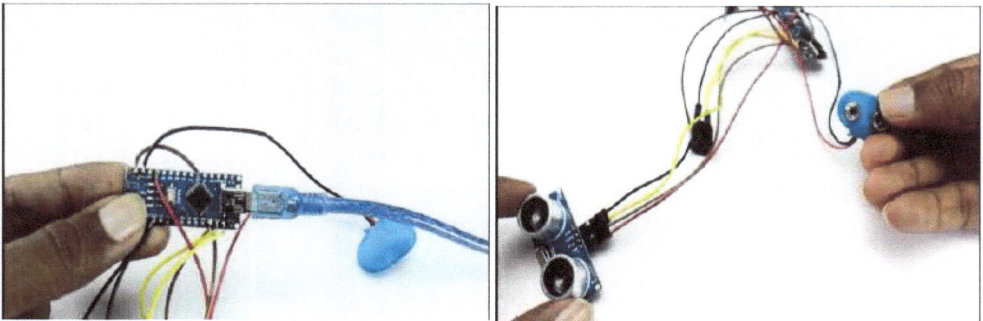

Fig. (8 a,b). Complete Prototype designing.

Benefits of Using Smart Shoe

This study shows the smart shoe system which is being interfaced with the Google mapping database and helps its users to reach his/her destination without actually hustling with mobile phones. Another thing here is the present technology of smartphones uses a battery or cell which either needs to be charged or replaced but in this study, we make sure that the power to the controller will be provided, and the user needs to be moving. The smart shoe uses navigational assistance while traveling, automatics re-routing alerts, and various user-controlled vibration patterns. Figs. (**9a** and **b**) show components inside the shoe.

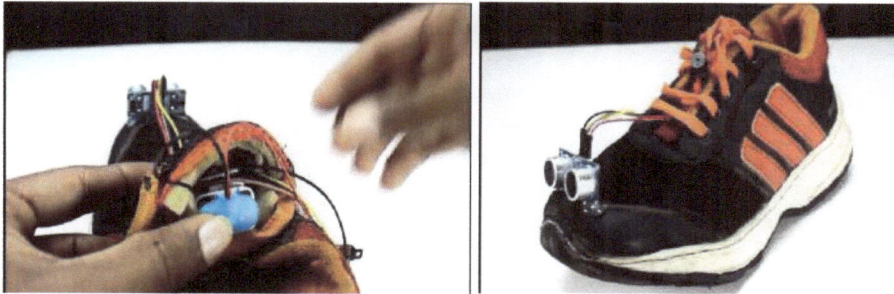

Fig. (9 a and b). Components inside shoe.

CONCLUSION

A smart shoe that helps a person while walking has been successfully created with added features. Our shoe designed with many features can be wear without affecting the feet while walking. The smart shoe was mainly designed with the minimum components and resources available. The circuitry designed was kept very simple with indeed features. This Shoe is nearly used for 3-4 hours of complete charging of the battery. The main feature of adjusting the range for detecting obstacles can be set up in the program, according to the size of the step and walking pattern of visually challenged people. GPS added is used for tracking the user's movement and to observe his/her position. Smart Shoe is mainly fitted for faintly populated zones such as college campuses and indoor and outdoor as well.

CONSENT FOR PUBLICATION

Not applicable.

CONFLICT OF INTEREST

The authors declare no conflict of interest, financial or otherwise.

ACKNOWLEDGEMENT

Declared none.

REFERENCES

[1]　Shubham Rastogi, Pankaj Sharma, Parth Dhall, Rishav Agarwal, and Shristhi Thakur, "Smart assistive shoes and cane:solemates for the blind people", *IJARECE,* vol. 6, no. 4, 2017.

[2]　Shlesha Khursade, Malavika Karunan, Ibtisam Sayyad, Saloni Mohanty, and Prof. B.B Gite, "Smart Shoes: A Safe Future for the Blind", *IJIRCCE,* vol. 6, no. 5, 2018.

[3]   SD Asha Mahesh, Supriya K. Raj , Pushpa Latha MVSSNK, Gowri P., Sonia T., and . Nani, B., "Smart assistive shoes and cane: solemates for the blind people", *International Journal of Engineering Science,* vol. 16, 2018.

[4]   D. Pascolini, and S. Mariotti, "Global estimates of visual impairment", *British Journal Ophtalmology Online.*

[5]   M. Divija, M. Rohitha, T. Meghana, and H. Monisha, "Integrated Smart Shoe for Blind People", *International Journal of Management Technology and Engineering,* vol. 8, pp. 748-752, 2018.

[6]   S. Rosen, "Algorithm for GPS Navigation, Adaptedfor Visually Impaired People", *International Journal of Biology and Biomedicine,* 2019.

[7]   A. Manjunathan, and C. Bhuvaneshwari, "Designing of smart shoe", *Mater. Today Proc.,* vol. 21, pp. 500-503, 2020.
       [http://dx.doi.org/10.1016/j.matpr.2019.06.645]

CHAPTER 41

Myxobacterial Metabolites: A Promising Resource for Big Pharma

Akshay Mohan[1], **Daljeet Singh Dhanjal**[1], **Chirag Chopra**[1] and **Reena Singh**[1,*]

[1] *School of Bioengineering and Biosciences, Lovely Professional University, Jalandhar-Delhi GT Road, Punjab, India*

Abstract: Abstract: The lives of people have been better since the discovery of medicines. Experts were able to discover a cure for even a few chronic diseases like cancer. Scientific studies from multiple areas, including biology, chemistry, and biotechnology, have created a diverse array of solutions to the issues faced by humans in health and medicine. The usage of microorganisms to make valuable products that are important to treat various diseases and improve people's lives by using it in many industrial levels like food, pharmaceutical, and agricultural sectors has made microbial species that include bacteria and fungus, *etc.*, ever-demanding. Research studies from the past few decades proved that many bacteria, including myxobacteria, have been used as a potent source for deriving lots of compounds that could be useful for humans like antibacterial, antifungal, anticancer agents, and enzymes that can degrade cellulose, protein, *etc.* Since it's a proteobacterium commonly found in organic dwells and swamps, it would be feasible to cultivate this bacterial species to produce chemical compounds with more incredible industrial applications. This review will explain why myxobacterium is considered a potential source for the production of industrially important enzymes and many other beneficial secondary metabolites. Also, this review will shed light on various ways to screen and characterize the myxobacterial population to produce cellulolytic enzymes.

Keywords: Cellulase, Enzyme, Myxobacteria, Proteobacteria, Secondary Metabolites.

INTRODUCTION

Myxobacteria are one of the widely distributed microbes ubiquitously found in aquatic and terrestrial ecosystems. A German botanist H.F. Link was the first scientist to discover the myxobacterium strain *P. vitellinium* in 1809. *Stigmatella aurantica* and *Chandromyces crocatus* were the other two genera discovered by M.J. Berkeley in 1857. These species were identified as myxobacteria for the first

---

* **Corresponding authors Reena Singh**: School of Bioengineering and Biosciences, Lovely Professional University, Jalandhar-Delhi GT Road, Punjab, India; E-mail: reena.19408@lpu.co.in

Dharam Buddhi, Rajesh Singh and Anita Gehlot (Eds.)

time in the year 1982 by R. Thaxter. Since the beginning of the 20th century, several studies have been made on myxobacteria. *Myxobacteria Xanthus* became an essential research tool of myxobactrial species [1]. Reichenbach and Coworkers by discovering a new field of biologically active secondary metabolites. Molecular biology, developmental biology, taxonomy, and especially biotechnology are the main focus nowadays in myxobacterial research. Following are some of the myxobacterial genera: *Sg. Stigmatella, H. Haploangium, Na. Nannocystis, Cc. Corallococcus, Cm. Chondromyces, M. Melittangium, Mx. Myxococcus, So. Sorangium* [1].

Two different types of myxobacterial cells are usually seen:

Cell type I: They have flexible rod-like structures usually slender 1 μm in diameter and belong to suborder: Cystobacterinae.

Cell type II: They have a cylindrical road-like structure, usually rigid one μm wide and belongs to suborder; Sorangineae.

Vegetative cells and myxospores have such a unique shape, which becomes a characteristic of the order Myxococcales. Fimbriae are involved in intercellular cohesion as well as in social behavior [2]. Myxobacteria show social behavior, which is characterized by their property of up-taking of food and cooperative motility [3]. They show gliding movement on surfaces made up of solid. This movement is accompanied by the secretion of slime and the bending of cells. 10–60 μm/min is their gliding speed. Flagella, an extracellular motility organelle are lacking [4].

Swarming behavior is a characteristic trait of myxobacterial colonies [5]. An increase in the colony diameter indicates swarming. The movement of myxobacteria is affected by many factors such as the density of cells, availability of nutrients, and temperature. Extracellular enzymes help in the digestion of food and increases feeding efficiency.

ISOLATION AND CULTIVATION OF MYXOBACTERIA

Isolation of myxobacteria is a process because myxobacteria grow slower than other bacteria present in the soil. The different methods for isolation of myxobacteria are [6];

• Baiting with dung pellets
• Baiting with *Escherichia coli* streaks
• Inoculation of filter paper with soil

Cultivation of isolated strains is equally difficult. Many different kinds of media like PDCY media, VY/2 media have been tried, but the best results were shown by CYE Media (in grams per liter): casitone at 28–32°C [1].

The isolation technique by rabbit dung pellets method mainly applies for isolation of *Myxococus Xanthus, Myxococus fulvus, Cystobacter fuscus*, and *Archangium gephyra, Cystobacter ferruginous, Stigmatella erecta, Corallococcus coralloides*. Saline-rich soil is completely unsuitable for myxobacterial growth, owing to which myxobacteria grow in a habitat that is rich in the organic matter having a pH of 6-8. Moreover, terrestrial habitat-specific myxobacteria cannot grow in an environment having above 1% salinity [7].

The soil property also plays a major role in the existence of myxobacteria. So they are mostly found in the soil, such as brown and black soil rich in nutrients, whereas sandy soil and decomposed rock soil lack myxobacteria [8,9].

Secondary Metabolites

Myxobacteria have become the most significant source of secondary metabolites. Many molecules are biologically active with antifungal, antibiotic, and anti-tumor. Chances for medical application have been found in many structures (Reichenbach and Holfle, 1990). Ambrucitin was the first chemical structure of a myxobacterial antibiotic. Then Myxothiazol formed by *M. fulvus* was found. Some essential secondary metabolites of myxobacteria are Corallophyronin C, Myxohiazol, Stigmatellin, Myxoniescin, Salranycine, Sorangicin, Aurachin E, Ambruticin, Nannochellin, Soraphen [10,11].

Epothilone

Epothilone was isolated from *S. cellulosum* epothilone in 1987, which acts upon the eukaryotic cytoskeleton (Jahn E,1911). The action spectrum of epothilone is narrow: rarely bacteria and fungi are affected. It treats cancer cells. Mainly it stops the growth of cancer cells in humans, *e.g.*, breast cancer, ovarian cancers. It is used for cancer treatment in clinical applications [12]. Epothilone has been FDA-approved as an anticancer drug. The structure of epothilone is shown in Fig. (**1**).

Geosmin

Geosmin is another secondary metabolite formed from streptomycetes and is mainly responsible for the typical odor of soil. Geosmin is formed from Nannocystisexedens [13]. The structure of geosmin is shown in Fig. (**2**).

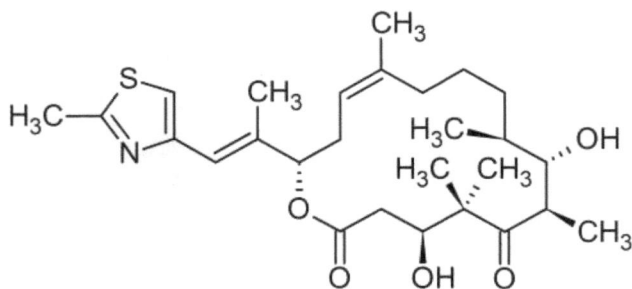

Fig. (1). Chemical structure of epothilone.

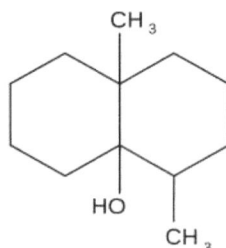

Fig. (2). Chemical structure of geosmin.

Other Important Secondary Metabolites By Myxobacteria

Myxothiazol and the Myxalamids

By combined PKS/NRPS systems of *S. aurantiaca* produces two electron transport inhibitors. Among these promising anticancer agents, Myxothiazol is an electron transport inhibitor. In the recently reported gene for Myxothiazol, aimed that during isolation of the combination of PKS/NRPS gene from myxobacteria, it was identified that a DNA fragment encoding β-ketoacyl-synthase [14].

Chondramides

These are the actin targeting cyclodepsipeptide obtained from the myxobacterium *Chondromycescrocatus*. During the Screening of strains of *Chondromyces,* there were four new antifungal depsipeptides isolated. Their macrocyclic structures comprise three amino acids and a polyketide. These compounds are similar in structure to jaspamide/jasplakinolide from marine sponges of the genus Jaspis. Severe shape malformations are caused by these chondramides when applied during growth [15].

Myxobacteria with Proteolytic Activity

Proteases, which catalyze the cleavage of peptide bonds in protein, offer the possibilities and potential for their biotechnology application and commercial importance. It is represented as one of the three most significant industrial enzymes and explained about 60% of the total enzyme market. In myxobacteria, the proteolytic enzyme has indicated considerable potential in the application area, especially for digestive diseases and food industries [16].

CONCLUSION

Myxobacteria offers a unique resource for several metabolites which can be useful for the pharmaceutical industry. The metabolite epothilone has been approved by the FDA as well. The need for effective secondary metabolites with medicinal or pharmacological properties is on the rise. Being saprophytes, the myxobacteria can break down complex polymeric substrates to synthesize different metabolites under stressful conditions. Owing to the complex life cycle as well as a complex genome, there is a need for more research into the biology of myxobacteria and their metabolites for the pharmaceutical industry. These metabolites have shown anti0-microbial, anti-cancer, antioxidant activities, among others, and are therefore worthy candidates for exploring unique ecological niches for myxobacterial diversity.

CONSENT FOR PUBLICATION

Not applicable.

CONFLICT OF INTEREST

The authors declare no conflict of interest, financial or otherwise.

ACKNOWLEDGEMENT

Declared none.

REFERENCES

[1]   L.J. Shimkets, M. Dworkin, and H. Reichenbach, "The Myxobacteria", In: *The Prokaryotes* Springer: New York, 2006, pp. 31-115.
       [http://dx.doi.org/10.1007/0-387-30747-8_3]

[2]   M. Dworkin, "Recent advances in the social and developmental biology of the myxobacteria", *Microbiol. Rev.,* vol. 60, no. 1, pp. 70-102, 1996.
       [http://dx.doi.org/10.1128/mr.60.1.70-102.1996] [PMID: 8852896]

[3]   D. Kaiser, M. Robinson, and L. Kroos, "Myxobacteria, polarity, and multicellular morphogenesis", *Cold Spring Harb. Perspect. Biol.,* vol. 2, no. 8, p. a000380, 2010.
       [http://dx.doi.org/10.1101/cshperspect.a000380] [PMID: 20610548]

[4]   H. Lunsdorf, and H. Reichenbach, "Ultrastructural details of the apparatus of gliding motility of Myxococcus fulvus (Myxobacterales)", *J. Gen. Microbiol.,* vol. 135, no. 6, pp. 1633-1641, 1989.
        [http://dx.doi.org/10.1099/00221287-135-6-1633]

[5]   D. Kaiser, and C. Crosby, "Cell movement and its coordination in swarms of myxococcus xanthus", *Cell Motil.,* vol. 3, no. 3, pp. 227-245, 1983.
        [http://dx.doi.org/10.1002/cm.970030304]

[6]   F. Gaspari, Y. Paitan, M. Mainini, D. Losi, E.Z. Ron, and F. Marinelli, "Myxobacteria isolated in Israel as potential source of new anti-infectives", *J. Appl. Microbiol.,* vol. 98, no. 2, pp. 429-439, 2005.
        [http://dx.doi.org/10.1111/j.1365-2672.2004.02477.x] [PMID: 15659197]

[7]   X. Zhang, Q. Yao, Z. Cai, X. Xie, and H. Zhu, "Isolation and identification of myxobacteria from saline-alkaline soils in Xinjiang, China", *PLoS One,* vol. 8, no. 8, p. e70466, 2013.
        [http://dx.doi.org/10.1371/journal.pone.0070466] [PMID: 23936436]

[8]   K.I. Mohr, "Diversity of Myxobacteria-We Only See the Tip of the Iceberg", *Microorganisms,* vol. 6, no. 3, p. 84, 2018.
        [http://dx.doi.org/10.3390/microorganisms6030084] [PMID: 30103481]

[9]   D. Kaiser, C. Manoil, and M. Dworkin, "Myxobacteria: cell interactions, genetics, and development", *Annu. Rev. Microbiol.,* vol. 33, no. 1, pp. 595-639, 1979.
        [http://dx.doi.org/10.1146/annurev.mi.33.100179.003115] [PMID: 115383]

[10]   S.C. Wenzel, and R. Müller, "Myxobacteria--'microbial factories' for the production of bioactive secondary metabolites", *Mol. Biosyst.,* vol. 5, no. 6, pp. 567-574, 2009.
        [http://dx.doi.org/10.1039/b901287g] [PMID: 19462013]

[11]   D. Krug, G. Zurek, O. Revermann, M. Vos, G.J. Velicer, and R. Müller, "Discovering the hidden secondary metabolome of Myxococcus xanthus: a study of intraspecific diversity", *Appl. Environ. Microbiol.,* vol. 74, no. 10, pp. 3058-3068, 2008.
        [http://dx.doi.org/10.1128/AEM.02863-07] [PMID: 18378661]

[12]   K. Gerth, N. Bedorf, G. Höfle, H. Irschik, and H. Reichenbach, "Epothilons A and B: antifungal and cytotoxic compounds from Sorangium cellulosum (Myxobacteria). Production, physico-chemical and biological properties", *J. Antibiot. (Tokyo),* vol. 49, no. 6, pp. 560-563, 1996.
        [http://dx.doi.org/10.7164/antibiotics.49.560] [PMID: 8698639]

[13]   J.S. Dickschat, H.B. Bode, T. Mahmud, R. Müller, and S. Schulz, "A novel type of geosmin biosynthesis in myxobacteria", *J. Org. Chem.,* vol. 70, no. 13, pp. 5174-5182, 2005.
        [http://dx.doi.org/10.1021/jo050449g] [PMID: 15960521]

[14]   B. Silakowski, H.U. Schairer, H. Ehret, B. Kunze, S. Weinig, G. Nordsiek, P. Brandt, H. Blöcker, G. Höfle, S. Beyer, and R. Müller, "New lessons for combinatorial biosynthesis from myxobacteria. The myxothiazol biosynthetic gene cluster of Stigmatella aurantiaca DW4/3-1", *J. Biol. Chem.,* vol. 274, no. 52, pp. 37391-37399, 1999.
        [http://dx.doi.org/10.1074/jbc.274.52.37391] [PMID: 10601310]

[15]   F. Sasse, B. Kunze, T.M.A. Gronewold, and H. Reichenbach, "The chondramides: cytostatic agents from myxobacteria acting on the actin cytoskeleton", *J. Natl. Cancer Inst.,* vol. 90, no. 20, pp. 1559-1563, 1998.
        [http://dx.doi.org/10.1093/jnci/90.20.1559] [PMID: 9790549]

[16]   P. Thakur, C. Chopra, P. Anand, D.S. Dhanjal, R.S. Chopra, J. Singh, D. Sharma, G. Kumar, and N. Sharma, Myxobacteria: Unraveling the potential of a Unique Microbiome Niche.*Microbial Bioprospecting for Sustainable Development.* 1st ed. Springer Singapore, 2018, pp. 137-163.
        [http://dx.doi.org/10.1007/978-981-13-0053-0_7]

CHAPTER 42

Modern Overtures in Biomaterials for Bone Tissue Engineering and its Applications

I. Shubhangi Das[1,*], Manisha Yadav[1] and Anjuvan Singh[1]

[1] *Department of Biotechnology, School of Bioengineering and Biosciences, Lovely Professional University, India*

Abstract: Bone tissue engineering is a stimulating way to directly repair and engineer the defected/damaged bone tissue by omitting the less efficient techniques used in conventional practices. Biomaterial plays a cardinal role in providing support and extracellular environment for cell proliferation, differentiation and enhances tissue regeneration. The traditionally used bone graft has been substituted by the tissue engineering technique, due to their minimal pathogen transmission property. As there have been some challenges and restrictions, bone tissue engineering has not been a hot topic in clinical practices. The aims and objective of this paper are to review the current approaches in the developing field of Bone tissue engineering and find a better substitute as an implant device for the welfare of society.

Keywords: Biomaterial, Bone Graft, Bone Tissue Engineering.

INTRODUCTION

The aspiration behind bone tissue engineering is to improve the current approaches for bone tissue regeneration. Bone being one of the significant issues that provide a structural backup for various organs in the human body also plays a vital role in mineral metabolism and provides a site for hematopoiesis [1]. There are chances of immune rejection and pathogen transmission during allograft an autogenous graft, so as a substitute to these bone grafts, metal and ceramics are being used [2]. Biomaterials are utilized for repair or replacement, treatment, augmentation, evaluation of tissue or specific parts of the human body. Scaffolds consist of biomaterials that act as a carrier of cells and signals playing a vital role in bone tissue engineering [3]. Then the bone tissue engineering started focusing on different alternative treatments which will overcome the current situation and issues for treatments available like immune rejection, donor site morbidity, and pathogen transfer [4]. Bone grafts performed each year cost $1.6 million [5].

---

[*] **Corresponding author I. Shubhangi Das:** Department of Biotechnology, School of Bioengineering and Biosciences, Lovely Professional University, India; E-mail: shubhangidas7@gmail.com

BONE TISSUE ENGINEERING

Bone tissue engineering is affined with the making of implantable bone exchange replacing defective critical skeletal which cannot heal on its own. In this scenario, priority is given to transplantation. The donor organ/tissue must be compatible with the recipient or it may result in graft rejection [6]. Nowadays, there have been many challenges in front of us related to preclinical, model business challenges; these issues have hampered clinical versions [7]. The technique involved in the making of a new substitute that can replace the damaged or lost bone tissues in a living body is referred to as bone tissue engineering [8]. Bone offers structure-designed formation, muscles binding place, tendons, and ligaments, and also the shield of vital tissues [9]. Some of the painful diseases regarded as musculoskeletal diseases include osteoporosis by losing bone mass, rheumatoid arthritis causing joint inflammation, and osteoarthritis affecting the joints in our body [10].

HISTORY OF BONE TISSUE ENGINEERING

Bone tissue restoration and replacement are bound to a significant amount of limitations and the need for quality life in the aging population have driven the field of bone tissue engineering at a demanding pace [11].

It all started with the introduction of limb transplants by Damien along with Saint Cosmos. Although the innovative idea was a result of mythology, science came from behind to prove this theory successfully. Soon, artificial analogs were used to replace the tissue and limbs and the tissue replacement concept became notable [12].

A small experiment performed in early 1970 by W. T. Green made it evident that future technologies could create biomaterials with great biocompatibility factors. His aim was to implant a mouse with chondrocytes in its bone spicules. The expected results were to form new cartilage. Although the experiment did not turn out as expected it invoked hope for future technologies [13].

BIOMATERIALS AND THEIR CLASSIFICATION

Any substance that can be engineered biologically into some material that posses therapeutic benefit and interact with biological systems which could also be utilized to repair or replace a specific portion of a living organism is termed as a biomaterial. The biomaterial chosen must have these three important properties bioinert, bioresorbable, and bioactive. In the present era, damaged tissues are either replaced or reconstructed. Both of these techniques require immense practice and compatibility. Though replacement or transplantations have aided a lot of people, it also has its setbacks [14].

Classification of Biomaterials

Metallic Biomaterial

The biomaterials falling under this category can be used as a support material for the biological tissues and are being used currently minute amounts. These can be used as fracture plates but usually, metals are susceptible to corrosion inside the body of an organism [15]. Metals being highly corrosive cannot be implanted for long inside the human body and this property limits the use of metallic biomaterials. Majorly, there are 3 sub-categories of metallic biomaterials as shown in Table **1** [16].

Table 1. Types of Metallic Biomaterial.

Metallic biomaterial	Implementation
Stainless steel	Screws, plates, nails, hip prosthesis
Co-based alloys	Joint prosthesis, dental fillings
Ti-based alloys	Cup for hip prosthesis, pacemakers

Ceramic Biomaterials

These can be used as an alternative to metal biomaterials as they are capable of strongly binding to the bone. Generally, bio-ceramics like alumina and zirconia are bio-inert which does not harm the living organism by causing the reaction [17]. Synthetic hydroxyapatite is remarkably one of the best biomaterials possessing great osteoinductive properties and accelerates mineralization for the rapid proliferation of bone tissues [23].

Polymeric Biomaterial

Polymer can be molded to the desired shape and is comparatively economical [8]. Apart from this, the fibrous feature of polymers allows manipulation at the time of scaffold fabrication so that the porosity and structure of scaffolds can be controlled [18].

Natural Polymer

Collagen, chitosan.

Synthetic Polymer

There is a wide variety of synthetic polymers like- Poly-Glycolic acid(PGA), Poly-Lactic acid(PLA), Poly-Lactide-co-glycolide.

Biodegradable Polymeric Biomaterial

Biodegradable polymeric substitutes are degraded gradually over time and also reduce the risk of causing hazards inside the host after years of implantation [19]. Another new property is seen in these biomaterials is that they also help in tissue regeneration inside the host as they get degraded in this process [20].

Composite/Hybrid Biomaterials

It is the mixture of two or more compounds that can bind together in different proportions and mimic the tissue functions and can be used to replace the damaged tissues [21].

APPLICATION OF BIOMATERIALS IN BONE TISSUE ENGINEERING

Alternatively, allogenic bone grafts can be utilized, merely they are prone to pathogenic infections and compromise biocompatibility [13]. Accordingly, alternative methods are being manipulated to provide effective and legitimate bone grafts for the ones who are unable to cope up with the clinical strategies [22]. Cartilage lesions are considered painful because of their slow degeneration process and conventional methods seem to show minimal benefits and switching to tissue engineering seems to be an effective alternative [23].

New technology allows *de novo* skeletal tissue formation [24]. When it comes to the repair of large bone defects, bone grafts are ascertained as the feasible technique. The instauration of a technique in which mesenchymal stem cells are gleaned from bone marrow and the cells are amplified *in vitro* followed by transplantation by loading the amplified cells onto a scaffold [25].

Modification of Hyaluronic Based Hydrogels for the Application in Bone Tissue Engineering

For Hyaluronic acid to be used in Bone tissue engineering, it must undergo chemical modifications and cross-link for alternation, so that it can be used in bone tissue engineering. To get the hydroxyl group to be modified generally, an ester bond is being used. For the carboxylic acid group, firstly it is being modified with hydrazide and then it is made to cross-linked with the ester bond [1].

Hydrogels have the property to respond to external stimuli and they are commonly called stimulation responsive hydrogels [26].

Fracture Healing Using Degradable Magnesium Fixation Plates and Screws

Fracture treatments often require some internal fixation materials like plates and screws to stabilize the fragments of bone throughout the process of healing. The plates and screw devices for fracture treatments are made up of non-degradable metals like alloys of Titanium. Due to the strength and property of being biocompatible, these metals have been selected to be used for fracture healing.

Magnesium alloy is also used as it has additional properties compared to other metal alloys and also it contains the property of being low in density, fracture toughness is high and it has strength similar to cortical bone [27].

Metallic Implants for Hip and Joint Replacements

Joint implants are usually composed of co alloys which are used for joint replacement in hips and shoulders commonly. The ball and socket model is designed in such a way that provides ease of movement to the recipient in Fig. (1) [16]. Harrington rods are used for the rectification of spine abnormalities and this is considered ideal for spine rectification because of its ability to restrict its invasion in the spinal canal, which could later result in neurological complications in Fig. (2) [28]. Pedicle screws on the other hand provide structural support to the healing tissues. Most commonly, this implant is used for cervical reconstruction in Fig. (3) [16].

Total Hip Prosthesis

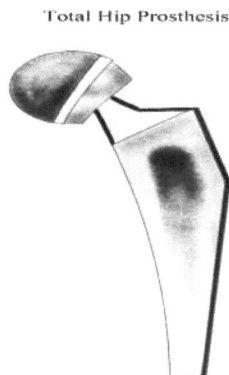

Fig. (1). Depicts the hip replacement biomaterial.

Harrington Rod

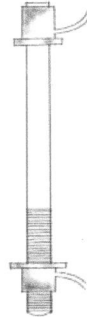

Fig. (2). Depicts the Harrington rod used in fixing spine curvature.

Pedicle Screw

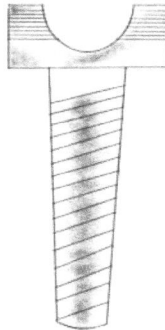

Fig. (3). Depicts a pedicle screw used in a wide range of applications to fix bone defects.

Present Advancements in Bone Tissue Engineering

Biological Method

The site suffering tissue damage has a slow healing tendency and could take months to heal. This is where tissue engineering takes over. Modern biological techniques allow the manipulation of tissues *in vitro* where cells are reared in an environment mimicking the condition similar to the natural one. While rearing, several growth factors are summed up and loaded into the tissues such as osteogenic growth factors, which helps in the fast recovery of the damaged tissue after implantation [29].

CONCLUSION

This article marks the schemes and different concepts of biomaterials and their derivatives required for bone tissue engineering. For instance, a magnesium-based alloy is used as biomaterials as it can dissolve in body fluid that makes implanted magnesium to get degraded during the time of healing process. Bone tissue engineering scaffolds are made up of different materials such as natural and synthetic; polymer sand bioactive ceramics. Novel current approaches for bone regeneration have been mentioned and their advancement in clinical applications of bone tissue engineering.

CONSENT FOR PUBLICATION

Not applicable.

CONFLICT OF INTEREST

The authors declare no conflict of interest, financial or otherwise.

ACKNOWLEDGEMENT

Declared none.

REFERENCES

[1]   K. A. Hing, "Bone repair in the twenty-first century: biology, chemistry or engineering?", *Philosophical Transactions of the Royal Society of London. Series A: Mathematical, Physical and Engineering Sciences,* 2004.
       [http://dx.doi.org/10.1098/rsta.2004.1466]

[2]   A.J. Salgado, O.P. Coutinho, and R.L. Reis, "Bone tissue engineering: state of the art and future trends", *Macromol. Biosci.,* vol. 4, no. 8, pp. 743-765, 2004.
       [http://dx.doi.org/10.1002/mabi.200400026] [PMID: 15468269]

[3]   A.R. Amini, C.T. Laurencin, and S.P. Nukavarapu, "Bone tissue engineering: recent advances and challenges", *Crit. Rev. Biomed. Eng.,* vol. 40, no. 5, pp. 363-408, 2012.
       [http://dx.doi.org/10.1615/CritRevBiomedEng.v40.i5.10] [PMID: 23339648]

[4]   J.P. Vacanti, and R. Langer, "Tissue engineering: the design and fabrication of living replacement devices for surgical reconstruction and transplantation", *Lancet,* vol. 354, suppl. Suppl. 1, pp. SI32-SI34, 1999.
       [http://dx.doi.org/10.1016/S0140-6736(99)90247-7] [PMID: 10437854]

[5]   R.J. O'Keefe, and J. Mao, "Bone tissue engineering and regeneration: from discovery to the clinic--an overview", *Tissue Eng. Part B Rev.,* vol. 17, no. 6, pp. 389-392, 2011.
       [http://dx.doi.org/10.1089/ten.teb.2011.0475] [PMID: 21902614]

[6]   Y. Ikada, "Challenges in tissue engineering", *J. R. Soc. Interface,* vol. 3, no. 10, pp. 589-601, 2006.
       [http://dx.doi.org/10.1098/rsif.2006.0124] [PMID: 16971328]

[7]   M. Ansari, and M. Eshghanmalek, "Biomaterials for repair and regeneration of the cartilage tissue", *Biodes. Manuf.,* vol. 2, no. 1, pp. 41-49, 2019.
       [http://dx.doi.org/10.1007/s42242-018-0031-0]

[8]   J. Henkel, M.A. Woodruff, D.R. Epari, R. Steck, V. Glatt, I.C. Dickinson, P.F. Choong, M.A. Schuetz, and D.W. Hutmacher, "Bone regeneration based on tissue engineering conceptions—a 21st century perspective", *Bone Res.,* vol. 1, no. 3, pp. 216-248, 2013.
       [http://dx.doi.org/10.4248/BR201303002] [PMID: 26273505]

[9]   J. Pajarinen, T. Lin, E. Gibon, Y. Kohno, M. Maruyama, K. Nathan, L. Lu, Z. Yao, and S.B. Goodman, "Mesenchymal stem cell-macrophage crosstalk and bone healing", *Biomaterials,* vol. 196, pp. 80-89, 2019.
[http://dx.doi.org/10.1016/j.biomaterials.2017.12.025] [PMID: 29329642]

[10]   S.M. Naghib, M. Ansari, A. Pedram, F. Moztarzadeh, A. Feizpour, and M. Mozafari, "Bioactivation of 304 stainless steel surface through 45S5 bioglass coating for biomedical applications", *Int. J. Electrochem. Sci.,* vol. 7, pp. 2890-2903, 2012.

[11]   E. Gentleman, and J.M. Polak, "Historic and current strategies in bone tissue engineering: do we have a hope in Hench?", *J. Mater. Sci. Mater. Med.,* vol. 17, no. 11, pp. 1029-1035, 2006.
[http://dx.doi.org/10.1007/s10856-006-0440-z] [PMID: 17122915]

[12]   J.P. Vacanti, and C.A. Vacanti, *Vacanti, J. P., & Vacanti, C. A. (2014). The history and scope of tissue engineering*, 2014.

[13]   R.R. Betz, "Limitations of autograft and allograft: new synthetic solutions", *Orthopedics,* vol. 25, no. 5, suppl. Suppl., pp. s561-s570, 2002.
[http://dx.doi.org/10.3928/0147-7447-20020502-04] [PMID: 12038843]

[14]   Y. Tabata, "Biomaterial technology for tissue engineering applications", *Journal of the Royal Society interface,,* vol. 6, no. no. suppl_3, pp. S311-S324, 2009.
[http://dx.doi.org/10.1098/rsif.2008.0448.focus]

[15]   K. Prasad, O. Bazaka, M. Chua, M. Rochford, L. Fedrick, J. Spoor, R. Symes, M. Tieppo, C. Collins, A. Cao, D. Markwell, K.K. Ostrikov, and K. Bazaka, "Metallic biomaterials: Current challenges and opportunities", *Materials (Basel),* vol. 10, no. 8, p. 884, 2017.
[http://dx.doi.org/10.3390/ma10080884] [PMID: 28773240]

[16]   Q. Chen, and G.A. Thouas, "Metallic implant biomaterials", *Mater. Sci. Eng.,* vol. 87, pp. 1-57, 2015.
[http://dx.doi.org/10.1016/j.mser.2014.10.001]

[17]   S.C.A. Pina, R.L. Reis, and J.M. Oliveira, *Ceramic biomaterials for tissue engineering.,* 2018.
[http://dx.doi.org/10.1016/B978-0-08-102203-0.00004-4]

[18]   P.N. Christy, S.K. Basha, V.S. Kumari, A.K.H. Bashir, M. Maaza, K. Kaviyarasu, M.V. Arasu, N.A. Al-Dhabi, and S. Ignacimuthu, "Biopolymeric nanocomposite scaffolds for bone tissue engineering applications–A review", *J. Drug Deliv. Sci. Technol.,* vol. 55, 2020.101452
[http://dx.doi.org/10.1016/j.jddst.2019.101452]

[19]   Z. Sheikh, S. Najeeb, Z. Khurshid, V. Verma, H. Rashid, and M. Glogauer, "Biodegradable materials for bone repair and tissue engineering applications", *Materials (Basel),* vol. 8, no. 9, pp. 5744-5794, 2015.
[http://dx.doi.org/10.3390/ma8095273] [PMID: 28793533]

[20]   R. Song, M. Murphy, C. Li, K. Ting, C. Soo, and Z. Zheng, "Current development of biodegradable polymeric materials for biomedical applications", *Drug Des. Devel. Ther.,* vol. 12, pp. 3117-3145, 2018.
[http://dx.doi.org/10.2147/DDDT.S165440] [PMID: 30288019]

[21]   E. Salernitano, and C. Migliaresi, "Composite materials for biomedical applications: a review", *J. Appl. Biomater. Biomech.,* vol. 1, no. 1, pp. 3-18, 2003.
[PMID: 20803468]

[22]   L. Meinel, V. Karageorgiou, R. Fajardo, B. Snyder, V. Shinde-Patil, L. Zichner, D. Kalpan, R. Langer, and G. Vunjak-Novakovic, "Bone Tissue Engineering Using Human Mesenchymal Stem Cells: Effects of Scaffold Material and Medium Flow", *Annals of Biomedical Engineering volume,* 2004.
[http://dx.doi.org/10.1023/B:ABME.0000007796.48329.b4]

[23]   B. Kristjánsson, T. Mabey, P. Yuktanandana, V. Parkpian, and S. Honsawek, "Mesenchymal stem cells for regeneration of cartilage lesions: focus on knee osteoarthritis", *Thai Journal of Orthopaedic Surgery,* vol. 37, no. 2-4, pp. 67-78, 2013.

[24]   C.R. Black, V. Goriainov, D. Gibbs, J. Kanczler, R.S. Tare, and R.O. Oreffo, "Bone tissue engineering", *Curr. Mol. Biol. Rep.,* vol. 1, no. 3, pp. 132-140, 2015.
[http://dx.doi.org/10.1007/s40610-015-0022-2] [PMID: 26618105]

[25]   W. Bensaïd, K. Oudina, V. Viateau, E. Potier, V. Bousson, C. Blanchat, L. Sedel, G. Guillemin, and H. Petite, "de novo reconstruction of functional bone by tissue engineering in the metatarsal sheep model", *Tissue Eng.,* vol. 11, no. 5-6, pp. 814-824, 2005.
[http://dx.doi.org/10.1089/ten.2005.11.814] [PMID: 15998221]

[26]   S. Yue, H. He, B. Li, and T. Hou, "Hydrogel as a Biomaterial for Bone Tissue Engineering: A Review", *Nanomaterials (Basel),* vol. 10, no. 8, p. 1511, 2020.
[http://dx.doi.org/10.3390/nano10081511] [PMID: 32752105]

[27]   A. Chaya, S. Yoshizawa, K. Verdelis, S. Noorani, B.J. Costello, and C. Sfeir, "Fracture healing using degradable magnesium fixation plates and screws", *J. Oral Maxillofac. Surg.,* vol. 73, no. 2, pp. 295-305, 2015.
[http://dx.doi.org/10.1016/j.joms.2014.09.007] [PMID: 25579013]

[28]   E. Ameri, H. Ghandhari, H. Hesarikia, H.R. Rasouli, H. Vahidtari, and N. Nabizadeh, "Comparison of harrington rod and cotrel-dubousset devices in surgical correction of adolescent idiopathic scoliosis", *Trauma Mon.,* vol. 18, no. 3, pp. 134-138, 2013.
[http://dx.doi.org/10.5812/traumamon.14663] [PMID: 24350172]

[29]   S. van Rijt, and P. Habibovic, "Enhancing regenerative approaches with nanoparticles", *J. R. Soc. Interface,* vol. 14, no. 129, 2017.20170093
[http://dx.doi.org/10.1098/rsif.2017.0093] [PMID: 28404870]

CHAPTER 43

A Review of Hybrid Effort Estimation Model for Agile Based Projects

Tina Bakshi[1] and **Mohit Arora**[1,*]

[1] *School of Computer Science And Engineering, Lovely Professional University, Phagwara, Punjab, India*

Abstract: Software effort estimation is an important part of the software development process as it strives to determine the success or failure of the project. The success of the project is entirely based on the prediction accuracy of the software effort estimation. There has been a major challenge in agile methodology adoption which is effort estimation but the conventional means of estimating the effort mainly results in inaccurate estimates so we need to follow an appropriate model approach. This paper is reviewed to focus on effort estimation in agile projects which is related to story points that help to prioritize user stories for faster deployment of the project and to review different hybrid models that are used to predict the effort. We reviewed the accuracy parameters based on three popular agile datasets and found that the Deep Belief Network-Ant Lion Optimizer (DBN-ALO) model works efficiently for all the datasets and outperforms all the other proposed hybrid models. Different techniques can be used for minimizing the effort estimation so that the tasks can be done efficiently.

Keywords: Agile Methodology, Ant Lion Optimizer, Deep Belief Network, Effort Estimation, Story Points.

INTRODUCTION

The software effort estimation is currently an important aspect of the software industry as the success or failure of a project depends upon how accurate an effort prediction of the software is. Various techniques are being used over the years for estimating the effort in both agile and non-agile environments [1]. In agile projects, two very commonly used metrics influence the project's growth efforts, one that defines the size and complexity of the project called the story points, and the other that defines the total number of story points that can be conveyed by the team in a sprint called the project team's velocity [2, 3].

---

* **Corresponding author Mohit Arora:** School of Computer Science And Engineering, Lovely Professional University, Phagwara, Punjab, India; E-mail: mohit.15980@lpu.co.in

Dharam Buddhi, Rajesh Singh and Anita Gehlot (Eds.)

We can estimate efforts that are required by the software project using the agile approach efficiently based on the two factors- story points and team velocity. This paper provides an overview of the hybrid software estimation techniques being used recently. The effort estimation process for user stories in the backlog is discussed in a sprint planning meeting, after which, the product owner prioritizes the item effectively based on the team's velocity. A very important factor in this process is to reduce any kind of influence on the team and to successfully practice the exercise [4, 5]. Different accuracy parameters can be used in effort estimation like MMRE, MdMRE, PRED which shows the difference between actual and the estimated effort [6, 7].

HYBRID ALGORITHMS USED IN AGILE-BASED PROJECTS

FLANN-WOA – FLANN-WOA – It is a hybrid algorithm that combines the FLANN (Functional Link Artificial Neural Network) which consists of three layers that are first, middle, and finally the last layer for getting the nodes, and then those nodes are being multiplied by the calculated weight vector by combining WOA (Whale Optimization Algorithm) [8].

RBFN-WOA- RBFN (Radial Bias Function Network) has 3 layers just like the FLANN algorithm. The first layer of RBFN has input neurons that provide input data whereas the Gaussian radial basis function generates the middle layer. The necessary output is then obtained by extracting the weighted sum which is calculated by WOA and then multiplied by the nodes of the middle layer [8].

DBN-ALO - The number of knots in the proposed DBN-ALO architecture has five RBM's for traditional inputs whereas 3 for agile inputs. Just one node is available to attempt the output layer. The input is linked and evaluated according to the training algorithm in the visible DBN layer. Three Restricted Boltzmann Machine (RBM) stacks were relocated. The effort is calculated as a linear amount of the final RBM output at the output layer [9].

SVR-RBF-The Support Vector Regression (SVR) has been designed to address regression problems. Oliveira initially investigated the application of SVR to estimate the cost of software projects. Kernel Learning algorithm uses a popular kernel function in machine learning called the radial base function which is also used for the classification of vector machines [9].

ABC-PSO - A novel method was proposed for the calculation of the commitment in agile development projects focused on velocity and the story points. A mixed variant of Artificial Bee Colony (ABC) and Particle Swarm Optimization (PSO) algorithms had been implemented as well for getting better results [11].

REVIEW METHOD

Based on the review method, two research questions were prepared and have been answered accordingly.

Below are the Research questions that were analyzed according to the review:

RQ1: What methods have been used to estimate effort or size in Agile Software Development (ASD)?

RQ2: What accuracy parameters and effort predictors are used in effort estimation for ASD?

RELATED WORK

Jørgensen *et al.* [8] presented a new method that was collaborated and applied to the software cost estimate model and it is being efficiently tested on the tera-PROMISE datasets. Kaushik *et al.* [9] proposed a DBN-ALO method. A prediction interval of effort to deal with uncertainty in estimation was also provided in the research work. Khan *et al.* [10] explored the characteristics of user stories that can influence the estimation of effort in ASD, which is useful for improving the effectiveness of the current techniques for estimation. Khuat [11] introduced a new technique that was based on the team velocity and story point factors. The parameters of these were then optimized by employing swarm optimization algorithms. Onkar Malgonde and Chari [12] took seven algorithms to predict the efforts of a story. They also carry out computer experiments to demonstrate that the ensemble-based in comparison with other ensemble-based benchmarking performed better. Mirjalili [13] proposed the algorithm was benchmarked in three stages. The ALO algorithm was initially compared in the first two test phases with a variety of algorithms that were from their literature survey. Nassif [14] initially performed the regression analysis. Results showed that model performance was affected by data heteroscedasticity. Ezghari and Zahi [15] proposed a strengthening of the FASEE, by imposing criteria of consistency to deal with the aforementioned disadvantages. Aditi Panda [16] improved the prediction accuracy of the agile-based software effort estimation. Different kinds of neural types for doing this were used. Pandey and Litoriya [17] proposed an analysis involving the use of five different estimates. They mapped the values of the correlation into fuzzy through linguistic values. Such fuzzy numbers state vector and adjacency are then used to represent the matrix. In recommending a software estimation method, they obtained a probability of success equal to 70%. Srikrishna *et al.* [18] introduced the Evolutionary Cost-Sensitive Deep Belief Network (ECS- DBN) model in this paper for the prediction of effort in any agile techniques. Rao and Rao [19] used seven machine learning-based classifiers for testing the model. Cláudio Ratke *et al.* [20] introduced an automated model based on narrative texts for estimating the software effort.

COMPARATIVE ANALYSIS OF DIFFERENT DATASETS AND ALGORITHMS USED

Tables **1** and **2** compare the hybrid models of Whale Optimization Algorithm (WOA) considering Company datasets I & II which were obtained on request by the company in Delhi. Company Dataset-1 is having 23, while Company Dataset-2 has 25 input points. On the other hand, the different hybrid techniques as shown in Table **3** compares the values of actual and predicted effort in terms of three commonly used accuracy metrics- MMRE, MdMRE, and PRED. As per the table, the DBN-ALO technique outperforms all the other techniques while considering Zia's Work [8 - 10].

Table 1. Hybrid Agile Models Used on Company Dataset-I [8].

Techniques	MMRE	MdMRE	PRED (25)
RBFN-WOA [8]	0.185	0.143	87.1
FLANN-WOA [8]	0.192	0.172	87.4

Table 2. Hybrid Agile Models Used on Company Dataset-II [8].

Techniques	MMRE	MdMRE	PRED (25)
RBFN-WOA [8]	0.160	0.182	87.0
FLANN-WOA [8]	0.138	0.185	89.3

Table 3. Hybrid Agile Models Used on Zia's Work [8 - 11].

Techniques	MMRE	MdMRE	PRED (25)
RBFN-WOA [8]	0.198	0.173	87.9
FLANN-WOA [8]	0.18	0.193	86.9
SVR-RBF [9]	0.0747	NA	95.9052
SVR-RBF-GS [9]	0.062	0.0426	100
ABC-PSO [11]	0.0569	0.0333	N. A
DBN-ALO [9]	0.0225	0.0222	98.4321

RESULTS

RQ1: Planning poker and expert judgment are the most common methods to estimate effort. In addition, various intelligent techniques are used to estimate effort to help the making of decisions.

RQ2: The most frequently used accuracy measures for effort prediction are Mean Magnitude of Relative Error (MMRE), Magnitude of Relative Error (MRE), and Prediction (PRED). The other factors or effort predictors in ASD are cost drivers and size metrics [5]. The most commonly used size metrics are use case stories and story points and these estimation techniques are used together to predict the effort obtained from velocity calculations and size estimated. The cost drivers also influence the effort estimation procedure. There are multiple cost drivers like a quality requirement, team skills, prior experience, task size, complexity, and work environment.

CONCLUDING REMARKS

This paper reviews different agile estimation papers and the development in error reduction through the years. A significant part of the software development process is the precise measurement of effort. The techniques for evaluating efforts are often focused on an expert judgment in Agile Software Development (ASD), but there has been an advancement in the area based on Machine Learning (ML) algorithms. A review of different hybrid algorithms has been presented in this paper. We have reviewed three agile datasets namely Zia, Company Dataset-I, and Company Dataset-II, and further on comparing we found that DBN-ALO outperforms all the other effort estimation hybrid models in the case of Zia's work, RBFN-WOA for Company Dataset-I and FLANN-WOA for Company Dataset-II.

CONSENT FOR PUBLICATION

Not applicable.

CONFLICT OF INTEREST

The authors declare no conflict of interest, financial or otherwise.

ACKNOWLEDGEMENT

Declared none.

REFERENCES

[1]   A. E. Akgün, "Team wisdom in software development projects and its impact on project performance", *Int. J. Inf. Manage,* vol. 50, pp. 228-243, 2019.
[http://dx.doi.org/10.1016/j.ijinfomgt.2019.05.019]

[2]   A. Ali, and C. Gravino, "Using Bio-Inspired Features Selection Algorithms in Software Effort Estimation: A Systematic Literature Review",
[http://dx.doi.org/10.1109/SEAA.2019.00043]

[3] A. Altaleb, M. Altherwi, and A. Gravell, "An Industrial Investigation into Effort Estimation Predictors for Mobile App Development in Agile Processes", *ICITM 2020 - 2020 9th Int. Conf. Ind. Technol. Manag,* 2020,pp. 291-296
[http://dx.doi.org/10.1109/ICITM48982.2020.9080362]

[4] S. Bilgaiyan, S. Mishra, and M. Das, "Effort estimation in agile software development using experimental validation of neural network models", *Int. J. Inf. Technol.,* vol. 11, no. 3, pp. 569-573, 2018.
[http://dx.doi.org/10.1007/s41870-018-0131-2]

[5] E. Dantas, A. Costa, M. Vinicius, M. Perkusich, H. Almeida, and A. Perkusich, "An effort estimation support tool for agile software development: An empirical evaluation", *Proc. Int. Conf. Softw. Eng. Knowl. Eng. SEKE,* 2019,pp. 82-87
[http://dx.doi.org/10.18293/SEKE2019-141]

[6] T. Foss, E. Stensrud, B. Kitchenham, and I. Myrtveit, "A Simulation Study of the Model Evaluation Criterion MMRE", *IEEE Trans. Softw. Eng.,* vol. 29, no. 11, pp. 985-995, 2003.
[http://dx.doi.org/10.1109/TSE.2003.1245300]

[7] M. Jørgensen, and T. Halkjelsvik, "Sequence effects in the estimation of software development effort", *J. Syst. Softw.,* vol. 159, 2020.
[http://dx.doi.org/10.1016/j.jss.2019.110448]

[8] A. Kaushik, D.K. Tayal, and K. Yadav, "The Role of Neural Networks and Metaheuristics in Agile Software Development Effort Estimation", *Int. J. Inf. Technol. Project Manage.,* vol. 11, no. 2, pp. 50-71, 2020. [IJITPM].
[http://dx.doi.org/10.4018/IJITPM.2020040104]

[9] A. Kaushik, D.K. Tayal, and K. Yadav, "A Comparative Analysis on Effort Estimation for Agile and Non agile Software Projects Using DBN-ALO", *Arab. J. Sci. Eng.,* vol. 45, pp. 2605-2618, 2020.
[http://dx.doi.org/10.1007/s13369-019-04250-6]

[10] Muhammad Ijaz Khan, "User Story Characteristics Affecting Software Cost in Agile Software Development: A Systematic Literature Review", *IJCSNS,* 2019.

[11] T. T. Khuat, and M. H. Le, "A Novel Hybrid ABC-PSO Algorithm for Effort Estimation of Software Projects Using Agile Methodologies", *J. Intell. Syst,* vol. 27, pp. 489-506, 2018.
[http://dx.doi.org/10.1515/jisys-2016-0294]

[12] O. Malgonde, and K. Chari, "An ensemble-based model for predicting agile software development effort, vol. 24, no. 2", *Empir. Softw. Eng.,* 2019.
[http://dx.doi.org/10.1007/s10664-018-9647-0]

[13] S. Mirjalili, "The ant lion optimizer", *Adv. Eng. Softw.,* vol. 83, pp. 80-98, 2015.
[http://dx.doi.org/10.1016/j.advengsoft.2015.01.010]

[14] A.B. Nassif, M. Azzeh, A. Idri, and A. Abran, "Software development effort estimation using regression fuzzy models", *Comput. Intell. Neurosci.,* vol. 2019, no. February, p. 8367214, 2019.
[http://dx.doi.org/10.1155/2019/8367214] [PMID: 30915110]

[15] D. Naud-Martin, C. Breton-Patient, F. Mahuteau-Betzer, and S. Piguel, ""Three-Component C–H Bond Sulfonylation of Imidazoheterocycles by Visible-Light Organophotoredox Catalysis". ", *Eur. J. Org. Chem.,,* vol. 2020, no. 42, 2020.
[http://dx.doi.org/10.1002/ejoc.202001219]

[16] A. Panda, S.M. Satapathy, and S.K. Rath, "Empirical Validation of Neural Network Models for Agile Software Effort Estimation based on Story Points", *Procedia Comput. Sci.,* vol. 57, pp. 772-781, 2015.
[http://dx.doi.org/10.1016/j.procs.2015.07.474]

[17] P. Pandey, and R. Litoriya, "Fuzzy Cognitive Mapping Analysis to Recommend Machine Learning-Based Effort Estimation Technique for Web Applications", *Int. J. Fuzzy Syst.,* vol. 22, no. 4, pp. 1212-1223, 2020.

[http://dx.doi.org/10.1007/s40815-020-00815-y]

[18]   H.M. Premalatha, and C.V. Srikrishna, "Effort estimation in agile software development using evolutionary cost- sensitive deep Belief Network", *Int. J. Intell. Eng. Syst.,* vol. 12, no. 2, pp. 261-269, 2019.
[http://dx.doi.org/10.22266/ijies2019.0430.25]

[19]   K.E. Rao, and G.A. Rao, "Ensemble learning with recursive feature elimination integrated software effort estimation: a novel approach", *Evol. Intell.,* vol. 14, no. 0123456789, pp. 151-162, 2021.
[http://dx.doi.org/10.1007/s12065-020-00360-5]

[20]   C. Ratke, H.H. Hoffmann, T. Gaspar, and P.E. Floriani, "Effort Estimation using Bayesian Networks for Agile Development", *2nd Int. Conf. Comput. Appl. Inf. Secur. ICCAIS 2019,* 2019,pp. 1-4
[http://dx.doi.org/10.1109/CAIS.2019.8769455]

CHAPTER 44

Accident and Theft Detection Using Arduino and Machine Learning

Paila Akhil[1,*], **Dibbyo Bhattacharjee**[1] and **B. Arun Kumar**[1]

School of Electronics and Electrical Engineering, Lovely Professional University Punjab, India, 144411

Abstract: Every human wants to be safe and secure health while driving. Here the big role comes for technology. Technology needs to improve a lot because we can't take risks in life. The life of human beings is very precious. The main reason for death during driving is attention. During driving, we all should be more careful and should be alert every time. The reason behind making this project is to rescue the people when they are in an emergency and to give them a medical rescue team *via* sensors and GPS, GSM, MPU6050 consists of gyroscope and accelerometer. Here Machine learning part is used for theft detection using accessing a local camera. And also this project will help to find our lost or stolen vehicles in India.

Keywords: Accident Detection, Emergency Services, Vehicle Tracking, Vehicle Theft Detection by accessing camera using machine learning coding.

INTRODUCTION

The use of vehicles has increased traffic congestion and thus led to an increase in road accidents. Lives are at stake, simply because of the urgent need for emergency supplies. Automatic microcontroller systems auto accident detection using GPS has received attention because it helps in an emergency. In this project, there is an Arduino microcontroller system that will detect the accident of a vehicle also detects the theft *via* Buzzer and Ultrasonic sensors. Hence we are introducing accident and theft detection using Arduino, Ultrasonic sensors, and Buzzers. With the help of GPS and GSM toolkit, we can find our lost and stolen vehicles also. This project can save the life of human beings from unplanned happenings. When we start the car engine a caution message is sent to the holder's

---

* **Corresponding author Paila Akhil:** School of Electronics and Electrical Engineering, Lovely Professional University Punjab, India, 144411; E-mail: 1akhilpaila6533@gmail.com

Dharam Buddhi, Rajesh Singh and Anita Gehlot (Eds.)

cell phone. When theft happens, the owner will be able to identify the theft and will be able for emergency services shortages *via* an app [1]. At last, the paper contains as follows. Section I Methodology. In Section II, we discuss the literature survey and block diagram. In Section III Block diagram of the GSM system for mobile communication work. In Section IV Block diagram of GPS working in accident and theft detection. In Section VI working of code with the camera using machine learning code. In Section V working on this project. At last, the conclusion of this project.

SECTION I: METHODOLOGY

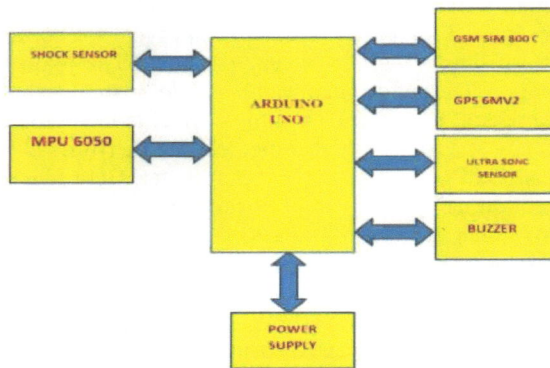

Fig. (1). Block diagram of this project [1].

Step 1: According to the Block diagram we have added a shock sensor, MPU 6050 that consists of an accelerometer and piezoelectric sensor, GSM, GPS, Ultrasonic sensor, and a buzzer.

Step 2: Firstly, a piezoelectric sensor will detect the accident, and the code written for MPU 6050 will be intimated.

Step 3: Latitudes and longitudes will be detecting on GPS and messages will be sent to nearby people with the help

of Buzzer and *via* GSM SIM800C.

Step 4: A message will receive that is a google maps IP number, that is prestored in Arduino Uno.

Step 5: If there is any false message an OFF switch is also provided (Fig. **1**).

SECTION II: LITERATURE SURVEY AND CIRCUIT DIAGRAM OF THE PROPOSED PROJECT

From the past circumstances, we come to know the Drawbacks that have been noted down

1. Detection of vehicle accidents plays a significant process in system projects.

2. After an unplanned accident, medical attention cannot be given to the needed person immediately.

After considering the drawbacks we have mapped out a proposed system that produce all the mentioned drawbacks.

1. A computerized system will be used once when an unplanned accident has happened.

2. This computerized system gives perfect output information about latitudes and longitudes in a system when any unplanned accident occurred in an area without taking any delays.

3. We can save human life using this computerized system when any accident happened.

4. This machine gives us attention when the theft happened.

Hardware Description

Arduino is a microcontroller device and is based on an open-source electronic platform prototyping board that can be easily programmed. In this project GSM accident detection, MPU6050 that consists of an accelerometer and gyroscope is used to detect or sense the posture of motion (Fig. **2**) [2]. A gyroscope sensor is a device that senses an angular velocity. In simple words, angular velocity is a change in rotational angle per unit of time. This computerized system is used to pitch of vehicle and because when any person is cross from an accident event, the vehicle tends to fall off to down, and the pitch angle is usually small (Fig. **3**) [3]. The work of receiving the pitch angle of the car also be executed by using a gyroscope that will provide the angular acceleration details of the GSM. Moreover, the tilt of pitch angle of the car is evaluated by using MPU6050 and in tie-up with the accelerometer. Gyroscope is used to help in computing the pitch angle which uses to help in executing the present algorithm (Fig. **4**) [4].

Arduino is a microcontroller device and it is based on an open-source electronic platform prototyping board that can be easily programmed.

Arduino supports C language and C++ language which is the authority of all other programmings.

Arduino especially provides the capability and flexibility to communicate with other systems or devices.

Arduino board is connected with the USB cable to a programming console. After done with coding, connect with the USB cable to system Port to Arduino board.

The Arduino board has 54 digital inputs and output pins (Fig. **5**) [5]:

1. 15 pins are PWM output.

2. One reset button and ICSP header button.

3. 4 were USARTS hardware serial port

4. 1 power jack

5. 1 USB connect

6. One 16 mega Hz crystal oscillator

7. 16 pins are ANALOG

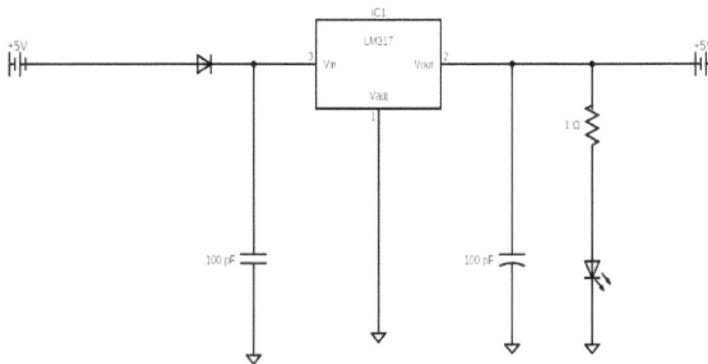

Fig. (2). Circuit Diagram of this project [2].

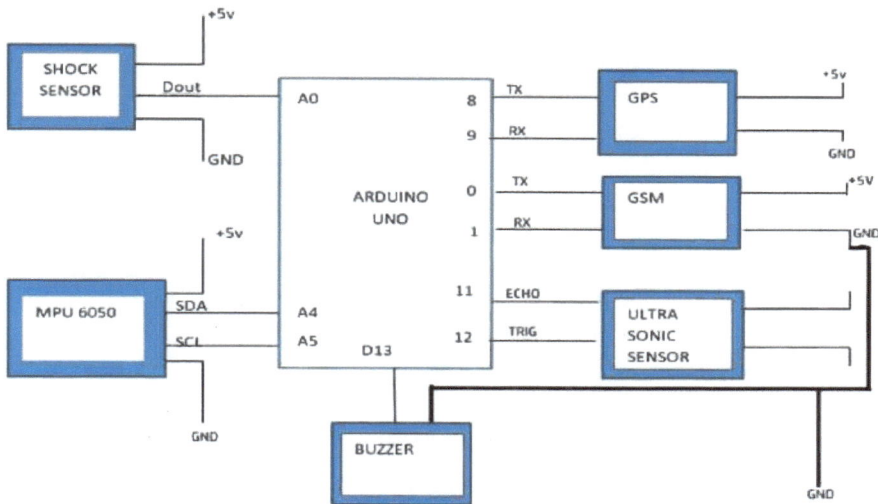

Fig. (3). Block diagram [3].

SECTION III: SIM 800C GSM

GSM SIM 800C is based on QUAD-BAND GPRS/GSM type of SMT (surface mount technology), it's a type of embedded system device.

GSM SIM 800C comes with QUAD-BAND 1900/1800/900/850 MHz that transmits data information, voice, and SMS with less power consumption (Fig. **6**) [6].

The size of this component is tiny 17.5 x 15.7 x 2.3 millimeters, and also it can be fitted into the compact and slim that demands custom design.

It has 4 ports that are tx, Rx, Vcc, and ground.

Specifications of SMS *via* GSM/GPRS

1. Point to point MO and MT.

2. SMS cell broadcast.

3. PDU & Text mode.

Fig. (4). Sim 800 C [4].

SECTION IV: GPS WORKING

In an orbit, there are GPS satellites that circle the earth twice a day. Each satellite forwards a unique signal output and the parameters of orbit will be allowing to GPS device to encode, decode, and compute the location of satellites.

After getting output from the satellite receiver the GPS will use this information to measure and calculates the user's accurate location.

The global positioning system receiver will calculate the distance to each satellite by the duration of time it will take to receive a transmitted output signal [6].

After getting information from the transmitted signal the output distance measurements from satellites, the receiver will decide a user's exact position and will display it on it.

To measure a 2D position that consists of longitude and latitudes, there should be needed 3 satellites and the signal must be locked and after this, we can track. A movement from the GPS receiver will receive and will be displayed on an LCD screen or Arduino software. To measure 3D position, there should be at least 8 or more than eight satellites locked, but that only depends on where we are on earth and what is the time of day on earth.

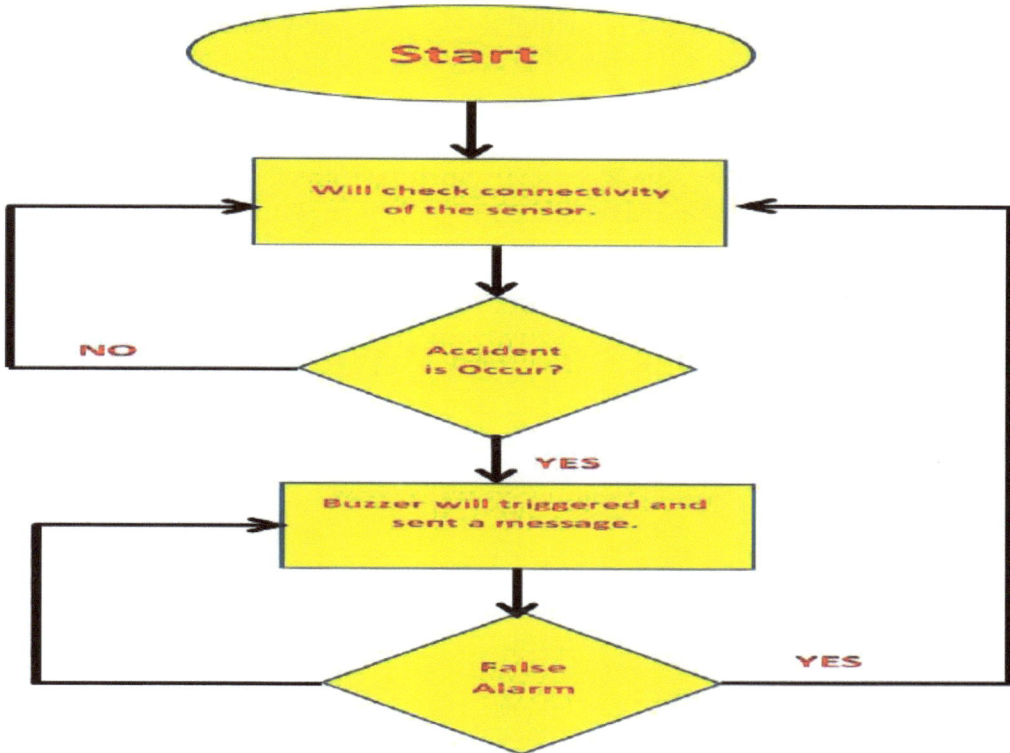

Fig. (5). Flow chart [5].

SECTION V: WORKING ON THIS PROJECT

1. After installing SIM into the GSM, but the USB cable for power supply.

2. Then, LED will be ON and the second LED will be blinking for 1 second and it OFF again because of network establishing.

3. There are 2 indications on GSM ie,

a) Network indication.

b) Power indication.

4. Here to ring with accident detection, 2 components are working in this project that is Ultrasonic sensor and Buzzer.

5. If an accident happened, the ultra-sonic sensors will sense the obstacle like

walls, other vehicles, *etc.*, and then it's immediately triggered the buzzer. People hear the sound of a buzzer and immediately come to see what happened here and they contact to rescue team for help (Fig. **7**) [7].

6. Now here about theft detection, go to Section VI.

Here an IP-based camera should be there to detect the face of the thief. After catching the thief we can raise a complaint to the police station.

Ip-based cameras should be attached or installed at traffic signals, garages, houses, *etc.* anywhere to check who has stolen the vehicles.

SECTION VI :WORKING OF CODE WITH A CAMERA USING MACHINE LEARNING CODE

Import package

In [1]: import cv2

Access stored video, laptop camera or any other ip based camera

In [2]: cap = cv2.VedioCapture(0) #Capture video from storage/Laptop camera/ Ip based camera

Keep displaying the each frame from video

In [3]; while True:

Ret, frame = cap.read() # read frame/ image one by one

Cv2.imshow("Frame", frame) # display frame/ image

Key = cv2.waitkey(1) #wait till key press

if key == ord ("q"): # exit loop on 'q' key press

break

cap.release() # release vedio capture object

cv2.destroyAllWindows() #destroy all frame windows

end of code

red color words = directions

black color words = code

blue color words = comments

Outputs

Fig. (6). Accessing real-time camera [6].

Fig. (7). After compiling the code, we can access the camera, see the theft, and file a complaint and FIR to the police station [7].

Advantages

1. You can locate your lost vehicle.

2. By ringing of loud Buzzer, people nearby can come to rescue you.

Limitations of this Project

1. A good network should be a must.

2. To catching the thief camera should be there.

Application

1. We can find lost and looted vehicles using latitudes, longitudes on google maps, and IP-based cameras.

CONCLUSION OF THIS PROJECT

The output is achieved through the making of this computerized project that is Accident and theft detection using an Ultrasonic sensor and Arduino.

The components like sensors, GPS, GSM, MPU6050, will be fitted on the project. It's a small idea that can save human life from unplanned accidents and theft happened.

DISCUSSION AND FUTURE SCOPE

In the future, we are planning to make this project at an advanced level *i.e.*, it will be completely working on smartphones. Nowadays a smartphone has certain features like a led touchscreen, fingerprint sensor, by adding google example OK GOOGLE voice recognition [8], Bluetooth, and Wi-Fi module.

It's safe from theft that happens.

This project is a trail-based model purpose for now but it will be implemented for a real-life purpose.

Hence, this project will save human's life from the unplanned accident and retrieve to find our vehicles easily *via* GSM and GPS, and IP based cameras.

CONSENT FOR PUBLICATION

Not applicable.

CONFLICT OF INTEREST

The authors declare no conflict of interest, financial or otherwise.

ACKNOWLEDGEMENTS

We have taken efforts in this project. However, it would not have been possible without the kind support and help of teachers and the organizations.

We are highly indebted to Lovely Professional University for their guidance and constant supervision as well as for providing necessary information regarding the capstone project and also for their support in completing this report. We would like to express our gratitude towards our Mentor that is B. Arun Kumar Professor of the University for their kind Cooperation and encouragement which helped us in the completion of this project.

We would like to express our special gratitude and thanks to the university persons for giving us such attention and time.

Our thanks and appreciation also goes to our HOD Dr. Shelej Khera and Dr. Vikram Kumar who allowed us to experience such knowledge and people who have willingly helped us out with their abilities.

REFERENCES

[1]   H. Bai, and C. Liu, "A hybrid license plate extraction method based on edge statistics and morphology", *18th International Conference on Pattern Recognition,* vol. 2, 2004, pp. 831-834

[2]   T. Celik, H. Demirel, H. Ozkaramanli, and M. Uyguroglu, "Fire Detection in Video Sequences Using Statistical Color Model", *IEEE International Conference on Acoustics Speed and Signal Processing Proceedings,* vol. 2, 2006, pp. II-213-II-216 [http://dx.doi.org/10.1109/ICASSP.2006.1660317]

[3]   M.S. Amin, M.B.I. Reaz, M.A.S. Bhuiyan, and S.S. Nasir, "Kalman filtered gps accelerometer-based accident detection and location system: A low-cost approach", *Curr. Sci.,* vol. 106, no. 11, pp. 1548-1554, 2014.

[4]   C. Prabha, R. Sunitha, and R. Anitha, "Automatic vehicle accident detection and messaging system using gsm and gps modem", *Inter- national Journal of Advanced Research in Electrical, Electronics and Instrumentation Engineering, ,* vol. 3, no. 7, pp. 10723-10727, 2014.

[5]   S. Sonika, K. Sathiyasekar, and S. Jaishree, "Intelligent accident identification system using gps, gsm modem", *Int. J. Adv. Res. Comput. Commun. Eng.,* vol. 3, no. 2, 2014.

[6]   S.R. Aishwarya, *An IoT Based Accident Prevention and Tracking System For Night Drivers,* vol. 3, no. 4, 2015.

[7]   R. Sahu, V. Rathore, S. Shukla, S. Maji, J. Gupta, and S. Karmakar, "Touch screen based security lock for bike", *International Journal of Scientific Research in Computer Science, Engineering and Information Technology,* vol. 2, no. 3, 2017.

[8]   D. Magar, S. Gadge, S. Gadakh, and S.S. Lavate, "Embedded based vehicle security using gsm-and gps system", *International Journal of Innovative Research in Electrical, Electronics, Instrumentation and Control Engineering,* vol. 5, no. 4, 2017. [http://dx.doi.org/10.17148/IJIREEICE.2017.5411]

CHAPTER 45

Review on Phyto-Remedy to Manage Diabetes Mellitus

Pratibha Kaushal[1] and **Sanjeev Singh**[1,*]

[1] *School of Bioengineering & Biosciences, Lovely Professional University, Punjab, India*

Abstract: Diabetes Mellitus is a heterogeneous metabolic disorder with various aetiologies that are characterized by chronic hyperglycemia. DM is a consequence of defective metabolism in the body was either cells fail to secrete insulin or insulin *per se* is inactive. Insulin is a pancreatic hormone responsible for glucose uptake and utilization by cells for energy production. Thus, insulin deficiency renders the cells unable to uptake and utilize glucose for metabolism. DM is majorly categorized into two conditions: Type 1, In this case, there is no enough insulin production by the pancreatic cells (Insulin-dependent). In type II, despite insulin production, the cells fail to respond to the hormone (Insulin Independent), the reason behind the former is not yet clear, but the latter is studied to be a result of obesity or inactivity. Thus, as a consequence of an error in glucose metabolism, it leads to comorbidities, with high glucose concentration in the blood causing damage to the circulatory and excretory system. This condition becomes life-threatening without proper medical care. Type 1 can be treated with an external source of insulin provided to the body. Type II is treated with hypoglycaemic drugs. Along with medicine, other options include an improved lifestyle with proper nutrition and exercise. The number of diabetes patients has been increased today. In the present review article, we have discussed the research reported to show medicinally important plants and plant-derived compounds which have been shown to affect insulin production and glucose intake and utilization in type 2 diabetes patients.

Keywords: Antidiabetic, Diabetes, Hyperglycemia, Medicinal Plants, Natural Treatment.

INTRODUCTION

Diabetes mellitus, a heterogeneous metabolic disorder, characterized by hyperglycemia condition because of the absence of insulin, and its decrement in concentration and its action mechanism. Hyperglycemia with chronic diabetes is related to several long-term microvascular complications affecting kidneys, ner-

* **Corresponding author Sanjeev Singh:** School of Bioengineering & Biosciences, Lovely Professional University, Punjab, India; E-mail: mohit.15980@lpu.co.in

Dharam Buddhi, Rajesh Singh and Anita Gehlot (Eds.)

ves, and damaging of the retina which is the subject of diagnostic criteria too for diabetes. In macrovascular, it is caused a high risk of cardiovascular disease (CVD) and tissue and organ damage [1]. Overall, it affected the patient's style and reduce the survival rate and increase the burden of diseases and cost of treatment. Conventional medicine and treatment are being utilized to suppress the diabetes effect but they have their side effect if used in long term. Plants and their products are natural and have many diverse chemicals which have been utilized for many purposes and one of them has medicinal potential. Antidiabetic medicinal property has already been observed in many plants and exploring in other plants too. Here, the information of some phyto-remedy is mentioned which can fight with type 2 diabetes to suppress and defeat diabetes with fewer side effects and better efficacy [2 - 4].

CLASSIFICATION AND DIAGNOSIS OF DIABETES

Long-time persisting hyperglycaemic condition is responsible to cause diabetes which may arise due to lack of insulin production responsible for type 1, insulin insensitivity, or resistance responsible for type 2, and during pregnancy cause gestational diabetes [5].

Epidemiology

Diabetes mellitus is no longer an international concept for the global community; it is increasingly prevailing in all countries [6], regardless of the volume of revenue. According to recent estimates, 629 million individuals are expected to suffer from DM by 2049 [7]. In children, T1DM is more predominant, especially in those going from birth to 14 years old, yet it can cause in people of all ages, with a man abundance found in youthful grown-ups [8]. Adults or obese individuals, on the other hand, are frequently affected by T2DM, but this can also occur in adolescents, according to the latest studies. T2DM is sometimes pointed to as the biggest donor to the diabetes community as a whole [9].

TREATMENTS

From ancient literature and to the present-day research, the use of plant varieties to extract medicinal valuable compounds for the treatment of several human diseases has been reported. In the present study, we have reported research demonstrating medicinally valuable plants and their products which have shown a positive effect in response to diabetes mellitus.

Plant Remedies in the Management of DM

From its roots on earth, civilization has used plants to cure illnesses and to promote general health. Plants have been the world's most important natural resource since ancient times [10,11]. The knowledge base concerning the therapeutics properties of herbs was primarily given by tribes and these plants are in high demand in both created and non-industrial nations. The World Health Organisation reports that 80% of the total populace relies on herbal drugs, most of which use plant remedies, to fulfill their primary health care needs. The current pharmacopeia also includes at least 25% of plant-inferred drugs and many others that are semisynthetic, because of plant-confined model mixtures. Recent studies indicate that among diverse cultures and nations, more than 9000 herbs have known medicinal uses [12].

Achyranthes Aspera

This herb belongs to the family *Amaranthaceae,* widely grown in India and called Chirchitta, Chirchira (Hindi). In indigenous medicine, *Achyranthes Aspera* is respected a great deal. *Achyranthes Aspera* ethanolic extract was tested for in-vitro anti-diabetic action against diabetic mice caused by alloxan. A substantial decrease in blood glucose levels was found observed after the treatment from the extract in diabetic mice [13].

Bael (Aegle marmelos)

Bael, a common name in Hindi, belongs to the genus of Rutaceae identified as a medicinal plant to cure various diseases [14]. The numerous bio-chemicals that are contained in the leaves of *A. marmelos* are alkaloids, heart glycosides, saponins, terpenoids, tannins, steroids, and flavonoids, *etc* even its fruit is also utilized for medicinal purposes [4] Ethanolic extract that contains *A. marmelos* has shown that diminishes blood sugar in alloxan-induced diabetic rodents. After continuous extract organization, glucose drops could be seen from the seventh day and sugar levels were discovered to be diminished by 54% on the 28th day. Alloxan-induced oxidative stress was observed to be greatly diminished when *A. marmelos* extract was administered. The Histopathological examination uncovered the recovery of β-cells in both the concentrate and glibenclamide treatment classes. The concentrate of ethanol given by the leaves of *A. marmelos* is a diabetic enemy of alloxan-incited rodents with promising antidiabetic action [15].

Aloe Vera (Aloe barbadensis Miller)

Aloe vera, a common name in Hindi, is being utilized globally due to its uncommon medicinal properties. It contains more than 70 active ingredients, including nutrients, minerals, proteins, amino acids, polysaccharides, and phenolic compounds [16]. The therapeutic effects of A. vera gel include, but are not limited to mitigating, corrective, antibacterial, cell reinforcement, malignancy, and antidiabetic properties [17, 18]. It's very significant to recognize compounds that prevent protein non-enzymatic glycation to help prevent the generation of secondary complications, taking into account the pathological consequences of glycation, which is why the objective of this study was to define the capacity of the Aloe vera (AVM) methanol extract by BSA/glucose assay to inhibit the protein glycation reaction. Methanol extricate prevents the development of AGEs and can repress the increase in glucose postprandial, meaning that AVM can prevent diabetes problems caused by the extract of aloe vera [19]

Annona Squamosa

Annona squamosa Linn., member of Annonaceae family, generally referred to as sitaphal (Hindi), and custard apples or sugar apples in India [20]. Its fruit is being consumed as cuisine and sweet preparation, as well as its different medicinal property, has been reported as anthelmintics, anti dysenteric, astringent and vermifuge as well as antidiabetic [21,22]. They produce huge quantities of flavonoids, so their antidiabetic activity can be because of flavonoids [23 - 25].

Azadirachta Indica

Azadirachta indica A. Juss., the deciduous tree that belongs to the Meliaceae is Juss., also known as neem [26 - 28]. Neem Leave extract (400mg/kg) dissolved in ethanol and even in chloroform, the solvent has been shown its different properties like anti-hyperglycemic, anti-lipid peroxidation and against hypercholesterolaemic action along with a decrease in serum fatty substance levels in diabetic rats induced with alloxan [29 - 35].

Biophytum Sensitivum

Biophytum belongs to the family Oxalidaceae provides many secondary metabolites; saponins, terpenes, flavonoids, and tannins, and have shown its antidiabetic effect. The hypoglycemic potential of *Biophytum sensitivum* has been studied in normal and diabetic rats induced with steptozotonic-nicotinamide where oral administration of this plant (200 mg/kg) for approximately 28 days was given [36]. Hypoglycemic reactions to *Biophytum sensitivum* are most likely to be related to the production or secretion of beta-cell insulin, according to published observations [37].

Cajanus Cajan

The native name of *Cajanus cajan* is Arhar which belonging to the family Fabaceae [38,39] have different medicinal as well as antidiabetic property [40]. Leave extract dissolved in methanol was used to reduce the blood glucose level in normal and diabetic rats. In other studies, ethanolic leave extract also has shown antidiabetic effect in mice which proves the potential of c.cajan's leaves to treat diabetes [41].

Cinnamomum Zeylanicum

Cinnamomum zeylanicum generally is used as a spice, inner bark part used, and called daalchini in Hindi [42]. During 12 weeks of treatment by the use of 1 gm of cinnamon in type 2 patients and observed the beneficial effect [43]. In a different research, along with antidiabetic antioxidant, cardioprotective, anti-inflammatory, anti-cancer and anti-microbial, *etc* have been reported [44].

Terminalia Chebula

Terminalia chebula is considered the king of medicine in Tibet and is extensively used in southern Asia [45]. Ethanolic extract of its fruit has been shown significant antidiabetic role and pancreatic cells proliferation enhancer among the Wistar diabetic rat [46].

Panax Ginseng

Panax ginseng is the part of Araliaceae family and has a wide variety of pharmacological properties to treat various diseases [47]. Ginseng has been proven its promising medicinal potential. The use of ginseng has been shown to reduce the fasting blood glucose, level of HbA1c, and systolic blood pressure comparatively placebo on the human study [48].

The Future Scenario of Phytoremedies for Diabetes Mellitus

Individuals use numerous types of natural medications, and numerous traditional drugs are frequently incorporated into modern therapeutics. Approximately 80 percent of the population in developed countries, particularly rural people, rely upon conventional clinical solutions for medical care needs. There has been a resurgence in interest in home-grown medications in non-industrial countries because of the inclination for a colossal scope of items from common sources. There is also a requirement for the distinction between homegrown treatments offered by a clinical specialist and certain natural medicines that are unreservedly accessible to people for self-drug. The seriousness of the human physical condition in the world is the rapidly increasing phenomenon of diabetes mellitus. New active plant medicinal drugs have recently been extracted and have anti-diabetic action that is more powerful than oral hypoglycaemic agents used in established therapy. The focus has been drawn to the exploration of plant's potential in current years that may be useful to people with anti-diabetic activity. Different experiment's results are provided by researchers for the betterment of medicine efficiency through the oral route to fight diabetes mellitus [49 - 51]

CONCLUDING REMARKS

Diabetes is associated with irregulating metabolism of glucose, lipid, and protein which consequently causes diminished insulin synthesis or rising resistance to its intervention. Diabetes is contributing significantly to morbidity and mortality cases across the world. The most commonly used pharmaceutical items in present-day medication, like ibuprofen, antimalarials, against malignancy meds, digitalis, and others, are collected from plant sources. Because of the purity, effectiveness, and dependable value in humans under various conventional drug regimes, plant sources are assumed to be safe. Plants in this manner give a characteristic substitute or a subordinate with fewer side effects results in comparatively conventional medicine. Deductively approved as powerful antidiabetics were flavonoids (queretin, and chrysin), alkaloids (berberberin and vindolin), glycosides and saponins, glycolipids, dietary strands, and imidazole compounds. Among these, in flavonoids, saponins, and alkaloids, various impacts have been seen. While ongoing advancement has been made in understanding the basic components and the various exercises of these plant-based medications, further investigations are required to use plant and combination of plant' components to use effectively for defeating the type 2 DM.

CONSENT FOR PUBLICATION

Not applicable.

CONFLICT OF INTEREST

The authors declare no conflict of interest, financial or otherwise.

ACKNOWLEDGEMENT

Declare none.

REFERENCES

[1]   Z. Punthakee, R. Goldenberg, and P. Katz, "Definition, Classification and Diagnosis of Diabetes, Prediabetes and Metabolic Syndrome", *Can. J. Diabetes,* vol. 42, no. 1, suppl. Suppl. 1, pp. S10-S15, 2018.
        [http://dx.doi.org/10.1016/j.jcjd.2017.10.003] [PMID: 29650080]

[2]   W. Kooti, M. Farokhipour, Z. Asadzadeh, D. Ashtary-Larky, and M. Asadi-Samani, "The role of medicinal plants in the treatment of diabetes: a systematic review", *Electron. Physician,* vol. 8, no. 1, pp. 1832-1842, 2016.
        [http://dx.doi.org/10.19082/1832] [PMID: 26955456]

[3]   A. Chaudhury, C. Duvoor, V.S. Reddy Dendi, S. Kraleti, A. Chada, R. Ravilla, A. Marco, N.S. Shekhawat, M.T. Montales, K. Kuriakose, A. Sasapu, A. Beebe, N. Patil, C.K. Musham, G.P. Lohani,

and W. Mirza, "Clinical Review of Antidiabetic Drugs: Implications for Type 2 Diabetes Mellitus Management", *Front. Endocrinol. (Lausanne),* vol. 8, no. 1, p. 6, 2017.
[http://dx.doi.org/10.3389/fendo.2017.00006] [PMID: 28167928]

[4]   S.R. Mudi, M. Akhter, S.K. Biswas, M.A. Muttalib, S. Choudhury, B. Rokeya, and L. Ali, "Effect of aqueous extract of Aegle marmelos fruit and leaf on glycemic, insulinemic and lipidemic status of type 2 diabetic model rats", *J. Complement. Integr. Med.,* vol. 14, no. 2, p. •••, 2017.
[http://dx.doi.org/10.1515/jcim-2016-0111] [PMID: 28284036]

[5]   N. Taneja, S. Kukal, and S. Mani, "Current treatments for type 2 diabetes, their side effects and possible complementary treatments", *Int. J.,* vol. 7, no. 3, pp. 13-18, 2017.

[6]   N.G. Forouhi, and N.J. Wareham, "Epidemiology of diabetes", *Medicine (Baltimore),* vol. 38, no. 18, pp. 602-606, 2010.
[http://dx.doi.org/10.1016/j.mpmed.2010.08.007]

[7]   E. English, and E. Lenters-Westra, "HbA1c method performance: The great success story of global standardization", *Crit. Rev. Clin. Lab. Sci.,* vol. 55, no. 6, pp. 408-419, 2018.
[http://dx.doi.org/10.1080/10408363.2018.1480591] [PMID: 30001673]

[8]   E.A.M. Gale, and K.M. Gillespie, "Diabetes and gender", *Diabetologia,* vol. 44, no. 1, pp. 3-15, 2001.
[http://dx.doi.org/10.1007/s001250051573] [PMID: 11206408]

[9]   K. Sahoo, B. Sahoo, A.K. Choudhury, N.Y. Sofi, R. Kumar, and A.S. Bhadoria, "Childhood obesity: causes and consequences", *J. Family Med. Prim. Care,* vol. 4, no. 2, pp. 187-192, 2015.
[http://dx.doi.org/10.4103/2249-4863.154628] [PMID: 25949965]

[10]   N. Chang, Z. Luo, D. Li, and H. Song, "Indigenous uses and pharmacological activity of traditional medicinal plants in Mount Taibai, China", *Evid. Based Complement. Alternat. Med.,* vol. 2017, no. 1, p. 8329817, 2017.
[http://dx.doi.org/10.1155/2017/8329817] [PMID: 28303162]

[11]   R.M.P. Gutierrez, Y.G.Y. Gómez, and M.D. Guzman, "Attenuation of nonenzymatic glycation, hyperglycemia, and hyperlipidemia in streptozotocin-induced diabetic rats by chloroform leaf extract of Azadirachta indica", *Pharmacogn. Mag.,* vol. 7, no. 27, pp. 254-259, 2011.
[http://dx.doi.org/10.4103/0973-1296.84243] [PMID: 21969798]

[12]   E. Salmerón-Manzano, J.A. Garrido-Cardenas, and F. Manzano-Agugliaro, "Worldwide research trends on medicinal plants", *Int. J. Environ. Res. Public Health,* vol. 17, no. 10, p. •••, 2020.
[http://dx.doi.org/10.3390/ijerph17103376] [PMID: 32408690]

[13]   J.L. Ríos, F. Francini, and G.R. Schinella, "Natural Products for the Treatment of Type 2 Diabetes Mellitus", *Planta Med.,* vol. 81, no. 12-13, pp. 975-994, 2015.
[http://dx.doi.org/10.1055/s-0035-1546131] [PMID: 26132858]

[14]   C.K. Pathirana, T. Madhujith, and J. Eeswara, "Bael (Aegle marmelos L. Corrêa), a Medicinal Tree with Immense Economic Potentials", *Advances in Agriculture,* vol. 2020, pp. 1-13, 2020.

[15]   A.N. Kesari, R.K. Gupta, S.K. Singh, S. Diwakar, and G. Watal, "Hypoglycemic and antihyperglycemic activity of Aegle marmelos seed extract in normal and diabetic rats", *J. Ethnopharmacol.,* vol. 107, no. 3, pp. 374-379, 2006.
[http://dx.doi.org/10.1016/j.jep.2006.03.042] [PMID: 16781099]

[16]   R. Bhavani, "Antidiabetic activity medicinal plant Aegle marmelos (linn.) on alloxan induced diabetic rats", *International Research Journal of Pharmaceutical and Biosciences,* vol. 1, no. 1, pp. 36-44, 2014.

[17]   B. Benzidia, B. Mohammed, H. Hind, B. Nadia, Z. Meryem, E. Hamid, A.D. Naima, B. Narjis, and H. Najat, "Chemical composition and antioxidant activity of tannins extract from green rind of Aloe vera (L.) Burm. F", *J. King Saud Univ. Sci.,* vol. 31, no. 4, pp. 1175-1181, 2019.
[http://dx.doi.org/10.1016/j.jksus.2018.05.022]

[18]   A. Muñiz-Ramirez, R.M. Perez, E. Garcia, and F.E. Garcia, "Antidiabetic Activity of Aloe vera

Leaves", *Evid. Based Complement. Alternat. Med.,* vol. 2020, no. 5, p. 6371201, 2020.
[PMID: 32565868]

[19]   A.A. Maan, N. Akmal, K.I.K. Muhammad, A. Tahir, Z. Rabia, M. Misbah, and A. Muhammad, "The therapeutic properties and applications of Aloe vera: A review", *J. Herb. Med.,* vol. 12, no. 01, pp. 1-10, 2018.
[http://dx.doi.org/10.1016/j.hermed.2018.01.002]

[20]   A. Bhattacharya, and R. Chakraverty, "The pharmacological properties of Annona squamosa Linn: A Review", *Int J Pharm Eng,* vol. 4, no. 2, 2016.

[21]   S. Gajalakshmi, R. Divya, V. Divya, S. Mythili, and A. Sathiavelu, "Pharmacological activities of Annona squamosa: A review", *Int. J. Pharm. Sci. Rev. Res.,* vol. 10, no. 2, pp. 24-29, 2011.

[22]   C. Ma, Y. Chen, J. Chen, X. Li, and Y. Chen, "A Review on Annona squamosa L.: Phytochemicals and Biological Activities", *Article in The American Journal of Chinese Medicine,* vol. 45, no. 5, pp. 933-964, 2017.
[http://dx.doi.org/10.1142/S0192415X17500501] [PMID: 28659034]

[23]   M. Kaleem, M. Asif, Q.U. Ahmed, and B. Bano, "Antidiabetic and antioxidant activity of Annona squamosa extract in streptozotocin-induced diabetic rats", *Singapore Med. J.,* vol. 47, no. 8, pp. 670-675, 2006.
[PMID: 16865205]

[24]   N. Kalidindi, N.V. Thimmaiah, N.V. Jagadeesh, R. Nandeep, S. Swetha, and B. Kalidindi, "Antifungal and antioxidant activities of organic and aqueous extracts of Annona squamosa Linn. leaves", *J. Food Drug Anal.,* vol. 23, no. 4, pp. 795-802, 2015.
[http://dx.doi.org/10.1016/j.jfda.2015.04.012] [PMID: 28911497]

[25]   N. Pandey, and D. Barve, "Phytochemical and Pharmacological Review on Annona squamosa Linn", *Int. J. Res. Pharm. Biomed. Sci.,* vol. 2, no. 4, pp. 1-30, 2021.

[26]   S. Pankaj, T. Lokeshwar, B. Mukesh, and B. Vishnu, "Review on neem (Azadirachta indica): thousand problems one solution", *International research journal of pharmacy,* vol. 2, no. 12, pp. 97-102, 2011.

[27]   J.F. Islas, A. Ezeiza, G. Zuca, L.D.J. Juan, G.M.T. María, E. Bruno, and E.M.C. Jorge, "An overview of Neem (Azadirachta indica) and its potential impact on health", *J. Funct. Foods,* vol. 74, no. 11, pp. 104-171, 2020.
[http://dx.doi.org/10.1016/j.jff.2020.104171]

[28]   J.J. Marín-Peñalver, I. Martín-Timón, C. Sevillano-Collantes, and F.J. Del Cañizo-Gómez, "Update on the treatment of type 2 diabetes mellitus", *World J. Diabetes,* vol. 7, no. 17, pp. 354-395, 2016.
[http://dx.doi.org/10.4239/wjd.v7.i17.354] [PMID: 27660695]

[29]   P. Khosla, S. Bhanwra, J. Singh, S. Seth, R.K. Srivastava, and R.K. Srivastava, "A study of hypoglycaemic effects of Azadirachta indica (Neem) in normaland alloxan diabetic rabbits", *Indian J. Physiol. Pharmacol.,* vol. 44, no. 1, pp. 69-74, 2000.
[PMID: 10919098]

[30]   R.M. Perez Gutierrez, and M. de Jesus Martinez Ortiz, "Beneficial effect of Azadirachta indica on advanced glycation end-product in streptozotocin-diabetic rat", *Pharm. Biol.,* vol. 52, no. 11, pp. 1435-1444, 2014.
[http://dx.doi.org/10.3109/13880209.2014.895389] [PMID: 25026338]

[31]   K. Satyanarayana, K. Sravanthi, I.A. Shaker, and R. Ponnulakshmi, "Molecular approach to identify antidiabetic potential of Azadirachta indica", *J. Ayurveda Integr. Med.,* vol. 6, no. 3, pp. 165-174, 2015.
[http://dx.doi.org/10.4103/0975-9476.157950] [PMID: 26604551]

[32]   S. Ponnusamy, S. Haldar, F. Mulani, S. Zinjarde, and H. Thulasiram, "Gedunin and azadiradione: Human pancreatic alpha-amylase inhibiting limonoids from neem (Azadirachta indica) as anti-diabetic agents", *PLoS ONE,* vol. 10, no. 10, 2015.

[33] M.A. Alzohairy, "Therapeutics role of azadirachta indica (Neem) and their active constituents in diseases prevention and treatment", *Evid. Based Complement. Alternat. Med.,* vol. 2016, no. 3, p. 7382506, 2016.
[PMID: 27034694]

[34] "K. I. Nwadike, N. J. I. Uwaezuoke, O. U. Eze, C. E. Anieze, and O. C. Nwoke. "Effects of fractionated neem leaf extract (IRC) on blood glucose level in alloxan induced diabetic wistar rats", *Int. J. Diabetes Clin. Res.,* vol. 6, no. 2, p. 105, 2019.

[35] U. Pingali, M.A. Ali, S. Gundagani, and C. Nutalapati, "Evaluation of the effect of an aqueous extract of azadirachta indica (Neem) leaves and twigs on glycemic control, endothelial dysfunction and systemic inflammation in subjects with type 2 diabetes mellitus – a randomized, double-blind, placebo-controlled clinical study", *Diabetes Metab. Syndr. Obes.,* vol. 13, no. 10, pp. 4401-4412, 2020.
[http://dx.doi.org/10.2147/DMSO.S274378] [PMID: 33244247]

[36] A.C. Bharati, and A.N. Sahu, "Ethnobotany, phytochemistry and pharmacology of Biophytum sensitivum DC", *Pharmacogn. Rev.,* vol. 6, no. 11, pp. 68-73, 2012.
[http://dx.doi.org/10.4103/0973-7847.95893] [PMID: 22654407]

[37] A. A. T. Pawar, and N. S. Vyawahare, *Int. J. Pharm. Pharm. Sci.,* vol. 6, no. 11, pp. 18-22, 2014.

[38] D. Pal, P. Mishra, N. Sachan, and A.K. Ghosh, "Biological activities and medicinal properties of Cajanus cajan (L) Millsp", *J. Adv. Pharm. Technol. Res.,* vol. 2, no. 4, pp. 207-214, 2011.
[http://dx.doi.org/10.4103/2231-4040.90874] [PMID: 22247887]

[39] I. Mohanram, and J.S. Meshram, "Treasures of Indigenous Indian Herbal Antidiabetics: An Overview", *Discovery and Development of Antidiabetic Agents from Natural Products: Natural Product Drug Discovery,* no. Oct, p. 271, 2016.

[40] D. Jaiswal, P.K. Rai, A. Kumar, and G. Watal, "Study of glycemic profile of Cajanus cajan leaves in experimental rats", *Indian J. Clin. Biochem.,* vol. 23, no. 2, pp. 167-170, 2008.
[http://dx.doi.org/10.1007/s12291-008-0037-z] [PMID: 23105745]

[41] S. Ariviani, D.R. Affandi, E. Listyaningsih, and S. Handajani, "The potential of pigeon pea (Cajanus cajan) beverage as an anti-diabetic functional drink", *IOP Conf. Ser. Earth Environ. Sci.,* vol. 102, no. 1, p. 12054, 2018.
[http://dx.doi.org/10.1088/1755-1315/102/1/012054]

[42] P. Ranasinghe, S. Pigera, G. S. Premakumara, P. Galappaththy, G. R. Constantine, and P. Katulanda, "Medicinal properties of 'true' cinnamon: A systematic review", *BMC Complementary and Alternative Medicine,* vol. 13, no. 1,275, pp. 13-275, 2013.

[43] A.S. Sahib, "Anti-diabetic and antioxidant effect of cinnamon in poorly controlled type-2 diabetic Iraqi patients: A randomized, placebo-controlled clinical trial", *J. Intercult. Ethnopharmacol.,* vol. 5, no. 2, pp. 108-113, 2016.
[http://dx.doi.org/10.5455/jice.20160217044511] [PMID: 27104030]

[44] P. Kawatra, and R. Rajagopalan, "Cinnamon: Mystic powers of a minute ingredient", *Pharmacognosy Res.,* vol. 7, no. 1, suppl. Suppl. 1, pp. S1-S6, 2015.
[http://dx.doi.org/10.4103/0974-8490.157990] [PMID: 26109781]

[45] A. Upadhyay, P. Agrahari, and D.K. Singh, "A review on the pharmacological aspects of Terminalia chebula", *Int. J. Pharmacol.,* vol. 10, no. 6, pp. 289-298, 2014.
[http://dx.doi.org/10.3923/ijp.2014.289.298]

[46] V.R. Kannan, R. Pandiyan, and R. Nachimuthu, "Anti-diabetic Activity on Ethanolic Extracts of Fruits of", *American Journal of Drug Discovery and Development,* vol. 2, no. 3, pp. 135-142, 2012.
[http://dx.doi.org/10.3923/ajdd.2012.135.142]

[47] N.T. Rokot, T.S. Kairupan, K.C. Cheng, J. Runtuwene, N.H. Kapantow, M. Amitani, A. Morinaga, H. Amitani, A. Asakawa, and A. Inui, "A Role of Ginseng and Its Constituents in the Treatment of

Central Nervous System Disorders", *Evid. Based Complement. Alternat. Med.,* vol. 2016, no. 10, p. 2614742, 2016.
[http://dx.doi.org/10.1155/2016/2614742] [PMID: 27630732]

[48]   W. Chen, P. Balan, and D.G. Popovich, "Review of ginseng anti-diabetic studies", *Molecules,* vol. 24, no. 24, p. 4501, 2019.
[http://dx.doi.org/10.3390/molecules24244501] [PMID: 31835292]

[49]   M. Modak, P. Dixit, J. Londhe, S. Ghaskadbi, and T.P.A. Devasagayam, "Indian herbs and herbal drugs used for the treatment of diabetes", *J. Clin. Biochem. Nutr.,* vol. 40, no. 3, pp. 163-173, 2007.
[http://dx.doi.org/10.3164/jcbn.40.163] [PMID: 18398493]

[50]   I. Oghogho Rosalie, "Antidiabetic potentials of common herbal plants and plant products: A glance", *Int. J. Herb. Med.,* vol. 4, no. 4, pp. 90-97, 2016.

[51]   B. Moradi, S. Abbaszadeh, S. Shahsavari, M. Alizadeh, and F. Beyranvand, "The most useful medicinal herbs to treat diabetes", *Biomed. Res. Ther.,* vol. 5, no. 8, pp. 2538-2551, 2018.
[http://dx.doi.org/10.15419/bmrat.v5i8.463]

Biometric Smart Card

Tahaab Maries[1,*], Tiwari Avnish[1], Parmar Deepak [1], Kumawat Devesh [1], Chouhan Mayank Singh [1] and Kumar Singh Dushyant[1]

[1] Lovely Professional University, Punjab, India

Abstract: It is a common saying in the security industry that the "S" in RFID stands for "Secure". An RFID Tag/Card dumps all the information it contains when brought close to the RFID Reader. The RFID Card is therefore considered unsuitable for storing sensitive data. Our project aims to find a way of adding a layer of security to the RFID Card using biometric authentication, which would widen the scope for the use of RFID technology in areas where they couldn't have been used before, such as contactless payments.

KEYWORDS: Biometric ATM Card, Biometric Authentication, Biometric Contactless Payment, Biometric Smart Card, Contactless Payment, RFID, Smart Card.

1. INTRODUCTION

The modern world is headed towards digitization at a pace that has never been seen before. It is therefore important for us to contribute towards this progress by doing our part in improving the technology that already exists and come up with innovative solutions to problems that haven't yet been solved.

The age of digitization has brought with it numerous ways which significantly reduce human effort. Technologies such as self-driving cars, which reduce the effort put in by the driver while driving. Online banking, which has eliminated the need for the customer to go to the bank for making a withdrawal/deposit. Home automation, which manages all the small and laborious tasks in a house, such as a temperature control, automatic motor control, gas leak detection, intruder alert system, *etc*. These are among the few examples of the part technology plays in improving the daily lives of people by eliminating the small tasks there by giving

* **Corresponding author Tahaab Maries:** Lovely Professional University, Punjab, India;
E-mail: mariestahaab @gmail.com

them more time to focus on improving their lives and spending time on the things that matter. As with every new technology introduced, there are several benefits and also certain drawbacks, the most significant of which is security. As smart devices make life more convenient for the user, the job of those who exploit these devices for personal gain also becomes easier.

One such example is contactless payment technology. It works by bringing the contactless payment-enabled card within 2cms of the post terminal and any amount within the predefined limit (Rs.2000) can be debited without requiring any form of authentication (PIN) from the user. This no doubt speeds up the payment process, but as is quite evident, it has the potential to be misused, and getting a refund from the bank would be quite a hectic and drawn-out process if at all the bank agrees to refund the amount. Hence, the need arises for a mode of authentication that is just as fluid and fast as the process of tapping the card on the POS terminal without entering the PIN. One of the solutions for this problem is adding a biometric authentication method to the card itself so that the contactless payment is disabled until a fingerprint sensor reads and authenticates the fingerprint of the cardholder. In this method, the cardholder would only have to hold the card by the fingerprint sensor to activate contactless payment and then tap the card on the POS terminal for the payment to go through. This method provides an additional layer of security, thereby drastically reducing the risks associated with contactless payments.

2. RESEARCH METHODOLOGY

The project that we aim to make different from any of those which are already available in the present market and are being taken into use by people in their day to day lives *e.g.*, contactless payment system using RFID technology is already available and is being readily used by the vast majority of people like at the toll plaza. A biometric attendance system is present in the market for a long time now and is very efficient and is used by people in several places like offices, institutes, universities just a name of a few. ATM cards with tap and go facilities are also readily available and are being by people everywhere around the globe. The project that we are making is a blend of all technologies mentioned above, just smarter, more reliable, more secure, portable, and efficient. Because in our project we plan to use RFID to store the concerned person's data who is supposed to use that particular card, like the fingerprint. We are also using a fingerprint scanner module to store the digital fingerprint and then later verify that it is being used by the same person which adds up additional safety. While all the above-mentioned systems are designed for a specific application, this smartcard is multifunctional and can also store important documents such as the ID card or the Driver's

License. It can be used for cashless payments just like a regular ATM card or as a contactless card with enhanced security.

3. OVERVIEW OF THE CURRENT BANKING SYSTEM

Many people from across the globe have turned towards contactless payments and the number just keeps on increasing. Especially in India, statistics show that the number of card payments has increased rapidly after the arrival of 4G refer (Fig. 1)) in 2016. Contactless payments are widely accepted by people and are also a preference in places where less time is required for the purchase. Instead of entering a pin that one tends to forget, one may simply prefer to "tap and go" [1]. Since the early 2000's many renowned companies from across the globe have launched their payment methods to make contactless payments, each one having its methodology to ensure that it will do the job without any security issue, as the only thing people were concerned about was "security" [2]. After that others also took part and launched their payment systems for their user base so that their users don't have to rely on somebody else's payment system to make payments [3]. Mobile payments increased rapidly across the globe and have proved to be very effective [4]. Some regions are not only accepting contactless payments but are also shifting towards cryptocurrencies, many countries have also opened banks just to support contactless payments [5], many people find it effective and have been saying that it is a much faster mode of payment than the conventional payment methods [7]. Places, where people tend to carry less amount of cash, are being dominated by contactless technology [8]. This technology is being supported and backed not only to make faster and safer payments but also to make the transition towards a cashless society [9].

Fig. (1). Value of card payments in india from financial year 2014 – 2020 [6].

4. POSSIBLE THREATS

The number of cyber-attacks and theft have also gradually increased, hackers are well-known for understanding the working principle of the technology and finding out the loopholes to divert the payments, leading to cyber theft (refer to Table **1**), the most common type of attack, that is being noticed nowadays is the Relay attack also known as Relay and Ghost attack Fig. (**2**) [10]. In this type of attack any person can access anyone's details from their RFID chip-enabled card with the help of an RFID reader by bringing the reader close to the card, to be safe and not fall prey to such techniques, the need for improved safety features is felt greatly. A more effective security field so that unauthorized individuals cannot enter it, and some precautions that can be taken from the user's side, such as using their network connection or mobile internet instead of using a public Wi-Fi. If payments are made using someone else's network, there are chances that someone might be able to see all the data traveling through the network which may compromise the user's confidential payment details such as their credit card information, bank account information, *etc.* [11].

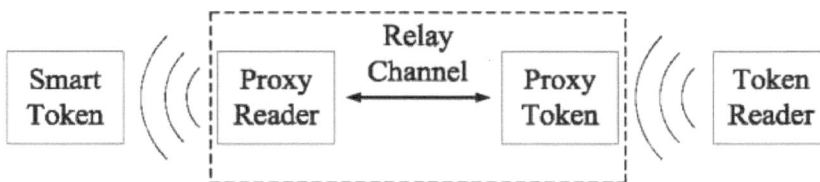

Fig. (2). Relay attack setup.

Table 1. Data on Frauds reported by scheduled commercial banks and select financial institutions on the category 'Card/Internet'.

Type of Cyber Fraud	Oct-19		Nov-19		Dec-19	
	No. of FMRs	Amount involved (Crores)	No. of FMRs	Amount involved (Crores)	No. of FMRs	Amount involved (Crores)
ATM/Debit Cards	3376	73.65	3533	10.55	4149	10.33
Credit Cards	1641	4.04	1711	4.77	2765	10.87
Internet Banking	360	7.01	2256	4.31	1250	2.27
Grand Total	5377	84.7	7500	19.63	8164	23.47

5. PROPOSED DESIGN

In this paper, we have proposed an efficient method of transaction whether it is done through an automated teller machine (ATM) or a POS terminal. In both types of transactions, biometric authentication will be used for verification before any transactions are done. In the case of contactless payment, there is a possible threat of illegal transactions if the card is not kept securely, so in our design, an RFID reader can only scan the details from the card after the user's biometric data has been verified. In the case of transactions through ATMs, the user doesn't have to remember the pin, and also if the card is lost, it cannot be misused. Figs. (**3** and **4**) show the activity flow diagrams of ATM and contactless payment-enabled POS terminal along with their possible interface system (Figs. **5** - **11**) for the transactions [12 - 14].

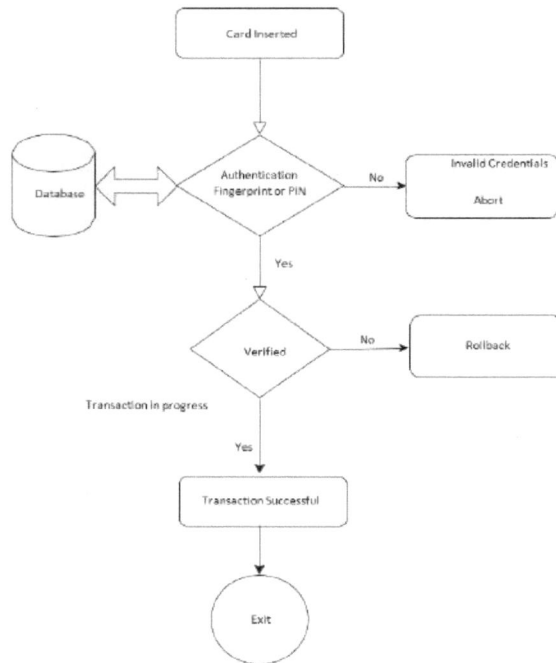

Fig. (3). Activity flow design for ATM transaction.

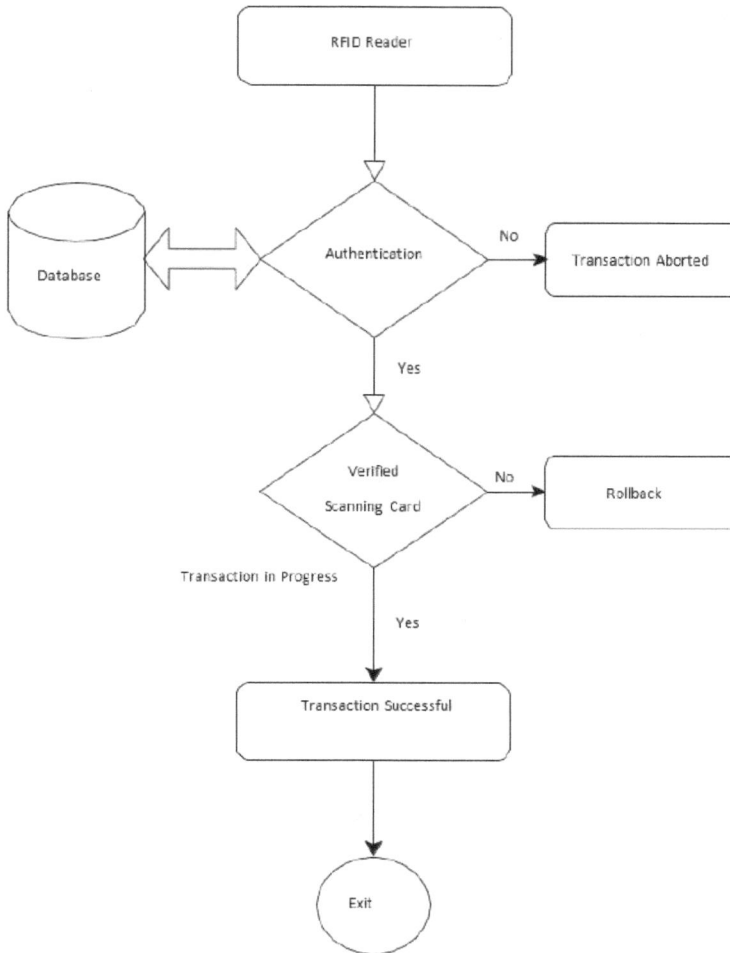

Fig. (4). Activity flow daigram for contactless transaction.

5.1. ATM Transaction Interface

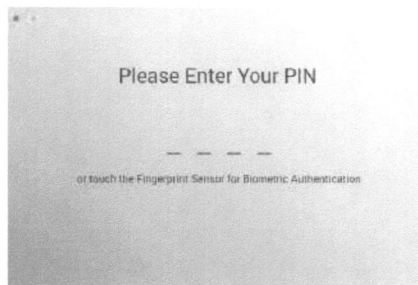

Fig. (5). Authentication interface.

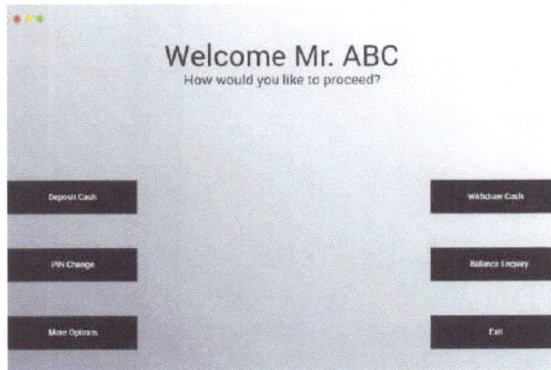

Fig. (6). Main menu.

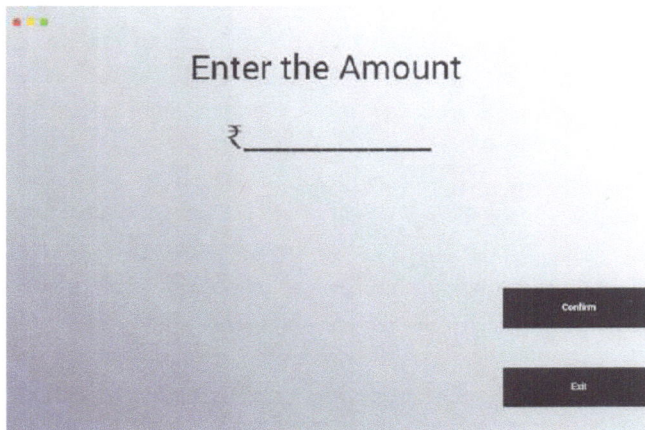

Fig. (7). Cash withdrawal interface.

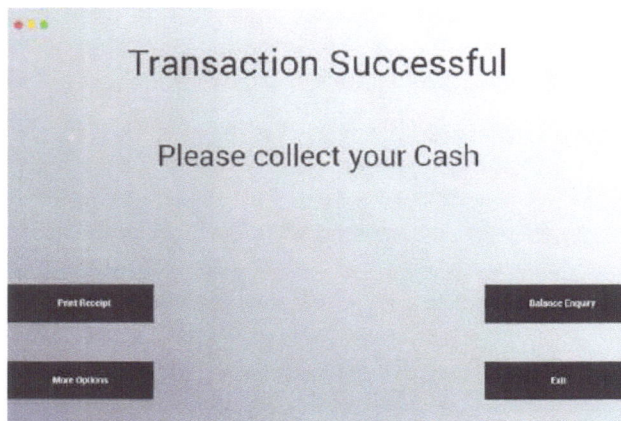

Fig. (8). Transaction completion.

6. CONTACTLESS PAYMENT POS INTERFACE

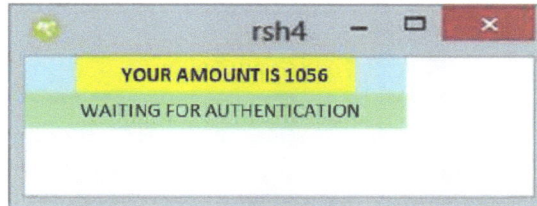

Fig. (9). This figure shows the interface waiting for a contactless card to be brought near it in order to initiate the transaction after the amount to be charged has been entered by the user.

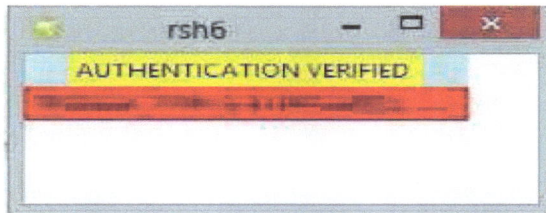

Fig. (10). After the biometric authentication has been verified and the card is active for the transaction, the interface sends over the transaction details to the bank and awaits confirmation.

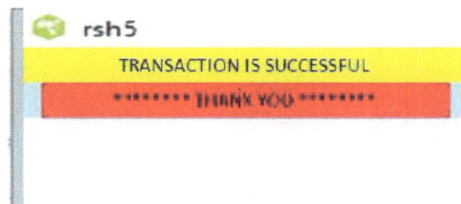

Fig. (11). After the transaction has been confirmed by the bank, the transaction is complete and a printed receipt of the same is provided to the customer.

7. APPLICATIONS AND FUTURE SCOPE

With the pandemic showing no sign of slowing down, it has become imperative for us to minimize unnecessary contact with people or surfaces that come into contact with other people. These surfaces can serve as Petri dishes for the virus and being used by a single infected individual may put everyone else that comes into contact with it to get infected. An example of such a surface is the POS terminal. It is out of the question that an infected individual will not touch his/her card and in doing so, will contaminate their card. When the same card is used at a POS terminal, it will cause everyone else who uses the terminal to become a potential carrier, thereby accelerating the spread of the virus. Now, imagine the same situation with a method of contactless payment. Even if there are traces of the virus on the card, it is highly unlikely that the post terminal would get contaminated since there is no contact between the card and the terminal. Since the authentication is biometric, there is no need for the user to enter the pin on the terminal as the fingerprint is verified on the card itself. This significantly reduces, if not eliminates the possibility of the POS terminal getting contaminated and leading to further infection of subsequent users.

Furthermore, with the country heading towards digitization, many of the documents required for identification have also been digitized and the need to carry original documents has greatly diminished. Documents such as the Aadhaar Card, Driving License, Vehicle Registration Certificate, *etc*. can now be stored on Digi-locker and can be accessed whenever needed. It is much more convenient to have access to all the documents on one's mobile phone than to carry physical copies of these documents. With the people beginning to realize the benefits of digitization, the need has arisen for a single card that would store all the documents of a person available on Digi-locker and make them available to the authorities for verification whenever required.

This paper proposes the integration of these two types of cards onto a single card, which uses contactless technology and cannot be accessed without biometric authentication. The final product could be used to make payments, withdraw cash at ATMs, and have all the functionalities of a conventional debit card along with being used as proof of identity.

8. RESULTS AND DISCUSSION

The global market has seen a great demand for biometric products (Fig. **13**) and if statistics are to be believed, the demand will continue to grow for the foreseeable future. Governments, private sectors, banks, and transit agencies are also implementing new contactless payment methods which are secure, reliable, and easy to use. This Radiofrequency enabled technology can deliver several benefits to an identity verification process such as speed, convenience, durability, reliability, and security. With these major factors, there will be improved positive levels for consumer acceptance. Furthermore, this technology is leading to a further upgrade to a smart card known as "Biometric Card". Table **2** shows the differences between our Biometric Smart Card and other existing technologies. As biometric authentication is the most popular and secure process these days in different sectors like healthcare, education, *etc*. (Fig. **12**) and it has also been very helpful in difficult times like Covid-19 in which there will be no need of touching any surface. Simply all the transactions will be easily done without any PIN or password the card will take the fingerprint data of the user and will match with the database, thereby making the transaction more secure and reliable.

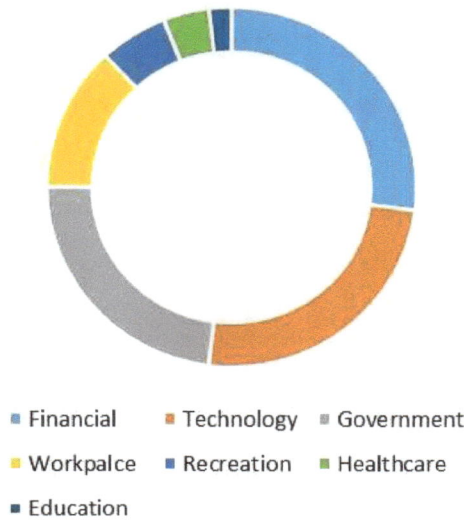

Fig. (12). After the transaction has been confirmed by the bank, the transaction is complete and a printed receipt of the same is provided to the customer.

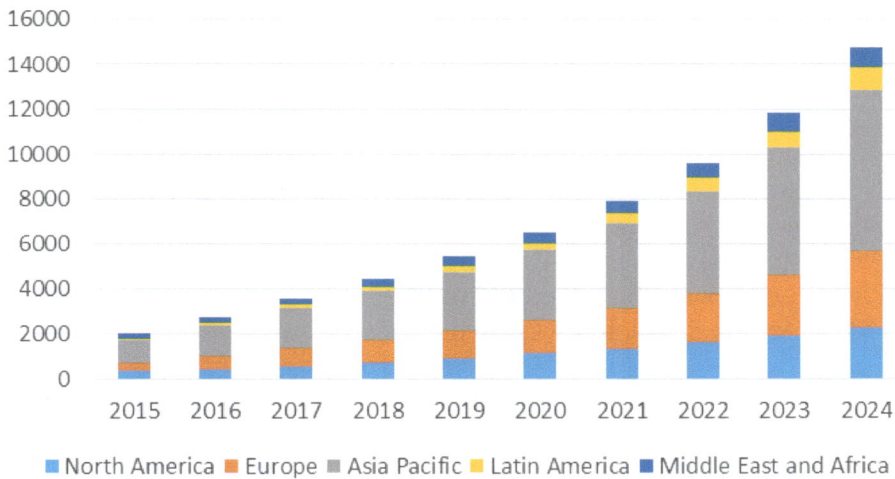

Fig. (13). After the transaction has been confirmed by the bank, the transaction is complete and a printed receipt of the same is provided to the customer.

Table 2. Comparison With other Technologies.

S.No.	Title	Description	Novelty in Biometric Smart Card
1.	**NFC and NFC Payments: A Review [15]**	Simple and safe bidirectional communication between NFC-enabled devices is made possible by this technology.	More secure than NFC system, as any such transaction requires biometric authentication.
2.	**Developing Secured Biometric Payments Model Using Tokenization [16]**	Secured Biometric Payments model using Tokenization	Check for the authentication of a user by verifying it from the database.
3.	**Digital Payment Systems [17]**	Conventional digital payments require pin authentication, which appears to be a challenge in some cases.	Biometric security eliminates the risk of money being lost or any such fraud.

CONCLUSION

With a drastic increase in the number of attacks against credit/debit card users, it is imperative to devise a method that can avert such malicious attempts and prevent the loss of one's hard-earned money. The idea of incorporating a fingerprint sensor with contactless payment technology not only enhances the security of the payment system but also increases the level of trust a customer has in the products offered by a financial institution. This in turn compels any skeptics of the aforementioned technology to reconsider their choices and may even turn them into potential customers. The ability to use the card on conventional POS terminals as well as contactless terminals with biometric authentication makes it versatile and hence suitable for a wide range of applications. In the future by adding the feature of receiving SMS alerts when the fingerprint is not recognized can also help the user to block the card before incurring any losses and know the location of the terminal which it was used.

CONSENT FOR PUBLICATION

Not applicable.

CONFLICT OF INTEREST

The authors declare no conflict of interest, financial or otherwise.

ACKNOWLEDGEMENT

Declared none.

REFERENCES

[1]   C. Chai, *Contactless 'tap-and-go' cards finally enter US market,* 2017.https://www.creditcards.com/credit-cardnews/contactless-tap-and-go-cards-us-market.php

[2]   C. Vinzant, *Contactless credit card technology now available in the USA,* 2007.https://www.creditcards.com/credit-card-news/contactless-creditcard-technology-1273.php

[3]   J. Wituschek, *Apple Card: Everything you need to know!,* 2019.https://www.imore.com/apple-goldmansachs-credit-card

[4]   B. Lovejoy, *Comment: China shows the future of mobile payment as it hits more than 50% of GDP,* 2017.https://9to5mac.com/2017/05/30/thefuture-of-apple-pay-mobile-payment-china/

[5]   A. Oxford, *FNB launches contactless payments from your phone,* 2016.https://www.htxt.co.za/2016/10/05/fnb-launchescontactless-payments-from-your-phone/

[6]   *Value of card payments in India from financial year 2014 to 2020,* 2021.https://www.statista.com/statistics/1034675/india-value-of-card-payments/

[7]   M. Vejacka, *Consumer acceptance of contactless payments in Slovakia,* 2015.https://www.researchgate.net/publication/286843735_Consumer_acceptanceof_contactless_payments_in_Slovakia

[8]   H. Griffiths, *New Zealand: Consumers move away from cash as digital and contactless payments rise,* 2017.https://www.rfigroup.com/rfi-group /news/newzealand-consumers-move-away-cash-digitl-and-contactlesspayments-rise

[9]   E. Ahmed, *The rising trend of cashless payments,* 2016.http://gulfnews.com/business/sectors/banking/the-rising-trend-ofcashless-payments-1.1863595

[10]   D. Ma, N. Saxena, T. Xiang, and Y. Zhu, "Location-Aware and Safer Cards: Enhancing RFID Security and Privacy via Location Sensing", *IEEE Trans. Depend. Secure Comput.,* vol. 10, no. 2, pp. 57-69, 2013.
[http://dx.doi.org/10.1109/TDSC.2012.89]

[11]   S. Brands, and D. Chaum, *"Distance-Bounding Protocols,"* in *Advances in Cryptology – EUROCRYPT '93.* vol. Vol. 765. Springer: Heidelberg, 1993, pp. 344-359.

[12]   *Hundreds of crores of rupees lost in card payment, internet banking frauds in just 3 months,* 2020.https://www.financialexpress.com/industry/banking-finance/hundreds-of-crores-of-r-pees-lost-in-card-payment-internet-banking-frauds-in-just-3-months/1886386/

[13]   R.L. German, and K.S. Barber, *Current Biometric Adoption and Trends,* 2018 .https://identity.utexas.edu/assets/uploads/publications/Current-Biometric-Adoption-and-Trends.pdf

[14]   *Global Biometric Market Analysis: Trends and Future Prospects.*https://www.bayometric.com/global-biometric-market-analysis/

[15]   N.S.S. Shobha, K.S.P. Aruna, M.D.P. Bhagyashree, and K.S.J. Sarita, "NFC and NFC payments: A review,", *2016 International Conference on ICT in Business Industry & Government (ICTBIG),* 2016, pp. 1-7
[http://dx.doi.org/10.1109/ICTBIG.2016.7892683]

[16]   R.K. Garg, and N.K. Garg, "Developing secured biometric payments model using Tokenization", *2015 International Conference on Soft Computing Techniques and Implementations (ICSCTI),* 2015, pp. 110-112
[http://dx.doi.org/10.1109/ICSCTI.2015.7489549]

[17]   J. Xu, *"Digital Payment Systems,"* in *Managing Digital Enterprise.* Atlantis Press: Paris, 2014, pp. 159-175.

CHAPTER 47

Bamboo Infoline in Himachal Pradesh-Analyzing the Current Status of the Furniture Industry and Artisans

Priyanka Shukla[1,*] and **Mahendra Joshi**[1]

[1] *Department of Architecture and Design, Lovely Professional University, Punjab, India*

Abstract: India is famous for its cultural legacy and its artisans and artistic heritage. Himachal Pradesh has always been known worldwide for its culture, styles, and workmanship. Bamboo is the most versatile and sprightly growing perennial grass. It is the fastest-growing resource that is getting recognition in all the states of India but has shown a very low pace in Himachal Pradesh. The state which is a hub of artists and bamboo resources is not able to utilize the resources and meet consumer demand. The prominent concerns in Himachal Pradesh are inadequate technological know-how & facilities; lack of awareness amongst artisans regarding the latest trend of bamboo furniture and design; lack of promotion of bamboo craft and furniture. The artisans and craftsmen are facing many challenges due to these impediments. Bamboo is an exciting design resource and has many positive attributes such as rapid growth, high strength, and ease of manipulation using simple tools with low investment and can act as a supportive measure to enhance the current status of the artisans in Himachal Pradesh. The removal of poverty and social tensions which Himachali craftsmen are facing, bamboo furniture development and productions can play a major stroke to protect the heritage legacy [1]. The focus of this paper is to find the current status of the bamboo handicraft and furniture industry in this highly tourist-oriented state of Himachal Pradesh, the gaps and challenges faced by all the stakeholders in the industry, and the development policies which are needed to produce a better source of handicraft and furniture to meet consumer demand.

Keywords: Bamboo, Bamboo Artisans, Craftsman, Furniture, Handicraft, Himachal Pradesh.

1. INTRODUCTION

The Indian bamboo and handicraft sector has significant importance in the economy-boosting. Besides promoting the cultural heritage and design of the state

---

* **Corresponding author Priyanka Shukla:** Department of Architecture and Design, Lovely Professional University, Phagwara, Punjab, India; E-mail: priyankas.archi@gmail.com

Dharam Buddhi, Rajesh Singh and Anita Gehlot (Eds.)

it is also generating an inflow of foreign exchange and interest in the foreign market.

The bamboo furniture and handicraft sector have been of importance since 1950 in promoting the creative skills and heritage of India. Himachal Pradesh is the main contributor to this development, it is a state with a geographical boundary of 55673 km². The ancient history described this state as the "Abode of Gods" located in the Northern part of India and boundaries by Jammu & Kashmir, Punjab, Uttar Pradesh, Haryana, and Tibet shown in Fig. (**1**). The state extended on 30^0 22' to $33^0 12'$ north latitude and 75^0 45' and 79^0 45'. Being a part of the Himalayan region, the state AMSL [average mean sea level] is 350 to 7000 meters [2].

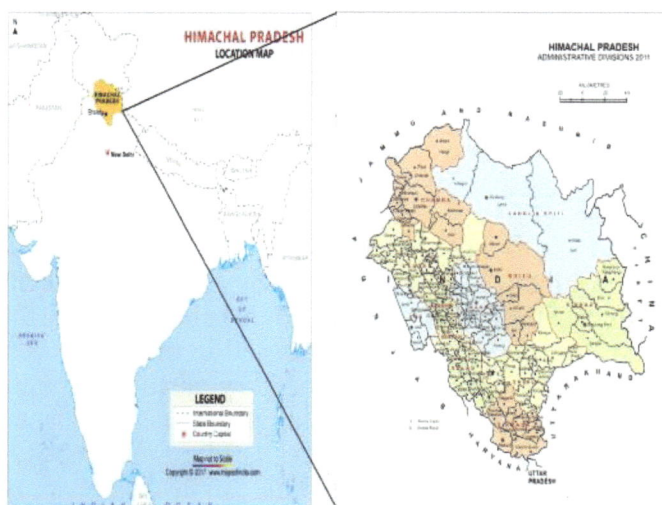

Fig. (1). Map showing location of study.

India is a very important supplier of bamboo in the world market. Since independence, with the legacy of Gandhian philosophy, the small-scale sector played an important role in boosting the Indian economy. It has been estimated that 40% of the total industrial output and 35% of the total direct export is been originated from the small-scale sector [3]. The bamboo and handicraft industry in the industry is highly labor-intensive. It is spread all over the rural and urban sectors of the country and provides a wide scale of employment to millions of artisan and large-scale women from the rural sector. The bamboo and furniture and handicrafts sector have been given importance since 1950, many policies were made to support the sector but policy implementation is facing lots of gaps and challenges. Our Rural and urban artisan and workers have given hefty contributions to the economy of the country. Bamboo, popularly been referred to as "poor man's timber" has been an integral part of life in the hilly states. Fig. (**1**) shows Map location of the Study.

The small scale industry of Himachal Pradesh which includes bamboo furniture industry and sawmills are less capital intensive compared to the other industry like paper, plywood, and other boards products. For centuries the local community is been using bamboo from food to construction. The paper is focusing on finding the current status of the bamboo handicraft and furniture industry in this highly tourist-oriented state, the gaps, and challenges faced by the industry and the artisans, and development policies that are needed to produce a better source of handicraft and furniture

2. BAMBOO CLUSTER HISTORY

The history of bamboo furniture and crafts has no specific records but it can be traced back to the 2nd century AD. During the ancient days, bamboo cutting was forbidden by natives because of its religious significance and auspicious value but the origin of bamboo crafts can be traced from the time when man started cultivating food crops thousands of years ago. The bamboo crafts and furniture industry is been one of the oldest cottage industries because of its versatility, strength, and easy workability with simply hand-handled tools [5]. The people of the state [Himachal Pradesh] are highly involved in the bamboo-related industry from making baskets to mats and this industry is highly contributing to the alleviation of rural poverty and women empowerment. These craftsmen are known by different names in different districts like Dumnas, Reharas, Bararas, Banjaras, Kolis, and Chamangas [6]. This industry is providing earning opportunities to many rural women and was majorly contributing to making them more financially independent. Most importantly this industry is playing a major role in environmental rejuvenation as bamboo absorbs 12 metric tonnes of carbon dioxide per hectare and produces 45% of more beneficial oxygen than other plants moreover bamboo needs less water than many other species which helps in maintaining soil stability. The industry was earlier confined to the household level but later it grew due to various central government policies. Many artisans were experts in the production of bamboo furniture and handicrafts. The cluster of bamboo working was practiced in many districts like Kangara, Pathankot, Dharamshala, Mandi, Kullu, and Baijnath [4].

3. METHODOLOGY

3.1. Questions

This study is attempted to highlight some issues like:

[a] Current state of the bamboo furniture and handicraft industry in Himachal?

[b] What are the current trends and preferences of the consumers?

[c] What are the challenges faced by the local artisan and workers?

[d] What are the major challenges faced in the promotion and sale of bamboo furniture and handicrafts?

[e] What are the major regulatory reforms and policies needed for promoting the bamboo furniture industry?

3.2. Methodology

The study is descriptive and compiled with the reference of qualitative and quantitative data. The primary data is being accumulated by the surveys. The secondary data is being sorted with the online documented literature, available articles, and government and private source of data to assess the clear picture of the bamboo furniture markets and the workers in Himachal Pradesh

4. KEY FINDINGS AND DISCUSSIONS

4.1. Bamboo Resources in Himachal Pradesh and its Status

In India, as per the Forest Survey of India [7], the total forest area is 67.7 million ha, from which bamboo acquires 11.4 million hectares. However, the contribution of India towards the global market is only 4%. Bamboo has always been the main source of the living hood for the rural class and around 8.6 million people rely on it for their daily living [8]. India is covering 16.0 million hectares of total bamboo area and Himachal owes 508km^2 area which caters around 3% of the total forest area. Bamboo plantation till 2008-2009 was accounted 8930 hac and till 2009-2010 and 2010-2011 it recorded 1242 and 754 hectares simultaneously has shown gradual rise after 2011 of 1096 Hac [2].The state has a total nine number of species under 5 genera like *Dendrocalamus hamiltonii, D. strictus, Arund-inaria falcata, Bambusa nutans, D. strictus, D. parishii, Phyllostachy aurea, P. bambusoides, Thamnocalamus spatifloras* which are widely available [2]. From these species, Arundinarifalcata and Thamnocalamus spatifloras are the two hill bamboo species which are locally called "Nirgal" or " Naghal" both these species have wide use in the state from roofing to fodder [9]. Bamboo has shown a wide variety of evidence all over the state with rich in raw material it has shown a wide variety of beautiful products each district has its style and work of art.

Apart from making baskets, bamboo is now given importance for turning bamboo into furniture with more modern innovation. The bamboo furniture industry has shown great potential for expansion and Himachal Pradesh has come out as a self-sufficient state in bamboo and its good money-making industry. Himachal Pradesh contributes a sizeable proportion to the Indian bamboo industry and provides livelihood to thousands of people.

Analyzing the market, it can be concluded that the bamboo market is segmented into two strata, firstly the domestic market and secondly the global market. The domestic market analysis concludes that bamboo being one of the vastly diverse resources and its applicability has shown the wide scope of social, environmental, and economic aspects but the industry has been held back because of a wide variety of issues in its value chains, improper channelization and regulatory and legislative barrier in cultivating of bamboo and promoting the bamboo products. The Global market analysis of bamboo has shown its high potential for high-value addition, the higher margin and enhanced profitability have transcended the image of bamboo as an outcast material. In the bamboo industry, most of the members which are associated or working are seen as socially backward who have taken up the skills of making handicrafts and furniture from bamboo.

4.2. Impediments in Domestic Market of Bamboo Handicrafts and Furniture

Most of the artisans were found that their prime sources of income were cultivation and making bamboo handicrafts while furniture is being focused as a part-time business only. Bamboo is being directly cultivated and obtained from the forest on a macro scale. The majority of the workers who are involved in this industry are suffering from a lack of elementary education shown in Fig. (**2**) and the appropriate techniques of working. Basic education qualification and knowledge regarding the concept and its value addition to the product is found to be very low.

Up to class 4	Class 5-8	Higher Secondary	Senior Secondary and above
4	8	5	3

Fig. (2). Education qualification of artisans in Himachal Pradesh [20 respondents].

The rural wages artisan's wages have shown significant changes after 2007-08. The Himachal Pradesh bamboo handicrafts and furniture industry is highly labor-intensive and has widespread in the rural and urban sectors of sub-regions. The state government has though introduced many schemes to increase the literacy rate of bamboo artisans, particularly the period after 2004-05 was the time of "jobless growth" with a growth rate of employment slowing down from 2.85% per annum and till 2011-12 it showed the increase of 0.46% per annum. But all the schemes to educate the local handicrafts artisan and bamboo workers were not able to increases the literacy rate of the bamboo workers.

Analyzing the data, shown in Fig. (3) it was found that only 30% of artisans work for 26 to 30 days whereas 45% of artisans work for almost 21 to 25 days and only 10% of artisans work for 10 to 15 days. This data states that a good percentage of Himachal artisans take the furniture crafts industry as their full-time occupation. It was also found that very few numbers of artisans are aware of new machines used in the production, and a large number of artisans working in the bamboo sector are unaware of the latest trends. It was also observed from the literature study that very few numbers of artisans are willing to know more about the new technology. This lack of awareness and low literacy rate lead to the decline of the bamboo sector in Himachal Pradesh. Most of the artisans are not aware of the latest trends and available working modules which is leading them towards a very low course of work and unemployment. The worker's and artisans' decline in employment has also partially contributed to the slowdown of revenue generation. Hence for the improvisation of the economic status, the Government has introduced many schemes to promote more working opportunities for artisans and bamboo workers [10]. Some schemes that are been governed by the government are shown in Fig. (4).

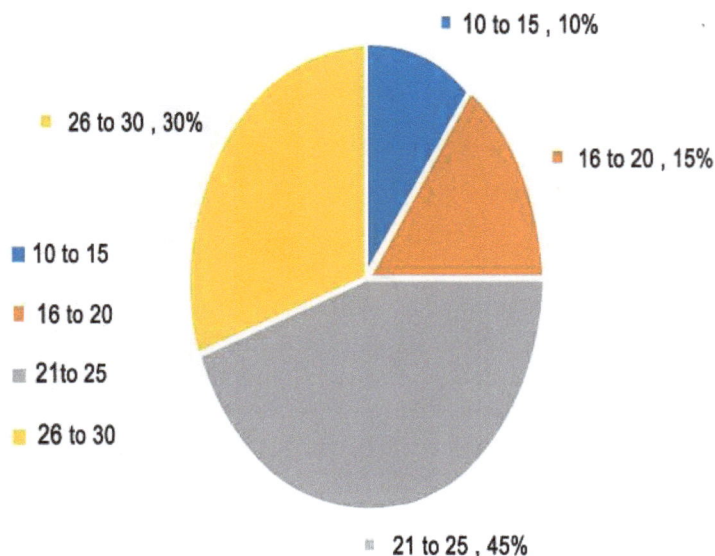

Fig. (3). No of working days of artisans (Source [6]).

S. No.	Scheme Initiated	Benefits
1	Direct Benefit to Artisans Scheme	Under this scheme financial support is provided to the old group and social security to the artisans who are the recipient of Shilp Guru Award/national award /state award not less 60 years financial assistance of Rs 3500/-pm shall be provided to Master Craftsman [11].
2	Margin Money	This scheme is included in 2018-19 and under this scheme concessional credit of 20% of Mudra loan the amount will be given as marginal support not exceeding to Rs 10000/- is provided to artisans [11].
3	Hathkargha Samvardhan Sahayata [HSS]	This scheme provides accessories to the weavers the enhance their earning through improved productivity. Intis scheme 90% of the cost is been borne by the government and 10% is borne by beneficiary [11].
4	Comprehensive Handicraft Cluster Development Scheme [CHCDS]	Under this scheme, the artisans /exporter/manufacturer are provided good quality raw material at subsidized rates [lower than market prices], and under the sub-scheme [11] a structure and technology support the raw material is provided at a subsidized rate of 8-10% less than market prices [11].

Fig. (4). Various Financial Schemes by Government of India (Source [11]).

Similarly, the National Bamboo Mission [NBM] under the Ministry of Agriculture Government India launched many integrated central schemes to proliferate the use of bamboo and the development of its industry [12]. NID's [National Institute of Design] Centre for Bamboo Initiatives [CFBI] introduced various activities related to the promotion of bamboo-based research design, development, and training students, young designers to experiment and innovate new innovative applications to build new resources for the bamboo sector in PAN India [13]. Such workshops have also been organized in Himachal Pradesh shown in Fig. (**5**). Setting up the furniture or handicraft workshops requires new motor skills from management to marketing. In addition, it has a perception that works shop based production need a formal premise, business plan and trading licensing whereas setting up home-based working doesn't need any financial assistance, huge capital, and labor investment and can still generate income from handicrafts sales with less disruption in their daily work and less overhead cost. But with these pros homestead working has to face some cons like distribution, sales of goods, and exploitation by the third parties like a middleman in the supply of their product to the consumers whereas these cons in case of home-based production get eliminated in the workshop production, production in the workshop is much easier and of better quality, as machine-based production is executed [14].

Fig. (5). Workshop for rural craftsperson's in Himachal Pradesh (Source : [4]).

4.3. Bamboo Furniture and Handicrafts Global Market

This sector has huge importance in the economy because of its employment potentials with less capital investment and demand in domestic and overseas markets [15]. As per the latest report by the Development commissioner handicrafts [DCHC], the total number of registered craftsmen's are 3837 including both male and female out of which 2834 are practicing their crafts in Himachal Pradesh. The state significantly contributes to the Indian export market.

The Indian handicrafts industry is being segmented into regional artisans and exporters or export houses. Indian bamboo handicrafts and furniture industry demonstrate the legacy and the culture of India in the global market. India's current demand for Bamboo is estimated at 27 million tons per year. Even though India has 30% of the world's Bamboo resources, we only meet 50% of our domestic demand; the rest has to be imported from China and other southeast Asian countries. Currently, the Bamboo sector has employed only 10 million people, much shorter than the potential that this sector has in our country which can range from 50 million going up to 129 million [4]. China accounts for 70% of the world's export of bamboo and rattan products, with a value of USD 1.18 billion [16]. Having one of the largest resources of bamboo in the world, India contributes just 4% to the global market. This shows lower productivity in comparison to other bamboo producing countries such as China, Malaysia, and Japan whose contribution is 80% shown in Fig. (**6**).

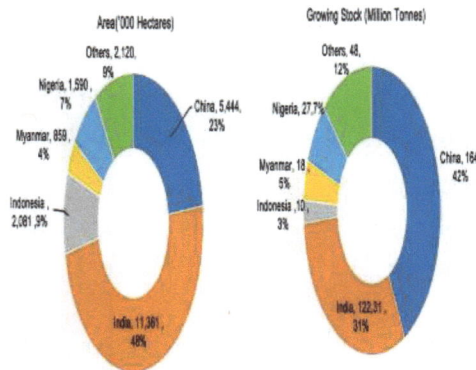

Fig. (6). Comparative analysis of the Bamboo area and growing stock of India with China (Source [16]).

4.4. Consumption Pattern of Bamboo Resources

Bamboo is capabe of thriving in an extreme range of climatic conditions and there are nearly 1,500 documented varieties of applications. Nearly 13.4 million tonnes of bamboo amounting to Rs. 2042 crores are used by various industries like pulp, paper, construction, cottage industry, and medicine.

The Indian demand for bamboo for industrial purposes is met from the state-owned forests resources, whereas for the other purpose we need to depend upon both private as well as government-owned resources [18]. The extent of commercial forestry is low and very few cultivators are interested in the cultivation of bamboo in Himachal Pradesh. The database of the resources utilization shown in Fig. (**7**), states that bamboo consumption by the furniture industry is only 1% [5]. This is leading to local and craftsmen being deprived of work and shifting to the other sector. This shifting of the craftsmen is affecting the cultural heritage and craft industry of Himachal Pradesh.

Uses	Consumption %
Pulp	35%
Housing	20%
Non - residential	5%
Rural uses	20%
Fuel(non - industrial)	8.5%
Packing,including basket	5%
furniture	1%
Others including ladders,mats etc	3%
Total	100%

Fig. (7). Consumption pattern of Bamboo in India (Source [17]).

4.4.1. Lack of Adequate Awareness

Management practices are a major problem in the bamboo sector. Farmers and artisans come with the traditional thinking that bamboo does not need cultural measures. On the other hand, harvesting bamboo culms from the natural plantation is a very dangerous job if we cannot harvest it carefully, it might be very dangerous to the life of people. Artificial plantation of bamboo with better management practices like maintaining the spacing, thinning makes it easier for harvesting. Farmers are also not aware of the harvesting season of bamboo culms. If they harvest bamboo culms in January/February, then there is less attack of insects/pests and diseases during storage. It was found that very few cultivators received training on bamboo cultivation and preservation. Whereas bamboo craftsmen are also not aware of the species that are adequate for furniture and handicrafts [19].

4.4.2. Inadequate Value Chain

There exists a huge untapped market for bamboo handicrafts and furniture items; however, the value chain is very weak. The two critical areas which are missing in a typical Indian value chain are dedicated designers and sales/ marketing. Additionally, the value chain in India suffers from the following problems:

4.4.2.1 Access to Raw Materials

• Bamboo as a resource for furniture is available relatively easily in rural areas but is quite difficult to find in urban areas.

Access to Markets

- Most of the artisans and craftsmen involved in the bamboo furniture industry are not aware of the markets.
- The existing manufacturers are scattered, with small-scale production; the transaction cost is high if a middle man wants to take the products to an urban area.
- Very little marketing or advertisement of bamboo-related products in any media whether TV, social media, or internet.

Technology and Skilled Human Resources Availability

- Small rural bamboo manufacturing establishments use simple tools, such as hand splitters, handsaws, knives, *etc*. The latest technologies which can produce high-quality furniture which is used in other Asian countries like China, Vietnam are not used in India.
- Very little design intervention to produce functional, aesthetic, and knock-down type furniture to excite new buyers.

Lack of Finance

- Most of the small artisans are not economically well-off and neither can invest on their own nor have access to financial aid.
- Bamboos, being an unexplored sector in India, banks are generally hesitant to provide finance.

The major problem is not how to produce bamboo but the main aspect which need to be focused on is "how to promote Bamboo" and stating a statement that if the bamboo building technology is made convenient to handle and promoted with standards, codes, and bye-laws it could go along way in creating the gateway all the alternative material and technologies [20].

4.4.2. Design Development Strategies for Rural based Economy Development

Once the local bamboo farms are stabilized and more farms are put in place there is a need to develop both local and external markets for the materials that are produced by the farmers. It will allow developing value-added strategies for both local use products as well as products for up the market customers in the vast Indian urban marketplace that can give large employment to local producer groups [19]. Each region of the country can choose a few locally appropriate bamboo species based on the nature of the soil, climate, and intended end uses and then invest in developing good practices for the systematic cultivation of these species in well-managed farms by using processes that comply with eco-friendly norms that would also need to be developed for each species. The post-harvest practices too will add value to the bamboo raw material and in some cases, primary local conversion of the bamboo stock and by-products of farming would create local employment for a large number of people. All parts of the bamboo are useful and the product design strategy should take this factor into account as well.

5. DEVELOPMENT DIRECTION NEEDED IN BAMBOO FURNITURE AND HANDICRAFT INDUSTRY

5.1. The Bamboo Furniture Design Should Detail its Culture

Culture has always been the pride of India and Himachal Pradesh is known for its old culture and heritage. There has always been a spiritual connotation of culture and furniture. Especially in today's world, it's difficult to meet societal needs. Furniture's material, form, and functionality are not able to meet the needs of the consumer. Similarly, China's bamboo culture has also long ancient history and is rich, and it has become a trend for the Chinese community to use Bamboo furniture and draw bamboo culture to their homes, but despite the huge bamboo production in India our cultural legacy we still are not able to cast this hidden treasure [19].

5.2. The Bamboo Furniture Design Should Focus on Information Technology

In this era of globalization where all countries are coming up with new technology every day the market is becoming fiercer with the competition. The constant development and introduction to new technology have created more competition in the furniture market. The old traditional design doesn't attract the customer's interest and the end-users are looking forward to the new latest technology and development in the product design. To meet the requirement of the end-users, the use of modern technology is important to connect designers, manufacturers, and end-users, so that each one can actively participate in designing, manufacturing, and developing the product to make India a more competitive producer of bamboo furniture globally [19].

5.3. The Bamboo Furniture Design Should be Personalized

The Indian bamboo furniture market needs to determine the needs of its main target consumers. Our young generation has always been community who like to do more experiment and single styled bamboo furniture is unable to attract this group who are mainstream line buyers. The young generation focuses on furniture convenience and individuality before buying it. Therefore, the design created should break the tradition and must be combined with considering ergonomics, taste, and design, to increase the aesthetic feel and individuality of the furniture to make the young generation a prominent buyer of this industry [19].

CONCLUSIONS

Bamboo can be an Economic Driver for the Rural Economy of Himachal

The key to this strategy is based on bamboo being used as a cultivated resource that can be intensively managed to produce high-quality resources for local consumption with the use of appropriate technologies and more importantly with the use of intensive knowledge inputs in innovation and design across all the stages of cultivation and utilization and to markets, both local and export. Such a strategy would give the rural inhabitants a perennial resource that can be fashioned into a whole range of end uses in an imaginative and economically and socially beneficial manner. The strategy included the possibility of at least three distinct types of cultivation to be adopted by the local farmers and it could extend to cultivation on large corporate farms where desired and it calls for a distinctly regulated method for the utilization of forest bamboo which has been the dominant form of access in India over the years [19]. The study states that the Himachal bamboo industry and its artisans can provide hefty support to the Indian economy, but local artisans and craftsmen need integrated development and a focused approach. They need a customized capacity development, Design, and product development, targeted market, and infrastructure development. The bamboo industry can add a new prospectus and a new dimension in the development of Himachal Pradesh will undoubtedly diversify the economic activities of the state [21].

CONSENT FOR PUBLICATION

Not applicable.

CONFLICT OF INTEREST

The authors declare no conflict of interest, financial or otherwise.

ACKNOWLEDGEMENT

Declare none.

REFERENCES

[1]   D.N. Tiwari, "A Monograph on bamboo", In: *Dehra Dun.: International Book Distributor,* , 1992.

[2]   "International Institute for Environment and Development", In: *Himachal Pradesh Forest Department, "Himachal Pradesh Forest Sector Review.* Department for International Development India, 2000.

[3]   R. Gupta, and I. Mukherjee, " "Scope of Cottage and Small Scale Industry in West Bengal in the Early 2000", *SSRN,* 2006.

[4]   A.U. Khan, and A. Hazara, *Industrialisation of the Bamboo Sector: Challenges and Opportunities.,* 2007.

[5]   "Katlamara Chalo: A Design for Development Strategy", In: *Centre for Bamboo Initiatives at National Insitute of Design (NID)* Development Commisioner Handicrafts, Govt of India: New Delhi, 2007.

[6]   G. Goraya, and M. Pal, "K. kapoor and V. Jishtu, "Hill Bamboo-Socio Economic significance and conservation Imperatives. A case study from Himachal Pradesh", *Indian For.,* 2008.

[7]   O. Singh, "Bamboo for Sustainable livelihood in India", *Indian For.,* 2008.

[8]   N.F. Fabeil, K.M. Marzuki, and J. Langgat, ""Dedicated Vs Modest Commercialization of Small-Scale Handicraft Production in Sabah," IRACST –", *Int. J. Commer. Bus. Manag.,* 2012.

[9]   "Bamboo: Green Construction Material", *Asia Pacific for global change research,* 2015.

[10]   N. Manjunath, "Contemporary Bamboo Architecture in India and its Acceptability", *10thWorld Bamboo Congress,* 2015.

[11]   N. Saini, C. Nirmala, and M. Bisht, "Bamboo Resource of Himachal Pradesh (India) and Potential of Shoots in Socio-economic Development of the State", *10th World Bamboo Congress,* 2015 Korea

[12]   "Integrated design and Technical Development project in cane and bamboo craft in balakurpi", *Export promotion council for handicraft,* 2017.

[13]   Himanshu, and S. Kundu, "Rural wages in India: Recent trends and determinants", In: *Indian Society of Labour Economics*, 2017.

[14]   K.K. Hazra, and A. Barman, "Prospect of traditional craft in present economy: a study of earthen doll of krishnagar, west bengal", *Int. J. Manag.,* 2017.

[15]   "Ministry of Environment & Forests", *FSI; Forest Survey of India,* 2017.

[16]   P. Hegde, "Greening India by planting bamboo", https://www.thehindubusinessline.com/opinion/greening-india-by-planting-bamboo-ep/article25009098.ece

[17]   C. R. Gupta, "Developing framework for designing craft based tourism circuits: a case of himachal pradesh",

[18]   "Financial Assistance to Handicraft and Textile Industry", *Government of India Ministry of Textiles,* 2019.https://pib.gov.in/Pressreleaseshare.aspx?PRID=1578541

[19]   *Bamboo Resource Status and Business Opportunities in Jharkhand.* Mukhyamantri Laghu Avam Kutir Udyam Vikas Board,Department of Industries, Government of Jharkhand, 2019.

[20]   T. Roy, "Intangible Heritage of Bamboo and Cane Crafts is in Stake: A Study of Selective Blocks of Malda", *International Journal of Scientific Research and Engineering Development,* 2019.

[21]   C. S. Susanth, "An Approach for Designing Solid Bamboo Furniture", *Business & Innovation - Product Development – Design.*

CHAPTER 48

Nutritional and Therapeutic Properties of Button Mushroom (*Agaricus Bisporus*)

Bhawna[1], Aditee Singh[1], Indresh Kumar Agnihotri[1], Jagjit Kaur[1] and **Manoj Kumar Jena[1,*]**

[1] *Department of Biotechnology, School of Bio-engineering and Biosciences, Lovely Professional University, Phagwara, Punjab, India, PIN- 144411*

Abstract: Medicinal plants have great importance these days and the better way to provide medicines is to give them as food. Mushrooms are one of the good sources of food that has good nutritional and therapeutic value. They are the constituents of a good diet which boosts up the immune system and helps to fight diseases like cancer. From ancient times people of China believed that the mushrooms have anti-aging properties and they had been consuming mushrooms as their important diet. Along with the anticancer and anti-aging properties, they increase brain activity. One major advantage of mushrooms over other vegetables is their growth period, *i.e.*, they can grow around the year while many vegetable plants are seasonal. The present review article is focused on the nutritional and therapeutic properties of button mushrooms in more detail, which will help the researchers working in this area to design further experiments and explore more on this plant.

Keywords: Anticancer Activity, Anti-Diabetic Activity, Button Mushroom, Nutritional Properties, Therapeutic Value.

INTRODUCTION

Mushrooms are the part of Basidiomycetes family and are used both as food and medicines. They are edible and poisonous too. The edible part of the Basidiomycetes family is known as "Mushrooms" while the non-edible part is "Toadstool" as they are poisonous. Unlike other plants, they do not perform photosynthesis as they have no chlorophyll in them and reproduce with the help of vegetative propagation. During the process of vegetative propagation, the mycelia of the mushroom secret the enzymes which are responsible for the breakdown of

* **Corresponding author Manoj Kumar Jena:** Department of Biotechnology, School of Bio-engineering and Biosciences, Lovely Professional University, Phagwara, Punjab, PIN- 144411, India; E-mail: manoj.20283@lpu.co.in

Dharam Buddhi, Rajesh Singh and Anita Gehlot (Eds.)

the compounds that are present in the substrate such as cellulose and lignin. The compounds that are degraded are again absorbed by the hyphae and allow mycelia to grow. Usually, mycelia grow in a lateral form but in some cases, the diameter enlarges [1].

NUTRITIONAL CONTENT IN *AGARICUS BISPORUS*

Proteins

A. bisporus is a protein-rich vegetable. The crude protein present in *A. bisporus* comes next to the protein content of animal meat but is superior as compared to milk protein [2]. *A. bisporus* contains most of the essential amino acids that can only be consumed from non-vegetarian products. *A. bisporus* contains many types of amino acids that are essential like histidine, leucine, valine, threonine, isoleucine, arginine, lysine, methionine, *etc*. The amount of amino acids is balanced in *A. bisporus*. Cysteine and threonine are limiting amino acids. *A. bisporus* is considered the best source of threonine, isoleucine, and tryptophan. In frozen and preserved white button mushrooms, there is an abundant amount of exogenous amino acids. The amount of aspartic acid, glutamic acid, and arginine is 8.1g, 6.2g, and 8.0g per 100g of the total amount of protein. By using various types of substrate, during the cultivation, protein content can be increased as various types of the substrate can provide nutrients to the fruiting body [3].

Carbohydrates and Dietary Fibers

Carbohydrates provide us with energy to do work. The carbohydrates content that we get from *A. bisporus* is not considered a major energy source for human beings. It was observed that based on dry matter percentage of carbohydrates present are mannitol (8.56), glycogen (5.34), reducing sugars (2.48), and hexuronic acid (0.32). Two forms of carbohydrates present are digestible carbohydrates, which are present in a small amount like glycogen that is 5-10% daily values whereas non-digestible carbohydrates which include non-starch polysaccharides like chitin, mannans which constitute major carbohydrate content in *A. bisporus*. Mannitol is the most abundant sugar in mushrooms as there is 80% mannitol out of total free sugars present in white button mushrooms [4]. Research shows that carbohydrate content is 3.26g per 100g of *A. bisporus* Dietary fibers consists of constituents of fungal cell wall like hemicelluloses, beta-glucan, *etc* which adds to the nutritional content of *A. bisporus* [5]. Dietary fibers are estimated to be 1g per 100g of *A. bisporus*.

Minerals

Mineral content vary according to diameter, type of substratum of fruiting body, and species of mushroom. *A. bisporus* is rich in Fe, K, Cu, Zn, Na, Co, Se, Ca, Cd and Mn [6]. Research shows that there is more amount of different types of minerals in the cap than they are in the stem but calcium and iron are more in an amount in the stem than in the cap. Phosphorous (10,400-11,230mg/kg) and potassium (38000-39500mg/kg) is present in maximum amount in *A. bisporus*. Mushrooms are grown on substrates rich in zinc, selenium, and copper. 56-70% of total ash content is produced by minerals like K, Na, Mg, and P whereas 45% of total ash is formed by potassium alone. *A. bisporus* is considered a poor source of selenium with a concentration of 1 microgram per gram on a dry basis.

Vitamins

A. bisporus contains a variety of vitamins like folate, pantothenic acid, riboflavin, thiamin, niacin, *etc*. It is reported that white button mushrooms are rich in vitamin B complexes [7]. *A. bisporus* is not usually rich in vitamin C and vitamin D. They are deficient in vitamin D because they are cultivated in a dark environment. There is a technique used to fix this deficient nature of vitamin D. When fruiting bodies are treated with ultraviolet rays. Perhaps this exposure leads to the production of vitamin D2 [8]. Ergosterol which is one of the constituents of the fungal cell wall is also a precursor of vitamin D2 [9]. Ergosterol from white as well as from brown button mushrooms. The antioxidant activities of both types of mushrooms are correlated with the ergosterol level of fruiting bodies [10].

THERAPEUTIC PROPERTIES OF BUTTON MUSHROOMS

Anticancer Activity

Lectin from the common mushroom *A. bisporus*, a popular edible species in western countries, has potent anti-inflammatory effects on human epithelial cancer cells, in addition to the apparent cytotoxicity. Lectin from *A. bisporus* has anti-inflammatory effects on various cell types. Selenium is an important chemical element in humans and animals. Selenium helps in the prevention of cancer. Selenium has a potential role in cancer protection by protecting antioxidants and/or increasing immune function [11]. The intervention trials also showed the benefits of selenium in reducing cancer, especially within the liver, prostate, colon, and lung, with excellent benefits for those with low selenium status. Aromatase is an enzyme that converts androgen into estrogen. Increased exposure of aromatase to breast tissue is considered a risk factor for carcinoma Ergothioneine, lectin, and beta-glucan found in white-buttoned mushrooms act as an antioxidant for breast cancer.

Antibacterial and Antifungal Activity

The antimicrobial effect *of A. bisporus* extract was tested and compared with Gram-positive and Gram-negative bacteria with the same type of yeast. The ethanol release of *A. bisporus* has shown anti-cancer activity. The release of Methanolic and acetone has shown strong antibacterial activity against tested bacteria. Methanolic extracts have shown a significant effect on inhibiting the growth of each experimental virus. There is a need for additional studies to differentiate and mark antibacterial in this fungus with effective disease control measures. Fragments from the bodies of fruit and mycelia of various mushrooms have been reported in the anti-bacterial activity of many infectious agents [12]. *A. bisporus* is the most widely grown mushroom and has 38% of the world's grown mushrooms. Therefore, the present study focused on testing the antimicrobial activity of the methanolic and acetone ingredients of *A. bisporus* using the agar distribution method against the separation of the two clinics *E.coli and Staphylococcus aureus*. White buttoned mushrooms are a rich source of saponins, flavonoids, tannins, and cardiac glycosides that are responsible for protecting against various viruses such as Escherichia coli, Bacillus subtilis, Staphylococcus aureus, Klebsiella pneumonia, and Proteus vulgaris [13].

Antidiabetic Activity

A. bisporus contains natural antidiabetic polysaccharides that can improve glucose tolerance and reduce the activity of hyperglycemia and α-amylase. So mushroom polysaccharides can help treat many syndromes of human diseases. Health conditions such as diabetes and heart disease have been linked to increased inflammation in the body. Mushrooms have anti-inflammatory properties that help control such conditions. Mushrooms are a great snack for people with diabetes as they have a low glycemic index as containing very low levels of carbs, which means they do not raise blood sugar levels like high carb foods like bread and pasta. Low-calorie foods with high water and fiber content that keep you full for a long time. Fresh mushrooms have both soluble and insoluble fibers, where soluble fiber is shown to maintain blood sugar levels. According to a study published in the Journal of Functional Foods, eating white mushrooms with a white button daily can act as a prebiotic by improving the bacterial community in the gut, which may improve glucose control in the liver. In the study, feeding white-buttoned mushrooms in mice altered the structure of the bacterium, microbiota- to produce more fatty acids, especially those derived from succinate, according to researchers.

Genoprotective Effect

A. bisporus is the most widely-eaten edible mushroom. *A. bisporus* fruit bodies protect the oxidative damage caused by H_2O_2 in cellular DNA *A. bisporus* is a heat-labeled protein, called FIIb-1, which is present inside the fruit body and found as tyrosinase. The genoprotective effect associated with the release of cold water from edible mushrooms, *Agaricus bisporus,* is related to the activity of tyrosinase found within the fruits of the mushroom fruit. The genoprotective effect of *A. bisporus* tyrosinase depends on the enzymic hydroxylation of tyrosine to L-DOPA and subsequent conversion of this metabolite to dopaquinone [14].

CONCLUSION

Mushrooms provide good quality food with a high level of proteins and can be easily produced from lignocellulosic wastes that are obtained from various origins. In the future, the improvement of strain and the advancement in cultivation technology will lead to an increase in the production of the presently consumed mushrooms. To break the yield barriers, modern biotechnological tools and computer-aided environmental control systems can be used. Medicinal mushrooms are special kinds of mushrooms that can be used for medical purposes and they are likely to spread all over the world due to their medicinal property.

CONSENT FOR PUBLICATION

Not applicable.

CONFLICT OF INTEREST

The authors declare no conflict of interest, financial or otherwise.

ACKNOWLEDGEMENT

Declared none.

REFERENCES

[1]   S. E. E. Profile, "comparative study of white button mushroom strains", , 2017.

[2]   A. Kakon, M.B.K. Choudhury, and S. Saha, "Mushroom is an Ideal Food Supplement", *J. Dhaka Natl. Med. Coll. Hosp.,* vol. 18, no. 1, pp. 58-62, 2012.
        [http://dx.doi.org/10.3329/jdnmch.v18i1.12243]

[3]   R. Gothwal, A. Gupta, A. Kumar, S. Sharma, and B.J. Alappat, *Feasibility of dairy waste water (Dww) and distillery spent wash (dsw) effluents in increasing the yield potential of pleurotus flabellatus (pf 1832) and pleurotus sajor-caju (ps 1610) on bagasse,* 2012.
        [http://dx.doi.org/10.1007/s13205-012-0053-9]

[4]   B.A. Wani, R.H. Bodha, and A.H. Wani, "Nutritional and medicinal importance of mushrooms", *J. Med. Plants Res.,* vol. 4, no. 24, pp. 2598-2604, 2010.
        [http://dx.doi.org/10.5897/JMPR09.565]

[5]   I. Golak-Siwulska, A. Kałuzewicz, S. Wdowienko, L. Dawidowicz, and K. Sobieralski, "Nutritional value and health-promoting properties of Agaricus bisporus (Lange) Imbach", *Herba Pol.,* vol. 64, no.

4, pp. 71-81, 2018.
[http://dx.doi.org/10.2478/hepo-2018-0027]

[6]   P. Mattila, K. Könkö, M. Eurola, J.M. Pihlava, J. Astola, L. Vahteristo, V. Hietaniemi, J. Kumpulainen, M. Valtonen, and V. Piironen, "Contents of vitamins, mineral elements, and some phenolic compounds in cultivated mushrooms", *J. Agric. Food Chem.,* vol. 49, no. 5, pp. 2343-2348, 2001.
[http://dx.doi.org/10.1021/jf001525d] [PMID: 11368601]

[7]   M.N. Owaid, A. Barish, and M.A. Shariati, "Cultivation of Agaricus bisporus (button mushroom) and its usages in the biosynthesis of nanoparticles", *Open Agric.,* vol. 2, no. 1, pp. 537-543, 2017.
[http://dx.doi.org/10.1515/opag-2017-0056]

[8]   J.S. Roberts, A. Teichert, and T.H. McHugh, "Vitamin D2 formation from post-harvest UV-B treatment of mushrooms (Agaricus bisporus) and retention during storage", *J. Agric. Food Chem.,* vol. 56, no. 12, pp. 4541-4544, 2008.
[http://dx.doi.org/10.1021/jf0732511] [PMID: 18522400]

[9]   S. Shao, M. Hernandez, J.K.G. Kramer, D.L. Rinker, and R. Tsao, "Ergosterol profiles, fatty acid composition, and antioxidant activities of button mushrooms as affected by tissue part and developmental stage", *J. Agric. Food Chem.,* vol. 58, no. 22, pp. 11616-11625, 2010.
[http://dx.doi.org/10.1021/jf102285b] [PMID: 20961043]

[10] F. Atila, M.N. Owaid, and M.A. Shariati, "The nutritional and medical benefits of Agaricus Bisporus: A review", *J. Microbiol. Biotechnol. Food Sci.,* vol. 7, no. 3, pp. 281-286, 2017.
[http://dx.doi.org/10.15414/jmbfs.2017/18.7.3.281-286]

[11] J. Marshall, *Selenium in cancer prevention.* Selenium Environ. Hum. Heal, 2013, pp. 55-56.
[http://dx.doi.org/10.1201/b15960-26]

[12] R. Balakumar, E. Sivaprakasam, D. Kavitha, S. Sridhar, and J. Suresh Kumar, *Antibacterial and antifungal activity of fruit bodies of Phellinus mushroom extract International Journal of Biosciences (IJB),* 2011.http://www.innspub.net

[13] M.V. Sharma, A. Sagar, and M. Joshi, "Study on Antibacterial Activity of Agaricus bisporus (Lang.) Imbach", *Int. J. Curr. Microbiol. Appl. Sci.,* vol. 4, no. 2, pp. 553-558, 2015.

[14] G. Dhamodharan, and S. Mirunalini, "A Novel Medicinal Characterization of Agaricus Bisporus (White Button Mushroom) Pharmacologyonline 2: 456-463 (2010) Newsletter Dhamodharan and Mirunalini", *Pharmacol. Newsl. Dhamodharan Mirunalini,* vol. 2, pp. 456-463, 2010. http:// pharmacology online.silae.it/ files/newsletter/2010/vol2/55.Mirunalini.pdf [Online]

CHAPTER 49

Phytochemical and Therapeutic Properties of Indian Bay Leaf (*Cinnamomum Tamala*) Plant: A Review

Sumit Bhattacharjee[1], Aeimy Mary Jose[1], Anuja Ajit More[1], Farhan Alam[1] and Manoj Kumar Jena[1,*]

[1] *Department of Biotechnology, School of Bioengineering and Biosciences, Lovely Professional University, Phagwara, Punjab, India 144411*

Abstract: The Indian bay leaf (*Cinnamomum Tamala*) plant is found in high altitudes along the tropical and subtropical regions of India. This plant is very useful to be used as a food ingredient and as a medicine for centuries to treat several ailments. The leaves are traditionally used in the Indian household as a spice in food and even as a mouth freshener and deodorant. This is because the leaves have a pepper-like aroma which is caused due to the essential oils present in the leaves and the bark. It is rich in alkaloids, flavonoids, terpenoids, polyphenols, *etc*. These phytochemicals are responsible for their aroma, flavor, and medicinal properties. The plant has therapeutic properties against various types of diseases. Due to its antimicrobial properties, it has been used as a preservative in the food industry. The present review discusses in detail the phytochemical and therapeutic properties of *C. Tamala,* along with its tremendous potential to be used as a medicine.

Keywords: Alkaloids, Bay Leaf Plant, Phytochemicals, Secondary Metabolites, Therapeutic Property.

INTRODUCTION

Nature is a treasure trove and it contains various treasures which help humanity grow and evolve. One of the most important and interesting treasures that can be found in it is plants, which provide us with minerals, vitamins, and even act as medicines. For thousands of years, humans have used plants not only as a source of food but also as a source of medicine for various ailments and disorders [1 - 3]. The advancement of technology prompted us to understand the therapeutic properties of various plants at a molecular level. The medicinal plants are rich in

---

* **Corresponding author Manoj Kumar Jena**: Department of Biotechnology, School of Bioengineering and Biosciences, Lovely Professional University, Phagwara, Punjab, India; E-mail: manoj.20283@lpu.co.in

Dharam Buddhi, Rajesh Singh and Anita Gehlot (Eds.)

phytochemicals which play a vital role in the treatment of different ailments. The phytochemicals are constitutive metabolites in plants that act with other nutrients and fibers to fight against diseases [4]. These chemicals on intake create a certain physiological action, suppress the growth of pathogens without causing any adverse effect to it, and have anti-mutagenic, anti-carcinogenic, and several other various attributes [5].

C. tamala is known as Indian bay leaf is a plant that belongs to the genus Cinnamomum under the family Lauraceae. In India, it is commonly known by names such as Tejpat, Tejpatta, and many more. Up to 350 species of this plant can be found worldwide [6, 7] and there are such species. In India, we can find around 20 such species. It is found in the states of Sikkim, Assam, Mizoram, Meghalaya, and on the North-Western parts of the Himalaya among its tropical and subtropical regions at an altitude of 900 -2500 m and is found mostly in shady and ravine slopes [8, 9]. The plant *C. tamala* is a medium-sized evergreen plant. Its trunk is grey-brown and its bark is soft and wrinkled in texture with abundant leaves present [6, 7]. The plant extract contains various phytochemicals used for various ailments such as diabetes, cardiac disorders, arthritis rheumatism, hyperlipidemia, inflammation, bladder disorders, hepatotoxicity, vomiting, diarrhea, and many more [10]. It is also used in the food industry as an ingredient for food items such as pickles, baked goods, sauces, some drinks, and many more, it is also used as a preservative for pineapple juice [8]. This review study focuses on the phytochemical and therapeutic properties of this plant.

Chemicals Found in *C. Tamala*

The frequently used part of *Cinnamomum Tamala* is its leaf, it has many components but the major component of bay leaf is the essential oils which consist of furanogermenone (59.5%), β- caryophyllene (6.6%), sabinene (4.8%), germacrene D (4.6%) and curcumenol (2.3%). It has other components such as eugenol, cinnamaldehyde which is present in the barks, it is responsible for the aroma [11]. The medicinal use of the oil is anti-flatulent, diuretic, and carminative properties [12]. The extract from Cinnamomum Tamala has terpenoids, alkaloids, phenol /polyphenols in extract fractions. The spice plant has phytochemicals that are used to develop antibacterial agents [13].

Phytochemical Properties of *C. Tamala*

Phytochemicals can be divided into seven different types which are polyphenols, alkaloids, tannins, terpenoid, phytosterols, saponins, and flavonoids [14].

Polyphenols

They are the secondary metabolites involved in the process of defending against UV rays and pathogens, plant polyphenol consumption in the diet prevents cancer formation, heart diseases, diabetes, and bone disorder, it has other properties like tests, aroma, smell, and against oxidative damage in food [15]. It is beneficial in reducing the adverse effect of aging on the brain or the nervous system.

Alkaloids

They are naturally occurring nitrogen-containing organic compounds. Alkaloids have complex chemical structures, it has been used for more than 3000 years and has various medicinal uses such as been used as purgative sedative and antitussive for fever, snakebite, and insanity. Alkaloids are known to have pharmacological effects, due to which it has been widely used in pharmaceuticals, such as stimulants and narcotics, but many alkaloids are toxic to different organisms so these are used as poisons [16, 17].

Flavonoids

It is a group of polyphenolic compounds that are classified into four groups, flavones, flavanones, anthocyanins, and catechins, and are mostly found in fruits, nuts, seeds, vegetables, stem, flower, honey, wine, tea, and propolis, and are known to have biochemical and pharmacological effects. antioxidant properties prevent cell damage, showing anti-tumor properties, improve blood circulation and decrease blood pressure [17].

Tannins

It is the most important secondary metabolite, tannins are described as the phenolic compound of high molecular weights, found mostly in plant leaves, barks, woods rots, *etc*. These provide the defense mechanism against insects, fungi, pathogens, herbivores, and help in regulating growth. Tannins have the ability to bound to proteins that form soluble or insoluble complexes, which gives the ability to form a complex with nucleic acid, alkaloids, saponins, polysaccharides, and steroids [18, 19].

Saponins

Saponins are secondary metabolites present in barks, stems, leaves, flowers, and roots. It has medicinal properties such as anti-inflammatory, antimicrobial, anti-tumor, anti-ulcer, hepatoprotective, decreases blood cholesterol level, and is used as an adjuvant in vaccines. It is used in the preparation of soaps, detergents, shampoos, beer, cosmetics, and fire extinguishers. Drugs based on Steroidal saponin have been used for disease treatment, saponins defend the plant from phytopathogenic microorganisms and insects [20].

Terpenoid

Terpenoids are primary metabolite which is present in a huge amount (maybe even thousands) in flowering plants. Terpenoids act as hormones, protein modification agents, components of the electron transfer system, membrane fluidity, anti-oxidant, determinants, defense against pathogens and herbivores, it also acts as signals to beneficial organisms like pollinators and so on. Herbs, spices, alcoholic drinks, and wine contain volatile terpenoids which give flavor and also preserves it due to its insecticidal and microbicidal properties [21, 22].

Phytosterols

It is an organic molecule having sterols and stanols are steroid compounds and is known as plant cholesterols. It has been seen that saturated phytosterol/ stanol has been more efficient in reducing the cholesterol level than unsaturated phytosterol/sterol. Health benefits are anti-inflammatory activities, reduced cholesterol level with decreased risk of heart diseases, and also prevents various diseases [23].

Therapeutic Properties

Anti-Diabetic and Antioxidant Effect

The extracts derived contain the compound procyanidin, which helps to protect the pancreatic β-cells and has also been shown to have a positive effect against diabetes [24]. The ethanolic extracts derived from the plant leaves have shown to be effective against both low blood sugar and high blood sugar [24]. It has also been seen to have a strong effect on cholesterol and has been shown to significantly increase antioxidant enzymes [25].

Anti-Tumour Effect

The extract of the *C. Tamala* leaf has shown anticarcinogenic effects against ovarian cancer in humans. Its extracts have shown strong cytotoxicity against cancerous cells under *in vitro* conditions due to the presence of the compound bornyl acetate in it [26]. It has also been seen to significantly reduce moderate inflammations [27].

Cardio-Protective Property

The extracts of the plants contain components such as cinnamaldehyde, benzoic acid, trans-cinnamic acid [28]. These components have shown to have positive results such as gastroprotective effects, anti-inflammatory effects, and effectiveness against cardiovascular diseases.

Anxiolytic and Antidepressant Effects

The extracts derived from *C.tamala* have shown anti-depressant activity like that of the drug imipramine and can be used for the treatment of conditions such as behavioral despair and learned helplessness. The plant C. Tamala has also shown significant anti-anxiety effects and thus can be used to manage psychological ailments [29].

Antimicrobial Activity

The extraction of *C. Tamala* is used for bio-synthesize iron nanoparticles. Degradation of extracellular material with the help of MFeNp leads to a reduction of proteins, polysaccharides, and water content. particles have shown that the significance of drainage from mud found out to be 85.9%, therefore, this water can be used for irrigating agricultural lands [30]. The extracts derived from the leaf have been shown to have a significant effect in the prevention of spoilage of food due to pathogens.

Immunomodulatory Activity

The bay leaf plant is rich in procyanidin which can suppress the immune cell function as required in many diseases [31].

CONCLUSION

Cinnamomum Tamala (Indian bay leaf) has several important phytochemicals which can be extracted from the leaves and barks of the plant. These chemicals have been shown to have various therapeutic properties that can be used to treat anxiety, depression, stress, diabetes, cardiac disorders, *etc*. The essential oils that are extracted from the plant have antimicrobial properties against several microorganisms and are thus used in food preservatives such as pineapple juice to prevent spoilage due to fungus or any other pathogen. This plant may have more hidden properties which could be used for designing drugs for several other diseases, thus proper research must be conducted to understand the full potential of the plant, for the development of proper medicine.

CONSENT FOR PUBLICATION

Not applicable

CONFLICTS OF INTERESTS

The authors declare no conflict of interest, financial or otherwise.

ACKNOWLEDGEMENT

Declared none.

REFERENCES

[1]   J. Choubey, A. Patel, and M.K. Verma, "Phytotherapy in the treatment of arthritis: a review", *IJ PSR,* vol. 4, no. 8, pp. 2853-2865, 2013.

[2]   H. Khan, "Medicinal plants in light of history: recognized therapeutic modality", *J. Evid. Based Complementary Altern. Med.,* vol. 19, no. 3, pp. 216-219, 2014.
        [http://dx.doi.org/10.1177/2156587214533346] [PMID: 24789912]

[3]   F. Jamshidi-Kia, Z. Lorigooini, and H. Amini-Khoei, "Medicinal plants: History and future perspective", *Journal of Herbmed Pharmacology,* vol. 7, no. 1, pp. 1-7, 2018.
        [http://dx.doi.org/10.15171/jhp.2018.01]

[4]   R.J. Molyneux, S.T. Lee, D.R. Gardner, K.E. Panter, and L.F. James, "Phytochemicals: the good, the bad and the ugly?", *Phytochemistry,* vol. 68, no. 22-24, pp. 2973-2985, 2007.
        [http://dx.doi.org/10.1016/j.phytochem.2007.09.004] [PMID: 17950388]

[5]   W. Hassan, S.N.Z. Kazm, H. Noreen, A. Riaz, and B. Zaman, "Antimicrobial Activity of Cinnamomum tamala Leaves", *Journal of Nutritional Disorders,* vol. 6, no. 2, pp. 2161-0509, 2016.

[6]   G. Sharma, and A.R. Nautiyal, "Cinnamomum tamala: A valuable tree from Himalayas", *Int. J. Med. Aromat. Plants,* vol. 1, no. 1, pp. 1-4, 2011.

[7]   R.K. Upadhyay, "Therapeutic and Pharmaceutical Potential of Cinnamomum Tamala", *Research Reviews: Pharmacy and Pharmaceutical Sciences,* vol. 6, no. 3, pp. 18-28, 2017.

[8]   S. Bisht, and S.S. Sisodia, "Assessment of antidiabetic potential of Cinnamomum tamala leaves extract in streptozotocin induced diabetic rats", *Indian J. Pharmacol.,* vol. 43, no. 5, pp. 582-585, 2011.
        [http://dx.doi.org/10.4103/0253-7613.84977] [PMID: 22022005]

[9]   V. Singh, A.K. Gupta, S.P. Singh, and A. Kumar, "Direct analysis in real time by mass spectrometric technique for determining the variation in metabolite profiles of Cinnamomum tamala Nees and Eberm genotypes", *ScientificWorldJournal,* vol. 2012, p. 549265, 2012.
        [http://dx.doi.org/10.1100/2012/549265] [PMID: 22701361]

[10]   S. Kumar, S. Sharma, and N. Vasudeva, "Chemical compositions of Cinnamomum tamala oil from two different regions of India", *Asian Pac. J. Trop. Dis.,* vol. 2, no. 2, pp. S761-S764, 2012.
        [http://dx.doi.org/10.1016/S2222-1808(12)60260-6]

[11]   P. Borhade, K. Lone, S. Joshi, A. Kadam, and P. Gaikwad, "Recent pharmacological review on Cinnamomum tamala", *Res. J. Pharm. Biol. Chem. Sci.,* vol. 4, no. 4, pp. 916-921, 2013.

[12]   M. Shah, and M. Panchal, "ETHNOPHARMACOLOGICAL PROPERTIES OF CINNAMOMUM TAMALA – A REVIEW", *Int. J. Pharm. Sci. Rev. Res.,* vol. 5, no. 3, pp. 141-144, 2010.

[13]   A.K. Mishra, B.K. Singh, and A.K. Pandey, "in vitro-antibacterial activity and phytochemical profiles of Cinnamomum tamala (Tejpat) leaf extracts and oil", *Rev. Infect.,* vol. 1, no. 3, pp. 134-139, 2010.

[14]   A.G. Bayir, H.S. Kiziltan, and A. Kocyigit, Plant Family, Carvacrol, and Putative Protection in Gastric Cancer.*Dietary Interventions in Gastrointestinal Diseases.* Academic Press: Cambridge, Massachusetts, 2019, pp. 3-18.
        [http://dx.doi.org/10.1016/B978-0-12-814468-8.00001-6]

[15]   K.B. Pandey, and S.I. Rizvi, "Plant polyphenols as dietary antioxidants in human health and disease", *Oxid. Med. Cell. Longev.,* vol. 2, no. 5, pp. 270-278, 2009.
[http://dx.doi.org/10.4161/oxim.2.5.9498] [PMID: 20716914]

[16]   T. Robinson, "Metabolism and function of alkaloids in plants", *Science,* vol. 184, no. 4135, pp. 430-435, 1974.
[http://dx.doi.org/10.1126/science.184.4135.430] [PMID: 17736509]

[17]   C. Jain, S. Khatana, and R. Vijayvergia, "BIOACTIVITY OF SECONDARY METABOLITES OF VARIOUS PLANTS: A REVIEW", *Int. J. Pharm. Sci. Res.,* vol. 10, no. 2, pp. 494-504, 2019.

[18]   S. Hassanpour, N. Maheri-Sis, B. Eshratkhah, and F.B. Mehmandar, "Plants and secondary metabolites (Tannins): A Review", *Int. J. For. Soil Eros.,* vol. 1, no. 1, pp. 47-53, 2011.

[19]   A.K. Das, M.N. Islam, M.O. Faruk, M. Ashaduzzaman, and R. Dungani, "Review on tannins: Extraction processes, applications and possibilities", *S. Afr. J. Bot.,* vol. 135, pp. 58-70, 2020.
[http://dx.doi.org/10.1016/j.sajb.2020.08.008]

[20]   E. Moghimipour, and S. Handali, "Saponin: Properties, Methods of Evaluation and Applications", In: *Annual Research & Review in Biology* vol. 5. , 2014, pp. 207-220.

[21]   A. Aharoni, M. Jongsma, T.Y. Kim, M.B. Ri, A.P. Giri, F.W.A. Verstappen, W. Schwab, and W. Bouwmeester, "Metabolic engineering of terpenoid biosynthesis in plants", *Phytochem. Rev.,* vol. 5, no. 1, pp. 49-58, 2006.
[http://dx.doi.org/10.1007/s11101-005-3747-3]

[22]   E. Pichersky, and R.A. Raguso, "Why do plants produce so many terpenoid compounds?", *New Phytol.,* vol. 220, no. 3, pp. 692-702, 2018.
[http://dx.doi.org/10.1111/nph.14178] [PMID: 27604856]

[23]   R.J. Ogbe, D.O. Ochalefu, S.G. Mafulul, and O.B. Olaniru, "A review on dietary phytosterols", *Asian Journal of Plant Science and Research,* vol. 5, no. 4, pp. 10-21, 2015.

[24]   P. Sun, T. Wang, L. Chen, B.W. Yu, Q. Jia, K.X. Chen, H.M. Fan, Y.M. Li, and H.Y. Wang, "Trimer procyanidin oligomers contribute to the protective effects of cinnamon extracts on pancreatic β-cells in vitro", *Acta Pharmacol. Sin.,* vol. 37, no. 8, pp. 1083-1090, 2016.
[http://dx.doi.org/10.1038/aps.2016.29] [PMID: 27238208]

[25]   U. Chakraborty, and H. Das, "Antidiabetic and Antioxidant Activities of Cinnamomum tamala", *Global Journal of Biotechnology & Biochemistry,* vol. 5, no. 1, pp. 12-18, 2009.

[26]   D. Shahwar, S. Ullah, M.A. Khan, N. Ahmad, A. Saeed, and S. Ullah, "Anticancer activity of Cinnamon tamala leaf constituents towards human ovarian cancer cells", *Pak. J. Pharm. Sci.,* vol. 28, no. 3, pp. 969-972, 2015.
[PMID: 26004731]

[27]   R.K. Dumbre, M.B. Kamble, and V.R. Patil, "Inhibitory effects by ayurvedic plants on prostate enlargement induced in rats", *Pharmacognosy Res.,* vol. 6, no. 2, pp. 127-132, 2014.
[http://dx.doi.org/10.4103/0974-8490.129031] [PMID: 24761116]

[28]   G.J. Kim, J.Y. Lee, H.G. Choi, S.Y. Kim, E. Kim, S.H. Shim, J.W. Nam, S.H. Kim, and H. Choi, "Cinnamomulactone, a new butyrolactone from the twigs of Cinnamomum cassia and its inhibitory activity of matrix metalloproteinases", *Arch. Pharm. Res.,* vol. 40, no. 3, pp. 304-310, 2017.
[http://dx.doi.org/10.1007/s12272-016-0877-7] [PMID: 28032317]

[29]   J. Ji, C. Lu, Y. Kang, G.X. Wang, and P. Chen, "Screening of 42 medicinal plants for in vivo anthelmintic activity against Dactylogyrus intermedius (Monogenea) in goldfish (Carassius auratus)", *Parasitol. Res.,* vol. 111, no. 1, pp. 97-104, 2012.
[http://dx.doi.org/10.1007/s00436-011-2805-6] [PMID: 22246367]

[30]   M. Hariri, and R. Ghiasvand, "Cinnamon and Chronic Diseases", *Adv. Exp. Med. Biol.,* vol. 929, pp. 1-24, 2016.

[http://dx.doi.org/10.1007/978-3-319-41342-6_1] [PMID: 27771918]

[31]   M. Kurokawa, C.A. Kumeda, J. Yamamura, T. Kamiyama, and K. Shiraki, "Antipyretic activity of cinnamyl derivatives and related compounds in influenza virus-infected mice", *Eur. J. Pharmacol.,* vol. 348, no. 1, pp. 45-51, 1998.
[http://dx.doi.org/10.1016/S0014-2999(98)00121-6] [PMID: 9650830]

CHAPTER 50

A Survey on FANET: Flying Ad-hoc Network (Situations & Model Functionality)

Azher Ashraf Gadoo[1] and Manjit Kaur[1,*]

[1] *Department of Computer Science & Engineering, Lovely Professional University, Punjab, India*

Abstract: Perhaps the contact that is key to coordination and interaction across UAVs (Un-named Aerial vehicle) has been some of the main marketing difficulties regarding mega-UAV (Interplanetary aerospace aircraft) platforms. If any of the UAVs were daisy-chained with the node or maybe a land base network, the UAV may interact *via* the infrastructure. Moreover, the mega-UAV device functions are limited by such a network-dependent interaction design. The issues of a completely infrastructural UAV system could be solved using ad-hoc networking within UAVs. The following study discusses flying ad hoc networks (FANET), an ad hoc network that links UAVs.

Keywords: Asymmetric key, Group key management, Key management, MANET, Routing protocols, Symmetric key.

INTRODUCTION

Owing to just the exponential developments throughout the development of electrical, detector, and connectivity capabilities, unmanned aerial vehicle (UAV) devices may be developed that would operate automatically or globally requiring management. Thanks to the range, mobility, simplicity of deployment and fairly low operational cost, the adoption of UAVs offers different opportunities for intelligence and defensive purposes including scan as well as kill operations [1], border monitoring [2], and the control of wildfires [3] when second-UAV structures are used for years, there are several benefits of a community of smaller UAVs, rather than designing and running a big UAV. Mega-UAV structures have special functions, and connectivity is the most critical architecture concern [4]. Flying Ad-Hoc Network (FANET) is a new program group that is effectively an ad hoc network for UAVs.

---

[*] **Corresponding author Manjit Kaur:** Lovely professional university, Punjab, India; E-mail: manjit.12438@lpu.co.in

Dharam Buddhi, Rajesh Singh and Anita Gehlot (Eds.)

The variations in the MANET, VANET, and FANET ad hoc cell tower are discussed and the key FANET implementation problems are identified [5] the FANET station's major technical problems. In addition to the advancement in microcontrollers as well as the move towards battery technology in hybrid propulsion machines, tiny or tiny UAVs were developed at a reasonable cost [6]. The potential of a specific UAV is therefore constrained. Different UAVs may be synchronized and combined in a framework that is above one UAV alone [7]. That effect of mega-UAV structures might be formulated as fare, ascendable, perseverance, discharge, micro transversal satellites. UAVs may also be connected to land or a drone in a mega-UAV network as in a standard UAV network. Components of such a policy cantered on star topology can exist [8]. Although only a UAV portion of soil bases or satellites may interact, all UAVs create an ad hoc web. It allows the UAVs to interact between the surfaces towards the other.

Good communication and communication with UAVs, FANET often requires third-party-to-peer communications. This also provides information either from the area, as is the case for embedded detector systems. FANET will also promote peer group-to-peer networking and simultaneously integrate recast congestion flow. There is a significantly greater standard disparity among the FANET networks than during the MANETs. MANET, VANET, FANET are displayed in Fig. (**1**). In the new section, we provided many FANET implementation situations and FANET interface features.

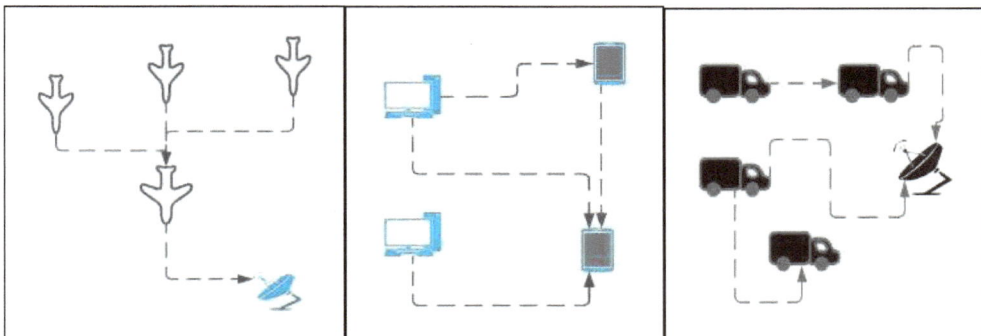

Fig. (1). View of FANET, MANET, VANET.

RELATED WORK

Ad hoc flying networks (FANETs) were discussed in their reports, an ad hoc structure linking UAVs [9]. Secondly, cellular informal channel differences between FANET and VANET are clarified, followed by significant impediments for the production of FANETs [10]. In keeping with both the new FANET protocol, the issues of research required are often dealt with [11]. They have essentially provided in this article a review of several FANET agility designs. Survivability is among FANET's several intractable problems [12]. The disparity among FANET as well as other ad hoc networks has been spoken about. As just an upcoming piece of research, they like to contrast such models with one route configuration pick the best with FANET [13]. A quest and destruction operation is carried out in a restricted area by several UAVs. The UAVs get a narrow band of sensors and are capable of carrying minimal assets to minimize demand. The UAVs do a job of finding objectives [14].

SITUATION WITH FANET

Extended Mega-UAV Operating Optimization

When a Mega-UAV transmission platform is installed on technology including a transmitter or a foundation, the field of service is restricted to architecture's connectivity range. Rather than UAV-to-footing connections, FANET is focused on the UAV-to-UAV ground stations, which can expand the service scope. (Fig. 2) shows an example.

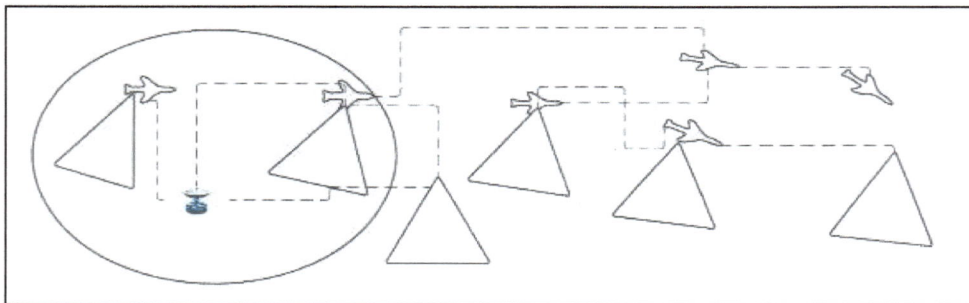

Fig. (2). A FANET circumstance to improve mega-UAV virtualization.

Many FANET architectures have been established to increase mega-UAV technologies interoperability. This was noted that perhaps the creation of a UAV connection string may stretch the operating range through mega-hop communications. FANET will also lead to the operation of mega-UAV systems under the hurdles and thus can improve underpinning.

Trustworthy Coordination for Mega-UAVs

Quad-UAV structures in many other situations are running in an extremely complex atmosphere. When no ad hoc link is available, all UAVs must be linked to a network, as seen in Fig. (**3**). Such communication function increases mega-UAV networks' reliability.

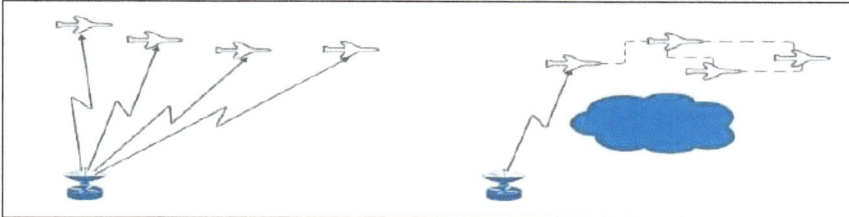

Fig. (3). A robust FANET dual-UAV contact node implementation example.

Hive of UAV

Micro UAVs are quite lightweight with narrow space for delivery. The group conduct of many tiny UAVs is capable of performing complicated assignments given its weak capabilities. The hive UAVs may avoid conflicts and communicate across UAVs with the aid of FANET algorithms. The FANET interaction structure is advocated for the Cooperative Autonomous Reconfigurable UAV Swarm (CARUS).

The aim of CARUS seems to be to track a variety of issues. Every UAV functions independently or every UAV decides in the sky but not on the surface. The centralized judgment and management framework was developed by Ben-Asher *et al*. For mega-UAV networks utilizing FANET. A UAV colony model focused on FANET was also suggested with collective judgment-making, to move UAVs to a remote site.

MODEL FUNCTIONALITY OF FANET

We attempted to clarify in the following segment FANET's four versatility Designs. Survivability simulations reflect the orientation of the network and how it can be position and pace as the period varies. To build a practical simulator setting usability systems can be needed. It shows very well how ad-hoc procedure efficiency would vary greatly with numerous usability designs.

Survivability Pattern Gauss-Markov

The survivability Template of Gauz Markov is required for the simulation of UAV behavior in a colony. Because of its motion rate, vertex location is often guided by its former location. The trajectory of an aircraft is dependent on the designer's data. Every unit is configured at a velocity within the Gauss-Markov Velocity Template that pace, position for the increasing connector is changed through specific duration periods. As seen in Fig. (**4**), the unit is shifted by the prior frame.

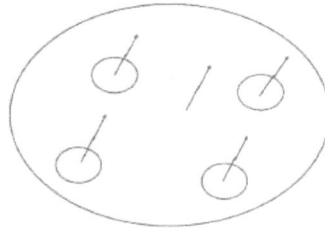

Fig. (4). Each unit shifts by all of the prior node's locations.

Curved Non-random Motion

This flexibility prototype is intended for UAV's rounded motion scenarios. This occurs in obtaining certain details for both the simulation of UAVs moving in a given location. Automobile selects throughout this template for instance the region in which the required subject in a squared zone is selected, as illustrated in Fig. (**5**).

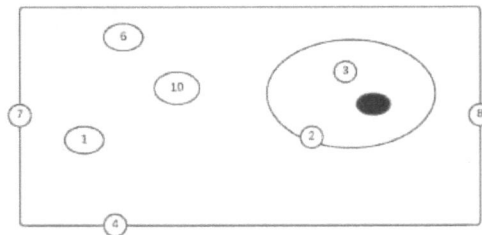

Fig. (5). Every aviation selects the required item at a tiny zone at a remote location.

Flexibility System Task Project Predicated

In the prototype MPB, aviation is indeed conscious about the full range of trajectories, normally pre-planned, meaning whether aviation is continuously moving across the pre-arranged course, and possible position data are accessible as shown in Fig. (**6**) at the physical plane.

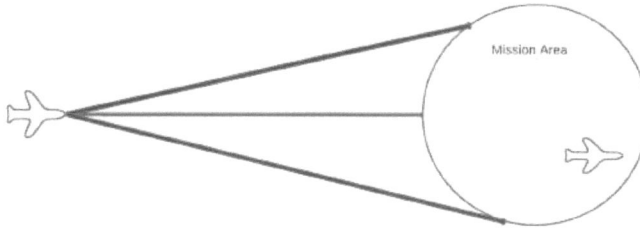

Fig. (6). Fixed road of aircraft.

The beginning and end position per aviation shall be specifically chosen when speed and duration of travel are defined.

CONCLUSION

Contact for dual- structures is among the key demanding problems of architecture. In this paper, the Moving Ad Hoc Platform (FANET) is indeed an ad hoc platform for UAVs. FANET has been explicitly described and many FANET possibilities are provided. We have also discussed the various types of stability, network length, a shift in neural networks, network transmission system, clock speed, computing status, and energy. Often analyzed as resilience, usability, responsiveness, UAV- limitations, and throughput are aspects of the FANET architecture. In this paper, we discussed briefly FANET, VANET, MANET, and separate FANET models.

CONSENT FOR PUBLICATION

Not applicable.

CONFLICT OF INTEREST

The authors declare no conflict of interest, financial or otherwise.

ACKNOWLEDGEMENTS

I thank Ms. Manjit Kaur for providing useful discussions and valuable supports.

REFERENCES

[1]   P. Olsson, J. Kvarnström, P. Doherty, O. Burdakov, and K. Holmberg, "Generating UAV communication networks for monitoring and surveillance", *2010 11th International Conference on Control Automation Robotics & Vision,* 2010pp. 1070-1077 [http://dx.doi.org/10.1109/ICARCV.2010.5707968]

[2]   T. Samad, J.S. Bay, and D. Godbole, "Network-Centric Systems for Military Operations in Urban Terrain: The Role of UAVs", *Proc. IEEE,* vol. 95, no. 1, pp. 92-107, 2007.

[http://dx.doi.org/10.1109/JPROC.2006.887327]

[3] A. Bürkle, F. Segor, and M. Kollmann, "Towards autonomous micro uav swarms", *J. Intell. Robot. Syst.,* vol. 61, no. 1, pp. 339-353, 2011.
[http://dx.doi.org/10.1007/s10846-010-9492-x]

[4] S. Chaumette, R. Laplace, C. Mazel, R. Mirault, A. Dunand, Y. Lecoutre, and J.N. Perbet, *CARUS, an operational retasking application for a swarm of autonomous UAVs: First return on experience,* 2011.

[5] Y. Ben-Asher, S. Feldman, P. Gurfil, and M. Feldman, "Distributed Decision and Control for Cooperative UAVs Using Ad Hoc Communication", *IEEE Trans. Control Syst. Technol.,* vol. 16, no. 3, pp. 511-516, 2008.
[http://dx.doi.org/10.1109/TCST.2007.906314]

[6] A.I. Alshbatat, and Q. Alsafasfeh, "Cooperative decision making using a collection of autonomous quad rotor unmanned aerial vehicle interconnected by a wireless communication network", *Global Journal on Technology.,* vol. 24, no. 1, pp. 212-218.

[7] J. George, P.B. Sujit, and J.B. Sousa, "Search strategies for multiple UAV search and destroy missions", *J. Intell. Robot. Syst.,* vol. 61, no. 1, pp. 355-367, 2011.
[http://dx.doi.org/10.1007/s10846-010-9486-8]

[8] I. Maza, F. Caballero, J. Capitán, J.R. Martínez-de-Dios, and A. Ollero, "Experimental results in multi-UAV coordination for disaster management and civil security applications", *J. Intell. Robot. Syst.,* vol. 61, no. 1, pp. 563-585, 2011.
[http://dx.doi.org/10.1007/s10846-010-9497-5]

[9] H. Xiang, and L. Tian, "Development of a low-cost agricultural remote sensing system based on an autonomous unmanned aerial vehicle (UAV)", *Biosyst. Eng.,* vol. 108, no. 2, pp. 174-190, 2011.
[http://dx.doi.org/10.1016/j.biosystemseng.2010.11.010]

[10] E. Semsch, M. Jakob, D. Pavlicek, and M. Pechoucek, "Autonomous UAV surveillance in complex urban environments", *2009 IEEE/WIC/ACM International Joint Conference on Web Intelligence and Intelligent Agent Technology,* vol. 2, 2009, pp. 82-85
[http://dx.doi.org/10.1109/WI-IAT.2009.132]

[11] H. Chao, Y. Cao, and Y. Chen, "Autopilots for small fixed-wing unmanned air vehicles: A survey", *2007 International Conference on Mechatronics and Automation,* 2007, pp. 3144-3149
[http://dx.doi.org/10.1109/ICMA.2007.4304064]

[12] E.W. Frew, and T.X. Brown, "Networking issues for small unmanned aircraft systems", *J. Intell. Robot. Syst.,* vol. 54, no. 1, pp. 21-37, 2009.
[http://dx.doi.org/10.1007/s10846-008-9253-2]

[13] M. Rieke, T. Foerster, and A. Broering, "Unmanned Aerial Vehicles as mobile multi-sensor platforms", *nThe 14th AGILE International Conference on Geographic Information Science,* vol. 1, 2011

[14] O.K. Sahingoz, "Mobile networking with UAVs: Opportunities and challenges", *2013 International Conference on Unmanned Aircraft Systems (ICUAS),* 2013, pp. 933-941
[http://dx.doi.org/10.1109/ICUAS.2013.6564779]

CHAPTER 51

Design and Implementation of Automated Electro-Mechanical Liquid Filling System

Abukar Ahmed Muse[1,*], Vankudoth Dinesh[1] and **Suresh Kumar Sudabattula[1]**

[1] *School of Electronics and Electrical Engineering, Lovely Professional University, Phagwara, Punjab, India*

Abstract: Automation hypothesis and strategies has increased global productivity in the last few years of worldwide competition. It influenced a great variety of enterprises on the manufacturing side through the decline of production time, superior system performance, and process control. The main purpose of today's manufacturing system is to upgrade productivity. The universe is full of technologies that pressure an elevated amount of production, particularly in industries where automation is needed. To proceed towards automation, the recent trends in all the industries are required to cope up with new technologies. The same vision is applied in water bottle filling plants, to meet the customer demands and to speed up the filling of bottles. Now the process of filling bottles is conducted by PLC in a large number of production units. Due to the high cost of the PLC machine, a filling is yet performed manually in small manufacturing units. The manual filling usually leads to imperfection in the operation of filling and spikes up the cost of labor. The prime purpose of our project is to design and develop an automated electro-mechanical bottle filling system, that can reduce the cost for small-scale en6terprises and assist them to build up an automated factory. The project proposes aspects of computer, electronics, and mechanical *i.e.*, designing and modeling, schematic circuit prototyping and programming, sensor and actuator application, project planning, and presentation skills.

Keywords: Arduino, Automation, Bottle Filling, Microcontroller, PLC.

INTRODUCTION

Overview

In evolving states, an advancement in liquid handling industries has been seen in the past years, as the production of beverages, milk, mineral water, cooking oil, and pharmaceutical liquid dosage forms are the main provision of the entire economy. Over the past years, industries have evolved in automation to adopt

---

* **Corresponding author Abukar Ahmed Muse**: School of Electronics and Electrical Engineering, Lovely Professional University, Phagwara, Punjab, India; E-mail: shifayare130@gmail.com

Dharam Buddhi, Rajesh Singh and Anita Gehlot (Eds.)

large-scale manufacturing and procedures. Mainly in large-scale industries Programmable Logic Controllers (PLC) are used to automate the process. PLC's found in the market are high initial cost that needs multifaceted changes for configuring the program and hardware [1]. The prime purpose of this project is to design and develop an automated electro-mechanical liquid filling system, that is low initial cost, small in size, easy to learn, and compatible with small and medium enterprises. A microcontroller is the brain of the system which controls the entire process and performs the concrete work in the process. Such microcontrollers are 8051, AVR microcontrollers, and Atmega328 [2].

Background

In the 19th century, the demand for bottled water refinement and beverage plants were fully-fledged and rapidly increased, due to the adulterated water supplied by the municipal corporation. Somalia is one of the evolving states that have seen burgeoning in the arena of liquid handling industries. Since 1999 the bottled water industries are promptly increasing in Somalia as the capital manufacturing machines and raw materials have been gradually imported. The number of small to medium enterprises (SMEs) has leveled up to around the capital city of Somalia, Mogadishu [3]. The prominent bottled water manufactures in Mogadishu are Caafi, Saafi, Zam-Zam, Jimmy, and Coca-Cola [4].

LITERATURE REVIEW

Kurkute *et al*. (2016) designed and developed an automatic liquid mixing and bottle filling using PLC and SCADA system. They proposed a mixing and filling management system for industries that are completely application of automation. They proposed a system that consists of three microcontrollers that controls all the various sections in the system [5]. Katre and Hora (2016), the researchers developed and designed a microcontroller-based automated solution filling module in which is used to fill a fixed amount or level that can reduce the manual filling or the imperfection in the operation of filling, which is a common problem in the manual filling [6]. Prajapati *et al*. (2019) developed a project of automatic bottle filling and capping system using PLC, it has high percentage output with minimum power consumption and high accuracy. It is capable to meet the requirements of the industries which can save time and increase productivity [7]. Darji and Parmar (2018) have designed and implemented an automatic bottle filling plant using the Geneva mechanism, they mainly focused on designing and develop a system that is easy to use, economically friendly to a Meghdev beverages plant located in Surendranagar, Gujrat, India [4]. Kumar *et al*. (2017), they simply developed an automatic bottle filling system using PLC, that follows a proposed methodology of sensing the bottle, dispensing the desired amount of liquid, and unloading the bottles [8].

PROPOSED WORKING METHODOLOGY

PLC and SCADA had a major role to play in an automated liquid filling system as a processing unit of handling and controlling the system. Mainly the automatic bottled water industries have been developed by using a microcontroller or Arduino nano based on Atmega328. The microcontroller-based system delivers acceptable performance with insignificant percentage error. The methodological approach to follow in the system is defined completely from loading the bottle to unloading the bottle in the below Fig. (**1**) that summarizes the procedure and entire flow chart of the system [9]. It would be carried out through system specification requirements that are concept studies related to the topic. Subsequently, we will come to the project planning *i.e.*, hardware specification requirements, software implementation, and system design. After taking different essential evaluation parameters, we will draw the block diagram and flow chart in Smart draw online software. Then the schematic diagram and PCB designing will be done by using Easyeda online software or proteus software. In the end, Arduino IDE will be utilized to write the program [10].

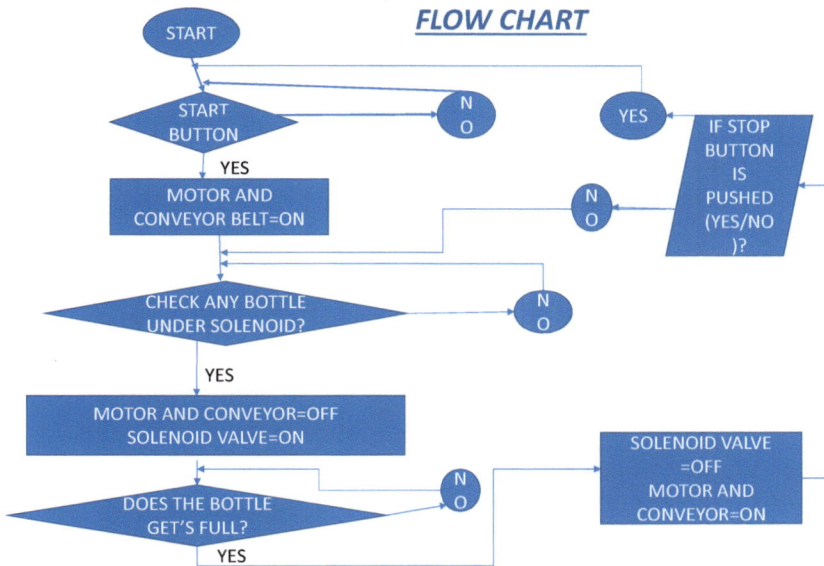

Fig. (1). Program flow chart of bottle filling system.

SYSTEM SPECIFICATION AND SOFTWARE DESIGN

Block Diagram

The block diagram shown below in Fig. (**2**) is described the complete working cycle of the jug filling system from loading to unloading the jug [11]. A microcontroller is the main part of the system and an interface between the program and the inputs. The microcontroller works on the inputs and outputs given [12].

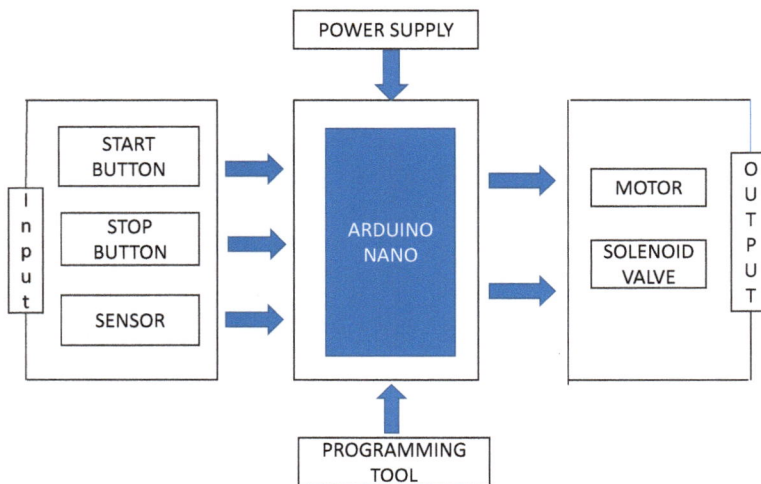

Fig. (2). Block Diagram of the system.

System Specification and Schematic Diagram

The schematic diagram and PCB are designed using Proteus software as shown in below Fig. (**3**). This plots the connection of all components used in the circuit as shown below. The Arduino development board used in the project is the centralized device, which controls the complete circuit or devices connected to it. We use Keil uvision software to code the program. The system uses a two-line, sixteen-character LCD to read the output and display it on the screen. We connected the Nema17 stepper motor to rotate the conveyor belt and the A4988 stepper motor driver to drive the Nema17 stepper motor. The solenoid valve is an electromechanically controlled valve that is used where fluid control is required automatically. We connected 7812 IC 12v regulator to drive the Nema17 motor and the water pump, 7805 IC 5v regulator is used to derive the Arduino and the LCD. We used two transistors to run the Buzzer and the water pump respectively [1,2]. The programming simulation is done in Keil micro vision software as you can see in the below mentioned Fig. (**4**) which encapsulates and demonstrates the code of the system [13].

Fig. (3). Circuit prototype of the system.

Fig. (4). Program code of the system.

RESULT AND DISCUSSION

The work analysis done to the system illustrates that the time required to fill a 1ltr bottle is 30 seconds. When the conveyor belt is moving at a speed of 417 mm/s, the length of the belt is 2.0712 m. Two ltr/sec is expected to be the discharge of water that comes out from the valve per unit time. Then the velocity of the liquid running into the solenoid valve will be 271 mm/s. the necessary time needed to fill one jug is 30 seconds, as the time delay of the motor that was defined in the Arduino code. When we load the empty jug into the conveyor belt, the proximity sensor and position sensor assures that the jug is empty and in a perfect position [13]. The solenoid valve will open for a specific time to fill the required amount of water in the jug. Fig. (5) below defines that as the solenoid valve upsurges the time needed to fill 200 ml water in the bottle declines. So as the valve opening upsurges the flow rate will increase. Hence for filling different sizes of the jug the filling will be fixed but the flow rate will be varying [8].

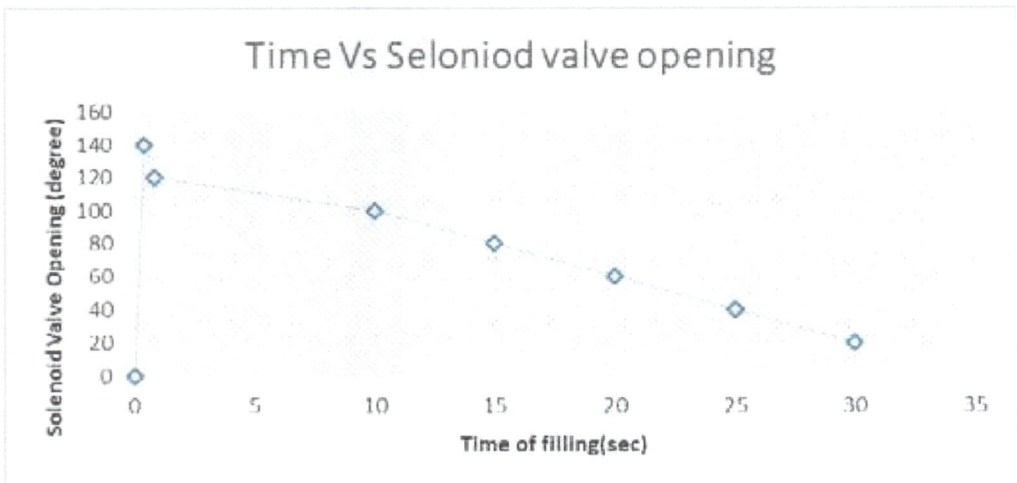

Fig. (5). Time and solenoid valve opening.

CONCLUSIONS

In this day and edge, industries are being confronted to ease the cost, misuse, and manufacturing time. This challenge has strained the older and traditional system which increased maintenance requirements and the pressure of continuous production volumes. However, the new technologies had a vital role to reduce the usage of water, upsurge the efficiency of the system and reduce the manufacturing time. In today's first moving, the competitive industrialized world PLC and SCADA are used to systematize all the industries. As the small and medium enterprises are not able to cope up with a system that is a high initial cost and maintenance requirement is supplementary. To automate and make it more accessible to numerous small and medium enterprises, we developed a system that is flexible, easy to use, and cost-effective for their survival in today's economical world. This system is cost efficiency ascribed to the decrease of labor cost and higher liquid filling rate. To resolve that problem, we used a microcontroller as the brain of the system which controls the entire system. The microcontroller is efficient, robust, and widely available in the market. We constructed a schematic diagram and PCB to precipitate the makeover of the system using Arduino nano. This reduces the cost of the machine to a great extent.

CONSENT FOR PUBLICATION

Not applicable.

CONFLICT OF INTEREST

The authors declare no conflict of interest, financial or otherwise.

ACKNOWLEDGEMENTS

We would like to express our special thanks of gratitude to our supervisor Mr. Suresh Kumar Sudabattula, for his support, cooperation, and motivation provided to us during the project for constant inspiration, presence, and blessings. We would also like to extend our sincere appreciation to Global Emerging Innovation Summit (GEIS-2021) for providing us a valuable opportunity to present in GEIS-2021. Lastly, we would like to thank the almighty and our friends for their moral support that we shared our day-to-day experience and received lots of suggestions that improved our quality of work. The success and outcome of this project required a lot of guidance from many people, I would like to express my thanks to all of them.

REFERENCES

[1]   B. Mashilkar, P. Khaire, and G. Dalvi, "Automated Bottle Filling System By Using", no. May 2018,

[2]   B. Mashilkar, P. Kumar, A. Chawathe, V. Dabhade, V. Kamath, and G. Pati, *Automated Bottle Filling System By Using,* no. May, p. 2018, 2019.

[3]   M.M. Mohamed, N.N. Isak, and D.H. Roble, "Private Sector Developments in Somalia : Analysis on Some Major Sectors", *Int. J. Econ. Commer. Manag.,* vol. VII, no. 12, pp. 938-951, 2019.

[4]   V.P. Darji, and B.G. Parmar, "Design and Modeling of Automatic Bottle Filling Plant Using Geneva Mechanism", *J. Manuf. Eng.,* vol. 13, no. 2, pp. 75-081, 2018.

[5]   P.S.R. Kurkute, M.A.S. Kulkarni, M.M.V. Gare, and M.S.S. Mundada, "Automatic Liquid Mixing and Bottle Filling - A Review", *Ijireeice,* vol. 4, no. 1, pp. 57-59, 2016.
[http://dx.doi.org/10.17148/IJIREEICE.2016.4114]

[6]   S. Katre, and J. Hora, "Microcontroller Based Automated Solution Filling Module", In: , 2016, pp. 708-710.

[7]   P. Prajapati, S. Singh, S. Gupta, and S. Srivastava, "Automatic bottle filling and capping system using PLC", *Int. Res. J. Eng. Technol.,* vol. 6, no. 11, pp. 1095-1104, 2019.

[8]   S. Suryawanshi, D. Shelke, A. Patil, and M. Tandale, "Automatic Bottle Filling System Using PLC", *Int. J. Trend Sci. Res. Dev,* vol. 2, no. 1, pp. 361-364, 2017.
[http://dx.doi.org/10.31142/ijtsrd4592]

[9]   K. Meah, "An automated bottle filling and capping project for freshman engineering students", *ASEE Annu. Conf. Expo. Conf. Proc.,* 2010
[http://dx.doi.org/10.18260/1-2--15723]

[10]   P. Das, K. Mandal, and P. Das, "Automatic Bottle Filling System Using PLC", *Int. J. Trend Sci. Res. Dev,* vol. 2, no. 1, pp. 361-364, 2017.
[http://dx.doi.org/10.31142/ijtsrd5953]

[11]   K. Manhas, M. Dogra, R. Tiwari, and J. Sharma, "Design and Implementation of Bottle Filling Automation System for Food Processing Industries using PLC", *Int. J. Power Electron. Control. Convert.,* vol. 4, no. 1, pp. 1-9, 2019.

[12]   *Function and Technology for Filling System,* vol. 03, no. 19, pp. 3947-3950, 2014.

[13]   A. Guha, A. Ganveer, M. Kumari, and A.S. Rajput, "Automatic Bottle Filling Machine", *SSRN,* 2019, pp. 1010-1015.
[http://dx.doi.org/10.2139/ssrn.3440463]

CHAPTER 52

Review of Mobile Ad-hoc Networks - Architecture, Usage, and Applications

Azher Ashraf Gadoo[1] and **Manjit Kaur**[1,*]

[1] *Department of Computer Science & Engineering, Lovely Professional University, Punjab, India*

Abstract: Advanced internet technology contributes to several new apps due to wireless communication technology. Mobile Ad-hoc networks are one of the most productive areas of wireless internet technology development. The cellular ad-hoc channels have now become one of the most competitive and successful communications and connections as their usage has increased considerably over the last few years. A remote ad hoc channel is an independent set of portable devices that interact through wireless networks and collaborate in a decentralized way, in the absence of fixed facilities, in an attempt to provide the communication capabilities required. This article addresses a summary of MANET and its use in devices. This article provides a brief analysis of the styles and benefits of MANET implementations in wireless transmission channels. This kind of network, which acts as an isolated channel or one or more places linked to cellular networks or the Internet, paves the way for various exciting new technologies. This article also gives an overview of the possible applications of ad hoc channels, complex assaults, and addresses the technical problems facing procedure architects and channel designers.

Keywords: Cellular Networks, Link Capacity, VANET, Wireless Transmission Channels.

INTRODUCTION

MANET is an ad hoc channel, where packet headers among two nodes require no engineering support. MANET is an ad hoc mobile network or known simply as an ad hoc cellular network which is also an active, self-ordained, and infrastructure-free network extender of portable devices [1]. Every machine or device in MANET design means that every system is both a router and an end-user. The MANET design's modules or machines are usually independent [2].

---
* **Corresponding author Manjit Kaur:** Lovely Professional University, Punjab, India;
E-mail: manjit.12438@ lpu.co.in

Every mobile phone is independent in a network [3]. Mobile phones can be carelessly moved and randomly arranged as shown in Fig. (**1**). Many implementations have been developed by MANET, such as strategic systems, wireless detector channels, cellular networks, computer networks, *etc* [4]. There are still some development difficulties and difficulties to address in many implementations. The main aim of ad hoc cellular networks is to broaden movement into the field of separate, mobile, remote territories, where adapters and users can be needed to create an ad hoc networking infrastructure [5]. Much exploitable vulnerability has been found in a mobile climate, such as MANET, and curb-measures have also been suggested. Nonetheless, only a handful of them gives a guarantee that solves the crucial safety issue [6].

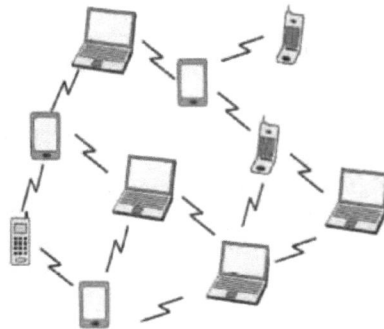

Fig. (1). Mobile Ad hoc Network.

Keeping this in mind, the key goal of wireless ad hoc communication is to promote the reliable and efficient operation of mobile networks through the implementation of routing capabilities in mobile channels. These networks are meant to have complex multi-hop and random routing protocols that often shift quickly, and generally consist of wireless links fairly restricted to throughput [7].

RELATED WORK

In this paper, the authors have given an insight into the possible implementations of ad hoc systems, various threats, and addresses software architects and system builders' face engineering challenges [8]. The author takes stock of the forms and strengths and drawbacks of MANET systems in conjunction with wireless transmission networks. MANET is an acronym for cellular ad hoc networks, also known as wireless ad hoc networks; it is an uncontrolled ego-configuring, an infrastructure-free channel of portable devices linked without the need for cables [9]. In this document, we will talk about the different main leadership systems for

MANETs. This academic research introduces a novel method for protected ethnicity-based data encryption using cryptographic and Knowledge Theoretical Technology [10]. The authors addressed the efficiency of some intrusion prevention devices in the MANET framework in this paper. It is suggested that SAODV be expanded to provide intrusion prevention mechanisms and a judge-based framework to foster cooperation between the cooperative nodes and severely punish the greedy nodes. Tests of the simulations were conducted to demonstrate the efficacy of their proposal as opposed to AODV and SAODV [11].

FEATURES OF MANET

In this section we aimed to give a general concept of some of the features of MANET although it has many characteristics, some of which are mentioned below and the given Fig. (**2**) gives a generalized idea of it:

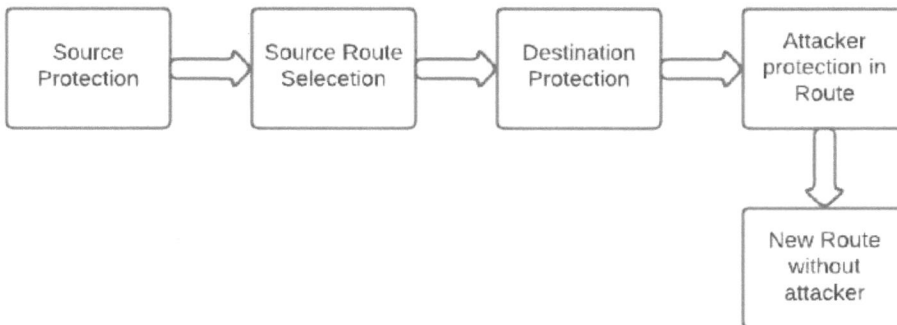

Fig. (2). Routing of MANET.

Multi-Hop Routing

Multi-hop networking is a type of wireless data communication in which the network link is wider than the ray range of the computing nodes. The node can use other clusters as relays to hit some endpoint.

Dynamic Network Topology

The channel is modeled, similar to the fixed scenario, as an unsupervised connected path, in which each node knows the peers but knows nothing more about the algorithm [12].

Partitioned Operation

There should be a collaboration between the nodes involved in MANET. Each node works as a relay.

Fluctuating Link Capacity

Most sessions share one direction of contact in MANET. The stream that communicates the networks has less capacity than a directed channel and is susceptible to interference, decay, and disruption.

Light Weight Terminal

MANET nodes have lower GPU, decreased cognitive ability, and reduced energy generation capacities. Such systems require optimized protocols and dynamics for computing and networking [13].

Autonomous Terminal

Each mobile interface in MANET technology is an independent device that can act as a server or a converter. There is also no means of distinguishing connection points and changes.

ARCHITECTURE OF MANET

This section includes the MANET framework and it possesses three central components, which are detailed as Fig. (**3**).

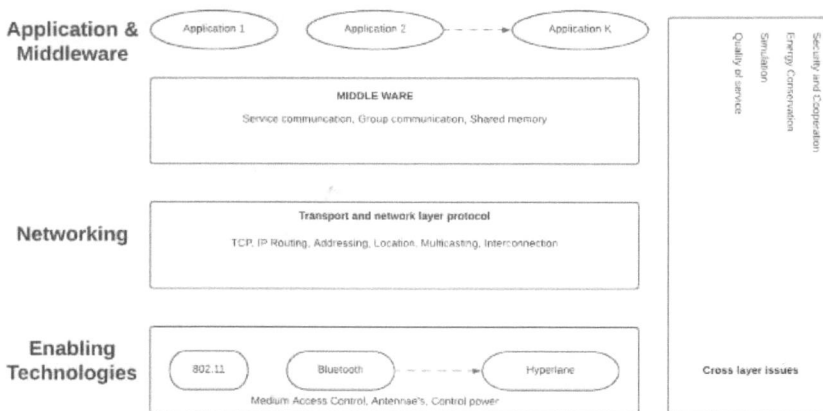

Fig. (3). Architecture of MANET.

Enabling Technology

The technology that enables is further differentiated by its scope. BAN's distance of contact is 1 to 2 meters. The distance of wireless communications is between 100 and 500 meters. A tower or a collection of structures may be linked [14].

Networking

The main aim of channel communications is to use the single-hop distribution services provided by the infrastructure to establish secure end-to-end applications *via* recipient to the receiver. The sender has to locate the target inside the channel to install a full contact. The main task of a locating provider is to map the receiver's address automatically to its current network position [15].

Middleware and Application

The use of ad hoc software and ad hoc communication systems is strongly promoted by the use of wireless devices, such as WIFI, Bluetooth, IEEE 802.11, WiMAX, or Hyper LAN, particularly in special fields of emergency response, business continuity, and atmosphere surveillance.

APPLICATION OF MANET

Mobile ad hoc networks are commonly used in the industry, military, and private companies. All MANET nodes are wireless and their communications are complex as opposed to backbone systems.

Commercial Industry

Ad hoc operations for relief efforts, for example, in the fire, flood or disaster may be used for crisis/evacuations. The work of rescuers was made more difficult by setting up an integrated mobile network with electronic equipment.

Military field

Military hardware now requires some form of electronic technology regularly. The ad hoc network strategies were derived from this field [16].

Data Network

Pervasive technology requires a business framework for MANETs. Machines can relay data to others by expanding data networks well beyond the normal range of the developed infrastructure [17].

Sensor Network

Its system is a network of very many tiny detectors. They can be used to detect several houses in a specific region [18].

Low Level

In domestic channels where machines can explicitly communicate to share information, an acceptable low-level application may be [19].

CONCLUSION

Mobile computing advancements are pushing a new hybrid route for wireless telephony where mobile devices form a cellular network that establishes itself, self-organizes, and self-administers itself, considered as Ad Hoc mobile network. Remote ad hoc networks are more sensitive than fixed or hardwired networks to physical security attacks. This article sheds light on numerous MANET's ideas that can enhance researchers to work. Its inherent simplicity, inadequate infrastructure, quick implementation, self-setting, reduced price, and possible applications make it a core aspect of future computing environments. In addition, the nodes in ad hoc networks would be tiny, simpler, quite efficient, and usable in any shape, particularly as the need for high penetration like battlefields and wearable devices persists. Although there is still a year from the expanded adoption of the Ad-Hoc networks, there will still be a very fruitful and innovative analysis in this area.

CONSENT FOR PUBLICATION

Not applicable.

CONFLICT OF INTEREST

The authors declare no conflict of interest, financial or otherwise.

ACKNOWLEDGEMENTS

I thank Ms. Manjit Kaur for providing useful discussions and valuable supports.

REFERENCES

[1]   A. Cho, J. Kim, S. Lee, and C. Kee, "Wind estimation and airspeed calibration using a UAV with a single-antenna GPS receiver and pitot tube", *IEEE transactions on aerospace and electronic systems.*, vol. 47, pp. 1-109, 2011.
[http://dx.doi.org/10.1109/TAES.2011.5705663]

[2]   I. Maza, F. Caballero, J. Capitán, J.R. Martínez-de-Dios , and A. Ollero, "Experimental results in

multi-UAV coordination for disaster management and civil security applications", *Journal of intelligent & robotic systems,* vol. 61, no. 1, pp. 563-85, 2011.

[3]   H. Xiang, and L. Tian, "Development of a low-cost agricultural remote sensing system based on an autonomous unmanned aerial vehicle (UAV)", *Biosyst. Eng.,* vol. 108, no. 2, pp. 174-190, 2011.
[http://dx.doi.org/10.1016/j.biosystemseng.2010.11.010]

[4]   E. Semsch, M. Jakob, D. Pavlicek, and M. Pechoucek, "Autonomous UAV surveillance in complex urban environments", *2009 IEEE/WIC/ACM International Joint Conference on Web Intelligence and Intelligent Agent Technology,* vol. 2, 2009, pp. 82-85
[http://dx.doi.org/10.1109/WI-IAT.2009.132]

[5]   H. Chao, Y. Cao, and Y. Chen, "Autopilots for small fixed-wing unmanned air vehicles: A survey", *2007 International Conference on Mechatronics and Automation,* vol. 5, 2009, pp. 3144-3149

[6]   B.S. Morse, C.H. Engh, and M.A. Goodrich, "UAV video coverage quality maps and prioritized indexing for wilderness search and rescue", *In2010 5th ACM/IEEE International Conference on Human-Robot Interaction (HRI),* 2010, pp. 227-234

[7]   P. Chouksey, "Comparative Study of AOMDV and AODV Routing based on Load Analysis in MANET", *International Journal of Scientific Research in Network Security and Communication.,* vol. 4, no. 5, pp. 12-19, 2016.

[8]   L Raja, "An overview of MANET: Applications, attacks and challenges", *International Journal of Computer Science and Mobile Computing,* vol. 3, no. 1, pp. 408-417, .

[9]   F. Albalas, M.B. Yaseen, and A.A. Nassar, "Detecting black hole attacks in MANET using relieff classification algorithm", *In Proceedings of the 5th International Conference on Engineering and MIS,* 2019 no. 22, pp. 1-6.
[http://dx.doi.org/10.1145/3330431.3330454]

[10]   A.K. Vatsa, P. Chauhan, M. Chauhan, and J. Sharma, "Routing mechanism for manet in disaster area", *Int. J. Netw. Mob. Technol.,* vol. 2, no. 2, 2011.

[11]   M. Fatima, R. Gupta, and T.K. Bandhopadhyay, "Route discovery by cross layer approach for MANET", *Int. J. Comput. Appl.,* vol. 37, no. 7, pp. 14-24, 2012.

[12]   E.W. Frew, and T.X. Brown, "Networking issues for small unmanned aircraft systems", *J. Intell. Robot. Syst.,* vol. 54, no. 1, pp. 21-37, 2009.
[http://dx.doi.org/10.1007/s10846-008-9253-2]

[13]   B. Kannhavong, H. Nakayama, N. Kato, Y. Nemoto, and A. Jamalipour, "A collusion attack against olsr-based mobile ad hoc networks", *nIEEE Globecom 2006,* vol. 26, no. 2, pp. 1-5, 2006.

[14]   Z. Karakehayov, "'Using REWARD to detect team black-hole attacks in wireless sensor networks," Wksp", *Real-World Wireless Sensor Networks.,* vol. 20, no. 2, pp. 20-21, 2005.

[15]   F. De Rango, and S. Marano, "Trust-based SAODV protocol with intrusion detection and incentive cooperation in MANET", *InProceedings of the 2009 International Conference on Wireless Communications and Mobile Computing: Connecting the World Wirelessly,* 2009, pp. 1443-1448
[http://dx.doi.org/10.1145/1582379.1582695]

[16]   C. Singh, V. Gupta, and G. Kaur, "A review paper on introduction to MANET", *International Journal of Engineering Trends and Technology,* vol. 1, pp. 38-43, 2014. [IJETT].
[http://dx.doi.org/10.14445/22315381/IJETT-V11P208]

[17]   S.B. Geetha, and C. Venkanagouda Patil, "BMWA: A Novel Model for Behavior Mapping for Wormhole Adversary Node in MANET", In: *In Intelligent Computing and Information and Communication,* 2018, pp. 505-513.

[18]   E. Yanmaz, C. Costanzo, C. Bettstetter, and W. Elmenreich, "A discrete stochastic process for coverage analysis of autonomous UAV networks", *In2010 IEEE Globecom Workshops,* pp. 1777-1782, 2010.

[http://dx.doi.org/10.1109/GLOCOMW.2010.5700247]

[19]   L. To, A. Bati, and D. Hilliard, "Radar cross section measurements of small unmanned air vehicle systems in non-cooperative field environments", *In2009 3rd European Conference on Antennas and Propagation,* 2009, pp. 3637-3641

CHAPTER 53

Face Expression Emoji as Avatar Creation and Detection

P. Heam Kumar[1], Pallala Akhil Reddy[1], Manukonda Manoj Kumar[1], Gade Vivek[1] and Manjit Kaur[1,*]

[1] *Department of Computer Science & Engineering, Lovely Professional University, Punjab, India*

Abstract: One of the latest emerging trends, the flexible way of communication in an entertainment manner is through emoji (smile emoji, sad emoji, etc), many top companies like Facebook, Snapchat, Instagram, and many more are using this type of trend but now also they are in progress of this trend for the creation of avatar for the virtual world. Therefore we will discuss the backend of this type of work creation, different technologies used for the implementation, all the set of algorithms, and past, present, and future of the Face emoji creation (FEC) and Face emoji detection (FED) with the review in avatar formation. We get to know different advantages and the disadvantages related to the above technology. One of the interesting things for everyone is the avatar of their face. So, we get to know the concepts of AC (avatar creation) with expression oriented and how it is going to be related with the virtual world, which is going to be experienced by the future generation.

Keywords: Avatar, Computer Vision, Convolutional Neural Networks, Face Emoji Creation, Face Emoji Detection.

1. INTRODUCTION

Social media would be boring, WhatsApp had been irritating, Snapchat could have blast our minds and many more without the implementation of emoji (or the set of emotions in a form of an image), different expression-based emojis are smile, anger, disgust, fear, etc. Expression/Emotions are often mediate and then facilitate the interactions among the different human beings that is why it had been annotated in a few Wordnets [1]. Thus, capturing the emotions although brings content to seeming bizarre ate and difficult social platform related communication. For the complex expressions, there is a mixture of different emotions that could be used for descriptors. Computer vision has grown at a large

---

[*] **Corresponding author Manjit Kaur:** Department of Computer Science & Engineering, Lovely Professional University, Punjab, India; E-mail: manjit.12438@lpu.co.in

Dharam Buddhi, Rajesh Singh and Anita Gehlot (Eds.)

scale and all the commercial products are using the benefits of a high speed image processing and machine learning to obtain useful details from the visual world to generate the virtual world, every company is in the race for the innovation as mentioned above one of the fastest-growing technology [2, 3]. Fig. (1) shows Convolutional neural network (CNN).

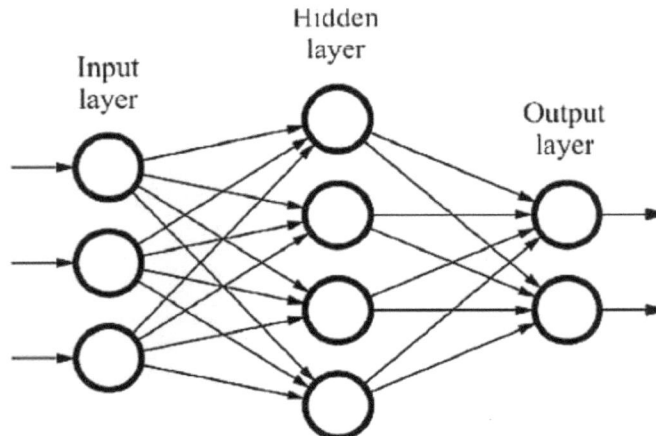

Fig. (1). Convolutional neural network (CNN) [1].

The emoji will be more interesting when it will be with the avatar of your face which would play a most interactive communication in social media and virtual space (from the real world to the virtual world). Avatar is represented in a set form like "switches" (parameters) which selected the different parts and modify it, as an instance, the shape of the nostril, the color of the attention, and the fashion of hair; all are taken from a predefined set of alternatives created *via* artists. Thus, we discuss all this with pros and cons. One of the important tasks is the detection of the face. After that expressions in the face then the creation of the required model like emoji or avatar. For these different techniques are there are fisher faces, eigenfaces, viola jones object detection framework, Hausdorff distance, etc. Many of the survey papers are there among which one of the papers by [4] provides a superb overview of the different states of computer work research as it directs to emotion detection from different primary sources are mentioned in this.

2. FACIAL ACTION CODING SYSTEM (FACS)

Ekman and Friesen had identified the facial muscles which are important in defining emotions and compiled finding to a system of 46 action units (AUs). These are for inner eyebrow and the raising of the outer eyebrow were immensely important in quantifying human expressions. It is important to note that automated detecting facial AUs is a difficult task, especially because some AUs are only visible from a certain point of angle [5, 6].

2.1. Facial Action Parameters (FAPs)

Ekman's Facial movement coding device became observed to be extremely comprehensive as researchers can identify over 7000 different combinations of the 46 atomic AUs [7, 8] however it is critical to note that real-life expressions are dynamic and there is more to them than still images of contracted muscles. The Motion Pictures Expert Group (MPEG) introduced a well-thought-out set of facial animation parameters (FAPs) This system develops a neutral face and a set of 84 different key facial characteristic points [9 - 11].

2.2. Emotion Detection from Frontal Facial Image

This segmentation of faces from the rest of the image is done by a skin color model to avoid noise [12]. Facial parts like the eye and lip and nose region are extracted from the image by the viola-jones algorithm. Emotion detection is then done by analyzing the position of the mouth and eyes in each of the test images.

3. NOVEL APPROACH TO FACE ANALYSIS

In this technique, image extraction is done from video. Image is segmented to different points of interest such as nose, eyes, mouth. Texture analysis is done on bass of cheek and forehead activity. Facial characteristic points are changes in the gradient, which is the vector sum of color differences in the neighborhood of a pixel. The texture is the brightness difference, color difference, shape, and smoothing factors. Now more techniques for the avatar improvement are pyramid histogram of gradients (PHOG) [13], AU conscious facial capabilities [14], boosted LBR descriptors [15], and RNNs [16]. The context for static pix all used deep CNNs, generating up to 62% test accuracy [17, 18].

(Discrepancy distance): Suppose 'Cap' be a category of capabilities from X1 to X2 and allow ℓ: B × B → R+ to be a loss feature over B. The discrepancy distance discC between distributions Dist1 and Dist2 is above the A is defined as:

discC(Dist1, Dist2) = sup$_{cap1,cap2 \in}$ Cap |RDist1 [cap1, cap2] − RDist2 [cap1, cap2]|,

where, RDist [cap1, cap2] = Ex~Dist [ℓ(cap1(x), cap2(x))]

We do not put into effect fashion losses in our technique as more substantially; the problem that we remedy differs. The literature of face comic strips additionally typically trains in a well-defined manner that requires correspondences among sketches and pictures.

4. PROPOSED METHODOLOGY

For the development of the project, we get to know that we required different types of the dataset to train the model the datasets that can be used are extended Cohn-Kanade dataset (CK+), AU-Coded facial features Database. These are small but it gives a nicely described facial feature in a controlled laboratory surrounding. We can also develop our datasets for the identification of the expressions [19, 20]. Then the VGG_S network for the image classification, but the use of the above datasets the image classification is not clear so that with the help of CNN's the image classification would be somewhat clear from the result of the past research paper [21, 22]. After the detection and the creation of the emoji the avatar creation process for which the TensorFlow module takes place as an approach as mentioned in Fig. (2). (TF) Tensor flow [64 bits], it is a ways an open-supply software program library for deep studying which modified into written in a python and C++ language. It changed into finished with the aid of way of Google thoughts team; an exceptional key aspect for the library is in a data go with the flow graph.

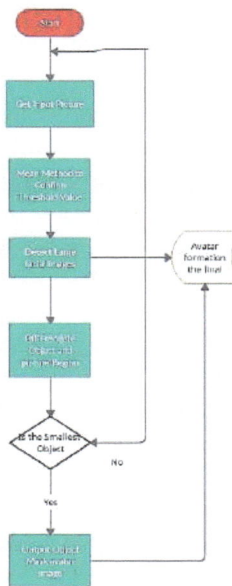

Fig. (2). Proposed Flowchart for facial expression.

5. EXPERIMENT DESCRIPTION

A normal digital camera-oriented Face expression relocation system consists of three predominant levels: pre-processing, feature extraction, and detection of the face. We get to know that in comparison popularity performance using the output of the automated face detector to overall performance on pictures with specific feature alignment the use of hand-labeled features. For manually aligned face photos, the faces are rotated so that the eyes ought to be horizontal and then warped so that the eyes and mouth had been aligned through every face. Fig. (3) shows the Cohn-Kanade dataset (CK+).

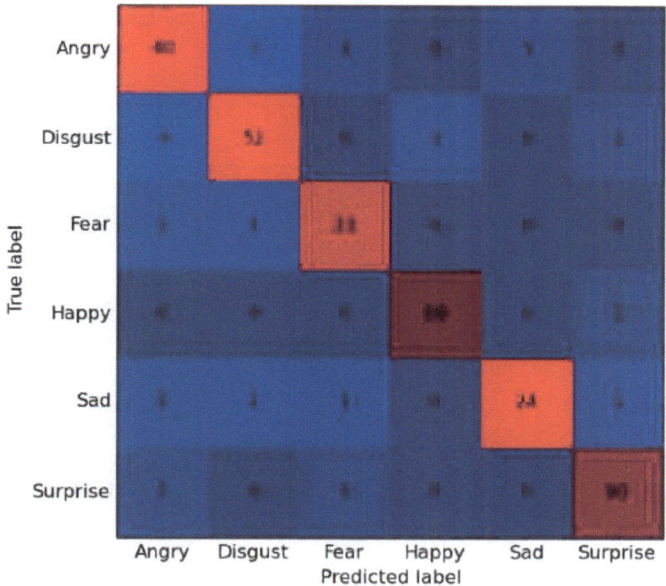

Fig. (3). Cohn-kanade dataset (CK+).

Then this SVM overall performance was compared to AdaBoost for the emotion class. The features carried out for an Adaboost emotion classifier are the man or woman Gabor filters. There were 48x48x40 = 92160 possible features. After this also many of the experiments are done by Google, Facebook but they are failed in that and new approaches and techniques are still in progress for the efficient implementation of the project for future technology.

CONCLUSION AND FUTURE SCOPE

In this paper, there is a proposal for a Facial expression recognition (FER) and face avatar creation algorithm one of the current and future projects for many companies because this as different fields like human interaction structures, intellectual disorder detection, and influence popularity. The goal for this research paper is for the implementation of a project on Face emoji creation and detection in avatar form. By using a custom educated VGG_S network with a face-detector provided *via* OpenCV, we have gone through we requirements to create a large dataset to improve the model generality. This required super lighting, digicam at eye stage, hassle going through digicam with an exaggerated expression, and the heavier pre-processing of the data is done. We get to know that deep learning and CNNs perform a critical position inside the advent of the undertaking for almost 90% accuracy.

Now the avatar which advent of higher computer photos engines and the plethora of available models, and the ability of neural networks to examine go-domain entities, and the lacking element for bridging between computer output and computer portraits is the capability to link picture records to an appropriate category. So, at last, we conclude that this type of the project or the technology are going to create a revolution in coming days with the implementation of artificial intelligent and deep learning concept which lead to a more security purpose for the safety and security.

CONSENT FOR PUBLICATION

Not applicable

CONFLICT OF INTEREST

The authors declare no conflict of interest, financial or otherwise.

ACKNOWLEDGEMENTS

We thank Ms. Manjit Kaur for providing useful discussions and valuable supports.

REFERENCES

[1]   M.V. Valueva, N.N. Nagornov, P.A. Lyakhov, G.V. Valuev, and N.I. Chervyakov, "Application of the residue number system to reduce hardware costs of the convolutional neural network implementation", In: *Mathematics and Computers in Simulation* vol. 177. , 2020, pp. 232-243.

[2]   V. Bettadapura, "Face expression recognition and analysis: the state of the art", *arXiv preprint arXiv,* 2012.

[3]   A. Krizhevsky, I. Sutskever, and G.E. Hinton, "ImageNet classification with deep convolutional neural

networks", *Communications of the ACM,* vol. 60, no. 6, pp. 84-90, 2017.
[http://dx.doi.org/10.1145/3065386]

[4]   A. Yao, J. Shao, N. Ma, and Y. Chen, "Capturing au-aware facial features and their latent relations for emotion recognition in the wild", *In Proceedings of the 2015 acm on international conference on multimodal interaction,* 2015, pp. 451-458
[http://dx.doi.org/10.1145/2818346.2830585]

[5]   C. Shan, S. Gong, and P.W. McOwan, "Facial expression recognition based on local binary patterns: A comprehensive study", *Image and vision Computing,* vol. 27, no. 6, pp. 803-816, 2009.
[http://dx.doi.org/10.1016/j.imavis.2008.08.005]

[6]   S. Ebrahimi Kahou, V. Michalski, K. Konda, R. Memisevic, and C. Pal, "Recurrent neural networks for emotion recognition in video", *Proceedings of the 2015 ACM on international conference on multimodal interaction,* 2015, pp. 467-474
[http://dx.doi.org/10.1145/2818346.2830596]

[7]   A. Radford, L. Metz, and S Chintala, "Unsupervised representation learning with deep convolutional generative adversarial networks", *arXiv preprint arXiv:1511.06434,* 2015.

[8]   H. Gammulle, ernando T, S Denman, Sridharan S, and Fookes C., "Coupled generative adversarial network for continuous fine-grained action segmentation", *2019 IEEE Winter Conference on Applications of Computer Vision (WACV),* 2019, pp. 200-209
[http://dx.doi.org/10.1109/WACV.2019.00027]

[9]   J.F. Cohn, Z. Ambadar, and P. Ekman, "Observer-based measurement of facial expression with the Facial Action Coding System", In: *The handbook of emotion elicitation and assessment* vol. 1. , 2007, no. 3, pp. 203-221.

[10]   M. Pantic, and I. Patras, "Dynamics of facial expression: Recognition of facial actions and their temporal segments from face profile image sequences", vol. 36 no. 2, pp. 433-449, 2006.
[http://dx.doi.org/10.1109/TSMCB.2005.859075]

[11]   J. Hrrigan, R. Rosenthal, K.R. Scherer, and K. Scherer, "New handbook of methods in nonverbal behavior research", In: *Oxford University Press,* 2008.

[12]   L. Malatesta, A. Raouzaiou, K. Karpouzis, and S. Kollias, "Towards modeling embodied conversational agent character profiles using appraisal theory predictions in expression synthesis", *Appl. Intell.,* vol. 30, no. 1, pp. 58-64, 2009.
[http://dx.doi.org/10.1007/s10489-007-0076-9]

[13]   A. Raouzaiou, N. Tsapatsoulis, K. Karpouzis, and S. Kollias, "Parameterized facial expression synthesis based on MPEG-4", In: *EURASIP Journal on Advances in Signal Processing,* vol. 10. , 2002, pp. 1-18.
[http://dx.doi.org/10.1155/S1110865702206149]

[14]   P. Viola, "Robust real-time object detection", *International journal of computer vision,* 2001.

[15]   Y. Wang, H. Ai, B. Wu, and C Huang, "Real time facial expression recognition with adaboost", In: *Proceedings of the 17th International Conference on Pattern Recognition* vol. 3. , 2004, pp. 926-929.
[http://dx.doi.org/10.1109/ICPR.2004.1334680]

[16]   "Emotion detection from frontal facial image", *Diss. BRAC University,* 2013..

[17]   N.J. Bailenson, "Real-time classification of evoked emotions using facial feature tracking and physiological responses", *Int. J. Hum. Comput. Stud.,* vol. 66, no. 5, pp. 303-317, 2008.
[http://dx.doi.org/10.1016/j.ijhcs.2007.10.011]

[18]   A. Dinculescu, "Novel approach to face expression analysis in determining emotional valence and intensity with benefit for human space flight studies", In: *E-Health and Bioengineering Conference (EHB)* IEEE, 2015.

[19]   S. Thuseethan, and S. Kuhanesan, "Eigenface Based Recognition of Emotion Variant Faces",

SSRN2752808, 2016.
[http://dx.doi.org/10.2139/ssrn.2752808]

[20]　T. Kanade, J.F. Cohn, and Y. Tian, "Comprehensive database for facial expression analysis", *In Automatic Face and Gesture Recognition, 2000. Proceedings. Fourth IEEE International Conference on IEEE,* 2000, pp. 46-53
[http://dx.doi.org/10.1109/AFGR.2000.840611]

[21]　S. Ouellet, "Real-time emotion recognition for gaming using deep convolutional network features", In: *CoRR, vol. abs/1408.3750,* 2014.

[22]　R.J. Williams, "Simple statistical gradient-following algorithms for connectionist reinforcement learning", In: *Machine Learning* vol. 8. , 1992, pp. 229-256.
[http://dx.doi.org/10.1007/978-1-4615-3618-5_2]

CHAPTER 54

Slice Isolation Through Classification: A New Dimension for 5G Network Slicing Security

Chandini[1] and **Atul Malhotra**[1,*]

[1] *Department of Computer Science and Engineering and Engineering, Lovely Professional University Lovely Professional University Phagwara, Punjab, India*

Abstract: Connectivity over the internet is drastically increasing which is the main factor for developing different generations in the mobile network. Earlier, the usage of the internet was primarily through smartphones but now with the invention of IoT devices, the shift has been changed into different sectors like healthcare, agriculture, infrastructures, and vehicles using this internet. With this shift the demand for bandwidth, connectivity has been increased. And can be resolved through 5G network slicing. In this paper, we proposed a slice isolation model for the security of 5G network slicing also it will help users utilize the characteristics of network slicing to the fullest. Our proposed model uses a Machine learning algorithm to perform slice isolation.

Keywords: Decision Tree, eMBB, k-Nearest Neighbor, Support Vector Machine, mMTC, Random Forest, uRLLC.

INTRODUCTION

Mobile devices, specifically smartphones, are increasingly computing and processed so that they overtake laptops and just become a leisure activity. Smartphones are increasingly used for everyday activities, such as shopping, reservation, transportation, and banking. All modern uses are establishing tremendous needs for mass communication. Mobile networks can play a central role in evolution and creativity to meet the dramatically rising needs of consumers and stuff about access and network availability. This results in the next generation of mobile devices, coined as "5G." 5G allows a new form of network to bring nearly all, including devices, objects, and appliances, together. Wireless 5G connectivity is expected to offer higher data rates of Multi Gbps, incredibly low latency, better reliability, vast network capability, greater availability, and much

---

[*] **Corresponding author Atul Malhotra:** Department of Computer Science and Engineering and Engineering, Lovely Professional University Lovely Professional University Phagwara, Punjab, India; E-mail: atul.18011@lpu.co.in

Dharam Buddhi, Rajesh Singh and Anita Gehlot (Eds.)

greater user experience for more devices [1]. The below Fig. (**1**) represents the requirement of 5G.

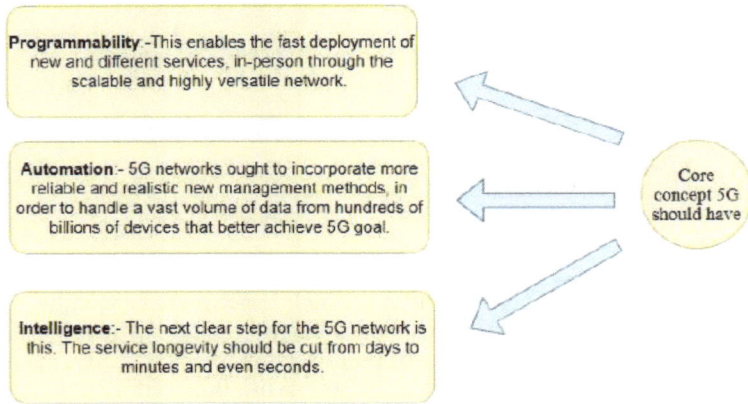

Fig. (1). 5G must have a concept.

5G will have a real effect on all sectors, rendering safe transport, remotely monitored health care, agriculture of precision, digital logistics, and so on. These challenging and ambitious goals must be met with the "one-fit-all" model [2]. This is why, in contrast with every previous transformation, the transition into 5G is even more sudden, meaningful, and demanding. To do this, 5G should have three core concepts [3]. The core concept underlying 5G is perhaps a single network, which is versatile enough to deal with a range of applications [4]. In 5G, network slicing is mainly of three types of slices as shown in Fig. (**2**), also known as "5G triangle" or "5G use-cases" [5, 6].

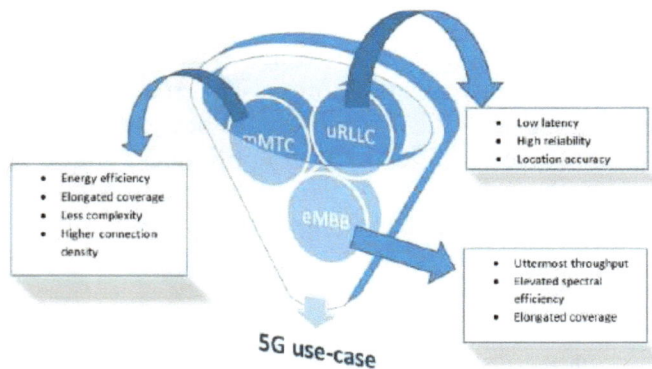

Fig. (2). 5G use-cases.

uRLLC

In additament to the improved mobile broadband implementations, the five-generation mobile networks are anticipated to provide essential URLLC services. URLLC covers a whole new case family of technologies by promoting new vertical market demands like autonomous driving, eHealth remote surgery, and industry 4.0 along with cloud robotics [7]. All the technologies clamor enhanced latency, reliability along with uttermost security and availability.

eMBB

The focus is on propping up an ever-expanding user rate as well as system capacity for enhanced mobile broadband (eMBB). eMBB unveils two significant technology improvements to meet these requirements:

- Scope shift with cmWave and mmWave to achieve significantly higher allocations for bandwidth.
- A state-of-the-art antenna array with ten or perhaps even thousands of TX/RX antenna arrays to encourage MIMO and beamforming.

The eMBB is a service extension allowed *via* 4G LTE networks first, in simplistic words, that allows the maximum data rate over a large coverage area [8]. For massive gatherings and end-users on the move, the EMBB would have the expanded capability to accommodate peak data rates.

mMTC

Concerning minimal cost and lengthy battery life, it embraces the huge density of devices including long-range transmission that makes it suitable for the Internet of Things. Without overburdening the network, mMTC aims at the cost-effective yet robust connection of thousands of devices [9]. Critical factors for performance involve -coverage, cost-effectiveness, reduced power usage, long-term availability. mMTC main focused is on IoT devices.

The below section deals with the research work of various scholars and researchers and their findings along with results is mentioned in Table **1**.

Table 1. Literature review.

S.No.	Author	Research Objective/Finding and Results
1.	Rabia Khan *et al.* [10]	• This paper provides an overview of the central and supporting technologies used in the development of the 5G security model from its security and privacy backgrounds. The paper also contains focuses on security surveillance and 5G network management. • This paper reviews the safety measures and requirements associated with core 5G technologies through recourse to various standardization bodies and a short description of security forces for 5G standardization
2.	Jianbing Ni *et al.* [11]	• In this article, the authors introduce an efficient and protected service-oriented authentication system that supports 5G-enabled IoT network slicing as well as fog computing. • In particular, *via* appropriate 5G infrastructure network sections identified by fog nodes centered on the slice/service form of access networks, users could efficiently create linkages *via* the 5G core network and access anonymous IoT services underneath its delegation.
3.	Ioannis P. Chochliouros *et al.* [12]	• The paper also recognized some such variables in methodology and carried out an initial assessment. Because of its design, the dynamic NS will help the 5G performance substantially.
4.	Dimitrios Schinianakis *et al.* [13]	• A security analysis of 5G networks with the utilization of trust zones within network slicing frameworks was addressed in this study. • Besides, a trust zone approach's efficiency, as well as isolation capabilities, are tested through a simulation process to identify as well as mitigate simulated risks.
5.	Xenofon Foukas *et al.* [14]	• The author starts with the analysis of the purpose built-in 5G NS as well as suggests a structure for holistic compilation and discussion of existing work. • They also assess the maturity of existing initiatives using this process and define a range of open research questionnaires.

LITERATURE REVIEW

Proposed Model-slice Isolation through Classification

In this session of the paper, we proposed our model, 5G aims to provide high connectivity through connecting billions of devices, heterogeneity of services such as smart city, e-health, Automotive, IoT devices, and many more over the similar 5G environment., high availability, low latency and several more attribute of 5G makes it the best fit for fulfilling the demands of the future era. With higher connectivity there comes the risk of higher attacks (*e.g* DOS/DDOS attack [15], API attack [3]) on the network.

To minimize the effect of such an attack we are presenting a model which will act as a preventive measure. Consider a scenario where two users want to access the

5G network to use services like smart- agriculture and smart healthcare. Among these two services smart healthcare contains highly sensitive data related to concerned users and any type of attack, modification of the data will cause severe loss. So security is a must requirement. And this will be achieved through Slice isolation [16 - 18].

Our proposed model will provide Isolation among different service slices through classification in terms of eMBB,uRLLC, and mMTC as every service will have different network requirements. This will make the slice that requires less security isolating from the one that requires high protection. Also attack in one will not affect the other performance as cyber breaches or defects damage the target slice alone and have no effect over the life span of any existing slices. Individual slice information is not exchanged between other slices. Here we are using five different ML algorithms stated in Table **2** to find the best slice type for the concerned service (Fig. **3**). Illustrate the flow of execution of the program.

ML Algorithm Used

Table 2. ML algorithm elaborated.

S.No.	ML Algorithm	Algorithm Description
1.	Random Forest	• The random forest itself is a technique for supervised learning. The 'forest' it constructs is a set of the decision-making trees, generally trained in the method of 'bagging.' • The idea behind the bagging method is generally to enhance the total result by combining learning models.
2.	Decision Tree	• The decision tree is such a mechanism for guiding decisions using a tree-like structure of decision making and its potential effects, including probabilities, resource costs, and efficiency. • Decision trees are also used to define a strategic approach most likely to achieve an objective during the operational study, particularly in decision analysis. • Every internal node in the decision tree refers towards an attribute query, every branch represents a test result, and each node in the leaf (terminal node) included a class label.
3.	k-Nearest Neighbor	• The KNN algorithm believes identical objects happen in the immediate vicinity. Related objects are close to each other, in other words. • For calculating the distance among the objects, the Euclidean formula is used which is given below. $$d(m,n) = \sqrt{\sum_{i=0}^{n}(m_i - n_i)^2}$$

(Table 2) cont.....

S.No.	ML Algorithm	Algorithm Description
4.	Logistic Regression	• Logistic Regression is simply an algorithm for classification. The objective variable (also refers as output), y, could only take separate values for a specified feature (referred to as inputs), X, in such a classification task. • The logistic regression theorem appears to reduce costs between 0 to 1.
5.	Support Vector Machine	• Inside this SVM algorithm, each of the data objects is traced as a point throughout n-dimensional space (in which n denotes the number of features that you have) with each function as its value. • We then classify by detecting that hyper-plane which very well distinguishes both groups. Support Vectors are just independent observation coordinates. • The SVM classification is a border that better separates all groups.

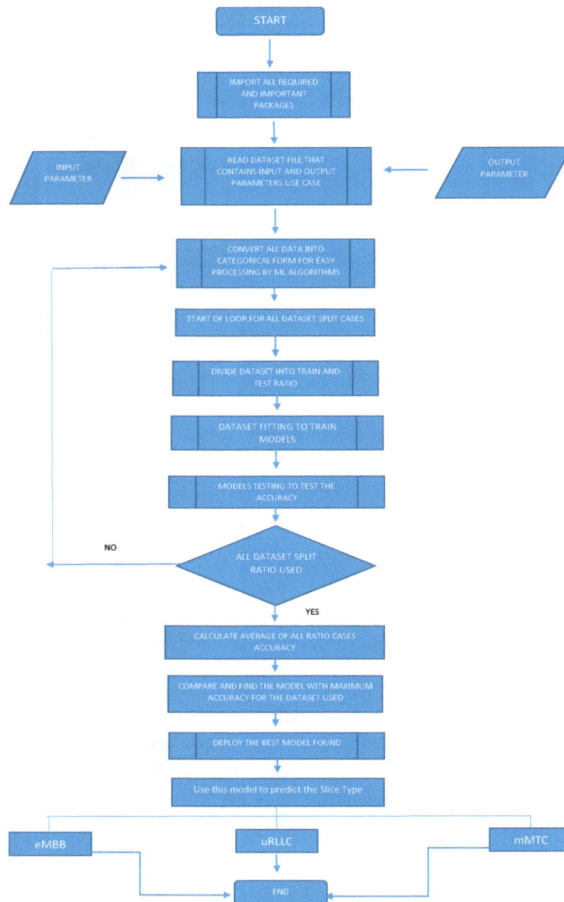

Fig. (3). Flow of execution.

RESULTS AND ANALYSIS

For the analysis of our proposed model, we have used this dataset from Kaggle [19]. About the authenticity, this dataset comes from a trustworthy source and is also been used by few researchers [20]. The reason for choosing this dataset is that as contains almost all the services that a 5G network slice would use along with the minimum criteria to be considered while classification for our slice isolation model. Few important points regarding our model are given below:

• In this proposed model the dataset is divided into 4 rations that are 90:10, 80:20, 70:30, and 60:40 for training and testing as the results are shown in Table **3**.

Table 3. Performance accuracy of ML algorithm.

Split-Ratio	Decision Tree	Random Forest	Logistic Regression	KNN	SVM
90/10	100%	100%	50%	50%	50%
80/20	100%	100%	50%	25%	25%
70/30	83%	100%	66.6%	50%	50%
60/40	62.5%	75%	75%	37.50%	50%

• The ML algorithm stated in Table **2** is used and for each ratio, the accuracy of each ML algorithm is calculated and stored as shown in Fig. (**4**). On taking the averages of all the cases for each ML algorithm we found the algorithm that has been consistent in terms of accuracy and better in terms of performance, so we choose that model as a slice type classifier among three different 5G network slices that is eMBB,uRLLC and mMTC.

Fig. (4). Performance measure.

• So based on the incoming request requirement our deployed ML algorithm will classify the request into a best-fit slice type category that will fulfill all the requirements of the request and hence the request will have access to the 5G network slice category in has been classified into preventing its interaction and access to other slice type category leading in slice isolation through classification and so improving the security of 5G network slices.

• The slice isolation model we are proposing is a preventive mechanism that leads to increased security at the earliest stage of interaction between a user and a slice Type.

CONCLUSION

The term "isolation" refers to distinguishing one object from another. In this paper our proposed model use isolation for implementing security in 5G network slicing. This work will add an extra layer of security by placing the service request in a different category, isolating one slice from another leading to the minimum or no interaction among them. In 5G network slicing, isolation can be performed on different bases such as physical isolation or virtual isolation but we have implemented isolation through classification. Our proposed model is cost-effective and easy to implement. It is not hardcoded and will efficiently work well with any 5G network slice dataset. We got to know that with all the utilized data set the above model delivers improved performance, but will get more strengthens after actual environment traffic has been deployed.

CONSENT FOR PUBLICATION

Not applicable.

CONFLICT OF INTEREST

The authors declare no conflict of interest, financial or otherwise.

ACKNOWLEDGEMENT

Declare none.

REFERENCES

[1]   G. Barb, and M. Otesteanu, "5G: An Overview on Challenges and Key Solutions", *2018 International Symposium on Electronics and Telecommunications (ISETC),* 2018pp. 1-4 Timisoara, Romania
[http://dx.doi.org/10.1109/ISETC.2018.8583916]

[2]   X. Li, M. Samaka, H.A. Chan, D. Bhamare, L. Gupta, C. Guo, and R. Jain, "Network Slicing for 5G: Challenges and Opportunities", *IEEE Internet Comput.,* vol. 21, no. 5, pp. 1-1, 2018.
[http://dx.doi.org/10.1109/MIC.2017.3481355]

[3]   G. Arfaoui, J. M. S. Vilchez, and J.-P. Wary, "Security and Resilience in 5G: Current Challenges and Future Directions", *2017 IEEE Trustcom/BigDataSE/ICESS,,* .
[http://dx.doi.org/10.1109/Trustcom/BigDataSE/ICESS.2017.345]

[4] I. Ahmad, T. Kumar, M. Liyanage, J. Okwuibe, M. Ylianttila, and A. Gurtov, "5G security: Analysis of threats and solutions", *2017 IEEE Conference on Standards for Communications and Networking (CSCN)*, 2017
[http://dx.doi.org/10.1109/CSCN.2017.8088621]

[5] A. Dutta, and E. Hammad, "5G Security Challenges and Opportunities: A System Approach", *2020 IEEE 3rd 5G World Forum (5GWF), ,* 2020
[http://dx.doi.org/10.1109/5GWF49715.2020.9221122]

[6] "Slice architecture for 5G core network", *2017 Ninth International Conference on Ubiquitous and Future Networks (ICUFN),* 2017.

[7] M. Parker, G. Koczian, S. Walker, K. Habel, V. Jungnickel, T. Rokkas, I. Neokosmidis, M. Siddiqui, E. Escalona, C. Canales-Valenzuela, A. Foglar, M. Ulbricht, Y. Liu, J. Point, D. Kritharidis, K. Katsaros, E. Trouva, Y. Angelopoulos, K. Filis, G. Lyberopoulos, E. Zetserov, D. Levi, P. Kralj, and P. Jenko, "Ultra-low latency 5G CHARISMA architecture for secure intelligent transportation verticals", *2017 19th International Conference on Transparent Optical Networks (ICTON),* 2017
[http://dx.doi.org/10.1109/ICTON.2017.8025024]

[8] A. Kostopoulos, I. Chochliouros, J. Ferragut, Y. Ma, M. Kutila, A. Gavras, S. Horsmanheimo, K. Zhang, L. Ladid, A. Dardamanis, and M-A. Kourtis, "Use Cases and Standardisation Activities for eMBB and V2X Scenarios", *2020 IEEE International Conference on Communications Workshops (ICC Workshops),* 2020.
[http://dx.doi.org/10.1109/ICCWorkshops49005.2020.9145377]

[9] C. Bockelmann, N.K. Pratas, G. Wunder, S. Saur, M. Navarro, D. Gregoratti, G. Vivier, E.D. Carvalho, Y. Ji, C. Stefanovic, P. Popovski, Q. Wang, M. Schellmann, E. Kosmatos, P. Demestichas, M. Raceala-Motoc, P. Jung, S. Stanczak, and A. Dekorsy, "Towards Massive Connectivity Support for Scalable mMTC Communications in 5G Networks", *IEEE Access,* vol. 6, pp. 28969-28992, 2018.
[http://dx.doi.org/10.1109/ACCESS.2018.2837382]

[10] R. Khan, P. Kumar, D.N.K. Jayakody, and M. Liyanage, "A Survey on Security and Privacy of 5G Technologies: Potential Solutions, Recent Advancements, and Future Directions", *IEEE Comm. Surv. and Tutor.,* vol. 22, no. 1, pp. 196-248, 2020.
[http://dx.doi.org/10.1109/COMST.2019.2933899]

[11] J. Ni, X. Lin, and X.S. Shen, "Efficient and Secure Service-Oriented Authentication Supporting Network Slicing for 5G-Enabled IoT", *IEEE J. Sel. Areas Comm.,* vol. 36, no. 3, pp. 644-657, 2018.
[http://dx.doi.org/10.1109/JSAC.2018.2815418]

[12] I.P. Chochliouros, A.S. Spiliopoulou, P. Lazaridis, A. Dardamanis, Z. Zaharis, and A. Kostopoulos, "Dynamic Network Slicing: Challenges and Opportunities," Artificial Intelligence Applications and Innovations. AIAI 2020 IFIP WG 12.5 International Workshops", *IFIP Adv. Inf. Commun. Technol.,* pp. 47-60, 2020.
[http://dx.doi.org/10.1007/978-3-030-49190-1_5]

[13] D. Schinianakis, R. Trapero, D.S. Michalopoulos, and B.G-N. Crespo, "Security Considerations in 5G Networks: A Slice-Aware Trust Zone Approach", *2019 IEEE Wireless Communications and Networking Conference (WCNC),* 2019.
[http://dx.doi.org/10.1109/WCNC.2019.8885658]

[14] X. Foukas, G. Patounas, A. Elmokashfi, and M.K. Marina, "Network Slicing in 5G: Survey and Challenges", *IEEE Commun. Mag.,* vol. 55, no. 5, pp. 94-100, 2017.
[http://dx.doi.org/10.1109/MCOM.2017.1600951]

[15] M.A. Javed, and S.K. Niazi, "5G Security Artifacts (DoS / DDoS and Authentication)", *2019 International Conference on Communication Technologies (ComTech),* 2019.
[http://dx.doi.org/10.1109/COMTECH.2019.8737800]

[16] R.F. Olimid, and G. Nencioni, "5G Network Slicing: A Security Overview", *IEEE Access,* vol. 8, pp. 99999-100009, 2020.

[http://dx.doi.org/10.1109/ACCESS.2020.2997702]

[17]   P. Porambage, and M. Liyanage, "Security in Network Slicing", In: *Wiley 5G Ref,* , 2020.

[18]   I. Ahmad, S. Shahabuddin, T. Kumar, J. Okwuibe, A. Gurtov, and M. Ylianttila, "Security for 5G and Beyond", In: *, " in IEEE Communications Surveys & Tutorials,* , 2019.
        [http://dx.doi.org/10.1109/COMST.2019.2916180]

[19]   https://www.kaggle.com/anuragthantharate/deepslice

[20]   A. Thantharate, R. Paropkari, V. Walunj, and C. Beard, "DeepSlice: A Deep Learning Approach towards an Efficient and Reliable Network Slicing in 5G Networks", *IEEE 10th Annual Ubiquitous Computing, Electronics & Mobile Communication Conference (UEMCON), New York, NY, USA,* , pp. 0762-0767, 2019.
        [http://dx.doi.org/10.1109/UEMCON47517.2019.8993066]

CHAPTER 55

Analyses of CSMA/CA Protocol Without Using Virtual Channel Sensing in DCF Mode

Harpreet Bedi[1,*] and **Kamal Kumar Sharma**[1]

[1] *School of Electronics and Electrical Engineering, LPU, Punjab, India*

Abstract: QoS is an important component of any operation, but it is particularly important in wireless transmission systems. To improve network strength and data transmission, we need to use a network that can provide the most benefits to its end users. This paper detail simulates the CSMA/CA protocol in DCF mode without using virtual channel sensing (*i.e.*, RTS/CTS frames). In this article, a comparison of the performance of an IEEE 802.11e network is made, and the results are compared to improve the network's efficiency.

Keywords: CSMA, CTS, CSMA, DCF, EDCF, LAN, RTS.

INTRODUCTION

In recent years, wireless networks have piqued people's attention. They are easy to set up and use. IEEE802.11 is the most well-known and commonly used WLAN protocol. Collisions can occur in wireless LANs, just as they can in wired networks when multiple stations communicate at the same time over a common transmission medium. In wireless LANs, the air is, of course, the shared medium [1] The medium access methods in the MAC layer are specified by the IEEE802.11 standard. The distributed coordination function (DCF) and the point coordination function (PCF) are their two primary functions (PCF). In the norm, DCF is needed, while PCF is not. Only DCF functionality is discussed in this study.

MAC LAYER

This is the most critical layer of the data link layer, and it is responsible for data transmission. It deals with the medium and gives it power over data access and transmission from the sender to the receiver. In multipoint transmission, this device makes use of an access control mechanism to provide a complete duplex

---

* **Corresponding author Harpreet Bedi:** School of Electronics and Electrical Engineering, LPU, Punjab, India; E-mail: singhharpreet08@gmail.com

Dharam Buddhi, Rajesh Singh and Anita Gehlot (Eds.)

structure [2]. This layer or channel may interact with each other in a single or dual direction.

DCF is the most widely used tool in IEEE 802.11 as shown in Fig. (**1**). It works on a transmission agenda with a first in first out algorithm and is based on the CSMA. It reallocates and integrates mac functionality and channel availability. If the channel is available or not. If the channel is busy, the MAC must ensure that the channel is free before transmitting the data. DCF uses CSMA/CA, which stands for carrier sense multiple access with collision avoidance. CSMA/CD (carrier sense multiple access with collision detection) [3] is not the same as this. Since wireless stations are incapable of transmitting and receiving at the same time, they are unable to detect collisions. Instead of detecting collisions, CSMA/CA attempted to stop them. The receiver can send a signal and collect data at the same time. This message is sent *via* SIFS, a smaller file system than DIFS. When SIFS is less than DIFS, the data is automatically shielded from other stations and transmitted, as shown in Fig. (**1**). If the CW file size is too small, the data will be lost. The backoff algorithm will be tested and analyzed if the size is doubled.

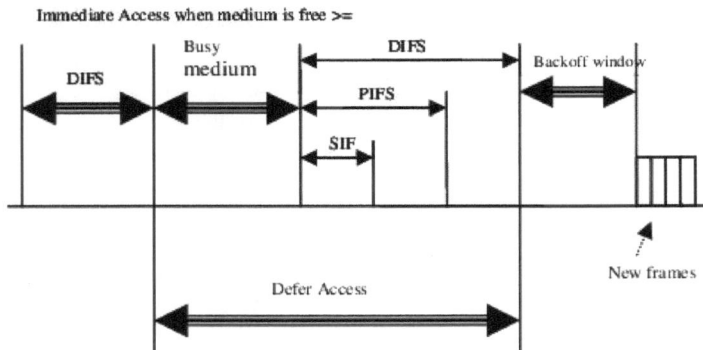

Fig. (1). Timing relationship for DCF.

It employs the "listen before speaking" mechanism, which requires each station to sense the medium for potential transmission [4]. The station transmits the packet if the medium has been idle for a certain amount of time. Otherwise, the station would delay transmission until the medium is free. After that, the backoff algorithm is invoked [5].

The CSMA/CA protocol has been documented in the literature for quite some time. Many articles have been published that address the protocol's performance problems and recommend various methods for improving it [6]. A lot of research has been done to improve the backoff procedure of the CSMA/CA protocol by changing the contention window adaptively [7, 8]. The backoff algorithm has a major effect on CSMA/CA efficiency because it specifies how long stations must wait before transmitting when the medium is sensed busy. As a result, if there are significant delays without a direct benefit, the channel will be underutilized, and bandwidth will be wasted. On the other hand, if the time between retransmission trials is quite short, there will be a growing number of collisions, resulting in low channel usage and bandwidth waste [9].

The work proposes a two-stage algorithm to improve the efficiency of the CSMA/CA protocol's backoff algorithm, and thus the IEEE802.11 WLAN's performance. This algorithm functions in the same way as the norm [10]. When a station has a collision, it raises the contention window, and when a station has a good transmission, it reduces it [11].

With the support of increasing DCF, EDCF is used to track and coordinate QOS in Fig. (2). To gain access to the access group [12, 13] the system will use the FIFO process, which will aid in the transmission of information to the stations. IEEE 802.1d [14] will be used for this. Various types of operations, such as video conferencing, online mail, and traffic, will be routed through a special AC [15, 16].

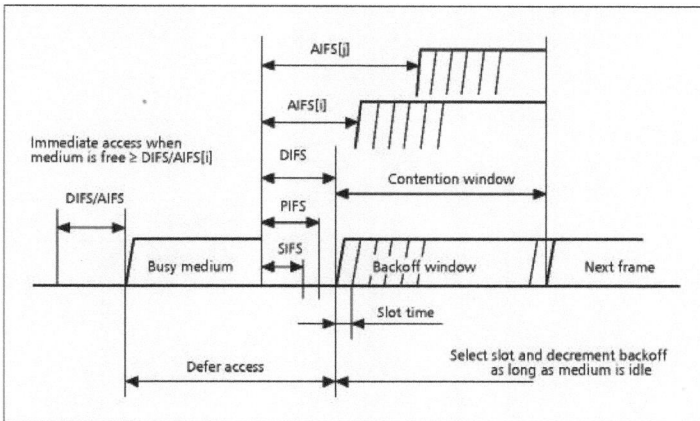

Fig. (2). Timing relationship for EDCF.

RESULTS AND DISCUSSIONS

Various simulator approaches will be explored in this report, and the network simulator will be used alongside other research operating devices as shown in Table **1**. The main goal is to test the system using a contention-based window whose size and variance are dependent on the Bit error rate and the number of users in the channel or medium. In this article, various comparisons were made concerning the number of stations, simulation time frame size with the successful transmission of the frame along efficiency.

Table 1. Various results taken at different stations and parameters.

Number of Stations	Simulation Time (Slot Time = 2)	Frame Size	Total Transmissions	Successful Transmission	Total Collisions	Unreachable Packets	Efficiency
2	600	5	18	18	0	0	1
2	700	5	23	14	0	9	0.687
2	800	5	24	24	0	0	1
2	1000	5	13	0	0	13	0
2	1500	5	43	28	4	11	0.6512
2	2000	5	51	35	2	14	0.6863
2	2500	5	51	37	0	14	0.7255
5	600	5	33	20	10	3	0.6061
5	700	5	38	24	8	6	0.6316
5	800	5	41	10	5	26	0.2439
5	1000	5	57	35	9	13	0.614
5	1500	5	82	61	14	7	0.7439
5	2000	5	92	41	5	46	0.4457
5	2500	5	131	71	14	46	0.542
7	600	5	43	18	14	11	0.4186
7	700	5	47	23	18	6	0.4894
7	800	5	56	32	13	11	0.5714
7	1000	5	65	42	16	7	0.6462
7	1500	5	98	41	21	36	0.4184
7	2000	5	123	50	23	50	0.4065
7	2500	5	150	83	30	37	0.553
10	600	5	49	25	13	11	0.5102
10	700	5	57	28	20	9	0.4912
10	800	5	64	32	27	5	0.5
10	1000	5	87	29	28	30	0.333
10	1500	5	118	47	44	27	0.3983
10	2000	5	155	58	42	55	0.3742
10	2500	5	180	64	47	69	0.3556
15	600	5	59	29	25	5	0.49

(Table 1) cont.....

Number of Stations	Simulation Time (Slot Time = 2)	Frame Size	Total Transmissions	Successful Transmission	Total Collisions	Unreachable Packets	Efficiency
15	700	5	69	27	26	16	0.3913
15	800	5	76	25	39	12	0.3289
15	1000	5	100	33	43	24	0.33
15	1500	5	146	43	53	50	0.2945
15	2000	5	172	60	58	54	0.3488
15	2500	5	218	80	76	62	0.367
20	600	5	78	20	37	21	0.2564
20	700	5	79	23	46	10	0.2911
20	800	5	87	26	45	16	0.2989
20	1000	5	118	25	45	48	0.2119
20	1500	5	162	46	83	33	0.284
20	2000	5	205	56	93	56	0.2732
20	2500	5	240	78	93	69	0.325

Fig. (**3**) and Graph 1 give the relationship between Total transmissions and successful transmission for different stations.

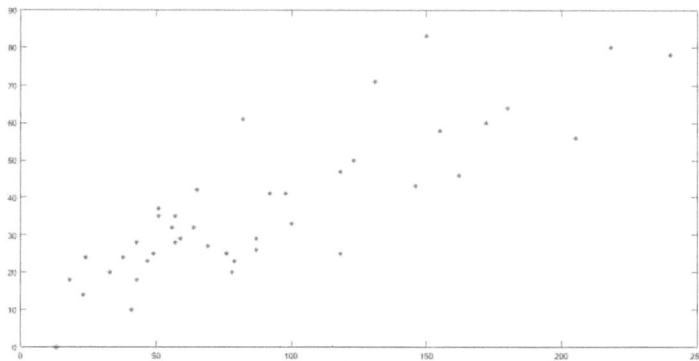

Fig. (3). Graph 1: Total transmissions and successful transmissions.

Fig. (**4**) and Graph 2 give the relationship between No of stations with Successful transmission and total collisions.

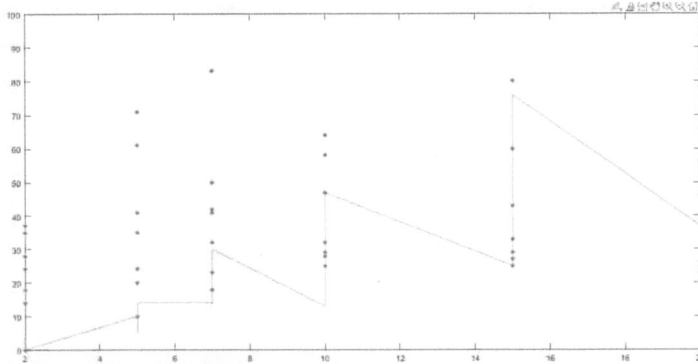

Fig. (4). Graph 2: No of stations with Successful transmission and total collisions.

Fig. (**5**) and Graph 3 give the relationship between No. of stations and total transmissions with the successful transmission.

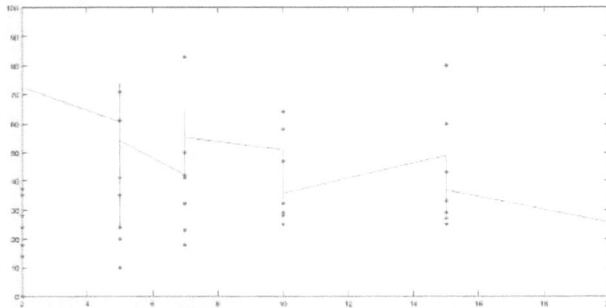

Fig. (5). Graph 3: No of stations and total transmissions with successful transmission.

Fig. (**6**) and Graph 4 give the relationship between No. of stations with Total transmissions and Successful Transmission and unreachable packets.

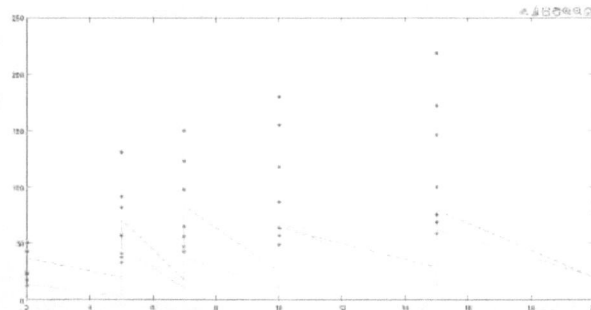

Fig. (6). Graph 4: No of stations with Total transmissions and Successful Transmission and unreachable packets.

CONCLUSION

The Mac layer and its most relevant protocols are discussed in this article, and an EDCA replica is created that covers all QoS features. Simulators are useful methods for assessing the performance of current and proposed protocols. They make this easier by removing the need to execute the protocol to be tested and providing a proper setup for it. However, due to reduced queuing, it could cause more delays to other waiting packets and, as a result, a higher number of dropped packets.

DECLARATION OF COMPETING INTEREST

The authors declare that they have no known competing financial interests or personal relationships that could have appeared to influence the work reported in this paper.

CONSENT FOR PUBLICATION

Not applicable

CONFLICT OF INTEREST

The authors declare no conflict of interest, financial or otherwise.

ACKNOWLEDGEMENT

Declared none

REFERENCES

[1]   W. Pattara-Atikom, P. Krishnamurthy, and S. Banerjee, "Distributed mechanisms for quality of service in wireless LANs", *IEEE Wirel. Commun.,* vol. 10, no. 3, pp. 26-34, 2003.
[http://dx.doi.org/10.1109/MWC.2003.1209593]

[2]   D. Chen, and P.K. Varshney, "QoS Support in Wireless Sensor Networks: A Survey", *International conference on wireless networks,* vol. vol. 233, 2004, pp. 1-7

[3]   K. Langendoen, and G. Halkes, "Energy-efficient medium access control", *Embedded systems handbook,* vol. 6000, 2005.

[4]   T. Wark, P. Corke, P. Sikka, L. Klingbeil, Y. Guo, and C. Crossman, "Transforming agriculture through pervasive wireless sensor networks", *IEEE Pervasive Comput.,* vol. 6, no. 2, 2007.
[http://dx.doi.org/10.1109/MPRV.2007.47]

[5]   O. Tsigkas, and F-N. Pavlidou, "Providing QoS support at the distributed wireless MAC layer: a comprehensive study", *IEEE Wirel. Commun.,* vol. 15, no. 1, pp. 22-31, 2008.
[http://dx.doi.org/10.1109/MWC.2008.4454701]

[6]   "Optimizing backoff procedure for enhanced throughput and fairness in wireless lans", In: *King Fahd University of Petroleum and Minerals,* 2009.

[7]   J.S. Vardakas, and M.D. Logothetis, "End-to-end delay analysis of the IEEE 802.11 e with MMPP

input-traffic", *International Symposium on Autonomous Decentralized Systems,* 2009, pp. 1-6

[8]   X. Liu, and W. Zeng, "Throughput and delay analysis of the IEEE 802.15. 3 CSMA/CA mechanism considering the suspending events in unsaturated traffic conditions", *6th International Conference on Mobile Adhoc and Sensor Systems,* 2009, pp. 947-952

[9]   Z.G. Khaki, and H.S. Bedi, "Transient correction using EDFA: in-line optical fiber with feedback", *International Conference on Computing Sciences,* 2012, pp. 233-238 [http://dx.doi.org/10.1109/ICCS.2012.76]

[10]   H.S. Bedi, G. Singh, T. Singh, and N. Kumar, "Overview of Performance of Coding Techniques in Mobile WiMAX Based System", In: *International Journal on Recent Innovation Trends in Computing Communication* vol. 1. , 2013.

[11]   N.S. Chauhan, and L. Kaur, "Implementation of QoS of different multimedia applications in WLAN", *Int. J. Comput. Appl.,* vol. 62, no. 8, 2013.

[12]   S. Singh, and A. Kumar, "Performance analysis of adaptive clipping technique for reduction of PAPR in alamouti coded MIMO-OFDM systems", *Procedia Comput. Sci.,* vol. 93, pp. 609-616, 2016. [http://dx.doi.org/10.1016/j.procs.2016.07.246]

[13]   H.S. Bedi, S. Verma, and B. Singh, "A Survey on Enhancing Quality of Service in Wireless Sensor Networks", *Int. J. Control Theory Appl.,* vol. 9, no. 41, pp. 37-42, 2016.

[14]   I. Syed, and B-h. Roh, "Delay analysis of IEEE 802.11 e EDCA with enhanced QoS for delay sensitive applications", *35th International Performance Computing and Communications Conference (IPCCC),* 2016, pp. 1-4

[15]   H. Bedi, I. Puri, and S. Verma, "Performance Comparison of Companding Techniques and New D-Cast Method for", *Int. J. Control Theory Appl.,* vol. 9, no. 24, pp. 271-222, 2016.

[16]   H.S. Bedi, K.K. Sharma, and R. Gupta, "A Review Paper on Performance Analysis of IEEE 802.11 e", *1st International Conference on Computing, Communications, and Cyber-Security (IC4S 2019),* 2020, pp. 47-56

CHAPTER 56

Sanjay: A Remote Vulnerability Scanning Framework

I. Abhinav Srivastava[1,*], Manish Kumar[1], Anshuman Das[1], Pratyaksh Kumar Singh[1], E. Pavan Kumar[1] and Atul Malhotra[1]

[1] *School of Computer Science & Engineering, Lovely Professional University, Phagwara 144411, India*

Abstract: Many techniques are laid out that try to protect the web application from unauthorized access with the growing concern for web security. To uncover the possible inconsistencies that can damage the application, new approaches have been introduced. Vulnerability scanning is the most widely used technique. By the term vulnerability, we meant the possible device flaws that make it vulnerable to attack. Scanning these vulnerabilities in the system provides a means of identifying and developing new strategies to protect the system from the risk of damage.

This paper focuses on the use of a remote vulnerability scanning framework-"SANJAY" which completely and comprehensively scans the given target most effectively and easily with the help of integrated tools like SLACK and VPS.

Keywords: Framework, SANJAY, Slack, VPS, Vulnerability Scanning, Web Security.

INTRODUCTION

Vulnerability is programming flaws or configuration issues that allow an attacker to achieve unauthorized access [1]. Vulnerability in a computer system can entail everything from a router's poor password to an unpatched programming defect in an exposed network service.

A vulnerability is a kind of technological error that makes a device or service vulnerable. Vulnerability is a warning sign, the more vulnerabilities associated with a device, the less secure it is. As per the Open Web Application Security Project (OWASP) [2], a list of the top vulnerabilities is been published based on the risk factor and threat level called OWASP TOP 10. These flaws possess a sig-

---

* **Corresponding author I. Abhinav Srivastava:** School of Computer Science & Engineering, Lovely Professional University, Phagwara 144411, India; E-mail: abhinav96fzd@gmail.com

Dharam Buddhi, Rajesh Singh and Anita Gehlot (Eds.)

nificant risk to the web application and may result in significant harm. Under the name OWASP Mobile Security Project, OWASP also announces a list for Mobile Vulnerability. In 2017, OWASP updated its list of the top vulnerabilities, illustrated in Table **1**.

Table 1. Shows the list of OWASP TOP 10 Vulnerabilities in 2017.

RANK	Vulnerability Name
1	Injection
2	Broken Authentication
3	Sensitive Data Exposure
4	XML External Entities (XXE)
5	Broken Access Control
6	Security Misconfiguration
7	Cross-Site Scripting (XSS)
8	Insecure Deserialization
9	Using Components with Known Vulnerabilities
10	Insufficient Logging & Monitoring

Some of the Web Application vulnerabilities are enlisted below [3].

Remote Code Execution (RCE)

Remote Code Execution occurs when an attacker can execute server commands on a remote server [4].

Cross-Site Scripting (XSS)

Cross-Site Scripting enables an intruder to insert untrusted javascript code into a web application with no authentication. When the victim enters the website, this script is executed, resulting in serious exploitation [5].

Cross-Site Request Forgery (CSRF)

In CSRF attackers can create forged HTTP requests by using CSRF. Using a forged HTTP message, the intruder attempts to trick the user to compromise his/her data [6].

Server-Side Request Forgery (SSRF)

It is a web security flaw that allows an attacker to force the server-side program to send HTTP requests to any domain of the attacker's choice [7].

Insecure Direct Object Reference (IDOR)

IDOR is essentially a defect in a web application in which specific database documents or directories are not fully secured and are readily exposed [8].

Injections

SQL Injection is a typical injection-based vulnerability found in database-driven web applications. An intruder sends basic text-based requests that take advantage of the syntax of the specific interpreter [9].

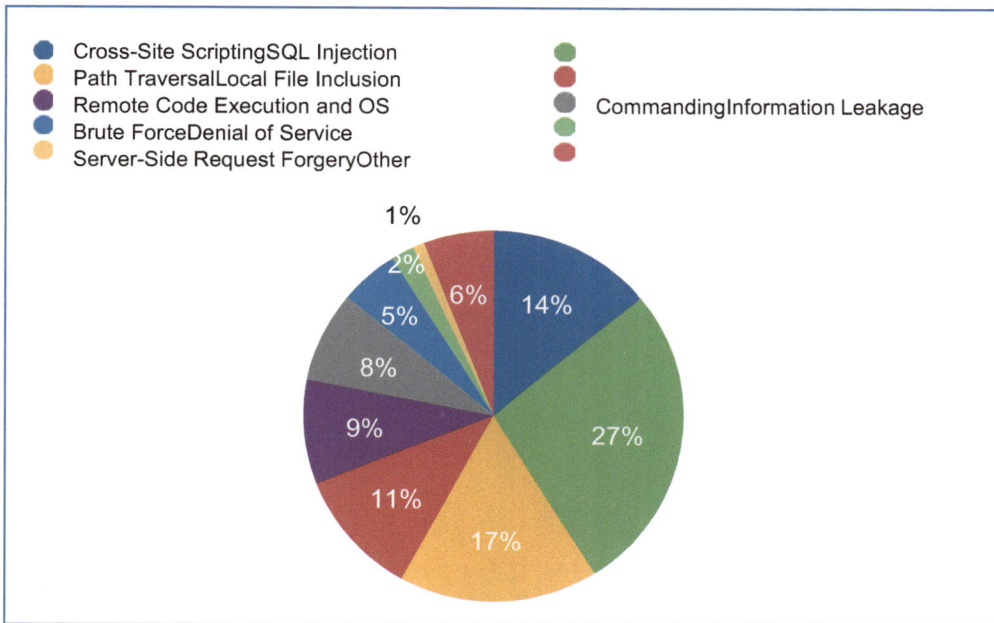

Fig. (1). Web application vulnerabilities.

Fig. (**1**) illustrated shows the round of percentage of Web Application based Vulnerabilities in the year 2019 in which 27 percent is of SQL Injection followed by Path Traversal having 17 percent and Cross-Site Scripting with 14 percent [10].

In this paper, we have proposed a comprehensive scanning framework called "SANJAY" which will make scanning and identifying the vulnerability in a system and make it more accessible to everyone.

In this framework, we have developed remotely accessible technology integrated with SLACK and VPS (Virtual Private Server). SANJAY will send the command through slack to our channel and it will start scanning the web application or system and generate the vulnerability report in the text format and send it over the slack chatbox.

LITERATURE REVIEW

Vulnerability Scanning

Vulnerability scanning is the method of finding loopholes/vulnerable points in a target structure that may cause an attacker to manipulate the target. It also describes how the flaws discovered could be exploited. This is a constructive strategy in which the flaws are discovered and addressed before someone else is aware of them. Vulnerability scanning is conducted not only on a single program, but also on the platform, middleware, and operating system on which the application is running. Generally, scanning is done in two ways Active scanning and Passive scanning.

Scanning a device or a network necessitates the use of a scanner, they can be categorized into different types:

Port Scanner

Port scanners are used to search the ports to determine which ports are open and which are closed, as well as the operating system and facilities available.

Application Scanner

Application scanners are used to test a single application on a network to detect bugs that may be abused to place the device at risk.

Vulnerability Scanner

Vulnerability scanners are programs that look for flaws in a system that, if exploited by a malicious person or intruder, might place the whole network system at risk. It notifies organizations about pre-existing vulnerabilities in their software and where they are based [11].

A comprehensive description of any application's weaknesses and the risks are detected with the help of Penetration testing and Vulnerability Scanning together [12]. Penetration testing aims to manipulate code vulnerabilities to assess whether unauthorized entry or other malicious behavior is feasible, as well as to detect which flaws pose a threat to the program [13]. To assess these flaws we go through six step vulnerability assessment approach as illustrated in Fig. (**2**).

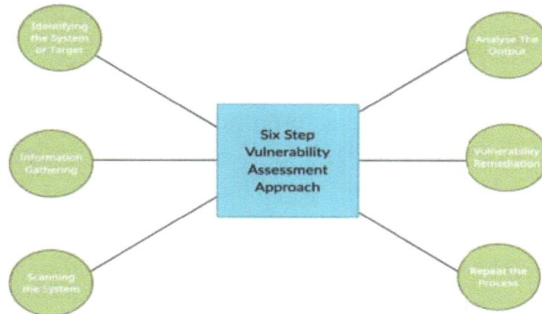

Fig. (2). Vulnerability assessment approach.

TYPES OF VULNERABILITY SCANS

In essence, a security evaluation employs a variety of approaches, techniques, and scanners to identify potential hazards and risks. Everything relies on how effectively the flaws in the assigned structures are identified to address the particular need we perform Vulnerability scans which are further categorized into various types. Some of the types are:

NETWORK-BASED SCANS

Its name suggests that it aids in the detection of potential network security threats. The scan identifies compromised devices on both wired and wireless networks [14].

Host-Based Scans

These scans will quickly find bugs in server mainframes or other network hosts. Ports and utilities are thoroughly reviewed as part of the procedure, it also offers excellent insight into configuration settings and patch records of the machine.

Application Scans

It's used to identify all suspected tech bugs on websites.

Wireless Network Scans

The architecture of a wireless network is scanned for bugs, which aids in the validation of a company's internal network [15].

Database Scans

Server scans help locate grey zones in a database, helping malicious actors to stop launching brutal attacks [16].

Web application-based attacks have remained a high concern in the last five years. High risk was 52 percent in 2017 and unexpectedly rose by 15% in 2018. Although it decreased by 17% in 2019, the risk remains at 50%. Fig. (**3**) shows the level of Web Application Vulnerabilities from 2015-2019 [17].

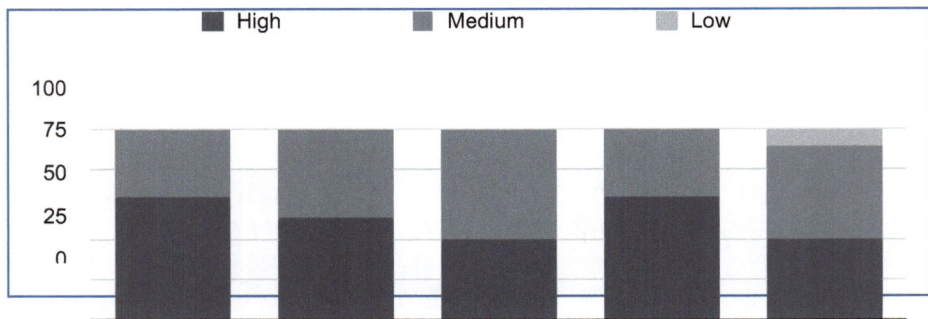

Fig. (3). Web Application Vulnerabilities.

Our Vulnerability Scanning Framework mainly focuses on finding the information related to the Web Application-based vulnerabilities. In a Web application, there are many phases through which a complete analysis can take place.

In March of 2019, one of the most serious web attacks occurred (Capital One Breach). Acting independently, an attacker obtained access to Capital One's secure network as well as over 100 million consumer accounts and credit card applications. The attacker also gained access to over 100,000 Social Security numbers, 80,000 bank accounts, and other emails, credit ratings, and records. According to the FBI agent, the attacker was thinking about publicly disclosing the Social Security numbers [18].

PROPOSED FRAMEWORK: SANJAY

SANJAY is an automated vulnerability scanning framework that makes scanning easy and reliable and it can be remotely accessible from all over the globe by any device by sending a single command through a slack integrated channel. This framework performs various tasks related to web application scanning and provides us the filtered false-positive result and sends it over the slack in the form of text files. All the processes are illustrated in Fig. (**4**).

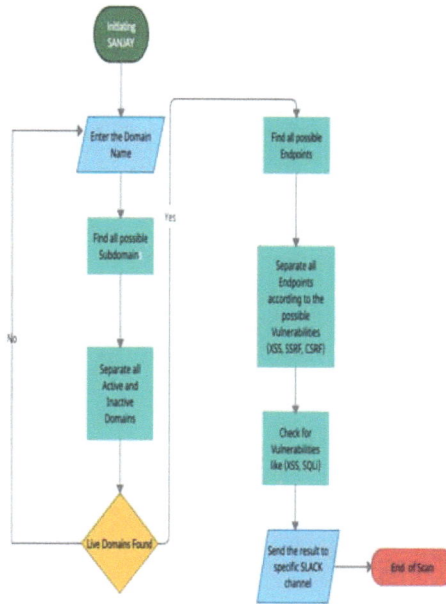

Fig. (4). Work-flow diagram of framework sanjay.

Firstly it will obtain a comprehensive list of all related subdomains for the domain given in the command. As we know all the subdomains are not active, some of them are inactive domains and are of no use, so SANJAY will filter out the list of subdomains and make a separate file of active domains/subdomains with the file name as [sublives.txt]. After getting the list of live domains further SANJAY will start crawling the subdomains from the internet and it will store the endpoints. SANJAY is integrated with various templates which are being used to find predefined OWASP-10 vulnerabilities and store the vulnerable endpoints in a categorized manner in the form of a text file. It also provides the option to customize templates according to the user's requirements. Now for the more precise result, it will again filter out the previous list of vulnerable endpoints and it will create a separate list with different vulnerabilities along with endpoints such as [domain].xss. txt, [domain name].csrf.txt, [domain].rce.txt, *etc*, and send those text files over the slack channel. Now after getting the resulting form our framework tester can easily trigger the vulnerabilities with the help of endpoints.

FEATURES OF THE PROPOSED FRAMEWORK

Remote Accessibility

SANJAY can be accessed remotely from any device that has SLACK in it.

Availability

VPS can be kept running 24x7 thus providing anytime, anywhere availability of SANJAY.

Usability

Using SANJAY as a vulnerability scanner can be easy as the user has to only provide the target name and the type of scan to perform.

Resource Consumption

Resource consumption can be drastically reduced onto a local system while using SANJAY as most of the heavy work is done over the VPS.

CONCLUSION

Vulnerability scanning is an integral component of the security of a system or a network. It helps you to be vigilant in closing loopholes and ensuring good protection for your applications, records, staff, and customers' details. Since data breaches are often the product of unpatched vulnerabilities, finding and closing these security loopholes prevents the attack vector.

Our proposed architecture, SANJAY, is critical in scanning the vulnerabilities for a given domain and providing the results in a sorted format over the slack channel in the form of a text file.

Since it is installed on a VPS, resource use on the local computer is drastically reduced, and the command's scanning operation is completed entirely over the VPS.

In the future, we would like to offer different applications for computers and smartphones, making it easier for users to use the framework.

CONSENT FOR PUBLICATION

Not applicable.

CONFLICT OF INTEREST

The authors declare no conflict of interest, financial or otherwise.

ACKNOWLEDGEMENT

Declare none.

REFERENCES

[1] B. Wang, L. Liu, F. Li, J. Zhang, T. Chen, and Z. Zou, "Research on Web Application Security Vulnerability Scanning Technology", *2019 IEEE 4th Advanced Information Technology, Electronic and Automation Control Conference (IAEAC), Chengdu, China,* 2019, pp. 1524-1528 [http://dx.doi.org/10.1109/IAEAC47372.2019.8997964]

[2] O.W.A.S.P. Top Ten, " [Online]. Available:", https://owasp.org/www-project-top-ten [Accessed: 20th February 2021].

[3] Common Web Application Vulnerabilities, https://www.veracode.com/security/web-applicatio--vulnerabilities [Accessed: 23rd March 2021].

[4] Saikat Biswas, M. Sohel, and M.D. Sajal, *A study on remote code execution vulnerability in web applications,* 2018.

[5] G. Wassermann, and Z. Su, "Static detection of cross-site scripting vulnerabilities", *2008 ACM/IEEE 30th International Conference on Software Engineering, Leipzig, Germany,* 2008, pp. 171-180 [http://dx.doi.org/10.1145/1368088.1368112]

[6] M.S. Siddiqui, and D. Verma, "Cross-site request forgery: A common web application weakness", *2011 IEEE 3rd International Conference on Communication Software and Networks, Xi'an, China,* 2011, pp. 538-543 [http://dx.doi.org/10.1109/ICCSN.2011.6014783]

[7] H. Luo, "SSRF Vulnerability Attack and Prevention Based on PHP", *2019 International Conference on Communications, Information System and Computer Engineering (CISCE),Haikou, China,* 2019, pp. 469-472 [http://dx.doi.org/10.1109/CISCE.2019.00109]

[8] https://portswigger.net/web-security/access-control/idor

[9] P.A. Sonewar, and S.D. Thosar, "Detection of SQL injection and XSS attacks in three-tier web applications", *2016 International Conference on Computing Communication Control and Automation (ICCUBEA),* 2016, pp. 1-4 Pune, India [http://dx.doi.org/10.1109/ICCUBEA.2016.7860069]

[10] https://www.ptsecurity.com/ww-en/analytics/web-application-attacks-2019/

[11] Y. Makino, and V. Klyuev, "Evaluation of web vulnerability scanners", *2015 IEEE 8th International Conference on Intelligent Data Acquisition and Advanced Computing Systems: Technology and Applications (IDAACS), Warsaw, Poland,* 2015, pp. 399-402 [http://dx.doi.org/10.1109/IDAACS.2015.7340766]

[12] M. Bishop, "About Penetration Testing", *IEEE Secur. Priv.,* vol. 5, no. 6, pp. 84-87, 2007. [http://dx.doi.org/10.1109/MSP.2007.159]

[13] P.S. Shinde, and S.B. Ardhapurkar, "Cyber security analysis using vulnerability assessment and penetration testing", *2016 World Conference on Futuristic Trends in Research and Innovation for Social Welfare (Startup Conclave), Coimbatore, India,* 2016, pp. 1-5 [http://dx.doi.org/10.1109/STARTUP.2016.7583912]

[14]   I. Sun-young, "S. Shin, Ki Yeol Ryu and Byeong-hee Roh, *"Performance evaluation of network scanning tools with operation of firewall"*, *2016 Eighth International Conference on Ubiquitous and Future Networks (ICUFN)*, 2016, pp. 876-881 Vienna, Austria
[http://dx.doi.org/10.1109/ICUFN.2016.7537162]

[15]   Y. Wang, and J. Yang, "Ethical Hacking and Network Defense: Choose Your Best Network Vulnerability Scanning Tool",
[http://dx.doi.org/10.1109/WAINA.2017.39]

[16]   D. Karabasevic, D. Stanujkic, M. Brzaković, M. Maksimović, and M. Jevtić, "Importance of vulnerability scanners for improving security and protection of the web servers", *Bizinfo (Blace)*, vol. 9, pp. 19-29, 2018.
[http://dx.doi.org/10.5937/bizinfo1801019K]

[17]   https://www.ptsecurity.com/ww-en/analytics/web-vulnerabilities-2020/

[18]   S. Nelson, and N. Anchises, *A Case Study of the Capital One Data Breach*.https://web.mit.edu/smadnick/www/wp/2020-07.pdf

CHAPTER 57

A Study of CRM Solution Using Salesforce

Ravi Prakash[1,*], **Shalabh Dwivedi**[1], **Sudarshan Sharma**[1] and **Jasvinder Pal Singh**[1]

[1] *School of Computer Science and Engineering, Lovely Professional University, Phagwara 144411, India*

Abstract: Customer Relationship Management (CRM) is a cloud-based technology that helps businesses to interact with their customers and potential customers. CRM makes easy access to shared computing resources with less management effort. Salesforce is an advanced and efficient CRM Solution. In this paper, we are discussing cloud computing, features of a good CRM tool, Salesforce as a CRM solution, analysis of Salesforce growth with other CRM providers, and the Salesforce CRM in COVID 19 insight. This paper aims to provide an extensive study of CRM using Salesforce and to substantiate the efficiency of Salesforce with other CRM providers like Adobe, Microsoft, and Oracle, *etc.*

Keywords: Cloud Computing, COVID 19, Customer Relationship Management, Good CRM Tool, Market Share, Salesforce.

INTRODUCTION

Traditional approaches to developing business applications, such as order and inventory management, software project management, and enterprise resource planning, are prohibitively expensive, complex, and time-consuming. To design, configure, install, test, run, and update such applications with traditional approaches, more resources and expertise are required. Managing all of these on a company's premises [1] can become a very time-consuming task. Because organizations need to set up on-premise infrastructures like data centers, servers and, hardware tools [2]. This required high investment and business risks in operation and maintenance. High infrastructure maintenance cost is an extra burden on these organizations. But over the last few decades, IT infrastructure has been evolving. From the measure of traditional hardware with high maintenance,

---

* **Corresponding author Ravi Prakash:** School of Computer Science and Engineering, Lovely Professional University, Phagwara 144411, India; E-mail: ravi140399@gmail.com

Dharam Buddhi, Rajesh Singh and Anita Gehlot (Eds.)

the IT industry has evolved to a point where infrastructure maintenance and operation are easily manageable. One solution in that journey has been the cloud, which alle*via*ted IT maintenance issues by handling core infrastructure requirements over the network. It provides on-demand shared resources [3] to the client. These resources include tools and services like storing data, access to databases and servers, networking and analytics, data security, and customer support [4]. The cloud services to individuals or companies are provided by cloud service providers.

Many technology and software services are directly deployed on the cloud. Out of these, one technology is CRM (Customer Relationship Management) [5]. It is used to manage and build customer relationships to increase sales, improve customer services and increase profitability. To grow a business at a good pace with up-to-date and reliable information on business progress, there is a need for CRM technology to get a clear view of customers. CRM systems offer several strategic advantages to companies. In the age of digitization, the scope and market of CRM are widening every day. Businesses are in a time of transition, as a technological shift has occurred from offline to online operations. To keep up with market changes, technology has been steadily improving the way businesses and their customers are managed. One of the most effective CRM technology is Saleforce.com. Salesforce is a cloud platform [6] that adheres to strict security requirements, allowing users to work in a secure environment. Salesforce CRM provides everything for users to grow their businesses. Users, for example, have no idea which CRM software is best for their applications. We compare and contrast Salesforce CRM with other CRM providers in this paper (like Adobe, Oracle, Microsoft, *etc*).

CLOUD COMPUTING

Cloud computing has become very prevalent and in high demand in businesses for several reasons, which include minimal cost, high productivity, high speed and efficiency, great performance, and maximum security. The main services provided on cloud computing include email, data storage, and backup, retrieval of data, creating and testing apps prototypes, streaming of audio and video, delivering software on demand, and also analyzing generated digital data [7]. To achieve consistency and cost advantages, cloud computing relies on resource sharing. Cloud computing has evolved as a result of the availability of good capacity networks, minimal cost computers and storage devices, as well as the global acceptance of hardware virtualization, utility, and autonomic computing, and service-oriented architecture [8, 9].

Fig. (**1**) illustrated shows the generalized cloud computing architecture. Cloud has deployment models which include public clouds, private clouds, hybrid clouds, and community clouds [10]. The public clouds deliver their services on servers and storage on the Internet. The private clouds deliver their services on a private network for a specific client such as an organization. The hybrid clouds are the blend of both public and private. One or more community groups, a third party, or a combination of them own, administer, and run the community cloud.

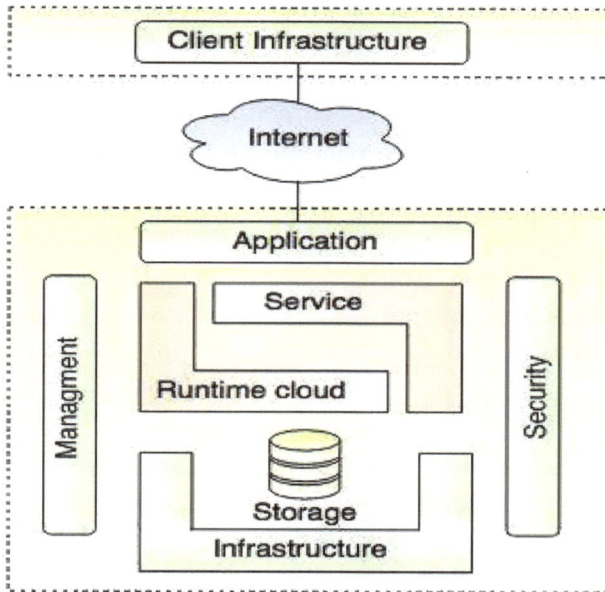

Fig. (1). Generalize cloud computing architecture [11].

Cloud providers offer their services through three basic models *i.e.* Infrastructure as a Service (IaaS), Platform as a Service (PaaS), and Software as a Service (SaaS) [12]. IaaS is a cloud computing service in which businesses rent or lease servers in the cloud for computing and storage. Cloud providers provide a computing platform in PaaS models, which usually includes an operating system, programming-language execution environment, database, and web server. Cloud providers install and run application software in the cloud, and cloud users access the software through cloud clients in the SaaS model [13].

OVERVIEW OF CRM

Customer Relationship Management (CRM) is a technology that helps in managing the relationship with customers and potential customers. It gathers and maintains a large volume of customer data. These data are processed for a better understanding of customers. We should use a two-pronged approach when selecting a CRM. The first one is the comparison of services of different CRM providers (like Adobe, Microsoft, Oracle, *etc*). The second one is, we should also consider how our company conducts business and how a CRM system can help and improve our workflow.

Some of the important considerations are listed below:

1. Minimal configuration is required, making it simple to get up and running quickly.
2. A good small business solution can document the sales processes and include dashboard-style reporting so we can see how our company is doing at a glance.
3. Consider a CRM that provides integrated customer support, marketing, and other business-critical apps.
4. Ease of use is promoted by a cloud-based CRM which allows us to log in from anywhere, *via* laptop or mobile app.
5. Choose a small business CRM that integrates with our company's other tools, such as email, file sharing, and document signing apps.

Key Features of a Good CRM Tool

A CRM tool, at its most fundamental level, is an application that aids in the positive interaction between a company and its customers. Such a broad definition means that a CRM tool can be anything used in setting and conducting business. These key features are the foundation of a good CRM tool [14].

Relationship Building

CRM's main goal is to help us to establish relationships with customers and get closer to them.

Automation of Work Processes

From capturing important business activity from our Inbox to allowing us to write and plan work emails.

Mobile access is Available

It is more important to set up our team to work from anywhere. A good CRM solution would have mobile apps that are compatible with both Android and IOS.

IMPACT OF COVID 19 ON CRM

The latest pandemic COVID-19 (Coronavirus) has prompted businesses to take all possible precautions to protect their workers and the general public. Regardless of the crisis, businesses must continue to pursue prospects, close sales, and resolve consumer complaints. As result, businesses are operated from remote area environments. As a consequence, demand for customer service CRM solutions is expected to grow. Most customer experience management features can be used by companies to improve employee efficiency, increase revenue, and serve consumers. To substantiate the growth of Salesforce in CRM, one of the fast-growing ed-tech company WhiteHat Jr [15] asserted that Salesforce helped them in scaling up their business and data operation at a global level. Salesforce helped them to increase their sales business by tenfold which in return increase the sales conversion rate from 6% to 15%.

SALESFORCE – A COMPLETE CRM SOLUTION

Salesforce CRM, also known as Salesforce Customer 360, is available in a range of models and configurations to meet the needs of companies of all sizes and industries. Customer 360 is the scope of Salesforce CRM technology - a single, integrated CRM framework that all of the teams can use to manage and expand customer relationships. All versions *i.e.* Essentials, Professional, Enterprise, and Unlimited, are developed on the same cloud platform, making it simple to update as the company expands. The most significant differences between versions are the maximum number of users we choose to serve, as well as the automation, advanced customization, and integration.

Salesforce Strength

Innovation

Artificial Intelligence

Salesforce's AI helps to make quicker decisions, make our workers more efficient, and make our customers happy.

Analytics

No matter what industry a consumer is in, augmented analytics will help them discover insights, predict results, and find suggestions to help their company.

Platform

One of the main assets is the Salesforce platform. Salesforce's platform sets it apart from other CRM solutions, from the AppExchange's expandability [16] and accessible integrations to the protection built into the cloud architecture.

Productivity

Salesforce has a track record of increasing productivity and revenue. Since introducing Salesforce, 38% of customers show an improvement in sales efficiency [17].

Mobile

Salesforce mobile app allows us to run the business from anywhere.

Community

The Salesforce Trailblazer network, which has over 2.3 million members, is a thriving community of CRM users who are promoting innovation in the Salesforce ecosystem.

ANALYSIS OF SALESFORCE WITH OTHER CRM APPLICATION PROVIDERS

Comparing Salesforce Customer 360 to other CRM applications is a key step in deciding Salesforce as a CRM solution best for our business. An easy way to view the analogy is to look at Salesforce's relative strengths vs the other CRM providers (like Adobe, Microsoft, Oracle, and SAP, *etc*).

Market Share

The top 10 CRM software vendors accounted for nearly 59.3% of the global CRM applications market in 2019 highlighted in Fig. (**2**) , which expanded 6.6 percent in license, maintenance, and subscription revenues. Salesforce was the clear leader in 2020, with a 30.2 percent market share and Adobe was ranked second, ahead of Oracle, SAP, and Microsoft.

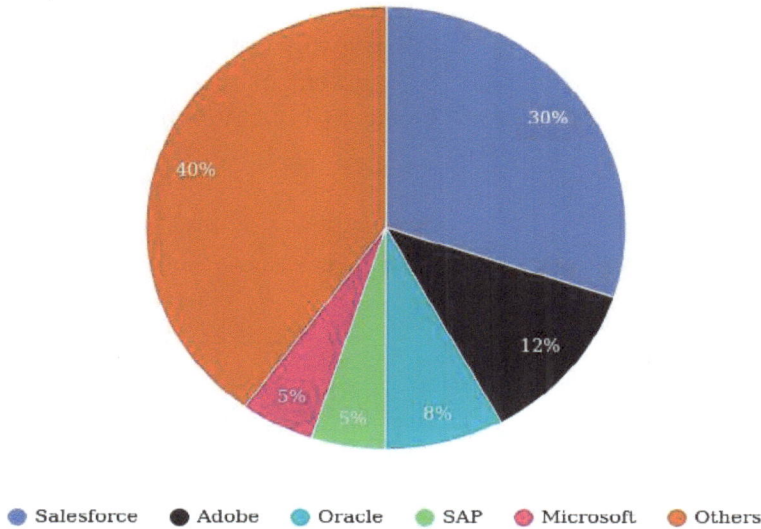

Fig. (2). CRM application market shares 2019 [18].

Percentage Increase in CRM Sales

Fig. (**3**) illustrates the increase in CRM sales over a period. It can be seen clearly that among the top 5 vendors, the share of Salesforce is high and at the top with a share of 18.3 percent in 2017, 19.5 percent in 2018, 19.6 percent in 2019, and 19.8 percent in 2020.

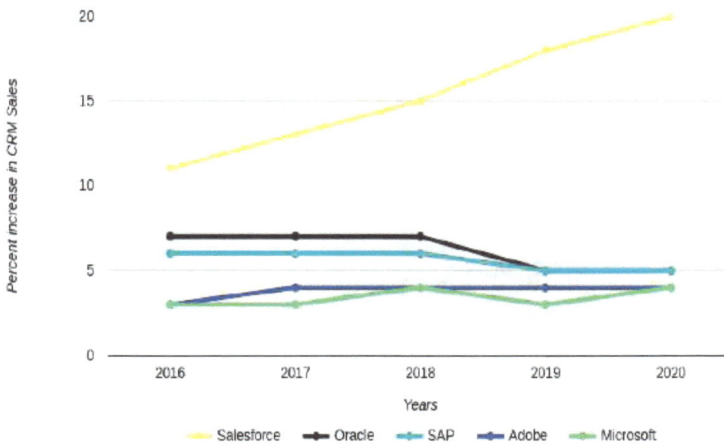

Fig. (3). Percent increase in CRM sales [19].

Revenue

Revenue is the most important indicator of a company's success. The CRM

software market is dominated by five firms, which control nearly 40% of the market as illustrated in Fig. (**4**) Even among top vendors, however, there is an unquestionable leader who controls 19.5 percent of the CRM software market, nearly equal to the combined market share of the top five: Salesforce made $9420.5 million in revenue in 2018.

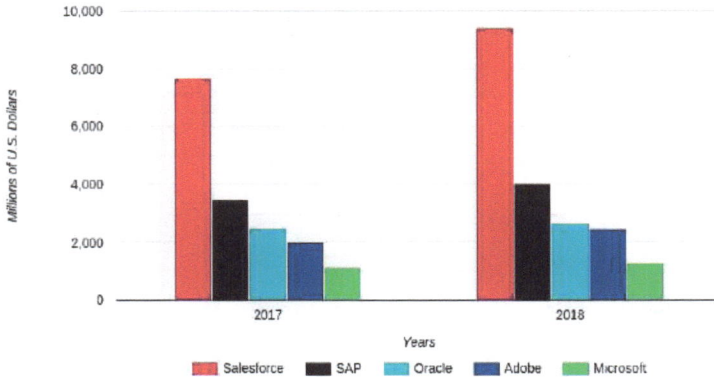

Fig. (4). Total software revenue worldwide (millions of U.S dollars) [20].

CONCLUSION

The transition to cloud computing is maybe the most important recent advancement in CRM systems. Organizations around the world are realizing the advantages of moving data, software, and services into a safe online environment since they are no longer required to install software on a large number of computers and mobile devices. Salesforce and other cloud-based CRM systems ensure that all users have access to the same information at all times.

This article examines Salesforce as a full CRM solution in depth. On different criteria, the features and capabilities of Salesforce CRM are addressed and compared to those of other CRM providers such as Adobe, Microsoft, and Oracle. Market share, revenue, and percentage increase in sales are among the parameters. The article outlines all of the main considerations when selecting CRM for a specific industry. This article explores the recent effect of COVID19 and how CRM is assisting companies in continuing and increasing their operations in the face of a global pandemic. This paper has offered valuable sources as well as a thorough review of Salesforce's features and CRM advantages. It will assist companies in accelerating their operations by allowing them to migrate to advanced platform-independent technology.

CONSENT FOR PUBLICATION

Not applicable.

CONFLICT OF INTEREST

The authors declare no conflict of interest, financial or otherwise.

ACKNOWLEDGEMENTS

The authors are grateful to the anonymous reviewers who made constructive comments.

REFERENCES

[1]   L. Björn, and B. Andrea, "Classifying Systemic Differences between Software as a Service and On-Premises Enterprise Resource Planning", *J. Enterp. Inf. Manag.,* vol. 28, no. 6, pp. 808-837, 2015.
        [http://dx.doi.org/10.1108/JEIM-07-2014-0069]

[2]   M. Ahmadreza, Y.M. Hossein, and L-G. Alberto, "Green Cloud Multimedia Networking: NFV/SDN Based Energy Efficient Resource Allocation", *IEEE Transactions on Green Communications and Networking,* vol. 4, no. 3, pp. 873-889, 2020.
        [http://dx.doi.org/10.1109/TGCN.2020.2982821]

[3]   D.C. Walden, "A Note on Interprocess in a Resource Sharing Computer Network", *IETF,* vol. 27, no. 2, pp. 765-234, 2020.

[4]   Q. Huang, and C. Yang, "Big Data and cloud computing: Innovation opportunities and challenges", *Int. J. Digit. Earth,* vol. 10, no. 1, pp. 13-53, 2017.
        [http://dx.doi.org/10.1080/17538947.2016.1239771]

[5]   E. Garbarino, and M.S. Johnson, "The different roles of satisfaction, trust, and commitment in customer relationships", *J. Mark.,* vol. 63, no. 4, pp. 70-87, 1999.
        [http://dx.doi.org/10.1177/002224299906300205]

[6]   L. Rao, *Everything You Need To Know About Salesforce Cloud 2,* 2009.https://techcrunch.com/2009/09/08/everything-you-need-to-know-about-salesforces-service-cloud-2/

[7]   E. Knorr, *What Cloud Computing Really Means,* 2018.https://www.infoworld.com/article/2683784/what-is-cloud-computing.html

[8]   R. Minvera, G.M. Lee, and N. Crespi, "Digital Twin in the IoT Context: A Survey on Technical Features Scenarios and Architectural Models", *Proc. IEEE,* vol. 108, no. 10, pp. 1785-1824, 2020.
        [http://dx.doi.org/10.1109/JPROC.2020.2998530]

[9]   P. Braun, A. Cuzzocrea, C.K. Leung, A.G.M. Pazdor, J. Souza, and S.K. Tanbeer, "Pattern Mining from Big IoT Data with fog Computing: Models Issues and Research Perspectives", *19th IEEE/ACM International Symposium on Cluster, Cloud and Grid Computing (CCGRID),* 2019, pp. 584-591
        [http://dx.doi.org/10.1109/CCGRID.2019.00075]

[10]   Timothy Grance, and Peter Mell, "The NIST Definition of Cloud Computing", *Tech. Report TR -80--145, National Institute of Standards and Technology,* 2011.

[11]   H. Wang, W. He, and F.K. Wang, "Enterprise cloud service architectures", *Inf. Technol. Manage.,* vol. 13, no. 4, pp. 445-454, 2012.
        [http://dx.doi.org/10.1007/s10799-012-0139-4]

[12]   Y. Duan, G. Fu, N. Zhou, X. Sun, N.C. Narendra, and B. Hu, "Everything as a Service (XaaS) on the Cloud: Origins, Current and Future Trends", *, 2015 IEEE 8th International Conference on Cloud Computing,,* 2015, pp. 621-628
        [http://dx.doi.org/10.1109/CLOUD.2015.88]

[13]   R. Du, and K. Xu, "Multiattribute Evaluation Model Based on the KSP Algorithm for Edge

Computing", *IEEE Access,* vol. 8, pp. 146932-146943, 2020.
[http://dx.doi.org/10.1109/ACCESS.2020.3015041]

[14]   X. Song, L. Liu, C. Li, and Z. Li, "A CRM-oriented model for evaluating intermediary", *2014 IEEE 7th Joint International Information Technology and Artificial Intelligence Conference,* 2014, pp. 497-502
[http://dx.doi.org/10.1109/ITAIC.2014.7065100]

[15]   G. Yadav, *WhiteHat Jr gears up to be India's fastest unicorn with Salesforce.*https://www.salesforce.com/in/customer-success-stories/whitehatjr/

[16]   "AppExchange", *AppExchange Features.*https://appexchange.salesforce.com

[17]   https:// www.macrotrends. net / stocks/charts/CRM/salesforce,-inc/financial-statements

[18]   C. Thompson, *Record Fourth Quarter And Full Year Fiscal 2019 Results CRM.*https://www.masonfrank.com/salesforce-blog/salesforce-q4-2019-financial-results/

[19]   https:// www.sec.gov/Archives/edgar/data/1108524/000110852416000053/crm-2016131x10k.htm

[20]   https://www.gartner.com/en/newsroom/press-releases/2019-06-17-gartner-says-worldwide-c-stomer-experience-and-relati [Accessed: 18th Januaary 2021]

CHAPTER 58

Using Renewable Electricity, Energy Storage, and Renewable Fuel to Produce Carbon-Neutral Process Heat and Power

Rhys Jacob[1,*], Martin Belusko[1,2], Shane Sheoran[2] and Frank Bruno[2]

[1] *Mondial Advisory, Malvern, SA 5061, Australia*

[2] *Future Industries Institute, University of South Australia, Mawson Lakes, SA 5095, Australia*

Abstract: In the pursuit of lower and more stable energy expenses coupled with lower carbon emissions, the industry is moving towards the integration of low-cost renewables. This can be achieved through a range of technologies and/or mechanisms, each with its own set of advantages and disadvantages. One potential solution to achieve low-cost, predictable, and low-carbon process heat and power is the coupling of renewable electricity with thermal energy storage backed up by renewable fuel. This combination takes advantage of the low-cost but variable renewable generation coupled with the dispatchability of thermal energy storage with robustness added by utilizing a higher cost but low volume renewable fuel. With this approach, a lower overall cost, dispatchable, robust, and carbon-neutral solution to process heat and power is realized. Therefore, in the current study, several combinations of renewable technologies coupled with electrically charged thermal storage (ECTES) and renewable fuel were designed and simulated to meet the process heat needs of two hypothetical scenarios. The solutions were then evaluated based on their economic (levelised cost of heat and power) and environmental (carbon emission avoided) merits to determine feasible solutions. Of the two studied cases, replacing diesel for heating at a mine site with solar PV, wind, ECTES, and renewable diesel delivered significant financial and environmental benefits. Unfortunately, at today's prices replacing natural gas heating with solar PV, ECTES, and biomethane is not economically feasible without further cost reductions and/or a carbon tax.

Keywords: Biofuels, Carbon-Neutral, Process Heat, Renewable Integration, Thermal Storage.

1. INTRODUCTION

Traditionally, commercial and industrial (C&I) heat has been generated using

---

* **Corresponding author Rhys Jacob:** Mondial Advisory, Malvern, SA 5061, Australia;
E-mail: rhys@mondiala dvisory.com

Dharam Buddhi, Rajesh Singh and Anita Gehlot (Eds.)

low-cost gas or coal. While the cost of coal continues to be erratic but largely unchanged for the last decade [1], companies are becoming increasingly aware of their carbon footprint and are starting to move away from coal. While a large number of companies were already using natural gas, the move to lower-carbon gas (over coal) encouraged more uptake. This was previously not an issue as domestic energy prices were low (around $5/GJ), however, since 2016 the domestic price of pipeline gas has doubled or tripled from this cost [2], while off-grid users are paying 3-5 times that cost in the form of LPG and/or diesel, putting pressure on companies' margins. Additionally, while gas as a heat source has a lower carbon intensity when compared to coal, it remains significantly higher than what is required to meet the Paris Accord targets. This has led to a growing number of consumer/end user calls for environmental accountability. For example, Woolworths (Australia) has announced their plan to go 100% green electricity by 2025 while requiring 100% of their 'own brand' products to be sustainable [3]. Similarly, the New South Wales government in Australia has recently launched a $750M fund to help decarbonise the process industry [4] highlighting the importance of decarbonising this sector.

Conversely, during this same time, the cost of generation from renewables (with limited carbon intensity) has sharply declined. This paradigm has led the way for companies to seek renewable electricity generation as a way of reducing their energy costs and carbon intensity. However, most renewable generation is variable; something most C&I processes are ill-adapted to cope with. Therefore, storage is needed to fill in these gaps. One such option is to use cheap renewable electricity to generate heat and store it thermally to be used when needed. There are a growing number of companies [5, 6] and demonstrations [7 - 9] highlighting the potential of electrically charged thermal energy storage (ECTES) as a means to provide heat and/or commercial and industrial power. These systems are designed to use excess and/or curtailed renewable electricity (and in some cases waste heat) to generate heat and store it thermally. When required the heat can be recovered and used directly and/or reconverted to electricity. These systems are ideally suited to process heat applications as long storage and usage durations make these systems significantly cheaper than chemical storage. Additionally, these systems deliver heat as air or steam and can be integrated into existing equipment, further saving costs. A non-exhaustive list of ECTES companies and demonstrations is given below highlighting the expanding market for ECTES (Table **1**).

Table 1. Planned or demonstrated ECTES technologies.

Company	System Configuration	Type	Media	Region
Siemens	Packed Bed	Sensible	Volcanic rock	Europe

Company	System Configuration	Type	Media	Region
RWE	2-tank	Sensible	Molten Salt	Europe
SaltX	Reactor	Thermochemical	Calcium Oxide	Europe
Lumenion	Packed Bed	Sensible	Steel	Europe
Storasol	Packed Bed	Sensible	Rocks	Europe
Malta	2-tank	Sensible	Molten Salt	North America
Brenmiller Energy	Coil-in-tank	Sensible	Crushed rocks	North America
EPRI	Coil-in-tank	Sensible	Concrete	North America
1414	Coil-in-tank	Latent	Silicon	Australia
CCT	Coil-in-tank	Latent	Silicon	Australia
Seas-NVE	Coil-in-tank	Sensible	Rocks	Europe
Energy Nest	Coil-in-tank	Sensible	Heatcrete	Europe
Eco-tech	Packed Bed	Sensible	Ceramic/waste	Europe
Azelio	Coil-in-tank	Latent	Metal PCM	Europe
Enesoon	2-tank	Sensible	Molten Salt	Asia
Moltex	Coil-in-tank	Sensible	Molten Salt	Europe
MGA	Coil-in-tank	Latent	Miscibility gap alloys	Australia
Graphite Energy/Solastor	Coil-in-tank	Sensible	Graphite	Australia
Kraftblock	Coil-in-tank/ Packed Bed	Sensible	Ceramic	Europe
Pebble Heater	Packed Bed	Sensible	Rocks	Europe
Team Solid	Iron Powder Reactor	Thermochemical	Iron Oxide	Europe

However, while it has been noted that variable renewable generation coupled with storage can have significant economic and environmental benefits, the amount of over-capitalization of generation and/or storage prevents this approach from effectively reaching 100% carbon neutrality and therefore a green fuel backup is required. For example [10], identified the least cost 100% renewable electricity grid for Australia resulted in 66% of the energy provided by variable renewable energy followed by 27% dispatchable renewable energy with the remaining 6% delivered using a renewable fuel. This result is consistent with recent 100% renewable energy studies for Germany such as those by [11] and [12] which focussed on the oversupply of variable renewable generation firmed up by energy storage with green fuel as the backup. Regarding green fuels, there are several options available such as green hydrogen, green diesel, and green methane. Green diesel or hydrotreated vegetable oil (HVO) diesel is made from vegetable oils, used cooking oil, or animal fats and is a direct drop-in diesel [13]. It possesses a longer shelf life than regular diesel, making it a potentially low-risk option, however, attracts prices around \$74/GJ. Alternatively, green hydrogen from renewably driven electrolysis has been attracting significant attention and will likely become a significant fuel source in the future. While the growing market and improvements will reduce the cost from around \$50/GJ today to \$10/GJ in 2050 [14], using 100% hydrogen will require significant upgrades to existing infrastructure and equipment. Therefore, the blending of renewable hydrogen with natural gas (up to 30%) may provide a useful compromise in the interim [15]. Lastly, green methane can be made from methanated green hydrogen and/or upgrading biogas. Like hydrogen, green methane has the advantage of being able to be gas grid injected essentially without concentration limits, however, is limited by supply. A recent report by Carlu *et al.* [16] highlighted that while the market for biomethane is large (500 PJ), this only represents 30% of current natural gas usage. In further support of this approach, most C&I companies already have significant fuel-based infrastructure, therefore, a 'drop-in' type green fuel makes best use of these potentially stranded assets. Additionally, these stand-alone fuel-based assets support the requirement of C&I processes to always be online providing a further layer of risk reduction.

Therefore, the current study aims to complement and extend existing studies investigating attaining 100% renewable process heat and power by economically and environmentally exploring the potential of converting existing C&I processes to be decarbonized using renewable generation, energy storage, and green fuel.

2. METHODOLOGY

The following section outlines the development of an in-house assessment model to determine the lowest cost of energy for C&I users. The technical, economical, and environmental assumptions are provided and summarised as well as the input weather and energy data as well as the baseline costs. As the assessed sites are assumed to have fully depreciated equipment, only operating costs are included with the suggested technology having to be integrated into existing infrastructure as shown in Fig. (1).

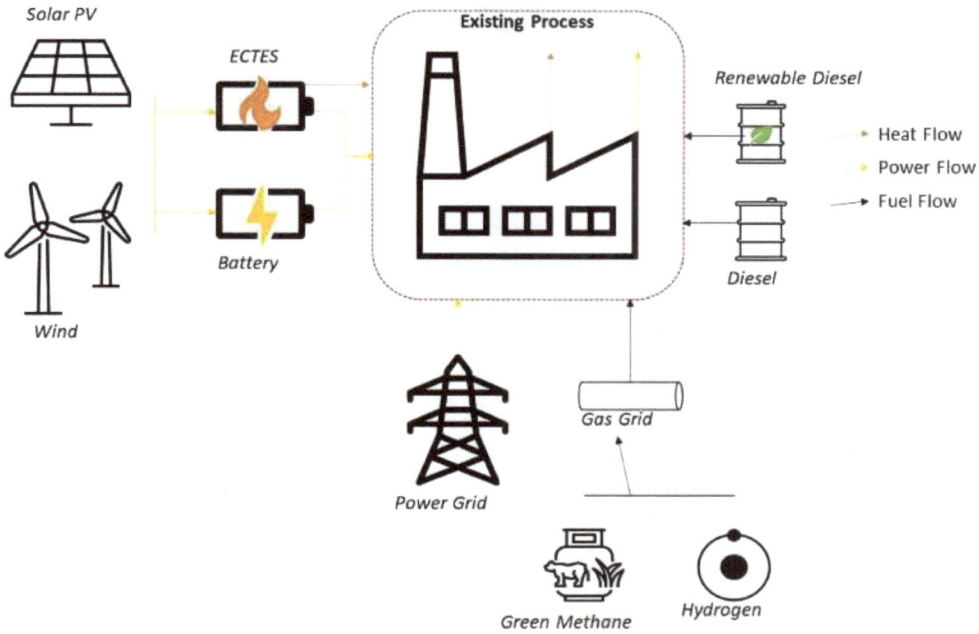

Fig. (1). Integration of renewable technologies for C &I customers.

2.1. Assessment Model

The performance of the renewable generation coupled with energy storage is simulated to provide energy for the process using PV, wind, and energy storage, with fuel as a backup. To determine the energy source for the process, the hourly generation of the renewable generation is estimated from the relevant weather data. The demand is initially met from the renewable generation converted to electricity and/or heat. If the generation exceeds demand the excess is sent to the store. If the generation is insufficient, the storage system is engaged. Finally, if the renewable generation and storage are insufficient to meet the demand, the fuel backup is used. The power and capacity of each of the technologies (*e.g.* solar, wind, energy storage, fuel) are determined by maximizing the net present value (NPV) of the site for a 30-year lifetime. Broadly, the model determines the 'optimal' technology mix resulting in a 'renewable energy hierarchy' as shown in Fig. (**2**).

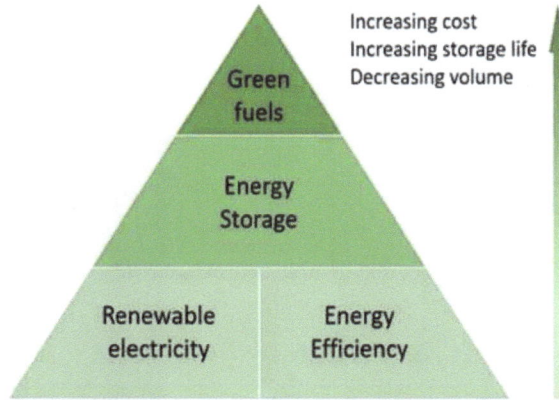

Fig. (2). Renewable energy hierarchy.

It should be noted that in addition to the aforementioned technologies, energy efficiency measures can reduce the demand, thereby reducing costs. This has no impact on the general trend of results but merely reduces the size of the 'renewable hierarchy' pyramid.

2.2. Cost and Performance of Assessed Technologies

The cost and performance of the technologies were determined using previous literature values [17] as well as engineering best practices and discussions with relevant personnel. The general assumptions of the study are described in Table **2**.

Table 2. General cost assumptions of current study.

Parameter	Value	Parameter	Value
Indirect Costs	19.7% Direct Costs	Contingency	10% Direct Costs
Inverter Cost	$300/kW$_e$	Discount Rate	20%
Loan Length	10 years	Interest Rate	4.0%
System Lifetime	30 years	Construction	2 years [Mine] 1 Year [Metro]

2.3. Cost and Performance of Renewable Generation

The cost and performance efficiencies used in the current study for the PV and wind systems are summarised in Table **3**. To determine the accurate performance of the wind turbine, technical and performance figures from the Vestas V162-5.6MW™ IEC S turbine [17] were used. The cost of the systems is based on work by [18], however, it should be noted that these values are general and will differ significantly from location and therefore are used as a guide only.

Table 3. Cost and performance of renewable generation.

Parameter	Value	Parameter	Value
PV Efficiency	15%	PV Lifetime	30 years
PV Cost	$1.52/$W_e$	PV O+M	$30/$kW_e$
Wind Turbine Cost	$1.975/$W_e$	Wind Turbine Height	166 m
Wind Turbine Lifetime	30 years	Wind Turbine O+M	$45/$kW_e$
Electrical-to-Heat Conversion Efficiency	98%	Electric-to-Heat Conversion	$0.23/$W_e$

2.4. Cost and Performance of Energy Storage

The type and cost of the energy storage are very dependent on the requirement of process heat fluid. In general, systems which use/generate steam (*e.g.*, SaltX, coil-in-tank, 2-tank molten salt) are better suited to process steam applications while air-based systems are better suited to process air (*e.g.*, Team Solid, packed bed). Similarly, modular systems are better suited to smaller applications while tank-based systems are suited for large applications. However, as this is a preliminary study, general cost and technical assumptions were applied with the cost of useful thermal storage set at $40/$kWh_t$[1] and useful life of 30 years. The O&M costs for the energy storage systems are 1% of the CAPEX.

2.5. Cost and Performance of Fuel Backup

The type and cost of fuel differ from the site; however, the performance of the backup boiler is assumed the same regardless of fuel. In the current study, two fuels and costs are considered for backup: diesel ($55/GJ) and natural gas ($15/GJ). The cost of these fuels is assumed to include all relevant costs such as contacting, excises, transport, on-site storage, *etc.* Additionally, only 'drop-in' fuels are considered to take advantage of existing infrastructure, therefore the 'green' version of these fuels includes green diesel ($70/GJ) and green methane ($18/GJ). Hydrogen is also included as a blended fuel for $25/GJ. The cost of the backup system is assumed to be a sunk cost while the system has a thermal conversion efficiency of 80% with a minimum run of 10%.

3. WEATHER DATA

Typical Meteorological Year (TMY) solar and wind data from the stated locations

(Geraldton, WA, and Adelaide, SA) was employed to determine the useful energy produced by renewable generation. The wind speed and global horizontal index (GHI) for Geraldton, WA is shown in Fig. (**3**) while the GHI of Adelaide, SA is shown in Fig. (**4**).

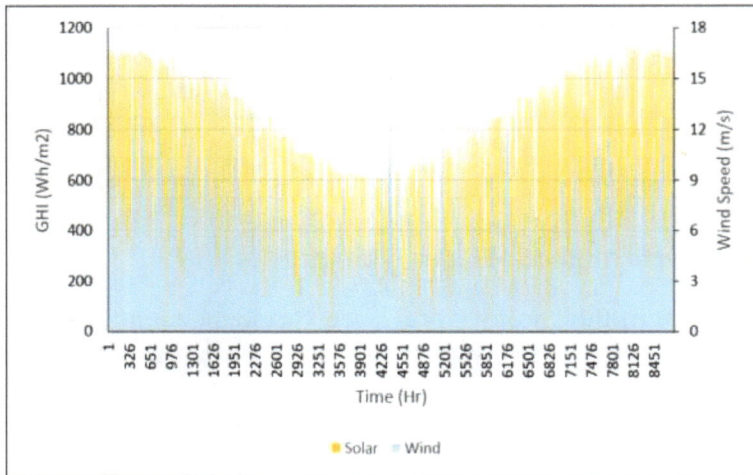

Fig. (3). Wind and Solar TMY Data for Geraldton, WA.

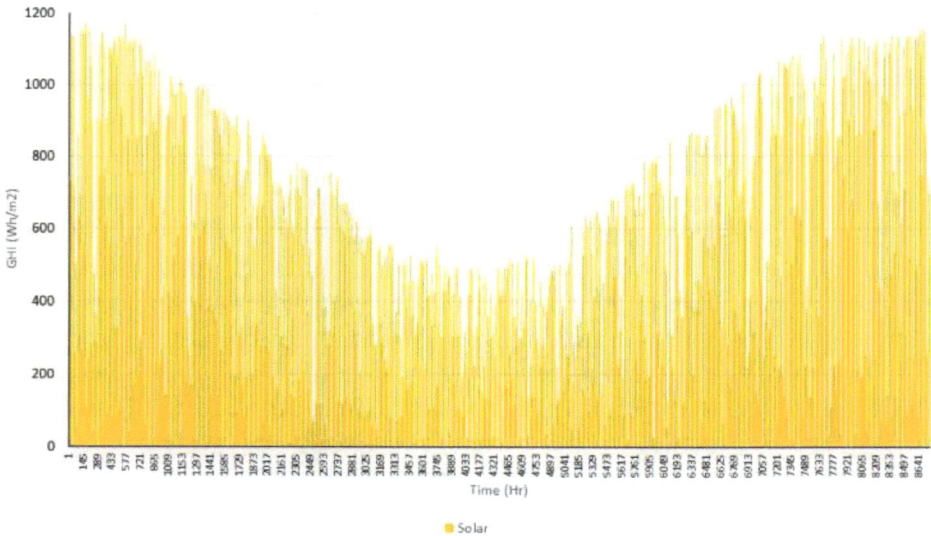

Fig. (4). Solar TMY data for Adelaide, SA.

3.1. Environmental Impact Assessment

The environmental impact of the assessed fuels and the power grid is determined using the Scope 1 and 3 emissions published in [13, 19, 20] and are summarised in Table **4**.

Table 4. Carbon intensities of fuel and power.

Parameter	Carbon Intensity	Parameter	Carbon Intensity
Diesel Fuel	73.8 kg $CO_{2\text{-}e}$/GJ	Natural Gas-SA Metro Grid	62.2 kg $CO_{2\text{-}e}$/GJ
Biomethane [upgraded municipal waste]	13.3 kg $CO_{2\text{-}e}$/GJ	Renewable Diesel	7.4 kg $CO_{2\text{-}e}$/GJ
Blended Hydrogen	43.6 kg $CO_{2\text{-}e}$/GJ	Blended Hydrogen and Biomethane	9.3 kg $CO_{2\text{-}e}$/GJ

For the current study, the operational carbon intensity of renewable generation and storage is assumed to be negligible and is ignored. Additionally, hydrogen is assumed to be produced from renewable energy and therefore also does not contribute to carbon emissions.

3.2. Assessed Sites

To assess the feasibility of converting process heat and power to 100% renewables, two hypothetical sites and loads were selected. For the current study the major markets assessed were:

- Off-grid mining
- On-grid manufacturing

These markets and operational data were based on previous conversations with businesses in these sectors and are therefore general. To maximize applicability, and declassify businesses, load data was assessed and averaged for the chosen market on an hourly basis throughout the year. Due to this, a more detailed study should be undertaken where promising commercial *via*bility exists. Nevertheless, the approach of the current study is valid for the intended purpose of highlighting the feasibility of renewable process heat.

3.3. Off-grid Mining

For the off-grid mining case, a mine site with a continuous load of 10 MW$_t$

located in Geraldton, WA, was assessed. This system was assumed to operate 365 days a year. In the current study, only the thermal load is analyzed which is currently being supplied by a diesel boiler. For the current study, the thermal energy is required to perform ore-preheating and can therefore be delivered as process air. The annual energy usage is estimated to be 394.2 TJ with an operational (fuel) cost of $21.7M/yr.

3.4. On-grid Manufacturing

The other analyzed case is based at a manufacturing plant in Adelaide, South Australia. This site has a continuous load of 2 MW_t between 6 AM and 10 PM supplied by a gas supply contract. The thermal energy is required as steam and is currently met by a gas boiler. The annual energy usage is estimated to be 52.6 TJ with an operational (fuel) cost of $788,619/yr.

4. RESULTS AND DISCUSSION

The following discussion provides the results of the aforementioned analyses as well as highlighting some important findings and early conclusions.

4.1. Off-grid Mining

In this scenario, the electricity supplied from the solar PV and wind turbines is fed into an ECTES system to generate heat which can be stored and/or discharged. As this site is assumed to not be connected to the power or gas grid, all power must be generated onsite while the diesel is transported in. Due to the site's current diesel operating equipment coupled with a lack of grid connection, renewable diesel has been selected as the 'green' fuel.

Under the assumptions and constraints presented earlier, it was found that 19.8 MW_e and 5.0 MW_e of wind and solar, respectively, would be required. This is fed into a 24.25 MW_t/138.25 MWh_t ECTES system providing 13.5 hours of storage. Lastly, approximately 75.5 TJ of renewable diesel is required, representing approximately 20% of the energy supplied. The annual energy flow is shown in Fig. (**5**) while the thermal energy source is shown in Fig. (**6**).

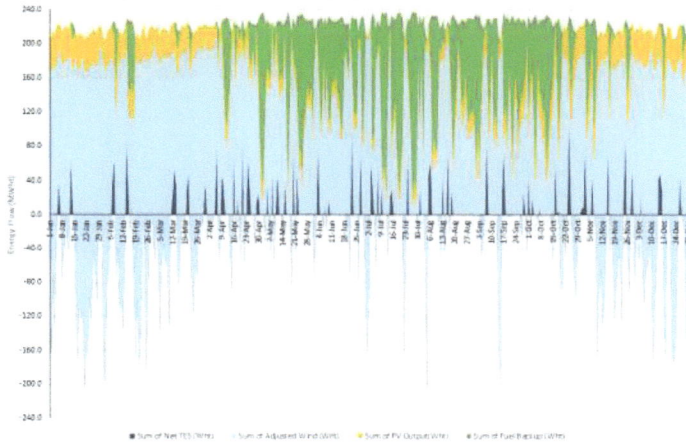

Fig. (5). Energy flow for an Off-grid mine site using renewables, energy storage.

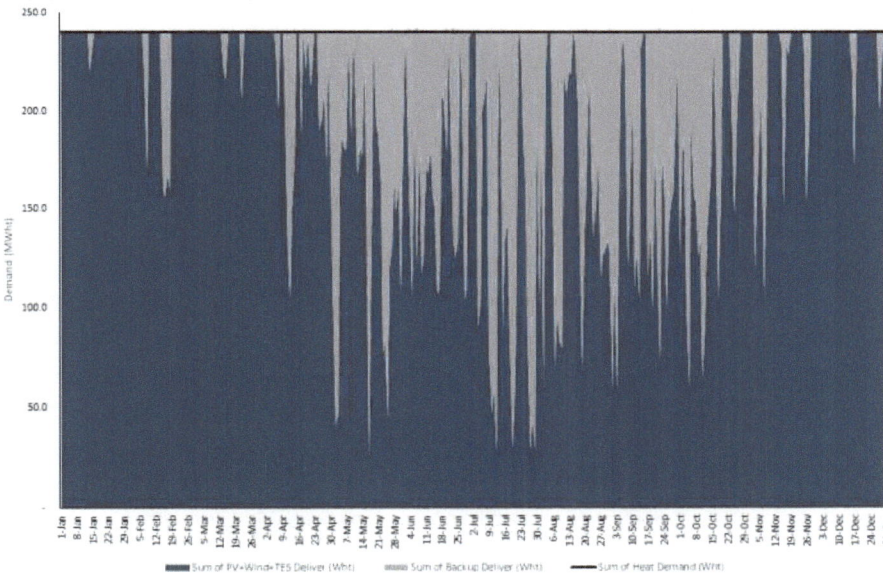

Fig. (6). Mine site thermal demand and energy source.

From Fig. (**5**) the majority of the energy is supplied from wind power with solar PV providing energy mainly in summer while the renewable diesel supplies energy mainly in winter. The ECTES is engaged throughout the year Fig. (**7**) but mainly through winter. However, due to the lower energy output of solar PV and wind throughout winter, it is not always able to completely charge.

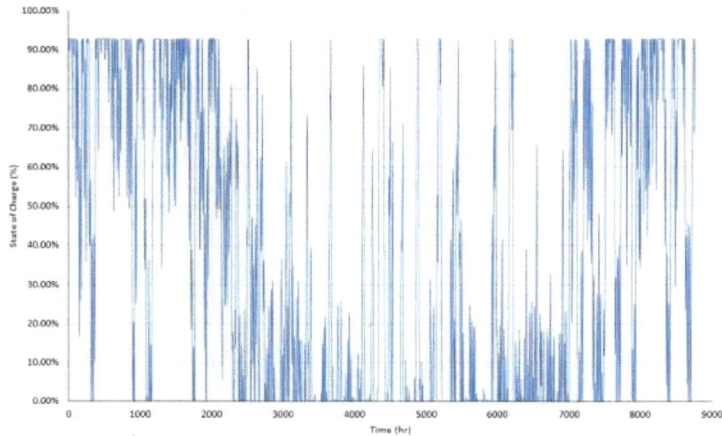

Fig. (7). ECTES annual state of charge.

The results are shown above highlight the value of all levels of the 'renewable' hierarchy in that the renewable generation does the majority of energy flow where possible while the energy storage can help daily fluctuations and shortfalls. Lastly, renewable fuel provides cover for when there is insufficient generation and storage rather than investing insignificant and severely underutilized renewable and storage overcapacity.

Financially, the site reduces operating expenses from \$21.7M/yr with diesel-only to \$15.5M/yr using renewables, energy storage, and renewable diesel. The simple payback for such a system is 11.2 years however, using loan-based repayments the site is cash positive in year 7 Fig. (**8**). The calculated NPV of the proposed system was determined to be \$18.1M with an internal rate of return (IRR) of 34.68%.

Finally, the proposed system is estimated to generate approximately 586.9 $t\text{-}CO_{2\text{-}e}$/yr in emissions which are significantly less than the 29,092 $t\text{-}CO_{2\text{-}e}$/yr under business-as-usual representing a 98% carbon reduction.

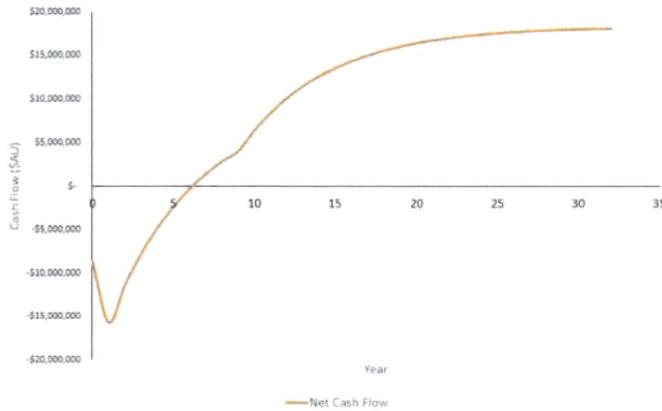

Fig. (8). Net cash flow for 100 % renewable mine site.

4.2. On-grid Manufacturing

In this scenario, a PV system is installed on the roof of the manufacturing plant to feed into an ECTES system which can generate steam for storage. The steam can then be discharged when required.

Under the assumptions of the current study, no economically *via*ble system could be designed (Fig. **9**). Despite deep cuts in emissions, the cost of capital outweighed any fuel savings due to the (relatively) low cost of thermal energy from the natural gas system.

It was also discovered that even with generous cost reductions in solar PV prices or higher than usual natural gas prices, the system would still struggle to be economically competitive. Therefore, an alternative system in which the renewable electricity was converted to heat *via* a heat pump ($1/W_e$, COP=3) was studied. These systems showed significantly better economic performance suggesting that this option may be better suited to this scenario if the delivered temperature can meet the required steam pressure and temperature. Other cost reduction strategies that could improve project economics include importing low-cost/curtailed grid electricity through a pull-through contract, importing off-site renewable electricity from wind (*e.g.* through a power purchase agreement (PPA)), and/or including a carbon tax.

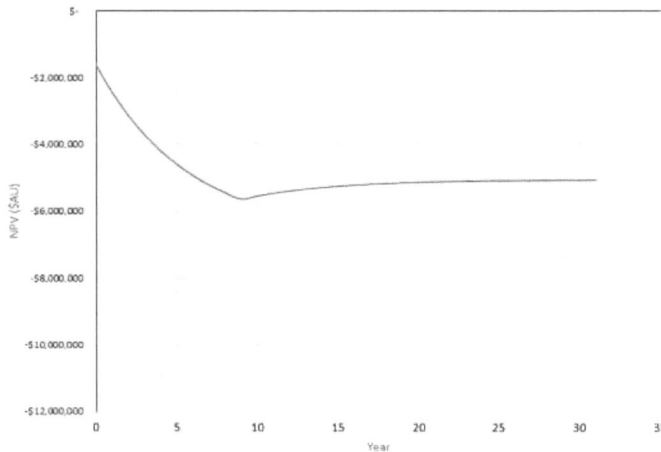

Fig. (9). NPV of solar PV, ECTES, and Green methane for manufacturing.

Lastly, it should be noted that in the current study there was no financial benefit in using 100% renewable blended hydrogen despite it having a slightly lower carbon impact. Given that for the studied case the fuel only represents 10% of energy and less than 2% of the original site emissions, this saving is not worthwhile (98% saving vs. 98.6% saving with blended H_2). This suggests that early adopters of biomethane will benefit more financially than those who use blended hydrogen and biomethane. Alternatively, blending the hydrogen with natural gas results in a similar price to biomethane ($17.5/GJ) but results in lower carbon emission reductions than biomethane (93.5% saving vs. 98.0%).

CONCLUSIONS

In the current study, two hypothetical sites were evaluated as to their potential in operating 100% renewably using renewable electricity generation, electrically charged thermal energy storage (ECTES), and a 'drop in' green fuel. Under the assumptions presented there was a strong financial and environmental case to deliver 100% renewable heat to mine sites using solar PV, wind, ECTES, and renewable diesel. Conversely, without significant cuts in solar PV price and the inclusion of a carbon tax, solar PV, ECTES, and biomethane struggle to financially compete with natural gas at today's prices in the manufacturing sector. While heat pumps present a better financial case, further cost decreases and cost pressure on emissions are required for significant deployment.

NOTES

[1] This is the cost of the system excluding integration and electrical charging equipment.

CONSENT FOR PUBLICATION

Not applicable.

CONFLICT OF INTEREST

The authors declare no conflict of interest, financial or otherwise.

ACKNOWLEDGEMENT

Declare none.

REFERENCES

[1]   Index Mundi, "Coal, Australian thermal coal Monthly Price-US Dollars per Metric Ton", 2020.

[2]   R. Jacob, M. Belusko, M. Liu, W. Saman, and F. Bruno, "Using thermal energy storage to replace natural gas in commercial/industrial applications",
        [http://dx.doi.org/10.1063/1.5067109]

[3]   A.H.F. Walmsley, *Sustainability Plan 2025.* Woolworths Group Australia, 2020, p. 22.

[4]   K. Fuller, "NSW puts $750 million on the table to help clean up big emitters", *ABC Illawarra,* 2021.

[5]   K. Erich, C. Dylan, and P. Alexander, *Powering our renewable energy future, MGA Thermal.,* 2021, .https://www.mgathermalstorage.com/about

[6]   E. Nest, *Thermal battery technology.,* p. 1, 2021, .https://energy-nest.com/technology

[7]   S. Gamesa, *Start of construction in Hamburg-Altenwerder: Siemens Gamesa to install FES heat-storage for wind energy.,* 2017 .https://www.siemensgamesa.com/en-int/newsroom/2017/11/start--f-construction-in-hamburg-altenwerder

[8]   A. Spence, "Renewable thermal solution provides green alternative for gas-hungry industries", https://reneweconomy.com.au/renewable-thermal-solution-provides-green-alternative-for-ga--hungry-industries-12919/

[9]   X. Salt, *Accelerating the energy transition with scalable and robust solutions.,* 2019 .https://saltxtechnology.com/references/

[10]   B. Elliston, I. Macgill, and M. Diesendorf, "Least cost 100% renewable electricity scenarios in the australian national electricity market", *Energy Policy,* vol. 59, pp. 270-282, 2013.
        [http://dx.doi.org/10.1016/j.enpol.2013.03.038]

[11]   T. Klaus, C. Vollmer, K. Werner, H. Lehmann, K. Müschen, R. Albert, M. Bade, T. Charissé, R. Herbener, and U. Kaulfersch, "Energy target 2050: 100% renewable electricity supply",

[12]   H-K. Bartholdsen, A. Eidens, K. Löffler, F. Seehaus, F. Wejda, T. Burandt, P-Y. Oei, C. Kemfert, and C.v. Hirschhausen, "Pathways for Germany's low-carbon energy transformation towards 2050", *Energies,* vol. 12, no. 15, p. 2988, 2010..
        [http://dx.doi.org/10.3390/en12152988]

[13]   https://www.neste.com/products/all-products/renewable-road-transport/neste-my-renewable-diesel

[14]   E. Taibi, H. Blanco, R. Miranda, and M. Carmo, *Green hydrogen cost reduction - Scaling up renewables to meet the 1.5°C climate goal.,* 2020 .https://www.aemo.com.au/-/media/Files/Gas/National_Planning_and_Forecasting/NGFR/2016/NGFR-Gas-Price-Review-Final-Report-October-2016.pdf

[15]   APGA, *Gas Vision 2050 - delivering a clean energy future.,* 2020 .https://www.apga.org.au/gas-vision-2050

[16]   E. Carlu, T. Truong, and M. Kundevsk, "Biogas opportunities for Australia", https://www.ieabioenergy.com/   wp-content/uploads/2019/07/ENEA-Biogas-Opportunities-for-Australia-March-2019_WebVersion-FINAL.pdf

[17]   https://nozebra.ipapercms.dk/Vestas/Communication/Productbrochure/enventus/enventus-produc--brochure/?page=12

[18]   P. Simshauser, and J. Gilmore, "On entry cost dynamics in Australia's National Electricity Market", *Energy J.,* vol. 41, no. 1, 2020.
[http://dx.doi.org/10.5547/01956574.41.1.psim]

[19]   S. Majer, K. Oehmichen, F. Kirchmeyr, S. Scheidl, and S. Proietti, *"Calculation of ghg emission caused by biomethane," Technical report.* BIOSURF, 2016.

[20]   G. Australia, *National Greenhouse accounts factors; Australian National Greenhouse Accounts,* 2020. https://www.industry.gov.au/sites/default/files/2020-10/national-greenhouse-accounts-facors-2020.pdf

CHAPTER 59

Techno-Economic Analysis of Thermal Energy Storage for Sensible and Latent Heat Systems as a Heat Source for CSP

Ming Liu[1,*], Soheila Riahi[1], Shane Sheoran[1], Frank Bruno[1], Rhys Jacob[2] and **Martin Belusko[2]**

[1] *Future Industries Institute, University of South Australia, Mawson Lakes, SA 5095, Australia*

[2] *Mondial Advisory, Malvern, SA 5061, Australia*

Abstract: With the advent of modern technologies in concentrated solar power(CSP), research focus has moved to cost reduction to make CSP a competitive alternative to the conventional heat source for power generation. One way to achieve this is by increasing the storage temperature so that size of thermal storage is lower which leads to reduced engineering and maintenance costs. To incorporate the CSP into power generation, one way to reduce the cost is to employ the higher temperature to improve the efficiency of the supercritical carbon dioxide Brayton power cycle. In the current study, a variety of phase change materials (PCMs) and graphite and their combination were assessed. For efficient heat transfer between PCM and heat transfer fluid, a shell and tube configuration was assessed as a suitable arrangement. The storage mediums can be contained in four indirect shell-and-tube configurations, including 3-PCM and 5-PCM cascade storage, PCM-graphite-PCM hybrid storage, and a single graphite storage tank. The sizing and design of the TES systems were performed by using a dynamic cycling methodology based on a transient 2D numerical model. The cost of these TES designs configurations was determined by using an economic model. This work also investigates the impact of some geometric parameters and cost assumptions on the techno-economic performance of the TES system. The analysis suggests a scenario exists whereby a low efficient storage system with less tube or lesser storage material could be more cost-effective. Overall, the cost of hybrid TES is the lowest among all studied systems, $26.96/kWh$_t$ and $21.49/kWh$_t$ for charging temperatures of 720 °C and 750 °C, respectively, followed by the 5-PCM storage of $28.06/kWh$_t$ and $21.82/kWh$_t$.

Keywords: Concentrated Solar Power, Design, Economic Analysis, latent heat, Sensible heat, PCM Cascade Storage, PCM Graphite Hybrid, TES Thermal Energy Storage.

---

[*] **Corresponding author Ming Liu:** Future Industries Institute, University of South Australia, Mawson Lakes, SA 5095, Australia; E-mail: ming.Liu@unisa.edu.au

Dharam Buddhi, Rajesh Singh and Anita Gehlot (Eds.)

1. INTRODUCTION

With the current advancement in renewable energy, the electricity market is evolving towards a carbon-free generation. The authors expect that this trend will continue to gather pace. However, some jurisdictions in the world are becoming constrained by where and what type of technology can be employed due to system strength or capacity issues [1 - 3]. Researchers are assessing ways to ensure how renewable technologies can deliver future electricity on demand. One such technology that can meet these requirements is concentrated solar power (CSP) with thermal energy storage (TES). However, CSP had limited success so far in fulfilling its potential as an on-demand energy producer [4]. CSP plant and cost per kWh are still seen as the biggest hurdle relative to solar photovoltaics and wind [5].

To make CSP competitive, its balance of plant, materials cost needs to be competitive. One way to reduce the cost of CSP is by higher temperature which helps to reduce the quantity of material and components sizes. This is especially prevalent for the receiver and two-tank sensible TES as the traditional heat transfer fluid (HTF) and storage material is a nitrate-based molten salt (60/40 wt% $NaNO_3/KNO_3$) which is limited to use in applications below 600°C [6, 7]. To take advantage of higher efficiency, the desired temperature of 700°C and above should be targeted [8, 9] as it will help to make the system more efficient. All this requires low-cost supercritical carbon dioxide (sCO_2) turbines, a new HTF, storage medium, and storage technology. One of the suitable high-temperature HTF fluids is liquid sodium. Liquid sodium has been heavily studied in past and a vast amount of research and data is available which may help in rapid development. The liquid sodium offers a low freezing point, high thermal stability, and high thermal conductivity, these properties can improve the performance and reduces the cost of the CSP receiver [10, 11]. Meanwhile, alternate high-temperature storage materials include solid and liquid sensible materials [12, 13], latent heat/phase change materials (PCMs) [14 - 19], or thermochemical materials [20, 21]. The techno-economic challenges associated with different TES technologies for use in CSP were addressed by Stekli, *et al.* [22]. This study focuses on sensible and latent TES technologies, which are currently viewed as the most deployable technologies.

The thermal performance and efficiency of the TES system are critical to determining the cost of the whole CSP plant. The PCM cascade system can achieve higher energy and exergy efficiencies compared to the single PCM system [23 - 25]. Mostafavi Tehrani, *et al.* [26] demonstrated that the hybrid storage of PCM-concrete-PCM provided the highest annual electricity output among the forty-five TES designs in a shell-and-tube configuration. Similar latent-sensible hybrid storage was studied by Jacob, *et al.* [27] in a high-temperature process heat application and hybrid storage was found to achieve the lowest storage cost. Both PCM cascade and latent-sensible hybrid TES were investigated in this study and graphite is of great interest as the sensible storage material due to its superior thermal conductive characteristics. A conventional and most-studied shell-and-tube configuration [28] was selected with PCM or graphite filling in the shell space.

For TES to be successfully adapted to CSP technology, TES must provide a technically viable and lower-cost economic solution. The TES capital cost of the current two-tank sensible system is $27/kWh$_t$ [29]. However, the U.S Department of Energy SunShot targeted cost for TES is $15/kWh$_t$ with a minimum discharge period of 6 h. Nithyanandam and Pitchumani [30] numerically evaluated the cost and performance of a CSP power tower with integrated encapsulated PCM packed-bed, PCM embedded with heat pipes, and two-tank sensible TES system. The effects of several storage system design parameters on the various performance metrics were analyzed and feasible operating regimes and design conditions were identified to meet the SunShot target of $15/kWh$_t$. The study of two-PCM cascade storage systems consists of 75 vol.% of high melting-temperature PCM and 25 vol.% of low melting-temperature PCM connected in series. Except for the height/length of the tank, all the other design parameters in both PCM systems are the same. It is believed by the authors in this study that there is still potential to optimize this cascade storage system. However, due to a large number of parameters, the design and optimization of cascade PCM and latent-sensible hybrid TES systems for CSP are more complicated than that of a direct sensible TES system.

It has been addressed in several studies [30 - 32] that the performance of a TES system needs to be evaluated annually in a CSP system model including power block and solar receiver considering the local climate conditions. The net annual electricity production is critical in examining the cost and efficiency of the CSP plant with TES at a geological location. However, the methodology of optimizing the cascade PCM and latent-sensible hybrid TES in system-level modeling has not been established. It is primarily due to a large number of parameters and the extremely intensive computing load required. Therefore, to improve the simulation speed, most previous system models employed a 1D model [31, 33] to predict the performance of the TES system. But it can be argued that the simplified 1D model has lower accuracy to capture the heat transfer in the PCM [34, 35].

This study suggests the design and optimization of the TES system can be carried out using a more detailed and accurate model and a simplified component model can be applied in the system model to predict the annual performance of the CSP plant. It was proposed [36] that the design and comparison of various TES systems can be achieved at the component level. Employing a validated 2D model, the methodology involves dynamically simulating 28 discharging-charging cycles at the design point and evaluating the design based on its performance at equilibrium conditions. It also addressed the necessity of economic analysis in design optimization and selection.

This work elaborates the design methodology and verifies the equilibrium condition by varying both the initial simulation condition and the cut-off condition in the discharging process. Four shell-and-tube TES configurations of the 3-PCM cascade, 5-PCM cascade, PCM-graphite-PCM hybrid, and single graphite were investigated and they were sized for a next-generation CSP plant under two temperature boundaries. An economic model was developed and employed to estimate the cost of the investigated TES systems. This study also investigated the impact of the tube spacing, amount of hot-end PCM, tube size, and cost assumption for the perspective of high-cost reduction potential. It is hoped that the design parameters obtained from this study will be used in the future system-level simulation to evaluate the benefit of integrating the investigated TES into the CSP plant.

2. METHODOLOGY

2.1. Numerical Modeling

A two-dimensional numerical model was developed to simulate the heat transfer performance of a shell-and-tube TES system as shown in Fig. (**1**). The system consists of multiple parallel tubes with an identical space between any two neighboring tubes and the tubes are contained inside a cylindrical tank. The HTF flows inside the tubes and the latent/sensible storage medium is placed in the shell, transferring energy through tube walls. The enthalpy method [37] was employed to formulate the energy conservation equation in the PCM region and the numerical model was validated by using the experimental data in Tay, *et al.* [38]. To further prove the validity of the model at high-temperature operating conditions with liquid sodium as the HTF, both the single PCM model and two-PCM cascade model were verified by using ANSYS FLUENT 17 and the verification results were reported in Liu, *et al.* [36]. By applying a hypothetically high melting temperature on the PCM (*e.g.* 10^4), the model can be extended to simulate the sensible energy storage system. The developed model allows the

HTF to enter into the TES system (single or cascade) in opposite directions for charging and discharging to maximize the heat transfer.

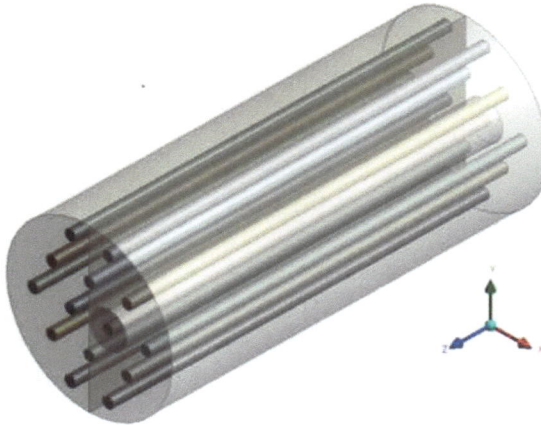

Fig. (1). Schematic diagram of a shell-and-tube TES system [36].

2.2. Methodology for Sizing Thermal Energy Storage Systems

The combination of the 3-PCM cascade, 5-PCM cascade, PCM-graphite-PCM hybrid, and single graphite as described in Liu, *et al.* [36] are further investigated in this work. Considering the availability of solar radiation, the charging process is limited to 6 h from 10 a.m. to 4 p.m. The discharging process starts directly after the charging and the power block only operates during the discharging. The storage system requirements are to provide a minimum of 1000 MWh thermal energy during 10 h of discharging process. The cascade and hybrid storage configurations and HTF flow direction are shown in Fig. (**2**). The charging and discharging cutoff temperatures are 580 °C and 700 °C, respectively. The detailed temperature boundaries were explained in Liu, *et al.* [36].

Fig. (2). Cascade storage configurations and HTF flow direction [36].

All the storage systems were sized at the hot liquid sodium temperatures of 720 °C and 750 °C, respectively, following the procedure illustrated in Fig. (**3**). All the PCMs considered in this study were experimentally proved and their thermo-physical properties are described in Liu, *et al.* [36]. The results are listed in Tables **1** and **2**, respectively. Firstly, the required amount of storage material is estimated to achieve the storage capacity of 1000 MWh$_t$, assuming (1) all the PCMs have the same equal storage capacity and the storage effectiveness (ratio of usable capacity to maximum storable capacity) is 50% and (2) graphite storage has an effectiveness of 70%. Stainless steel pipe DN 10 Sch 40S (OD: 17.51 mm; ID: 12.53 mm) is assumed for all the storage systems in the preliminary design. Using the effectiveness-number of transfer units (ε-NTU) method [39, 40] as a guideline, preliminary sizing is conducted to identify the other design parameters, such as the tube spacing, number of tubes, and tube length. A tube spacing of 60 mm is selected to give reasonable effectiveness for all the PCM storage. A tube spacing of 120 mm is applied on the graphite storage due to its high thermal conductivity. A parametric study of tube dimension and tube spacing has been carried out in this study and the results are presented later.

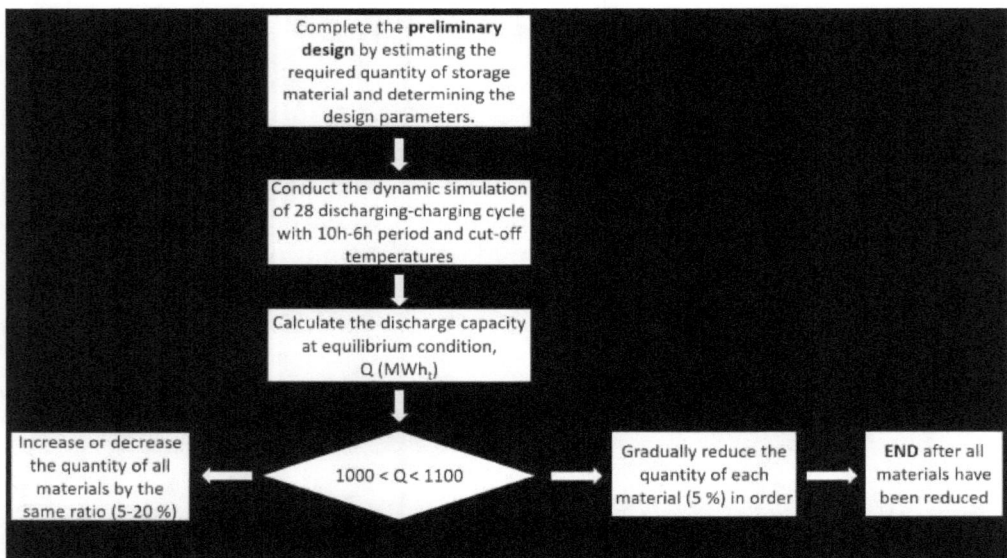

Fig. (3). Flow chart of the design and sizing procedure.

Table 1. Material costs in the current study.

Material	Cost ($US)	Material	Cost ($US)	Material	Cost ($US)
HTF	$2/kg	*PCM710*	$0.51/kg	*PCM662*	$0.19/kg
PCM635	$0.12/kg	*PCM597*	$0.10/kg	PCM569	$0.13/kg

(Table 1) cont.....

Material	Cost ($US)	Material	Cost ($US)	Material	Cost ($US)
Graphite	$0.7/kg	Vessel Material	$3.5/kg	*Tube Material*	$3.5/kg
External Insulation	$256/m^2	*Foundation and Footings*	$1,320/m^2	Installation	30% Material Cost

Table 2. Simulation results when the hot inlet sodium temperature is 720 °C.

Design No.	Storage configuration	Storage material	Required quantity (tonnes)	Discharge capacity (MWh)	The energy density (KWh/tonne)	Storage Effectiveness	No. of tubes	Total Tube length (m)	Tank Size [r × L] (m × m)	No. Tanks
1.	3-PCM cascade	PCM705	8768	294	44.4	25.9%	8040	210	2.99 x 15	14
		PCM635	2869	223		46.0%		66.4	2.99 x 13.28	5
		PCM569	11273	501		29.5%		270	2.99 x 15.00	18
		total	22910	1018		30.6%		546.4		37
2.	5-PCM cascade	PCM705	3087	119	59.4	32.8%	8290	71.7	3.03 x 14.34	5
		PCM662	2770	119		32.1%		64.4	3.03 x 12.88	5
		PCM635	2368	192		51.9%		53.2	3.03 x 13.30	4
		PCM597	2947	225		61.0%		68.5	3.03 x 13.70	5
		PCM569	5941	360		43.5%		138	3.03 x 13.80	10
		total	17113	1014		44.1%		395.7		29
3.	PCM-graphite-PCM hybrid	PCM705	1546	43	60.9	24.0%	8135	36.6	3.01 x 12.20	3
		graphite	15152	972		80.2%		93.4	6.01 x 13.34	7
		PCM569	845	53		45.3%		20.0	3.01 x 10.00	2
		total	17543	1069		70.7%		150.0		12

(Table 2) cont.....

Design No.	Storage configuration	Storage material	Required quantity (tonnes)	Discharge capacity (MWh)	The energy density (KWh/tonne)	Storage Effectiveness	No. of tubes	Total Tube length (m)	Tank Size [r × L] (m × m)	No. Tanks
4.	Graphite	Graphite	22542	1065	47.3	59.0%	7905	143	5.93 x 14.30	10

After determining the preliminary design parameters, firstly the simulation is carried out by dynamically repeating a consecutive 28 discharging-charging cycles. The simulation commences from the discharging process with a fully charged system with a uniform temperature of 720 °C or 750 °C. During the repeating cycles, the discharging-charging process reaches an equilibrium condition, when the difference of state of charge at the end of both discharging and charging between the two consecutive cycles is less than 1% [36]. The discharge capacity is calculated at this equilibrium condition. Next, the volume of all storage media will be increased/decreased by 5-20% at the same ratio until the re-stimulated capacity falls in between 1000 MWh_t and 1100 MWh_t (coarse sizing). Then, the volume of each storage material will be reduced in order gradually until the volume of every PCM is optimized and eventually the design is completed.

2.3. Verification of Equilibrium Condition

In the current study, two cases were studied to verify the equilibrium condition achieved with the initially fully-charged system (T_{init} = 750 °C) for the 3-PCM cascade storage system (Design 1 in [36]). For case 1, the simulation starts with the initially fully-discharged system (T_{init} = 540 °C) and for case 2 the discharging period is extended beyond 10 h and it ends only until the cut-off temperature is reached. Figs. (**4** and **5**) show the HTF outlet temperatures, PCM average temperatures, and PCM liquid fractions (LF) overtime during a consecutive of 10 discharging-charging cycles for case 1 and case 2, respectively. Both case 1 and case 2 take longer to achieve the equilibrium condition than Design 1 [36] (10 cycles and 9 cycles *vs.* 5 cycles). However, the equilibrium conditions are identical for all three cases, which is demonstrated by the identical HTF outlet temperatures, PCM average temperatures, and PCM liquid fractions presented in Fig. (**6**). The same study was conducted on the 3-PCM cascade storage with hot liquid sodium temperature of 720 °C and the hybrid storage with both liquid sodium temperatures, and the equilibrium condition was verified in all the cases. This reveals the potential of operating the storage system in varying scenarios.

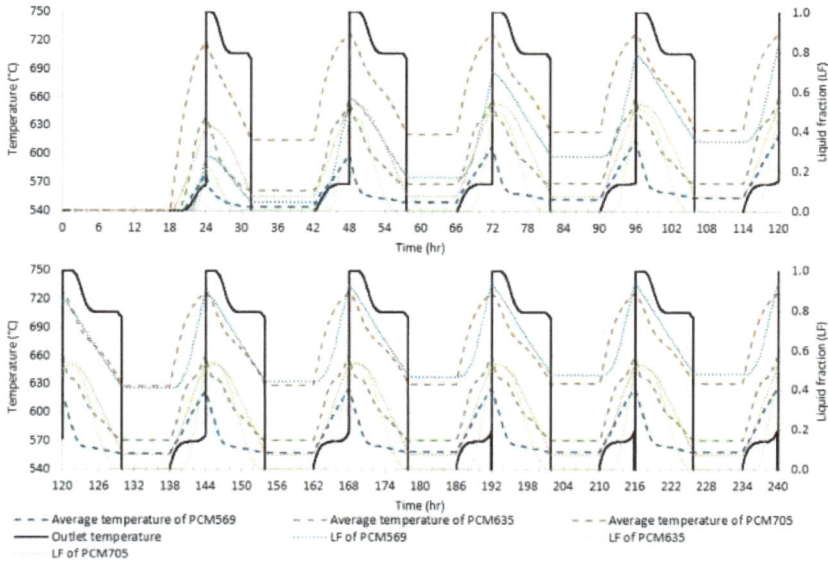

Fig. (4). HTF outlet temperature, PCM average temperature and LF over time for Case 1.

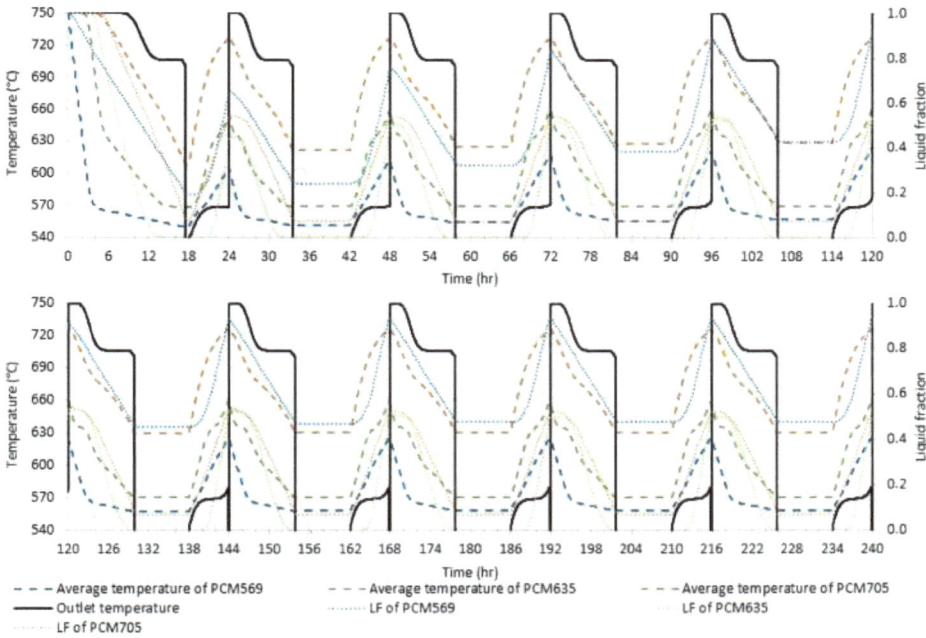

Fig. (5). HTF outlet temperature, PCM average temperature, and LF over time for Case 2.

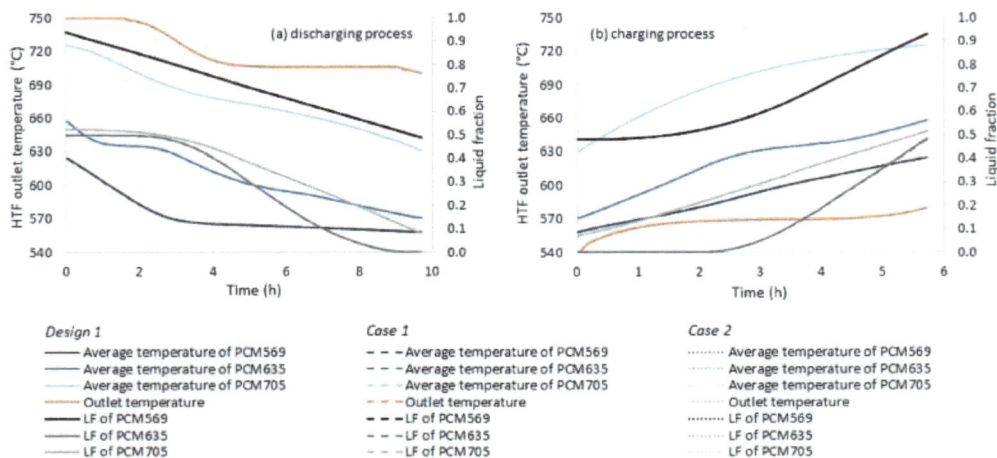

Fig. (6). HTF outlet temperature, PCM average temperature, and LF over time for Design 1 [36], Case 1, and Case 2.

2.4. Estimating the Cost of Thermal Energy Storage Systems

To investigate the economic impact of different designs in TES systems, an estimate of the cost of the studied designs was undertaken. The cost of the system was determined by estimating the material requirements for each design to meet the stated energy storage power and capacity. This methodology has previously been used to estimate the cost of encapsulated PCM storage systems [27, 41] and other PCM shell-and-tube systems [42]. Material costs and assumptions were assigned based on average bulk costs of raw materials from various vendors and/or literature values [41 - 44] and are summarized in Table **1**. It should be noted that these costs are bulk material costs averaged across a range of vendors and locations and may not reflect true system cost, but ultimately provide data from which comparison of systems can be made. To better determine how these cost assumptions may affect the conclusions of the current study, further analysis of several of the designs is given in Section 3.3.3. All costs have been converted to US_{2019}$ unless otherwise stated.

2.4.1. Storage Vessel Size Estimation

The simulation of the system assumes a continuous length of tube with the storage material on the outside. In reality, it is unfeasible for this to occur. As such, the length will need to be split across several tanks in series. In the current study, the maximum tank length was taken as 15 m. Therefore, the number of tanks ($\#_{tanks}$) could be determined as:

$$\#_{tanks} = \frac{L_{tank}}{x}; \quad 0 < x \leq 15 \text{ m} \tag{1}$$

where, L_{tank}, and x are the total length of one tube and the length of one tank, respectively. Subsequently, the tank radius can be calculated using Eq. (2):

$$r_{tank} = \frac{\sqrt{\frac{\#_{tubes} \cdot s_{tubes}^2}{4}}}{0.9} \tag{2}$$

where $\#_{tubes}$ and s_{tubes} relate to the number of tubes and the spacing of the tubes, respectively. The value of 0.9 is the assumed packing factor of the tubes.

2.4.2. Storage Vessel Material Requirement

Once the size of the TES vessels is calculated, the material required for the storage vessel can be estimated. To improve the accuracy of the cost estimation, the methodology used previously [42] has been modified to include the vessel header and tubesheet as shown in Fig. (7). As the current study is comparative, the shell, tubing, and tube sheet thickness are assumed fixed at 12 mm. This thickness was assumed to be sufficient to meet mechanical loading requirements which are small however, the mechanical validity of this assumption requires further analysis but is outside of the scope of the current study.

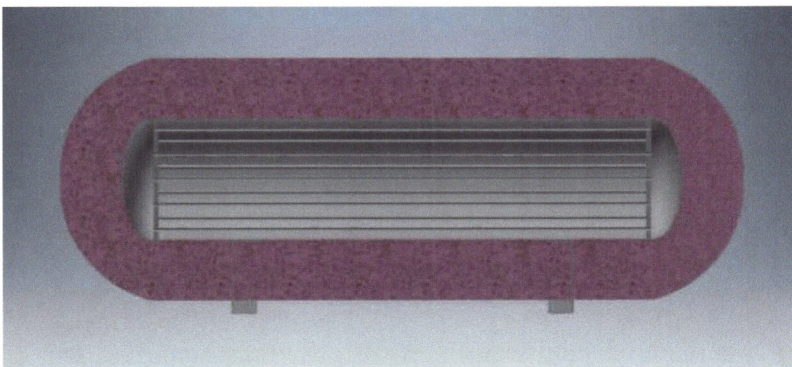

Fig. (7). Schematic of a shell-and-tube storage vessel showing the shell, tube sheet, header, insulation, and footings.

To estimate the steel required for such a vessel, Eq. (3) was used:

$$Steel\ Mass_{shell} = (\pi \cdot (r_{tank} + w_{tank})^2 \cdot L_{tank} - (\pi \cdot (r_{tank})^2 \cdot L_{tank})) \cdot \rho_{steel} \quad (3)$$

where w_{tank} and ρ_{steel} are the tank wall thickness (12 mm) and density of steel (7,900 kg/m³), respectively.

It is envisioned that for these types of systems, elliptical headers with a volume and steel requirement according to Eq. (4) will be utilized to reduce thermal stresses and ensure even flow through the tubes.

$$Steel\ Mass_{Header} = (\frac{\frac{4}{3}(d_{tank} + w_{tank})^3}{16} - \frac{\frac{4}{3}(d_{tank})^3}{16}) \cdot 1.05 \cdot \rho_{steel} \quad (4)$$

where d_{tank} is the diameter of the storage vessel.

The tube sheet material requirement can be estimated using Eq. (5) assuming a tube sheet thickness of 12 mm.

$$Steel\ Mass_{Tubesheet} =$$

$$(\pi \cdot (r_{tank} + w_{tubesheet})^2 \cdot w_{tank} - \pi \cdot \frac{S_{tubes}^2}{2} \cdot w_{tubesheet} \cdot \#_{tubes}) \cdot \rho_{steel} \quad (5)$$

The foundation area of the storage vessels was assumed to be a rectangular block with the dimensions of the width of the vessel ($d_{tank} + w_{tank}$) and length. The surface area of each storage vessel was determined using Eq. (6):

$$= 2 \cdot \pi \cdot (r_{tank} + w_{tank}) \cdot (L_{tank} + 2 \cdot (\frac{d_{tank}}{4} + w_{tank}) + 2 \cdot \pi \cdot (r_{tank} + w_{tank})^2 \quad (6)$$

During operation, there will be HTF in the header and tubes which is in addition to the HTF in the system. While the storage capacity of the stored HTF is ignored in the current study, the cost has been included to better reflect the actual cost. To estimate the mass of additional HTF required, the volume of the tubes and headers should be calculated (Eq. 7).

where the r_{tube} is the inner radius of the tubes (12.89 mm) and ρ_{sodium} is 782.5 kg/m³.

2.4.3. Storage Vessel Cost Estimate

In the current study, the purchased cost of each system was estimated as being the sum of the material requirements while the direct cost includes the cost of installation (30% of the purchased cost) [43]. While this 30% additional cost is intended to account for additional processing, manufacturing, *etc.*, no further allowances have been made to differ between manufacturing processes or complex issues that may arise.

3. RESULTS AND DISCUSSION

3.1. Sizing Thermal Energy Storage Systems

The results of sizing the TES system in four configurations with a hot liquid sodium temperature of 720 °C and 750 °C are listed in Tables **2** and **3**, respectively. The amount of storage material, volume of tube material, energy density, and storage effectiveness of the four configurations are compared and plotted in Fig. (**8**) using an unfilled column with blue borders. The TES system size for sodium inlet temperature at 750 °C drops significantly compared to the 720 °C inlet temperature. It is due to the increased temperature difference between the hot HTF and the melting temperature of the hot end storage material in the charging process, leading to an improved heat transfer performance. Also at 720 °C, the required storage material quantity decreased by 27.3%, 19.2%, 17.4%, and 27.3% for 3-PCM cascade, 5-PCM cascade, hybrid and single graphite design, respectively, while the volume of tube material presents a similar reduction for those four investigated designs (27.3%, 19.3%, 22.7%, and 27.3%, respectively). Accordingly, the energy density is dramatically improved by 39.6%, 31.7%, 18.2%, and 37.4% by increasing the hot HTF temperature from 720 °C to 750 °C. The storage effectiveness of the 3-PCM cascade, 5-PCM cascade, and single graphite design is increased as well, however, that of the hybrid design remains similar.

Table 3. Simulation results when the hot inlet sodium temperature is 750 °C.

Design No.	Storage configuration	Storage material	Required quantity (tonnes)	Discharge capacity (MWh)	The energy density (KWh/tonne)	Storage Effectiveness	No. of tubes	Total Tube length (m)	Tank Size [r × L] (m x m)	No. Tanks
1.	3-PCM cascade	PCM705	8768	535		47.1%		210	2.99 x 15.00	14
		PCM635	2869	222	62.0	45.7%	8040	66.4	2.99 x 13.28	5
		PCM569	5010	275		36.3%		120	2.99 x 15.00	8
		total	*16647*	*1031*		*43.4%*		*396.4*		*27*
2.	5-PCM cascade	PCM705	3087	198		49.7%		71.7	3.03 x 14.34	5
		PCM662	2770	196		49.1%		64.4	3.03 x 12.88	5
		PCM635	2368	259	78.2	64.8%	8290	53.2	3.03 x 13.30	4
		PCM597	2947	221		55.3%		68.5	3.03 x 13.70	5
		PCM569	2649	206		51.3%		61.5	3.03 x 12.30	5
		total	*13821*	*1081*		*54.0%*		*319.2*		*24*
3.	PCM-graphite-PCM hybrid	PCM705	760	47		24.9%		18	3.01 x 9.00	2
		graphite	12978	972	72.0	80.3%	8135	80	6.01 x 13.33	6
		PCM569	760	25		41.3%		18	3.01 x 9.00	2
		total	*14499*	*1044*		*73.3%*		*116*		*10*
4.	Graphite	Graphite	*16394*	*1065*	*65.0*	69.6%	7905	104	5.93 x 14.86	7

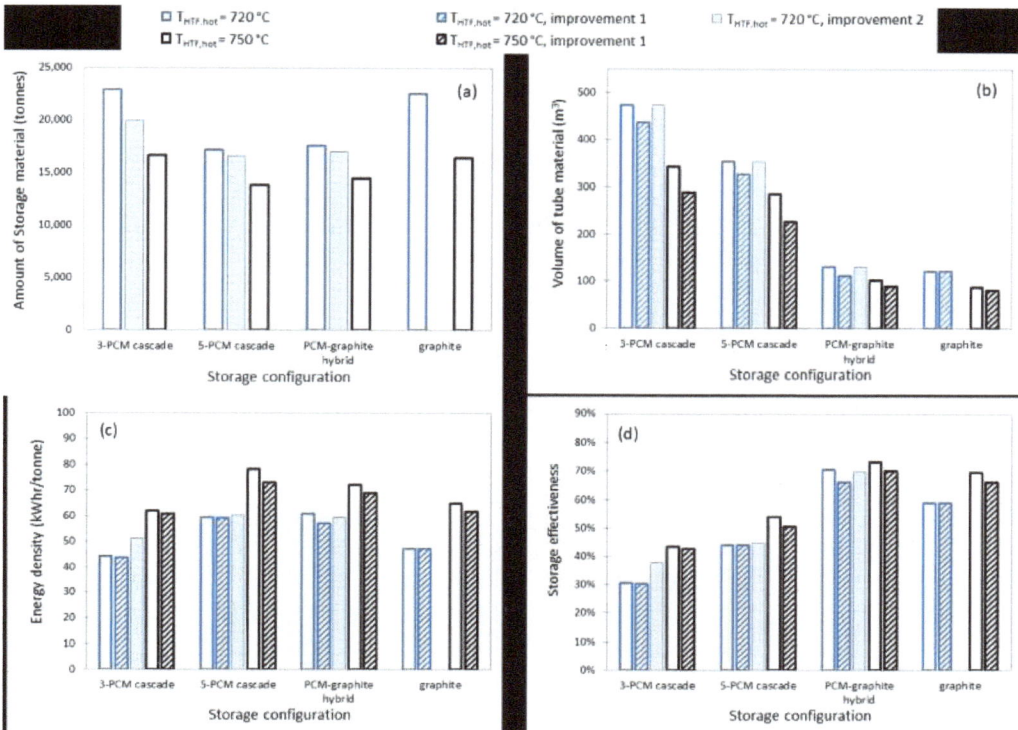

Fig. (8). Comparison of (a) amount of storage material, (b) volume of tube material, (c) energy density, and (d) storage effectiveness for four storage configurations under two design conditions and improvements 1 & 2.

It should be noted that both the required amount of PCM569 and tube material in the PCM569 storage tank is reduced by more than 55% for both PCM cascade designs when the hot HTF temperature rises from 720 °C to 750 °C. However, the other PCM tanks remain the same size, while most of them have higher storage effectiveness, meaning more useful energy is extracted. This result indicates that when the temperature difference between the hot HTF and melting temperature of the hot end PCM (PCM705) becomes larger, even though the storage effectiveness of the hot end PCM is improved, its size is not reduced as expected. However, the size of the cold end PCM (PCM569) is reduced. Therefore, the transient simulation on multiple charging-discharging cycles is very critical in designing a PCM cascade system.

Overall, the total amount of storage material required is for the 5-PCM cascade and hybrid design is 22.2% - 25.3% and 11.6% - 17.0% less than that of the single graphite and 3-PCM cascade design for hot HTF temperature of 720 °C and 750 °C, respectively. The amount of tube material used in hybrid and graphite design is similar, but it is dramatically less than that in the other two designs. Both 5-PCM cascade and hybrid designs present superior energy density of 59.4 kWh/ton and 60.9 kWh/ton, respectively when the hot HTF temperature is 720 °C. The 5-PCM cascade design achieves the highest thermal energy density of 78.2 kWh/ton when the hot HTF temperature increases to 750 °C, following by the hybrid of 72.0 kWh/ton, graphite of 65.0 kWh/ton, and the 3-PCM cascade of 62.0 kWh/ton, respectively. The hybrid and 3-PCM cascade have the highest and lowest storage effectiveness, respectively, regardless of the hot HTF temperature.

3.2. Cost of Thermal Energy Storage Systems

Using the storage system design requirements from Tables **2** and **3**, their cost and the cost breakdown of each system were determined and presented in Fig. (**9**). The 3-PCM cascade and hybrid configuration have the highest ($41.35/kWh$_t$ and $30.86/kWh$_t$) and lowest cost ($26.96/kWh$_t$ and $22.74/kWh$_t$) among all the investigated configurations for both hot HTF temperatures of 720 °C and 750 °C, respectively. At 750 °C, the cost difference among the 5-PCM cascade, hybrid, and single graphite design is very small. The cost reduction when applying higher HTF temperature is approximately 25% for PCM cascade and single graphite designs and 15.6% for hybrid design, respectively, compared to applying lower HTF temperature. The breakdown of the cost revealed that the tube material contributes to most of the cost (nearly one-third) for both the PCM cascade designs, followed by the storage material (11.4% - 17.2%). On the other hand, for the hybrid and single graphite designs, the cost of storage material accounts for the highest proportion of 40.0% - 45.4% followed by the cost of the tube material (9.8% - 12.9%).

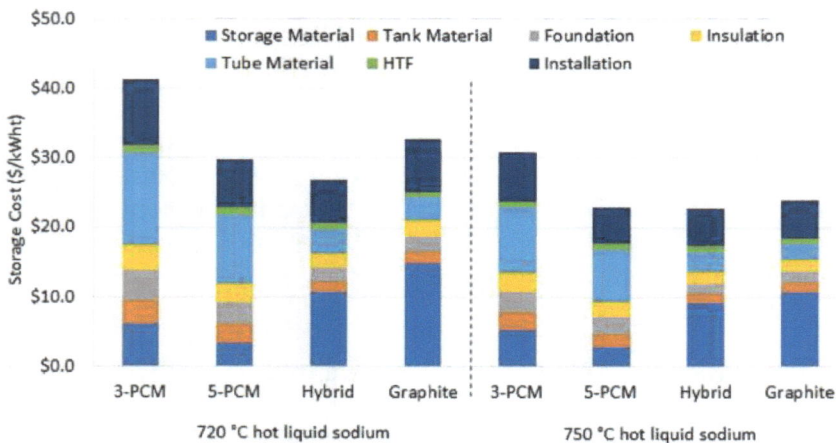

Fig. (9). Comparison of the studied TES system cost and the cost breakdown.

3.3. Parametric Analysis

3.3.1. Impact of TES System Size

The quantity of storage material and tube material is critical to determine the cost of the TES system. The parametric study was carried out to improve the design to reduce the amount of material, hence improving the techno-economic performance of the TES system. The first attempt is to decrease the tube length by increasing the tube spacing in each storage tank for the given amount of storage material and number of tubes (known as improvement 1). This study was conducted for HTD at both 720 °C and 750 °C. It starts from the cold end storage material and moves to the hot end material in order and the increment in tube spacing is 5 mm until the discharge capacity at equilibrium drops below 1000 MWh$_t$. It was noted from Table **2** that the storage effectiveness of PCM705 is quite low (24% - 32.8%) when the entering hot HTF temperature is 720 °C due to the small temperature difference. The hypothesis is that the amount of PCM705 can be lessened if its effectiveness is improved. The second trial is to reduce the tube spacing in PCM705 (while the amount of tube material is maintained the same) and hence to reduce the thermal resistance and improve the effectiveness (improvement 2). The decrease is 5 mm until the discharge capacity at equilibrium drops below 1000 MWh$_t$. It is also attempted to improve the effectiveness of all storage tanks using the same method, however, ending up with much less discharge capacity of around 700 MWh$_t$. Therefore, this approach is not pursued. The results of the above two case studies and the comparison with the results in Section 3.1 were plotted in Fig. (**8**).

The amount of tube material in all the configurations can be reduced by 7.5% - 20.6%, among which the 5-PCM cascade design in improvement 1 achieves the highest reduction when the hot HTF temperature is 750 °C. In the case of lower HTF temperature, it was found that (1) the reduction is lower and even no reduction is observed in the single graphite design; (2) the energy density and storage effectiveness marginally decrease in improvement 1.

For the 3-PCM cascade configuration, the required amount of PCM705 drops from 8768 tonnes to 5850 tonnes in improvement 2, contributing to a significant reduction of 12.7% in the total amount of storage material. Thus, the energy density and storage effectiveness increased to 51.2 kWh/tonne and 37.9%, respectively. In the designs of 5-PCM cascade and hybrid storage, the impact of PCM705 reduction is minimal due to its low proportion. When focusing on the PCM750 storage tank, its storage effectiveness is improved from 25.9% to 35.9%

and from 32.8% to 37.6% for 3-PCM cascade and 5-PCM cascade configuration, respectively.

3.3.2. Impact of Tube Size

The ultimate improvement of designing and sizing the TES system is the cost reduction, therefore, an economic analysis was conducted using the methodology aforementioned in Section 2.4. The lowest cost option from each storage configuration was selected and further attempted improvement (improvement 3) was conducted to investigate the impact of tube size on the cost and performance of the TES system. So far, the most cost-effective designs are improvement 1 for 3-PCM and 5-PCM cascade storage and original design for hybrid and graphite storage under hot HTF temperature of 750 °C and improvement 2 for 3-PCM cascade storage, improvement 1 for 5-PCM cascade storage and original design for hybrid and graphite storage under HTF temperature of 720 °C. One smaller tube size of DN 8 Sch 40S (OD: 13.72 mm; ID: 9.24 mm) and one larger size of DN 15 Sch 40S (OD: 21.34 mm; ID: 15.8 mm) were applied to the above designs with the same number of tubes and tube length and the same amount of storage material. The impact of the tube diameter on pumping power is ignored in the current study, due to it having a similar effect on all the investigated systems and only a marginal impact on the overall thermal power and system cost [45].

Replacing DN 10 tube with DN 15 tube increases the heat transfer area by 26.1% and the tube volume by 50%, which potentially increases the cost. However, the energy density and storage effectiveness are only marginally improved and hence this option was excluded. Both of the 3-PCM and 5-PCM cascade designs with a tube size of DN 8 could not achieve the discharge capacity of 1000 MWh$_t$. With a slight decrease in energy density and storage effectiveness, the tube size of DN 8 can be employed in both hybrid and graphite storage designs and they benefit from a 25% tube volume reduction. The improved designs for all the storage configurations and both hot HTF temperatures of 720 °C and 750 °C are listed in Table **4** and the storage system cost and storage effectiveness are presented in Fig. (**10**). The steel required for each design and inlet temperature is shown in Tables **5** and **6**.

Table 4. Cost-effective/improved design for all the storage configurations.

Storage configuration	Storage material	Hot HTF temperature = 720 °C							Hot HTF temperature = 750 °C						
		Required quantity (tonnes)	Tube length (m)	Tube spacing (mm) and size	The energy density (KWh/tonne)	Storage effective-ness	Tank Size [r x L] (m x m)	No. Tanks	Required quantity (tonnes)	Tube length (m)	Tube spacing (mm) and size	Energy density (KWh/tonne)	Storage effective-ness	Tank Size [r x L] (m x m)	No. Tanks
3-PCM cascade	PCM705	5850	210	50			2.49 x 15.00	14	8768	210	60			2.99 x 15.00	14
	PCM635	2869	66.4	60	51.2	37.9%	2.99 x 13.28	5	2869	66.4	60	60.9	42.8%	2.99 x 13.28	5
	PCM569	11273	270	60			2.99 x 14.19	16	5010	57.2	85			4.23 x 14.30	4
	total	*19992*	*546.4*	DN 10				*35*	*16647*	*333.6*	DN 10				*23*
5-PCM cascade	PCM705	3087	71.7	60			3.03 x 14.34	5	3087	71.7	60			3.03 x 14.34	5
	PCM662	2770	34.8	80			4.05 x 11.60	3	2770	30.7	85			4.30 x 10.23	3
	PCM635	2368	53.2	60	59.0	44.0%	3.03 x 13.30	4	2368	53.2	60	73.1	50.5%	3.03 x 13.30	4
	PCM597	2947	68.5	60			3.03 x 13.70	5	2947	68.5	60			3.03 x 13.70	5
	PCM569	5941	138	60			3.03 x 13.80	10	2649	29.4	85			4.30 x 14.70	2
	total	*17113*	*366.2*	DN 10				*27*	*13821*	*253.5*	DN 10				*19*
PCM-graphite-PCM hybrid	PCM705	1546	36.6	60			3.01 x 12.20	3	760	18	56.21			2.82 x 9.00	2
	graphite	15152	93.4	120	60.9	70.7%	6.01 x 13.34	7	12978	80	116.21	71.3	72.6%	5.82 x 13.33	6
	PCM569	845	20.0	60			3.01 x 10.00	2	760	18	56.21			2.82 x 9.00	2
	total	*17543*	*150*	DN 10				*12*	*14498*	*116*	DN 8				*10*
Graphite	Graphite	*22542*	*143*	120 DN 10	*47.3*	*59.0%*	*5.93 x 14.30*	*10*	*16394*	*104*	120 DN 10	*65.0*	*69.6%*	*5.93 x 14.86*	*7*

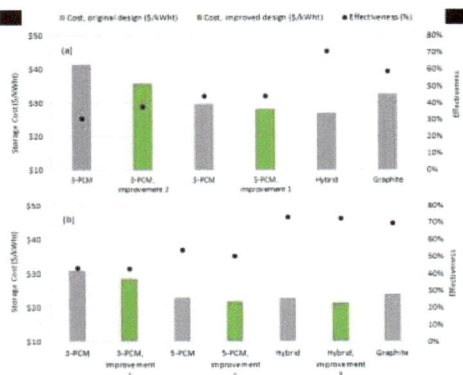

Fig. (10). Comparison of storage cost and effectiveness on the original and improved designs with a sodium Inlet temperature of 720°C (a) and 750°C (b).

Table 5. Steel requirement for assessed storage systems for a sodium inlet temperature of 720°C.

-		Original Design				Improved Design				% Mass Saving
Storage configuration	Storage material	Total Tube Length (km)	Tube Mass (t)	Shell+Header/Skirt+ Tubesheet Mass (t)	System Steel Mass (t)	Total Tube Length (km)	Tube Mass (t)	Shell+Header/Skirt+ Tubesheet Mass (t)	System Steel Mass (t)	
3-PCM cascade	PCM705	1,688.40	1,471.32	382.72	1,854.05	1,688.40	1,471.32	317.88	1,789.21	9.2
	PCM635	533.86	465.22	126.56	591.78	533.86	465.22	126.56	591.78	
	PCM569	2,170.80	1,891.70	489.76	2,381.46	1,825.88	1,591.13	413.23	2,004.36	
	total	4,393.06	3,828.24	999.04	4,827.28	4,048.14	3,527.67	857.67	4,385.34	
5-PCM cascade	PCM705	594.39	517.97	138.23	656.21	594.39	517.97	138.23	656.21	6.6
	PCM662	533.88	465.24	125.01	590.25	288.49	251.40	98.99	350.39	
	PCM635	441.03	384.33	104.72	489.05	441.03	384.33	104.72	489.05	
	PCM597	567.87	494.85	132.44	627.29	567.87	494.85	132.44	627.29	
	PCM569	1,144.02	996.93	258.33	1,255.26	1,144.02	996.93	258.33	1,255.26	
	total	3,281.18	2,859.32	758.73	3,618.05	3,035.80	2,645.49	732.71	3,378.20	
PCM-graphite-PCM hybrid	PCM705	297.74	259.46	73.88	333.34	N/A				
	Graphite	759.81	662.12	368.14	1,030.26					
	PCM569	162.70	141.78	44.09	185.87					
	total	1,220.25	1,063.36	486.11	1,549.47					
Graphite	Graphite	1,130.42	985.08	537.73	1,522.81	N/A				

Table 6. Steel requirement for assessed storage systems for a sodium inlet temperature of 750°C.

		Original Design				Improved Design				% Mass Saving
Storage configuration	Storage material	Total Tube Length (km)	Tube Mass (t)	Shell+Header/Skirt+ Tubesheet Mass (t)	System Steel Mass (t)	Total Tube Length (km)	Tube Mass (t)	Shell+Header/Skirt+ Tubesheet Mass (t)	System Steel Mass (t)	
3-PCM cascade	PCM705	1,688.40	1,471.32	382.72	1,854.05	1,688.40	1,471.32	382.72	1,854.05	14.3
	PCM635	533.86	465.22	126.56	591.78	533.86	465.22	126.56	591.78	
	PCM569	964.80	840.76	222.17	1,062.93	459.89	400.76	160.90	561.66	
	total	3,187.06	2,777.30	731.45	3,508.75	2,682.14	2,337.30	670.18	3,007.48	

(Table 6) cont.....

Storage configuration	Storage material	Original Design				Improved Design				% Mass Saving
		Total Tube Length (km)	Tube Mass (t)	Shell+Header/Skirt+Tubesheet Mass (t)	System Steel Mass (t)	Total Tube Length (km)	Tube Mass (t)	Shell+Header/Skirt+Tubesheet Mass (t)	System Steel Mass (t)	
5-PCM cascade	PCM705	594.39	517.97	138.23	656.21	594.39	517.97	138.23	656.21	18.2
	PCM662	533.88	465.24	125.01	590.25	254.50	221.78	95.67	317.46	
	PCM635	441.03	384.33	104.72	489.05	441.03	384.33	104.72	489.05	
	PCM597	567.87	494.85	132.44	627.29	567.87	494.85	132.44	627.29	
	PCM569	509.84	444.29	119.76	564.04	243.73	212.39	92.34	304.73	
	total	2,647.00	2,306.67	620.16	2,926.83	2,101.52	1,831.32	563.41	2,394.73	
PCM-graphite-PCM hybrid	PCM705	146.43	127.60	40.50	168.11	146.43	93.45	37.51	130.97	19.4
	Graphite	650.80	567.13	320.10	887.22	650.80	415.35	309.04	724.39	
	PCM569	146.43	127.60	40.50	168.11	146.43	93.45	37.51	130.97	
	total	943.66	822.33	401.10	1,223.43	943.66	602.26	384.07	986.33	
Graphite	Graphite	822.12	716.42	399.90	1,116.32	822.12	524.69	386.35	911.04	18.4

As shown in Fig. (**10**), the largest cost reduction of 13.8% results from the significant decrease in the amount of PCM705 in the 3-PCM cascade design when the HTF temperature is 720 °C. For all the other cases, marginal cost savings of below 7.3% was found by applying the three trial improvements. In addition, the effectiveness does not necessarily lead to lower-cost designs. Despite significant reductions in steel usage for the graphite designs, none of the studied 'improvements' was able to lower the cost. This is most likely due to the higher cost of graphite when compared to PCMs, suggesting that improvements to effectiveness would result in lower system costs. However, when this was attempted no further cost reductions from the original design could be found, suggesting that the original design of the current study was balanced in this regard.

3.3.3. Impact of Cost Assumptions

To determine if and how the cost of the original and 'improved' designs may change with differing cost assumptions, a separate analysis was performed on the cost of the PCMs, graphite, and steel for the 3-PCM, hybrid, and single graphite systems for a sodium inlet temperature of 750°C. This was to better understand the impact of previously excluded costs such as purification, melting, transport, *etc.* which may impact the final cost of material. For example, Kelly and Kearney [46] suggest that a cost of $0.07/kg be added to the purchased salt price to cover the costs of melting and filling the salt, while metal costs could be as high as $13.23-88.18/kg [47]. Furthermore, the cost of the graphite is impacted by the grade, size, and form (*i.e.* powder *vs.* block) required and can differ from $0.37-15/kg [48]. The graphite price may also be impacted by increased demand due to the proliferation of batteries and electric vehicles [49] which may impact the cost of graphite used in TES. The alternate cost assumptions for this study are summarized in Table **7** and uses the storage requirements for the improved design of the 3-PCM cascade and hybrid, and the original design for the graphite system.

Table 7. Alternate cost assumptions for the sensitivity analysis.

Material	Original Cost ($/kg)	Alternate Cost ($/kg)	% change
PCM710	0.51	1.01	98.0
PCM635	0.12	0.62	416.7
PCM569	0.13	0.63	384.6
Graphite	0.70	3.00	328.6
Steel	3.5	8.00	128.6

The cost of these systems is given in Fig. (**11**) for comparison. From Fig. (**11**), it can be seen that for the original and improved designs the hybrid and graphite systems represent the lowest cost, however, when the cost of the storage materials is increased the 3-PCM cascade becomes a much cheaper option (15.8-16.2%). This result seems to suggest that while the steel cost has the most significant impact on the stated cost, the choice of the lowest cost technology is more sensitive to the storage material cost assumptions. However, across all scenarios, the cost difference is relatively small, and the selection of technology is more likely to be dictated by which system is more functional.

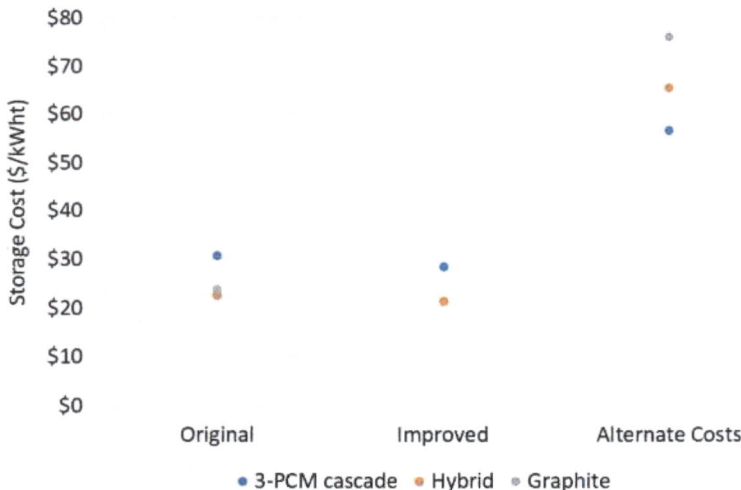

Fig. (11). Impact of cost assumptions on the storage cost.

CONCLUSION

In this study, the techno-economic performance of TES systems deployable to next-generation CSP was investigated and assessed at the component level. Four promising TES systems which include cascade storage with 3 and 5 PCMs, hybrid storage with PCM-graphite-PCM, and single graphite sensible storage in a simple shell-and-tube configuration were studied. By using a previously validated and verified 2D numerical model, the aforementioned storage systems were sized based on a storage capacity of 1000 MWh$_t$, charging/discharging period of 6 h/10 h, cut-off temperatures in both charging and discharging, and two hot sodium (charging) temperatures. The cost of the studied designs was undertaken, and a parametric analysis was conducted to improve their techno-economic performance. The following conclusions can be drawn:

• It is a necessity to apply the methodology of dynamic simulation of multiple charging-discharging cycles and size the TES system based on the performance at equilibrium condition, though heavy computing load is required. For example, with the PCM cascade storage, the increase of difference between the hot HTF temperature and the melting temperature of hot end PCM leads to the reduction of the amount of cold end PCM, which is not expected in the static design and single charging or discharging process simulation.

• Regardless of the charging temperature, the cost of the studied designs increases in the order of hybrid, 5-PCM cascade, single graphite, and 3-PCM cascade, among which the difference between hybrid and 5-PCM cascade is small.

• Three promising improvements were investigated to reduce the cost of the TES system. A notable cost reduction of 13.8% was found with the 3-PCM cascade design when the effectiveness of PCM705 is improved as a result of reducing the tube spacing. The graphite system did not benefit from those improvements, while the other two systems received the minor benefit.

• Higher hot HTF temperature for charging is preferable to enable a more cost-effective storage system. A cost reduction of approximately 20% was found with the hot HTF temperature increasing from 720 °C to 750 °C. However, this inlet boundary condition is affected by the operating performance of other CSP components, *e.g.* receiver. Therefore, system-level assessment is critical to determine the boundary conditions to achieve the economic viability of the CSP technology.

CONSENT FOR PUBLICATION

Not applicable

CONFLICT OF INTEREST

The authors declare no conflict of interest, financial or otherwise.

ACKNOWLEDGEMENTS

This research was funded by the Australian Solar Thermal Research Institute (ASTRI), which is supported by the Australian Government, through the Australian Renewable Energy Agency (ARENA).

REFERENCES

[1] *Power System Limitations in North Western Victoria and South Western New South Wales.*.https://www.aemo.com.au/-/media/Files/Electricity/NEM/Network_Connections/Power-System-Limitations-December.pdf

[2] P. Denholm, M. O'Connell, G. Brinkman, and J. Jorgenseon, "Overgeneration from Solar Energy in California: A Field Guide to the Duck Chart", *National Renewable Energy Laboratory (NREL),*, 2015.https://www.nrel.gov/docs/fy16osti/65023.pdf
[http://dx.doi.org/10.2172/1226167]

[3] *Managing system strength during the transition to renewables,* 2020.https://arena.gov.au/assets/2020/05/managing-syste--strength-during-the-transition-to-renewables.pdf

[4] M. Mehos, *Concentrating Solar Power Best Practices Study,* 2020.https://www.nrel.gov/docs/fy20osti/75763.pdf
[http://dx.doi.org/10.2172/1665767]

[5] *Renewable power generation costs in 2019,* 2020.https://www.irena.org/-/media/Files/IRENA/Agency/Publication/2020/Jun/IRENA_Power_Generation_Costs_2019.pdf

[6] T. Bauer, N. Pfleger, D. Laing, M. Eck, and S. Kaesche, "High-Temperature Molten Salts for Solar Power Application", In: *in Molten Salts Chemistry, F. G. Lantelme, H., Ed., ed. Burlington, MA, USA: Elsevier Inc*, 2013, pp. 415-438.
[http://dx.doi.org/10.1016/B978-0-12-398538-5.00020-2]

[7] W. Benaissa, and D. Carson, "Oxidation properties of "Solar Salt"." presented at the AIChE Spring Meeting 2011 & 7", *Global Congress on Process Safety (GCPS), Chicago, United States, Mar 2011*, 2011.https://hal-ineris.archives-ouvertes.fr/ineris-00976226/document

[8] M. Mehos, *Concentrating Solar Power Gen3 Demonstration Roadmap,* 2017 .https://www.nrel.gov/docs/fy17osti/67464.pdf
[http://dx.doi.org/10.2172/1338899]

[9] *ASTRI Public Dissemination Report,* 2019 .https://arena.gov.au/assets/2013/01/astri-publi--dissemination-report.pdf

[10] N. Boerema, G. Morrison, R. Taylor, and G. Rosengarten, "Liquid sodium versus Hitec as a heat transfer fluid in solar thermal central receiver systems", *Sol. Energy,* vol. 86, no. 9, pp. 2293-2305, 2012.
[http://dx.doi.org/10.1016/j.solener.2012.05.001]

[11] J. Pye, M. Zheng, C-A. Asselineau, and J. Coventry, "An exergy analysis of tubular solar-thermal receivers with different working fluids", *International Conference on Concentrating Solar Power and Chemical Energy Systems, SolarPACES 2014,* 2014 Beijing, China

[12] S. Khare, M. Dell'Amico, C. Knight, and S. McGarry, "Selection of materials for high temperature sensible energy storage", *Solar Energy Materials and Solar Cells,* vol. 115, pp. 114-22, 2013.

[http://dx.doi.org/10.1016/j.solmat.2013.03.009]

[13]   G. Mohan, M. B. Venkataraman, and J. Coventry, "Sensible energy storage options for concentrating solar power plants operating above 600 °C", *Renewable and Sustainable Energy Reviews*, vol. 107, pp. 319-337, 2019.
[http://dx.doi.org/10.1016/j.rser.2019.01.062]

[14]   M.M. Kenisarin, *High-temperature phase change materials for thermal energy storage*, 2010 .http://www.sciencedirect.com/science/article/pii/S1364032109002731
[http://dx.doi.org/10.1016/j.rser.2009.11.011]

[15]   B. Xu, P. Li, and C. Chan, "Application of phase change materials for thermal energy storage in concentrated solar thermal power plants: a review to recent developments", *Appl. Energy*, vol. 160, pp. 286-307, 2015.
[http://dx.doi.org/10.1016/j.apenergy.2015.09.016]

[16]   Y. Jiang, Y. Sun, M. Liu, F. Bruno, and S. Li, "Eutectic Na2CO3–NaCl salt: a new phase change material for high temperature thermal storage", *Sol. Energy Mater. Sol. Cells*, vol. 152, pp. 155-160, 2016.
[http://dx.doi.org/10.1016/j.solmat.2016.04.002]

[17]   M. Liu, "Review on concentrating solar power plants and new developments in high temperature thermal energy storage technologies", *Renewable and Sustainable Energy Reviews*, vol. 53, pp. 1411-1432, 2016.
[http://dx.doi.org/10.1016/j.rser.2015.09.026]

[18]   R. Jacob, M. Liu, Y. Sun, M. Belusko, and F. Bruno, "Characterisation of promising phase change materials for high temperature thermal energy storage", *J. Energy Storage*, vol. 24, 2019.100801
[http://dx.doi.org/10.1016/j.est.2019.100801]

[19]   C. Prieto, and L. F. Cabeza, "Thermal energy storage (TES) with phase change materials (PCM) in solar power plants (CSP). Concept and plant performance", *Applied Energy*, vol. 254, p. 113646, 2019 .
[http://dx.doi.org/10.1016/j.apenergy.2019.113646]

[20]   X. Chen, Z. Zhang, C. Qi, X. Ling, and H. Peng, "State of the art on the high-temperature thermochemical energy storage systems", *Energy Convers. Manage.*, vol. 177, pp. 792-815, 2018.
[http://dx.doi.org/10.1016/j.enconman.2018.10.011]

[21]   D. A. Sheppard, and C. E. Buckley, "The potential of metal hydrides paired with compressed hydrogen as thermal energy storage for concentrating solar power plants", *International Journal of Hydrogen Energy*, vol. 44, no. 18, pp. 9143-9163.
[http://dx.doi.org/10.1016/j.ijhydene.2019.01.271]

[22]   J. Stekli, L. Irwin, and R. Pitchumani, "Technical Challenges and Opportunities for Concentrating Solar Power With Thermal Energy Storage", *J. Therm. Sci. Eng. Appl.*, vol. 5, no. 2, 2013.021011
[http://dx.doi.org/10.1115/1.4024143]

[23]   Z.-X. Gong, and A. S. Mujumdar, "Thermodynamic optimization of the thermal process in energy storage using multiple phase change materials", *Applied Thermal Engineering*, vol. 17, no. 11, pp. 1067-1083.
[http://dx.doi.org/10.1016/S1359-4311(97)00012-4]

[24]   H. Michels, and R. Pitz-Paal, *Cascaded latent heat storage for parabolic trough solar power plants*, 2007 .http://www.sciencedirect.com/science/article/B6V50-4M87BS3-1/2/3245f9652a68 ec4a2703ea 2116aafe1e
[http://dx.doi.org/10.1016/j.solener.2006.09.008]

[25]   H. J. Xu, and C. Y. Zhao, "Thermal performance of cascaded thermal storage with phase-change materials (PCMs). Part II: Unsteady cases", *International Journal of Heat and Mass Transfer*, vol. 106, pp. 945-957, 2017.
[http://dx.doi.org/10.1016/j.ijheatmasstransfer.2016.10.066]

[26]   S. S. Mostafavi Tehrani, Y. Shoraka, K. Nithyanandam, and R. A. Taylor, "Cyclic performance of cascaded and multi-layered solid-PCM shell-and-tube thermal energy storage systems: A case study of the 19.9 MWe Gemasolar CSP plant", *Applied Energy,* vol. 228, pp. 240-253, 2018. [http://dx.doi.org/10.1016/j.apenergy.2018.06.084]

[27]   R. Jacob, M. Belusko, M. Liu, W. Saman, and F. Bruno, "Using renewables coupled with thermal energy storage to reduce natural gas consumption in higher temperature commercial/industrial applications", *Renew. Energy,* vol. 131, pp. 1035-1046, 2019. [http://dx.doi.org/10.1016/j.renene.2018.07.085]

[28]   F. Agyenim, N. Hewitt, P. Eames, and M. Smyth, "A review of materials, heat transfer and phase change problem formulation for latent heat thermal energy storage systems (LHTESS)", *Renew. Sustain. Energy Rev.,* vol. 14, no. 2, pp. 615-628, 2010. [http://dx.doi.org/10.1016/j.rser.2009.10.015]

[29]   G. Glatzmaier, *Developing a cost model and methodology to estimate capital costs for thermal energy storage.* National Renewable Energy Laboratory, 2011. [http://dx.doi.org/10.2172/1031953]

[30]   K. Nithyanandam, and R. Pitchumani, "Cost and performance analysis of concentrating solar power systems with integrated latent thermal energy storage", *Energy,* vol. 64, pp. 793-810, 2014 . [http://dx.doi.org/10.1016/j.energy.2013.10.095]

[31]   P. Sharan, C. Turchi, and P. Kurup, "Optimal design of phase change material storage for steam production using annual simulation", *Sol. Energy,* vol. 185, pp. 494-507, 2019. [http://dx.doi.org/10.1016/j.solener.2019.04.077]

[32]   S. S. Mostafavi Tehrani, Y. Shoraka, K. Nithyanandam, and R. A. Taylor, "Shell-and-tube or packed bed thermal energy storage systems integrated with a concentrated solar power: A techno-economic comparison of sensible and latent heat systems", *Applied Energy,* vol. 238, pp. 887-910, 2019 . [http://dx.doi.org/10.1016/j.apenergy.2019.01.119]

[33]   H. Chirino, and B. Xu, "Parametric Study and Sensitivity Analysis of Latent Heat Thermal Energy Storage System in Concentrated Solar Power Plants", *J. Sol. Energy Eng.,* vol. 141, no. 2, 2019. [http://dx.doi.org/10.1115/1.4042060]

[34]   J. Virgone, and A. Trabelsi, ""2D conduction simulation of a PCM storage coupled with a heat pump in a ventilation system," Applied Sciences (Switzerland)", *Article,* vol. 6, no. 7, pp. 1-17, 2016. [http://dx.doi.org/10.3390/app6070193]

[35]   S. Farah, M. Liu, and W. Saman, "Numerical investigation of phase change material thermal storage for space cooling", *Applied Energy,* vol. 239, pp. 526-535, 2019 . [http://dx.doi.org/10.1016/j.apenergy.2019.01.197]

[36]   M. Liu, S. Riahi, R. Jacob, M. Belusko, and F. Bruno, "Design of sensible and latent heat thermal energy storage systems for concentrated solar power plants: Thermal performance analysis", *Renewable Energy,* vol. 151, pp. 1286-1297, 2020. [http://dx.doi.org/10.1016/j.renene.2019.11.115]

[37]   V.R. Voller, "Fast implicit finite-difference method for the analysis of phase change problems", *Numer. Heat Transf. B,* vol. 17, pp. 155-169, 1990. [http://dx.doi.org/10.1080/10407799008961737]

[38]   N. Tay, M. Belusko, M. Liu, and F. Bruno, Static concept at University of South Australia.*High temperature thermal storage systems using phase change materials.,* L. Cabeza, N. Tay, Eds., Academic Press: US, 2017.

[39]   N.H.S. Tay, M. Belusko, and F. Bruno, "An effectiveness-NTU technique for characterising tube-i--tank phase change thermal energy storage systems", *Appl. Energy,* vol. 91, no. 1, pp. 309-319, 2012. [http://dx.doi.org/10.1016/j.apenergy.2011.09.039]

[40]   M. Liu, N.H.S. Tay, M. Belusko, and F. Bruno, "Investigation of cascaded shell and tube latent heat

storage systems for solar tower power plants", *International Conference on Concentrating Solar Power and Chemical Energy Systems, SolarPACES 2014,* 2014 Beijing, China

[41] K. Nithyanandam, and R. Pitchumani, "Optimization of an encapsulated phase change material thermal energy storage system", *Solar Energy,* vol. 107, pp. 770-788, 2014 . [http://dx.doi.org/10.1016/j.solener.2014.06.011]

[42] R. Jacob, M. Belusko, A. Inés Fernández, L. F. Cabeza, W. Saman, and F. Bruno, "Embodied energy and cost of high temperature thermal energy storage systems for use with concentrated solar power plants", *Applied Energy,* vol. 180, pp. 586-597, 2016 . [http://dx.doi.org/10.1016/j.apenergy.2016.08.027]

[43] M.S. Peters, and K.D. Timmerhaus, *Plant design and Economics for Chemical Engineers.* McGraw-Hill: New York, 1991.

[44] MEPS, *"World Stainless Steel Prices- Stainless Steel 316."*.https://www.meps.co.uk/gb/en/products/world- stainless-steel-prices (accessed 25 June, 2020).

[45] S. Riahi, Y. Jovet, W. Y. Saman, M. Belusko, and F. Bruno, "Sensible and latent heat energy storage systems for concentrated solar power plants, exergy efficiency comparison", *Solar Energy,* vol. 180, pp. 104-115, 2019 . [http://dx.doi.org/10.1016/j.solener.2018.12.072]

[46] B. Kelly, and D. Kearney, *Thermal storage commercial plant design for a 2-tank indirect molten salt system.* National Renewable Energy Laboratory, 2006.

[47] J. Moore, *Testing of a 1 MW$_e$ supercritical CO$_2$ test loop* .https://www.energy.gov/ sites/ prod/files /2019/11/f68/DOE%20sCO2%20Workshop %2C %20SWRI%2C%20Jeff%20Moore.pdf (accessed 1 July, 2020).

[48] Syrah Resources, *"Diggers and Dealers Mining Forum- Syrah Resources"* .http://www.syrah resources.com. au /asx-announcements/august-2019 (accessed 4 August, 2020).

[49] B. Ballinger, M. Stringer, D.R. Schmeda-Lopez, B. Kefford, B. Parkinson, C. Greig, and S. Smarta, "The vulnerability of electric vehicle deployment to critical mineral supply", *Applied Energy,* vol. 255, p. 113844, 2019. [http://dx.doi.org/10.1016/j.apenergy.2019.113844]

SUBJECT INDEX

A

Abraham Maslow's theory 347
Accuracy of packet-level classification
     process 20
Acid(s) 246, 247, 249, 266, 369, 430, 431,
     437, 439
  aspartic 266, 430
  glutamic 266, 430
  hexuronic 430
  hyaluronic 369
  nucleic 437
  oleic 246, 249
  pantothenic 431
  phosphorous 5
  poly-glycolic (PGA) 369
  stearic 247
  synthetic polymers like- poly-glycolic 369
  trans-cinnamic 439
Acoustic environment, time-varying 23
Action 249, 395, 397, 436
  anti-diabetic 395, 397
  anti-oxidative 249
  hypercholesterolaemic 396
  physiological 436
Active noise 23, 24, 31
  cancellation system 31
  control 23, 24
Activity 40, 101, 186, 190, 229, 249, 328,
     345, 364, 397, 422, 429, 431, 432, 433,
     438, 439, 474
  anti-bacterial 432
  anti-cancer 432
  anti-depressant 439
  anti-diabetic 397, 429
  anti-estrogenic 328
  anti-inflammatory 438
  antimicrobial 432, 439
  antioxidant 249, 364, 431
Ad hoc 444, 459, 462
  communication 459
  flying networks 444

networking infrastructure 459
  software 462
ADMET information 35
Agile software development 377, 379
Air cushion vehicle (ACVs) 6
Algorithms 204, 216, 254
  computer-based 216
  cryptographic 204, 254
Alkaline fuel cells (AFC) 5
Alloxan-induced oxidative stress 395
Alzheimer's disease 52
Anti-noise control (ANC) 23, 24, 26, 31
Applications of wearable devices 276
Applying machine learning 147
Artificial 51, 54, 56, 57, 58, 59, 125, 131, 132,
     133, 134, 158, 163, 376
  bee colony (ABC) 376
  intelligence techniques 158, 163
  neural networks (ANN) 51, 54, 56, 57, 58,
     59, 125, 131, 132, 133, 134
ATM transaction 407, 408
  Interface 408
Attention deficit hyperactivity disorder
     (ADHD) 332, 333, 334, 335, 337, 338
Autism spectrum disorder (ASD) 335, 337,
     377, 379
Automated 126, 219, 406, 407, 411
  machine learning process 219
  segmentation-based detection technique
     126
  teller machine (ATMs) 406, 407, 411
Automation of work processes 505
Autonomous underwater vehicle (AUV) 7
Autoregressive neural network 256
Azure machine learning studio (AMLS) 194

B

Bacillus subtilis 432
Bamboo 417, 418, 423, 426
  cluster history 417
  consumption 423

www.ingramcontent.com/pod-product-compliance
Lightning Source LLC
Chambersburg PA
CBHW050519240326
41598CB00086B/37